Bjørn Olav Tveit

A Birdwatcher's Guide to Norway

Where, When and How to find Scandinavia's most sought-after birds

Featuring Varanger, Svalbard, Lofoten, Oslo, Bergen & more

Enhanced 2nd edition

WWW.ORNFORLAG.NO

Co-published in 2024 by

Ørn Forlag
Veståsen 4, 1362 Hosle, Norway
www.ornforlag.no

Pelagic Publishing
20–22 Wenlock Road
London N1 7GU, UK
www.pelagicpublishing.com

ISBN Norway: 9788293314035
ISBN International: 9781784275082

Layout: Bjørn Olav Tveit
Print: Livonia Print, Latvia

The author of this book acknowledges financial support from the Norwegian Non-fiction Writers' and Translators' Association.

The publishers and author of this book accept no responsibility for any loss, damage, injury or death sustained while using this book as a guide.

Reference to this book:
Tveit, B. O. 2024. *A Birdwatcher's Guide to Norway. Where, When and How to find Scandinavia's most sought-after birds. Featuring Varanger, Svalbard, Lofoten, Oslo, Bergen & more.* Enhanced 2nd edition. Ørn Forlag, Hosle, Norway

Front cover image: Atlantic Puffin. *Photo: Kjetil Schjølberg*
Back cover images: Slavonian (Horned) Grebe. *Photo: Kjetil Schjølberg*
Yellow-browed Warbler. *Photo: Bjørn Fuldseth*
Tisnes, Troms. *Photo: Bjørn Olav Tveit*
Image p. 3, Great Grey Owl. *Photo: Bjørn Fuldseth*

CONTENTS

PREFACE TO 1ST EDITION

Every respectable nation should have at least one birdwatching site guide. Frankly, in my opinion, a birdwatching site guide is the ultimate sign of an advanced civilization! When I grew up as a young birdwatcher in the Oslo area in the 1980s I was embarrassed that my own country didn't have a site guide and I desperately hoped someone would write one. It was not just a matter of filling the gap in my self esteem on behalf of The Kingdom of Norway. I was also badly in need of such a guide as the years went by and my home range gradually expanded due to my realization of how many wonderful birdwatching experiences I had missed out on simply by limiting my efforts to the Oslo region.

At one point, in desperation perhaps, I even wrote a paper about the sites of the Oslo region in the national birdwatcher's magazine *Vår fuglefauna*, hoping to spark off a series of papers by other authors dealing with their local sites, ultimately covering all parts of the country. The project was more or less futile and so, after 30 years of waiting, I realized I had to take on the job of writing a site guide covering the whole country by myself.

I quickly understood that this was not going to be a desk-top job only, even though I did a lot of research ploughing through local ornithological magazines and websites. I had to get out there and see for myself how each site was best approached in order to maximize the birdwatching results. I had done a lot of travelling abroad and knew perfectly well the value of extremely detailed road descriptions. I wouldn't be able to write those from the parameters of my living room. And so I soon waved goodbye to my wife and three kids and travelled to all corners of the country, roaming the land for weeks on end. My expeditions through Norway were not only extremely useful in terms of collecting material for the book, it was also a truly magnificent experience, just me, my Jeep, the birds and the countryside. Even though I had travelled a lot in Norway in advance, it felt different experiencing it all in such a condensed period of time. I was truly stunned not only by the birds, but also by the shear beauty of the country, ranging from the deep fjords, vast mountain plateaus, pristine taiga forests, gleaming glaciers mirrored in bright turquoise blue lakes, a blood red sun just touching the horizon before rising again in the northern summer, and white sandy beaches contrasting with black and rugged cliffs. I realized that in terms of nature, Norway is truly the crown jewel of Europe.

It took me three years to complete the Norwegian edition of this book, and another year translating it to English and adapting it to a foreign audience. I also changed from the original county-based presentation of the sites, to a region-based, and added a Selected species-section. In the process with the English version, I have had extensive and invaluable help from Oslo birder Simon Rix, whose native English tongue, expertise on birds and experience as a foreign birdwatcher living in Norway has graced the language as well as the general content of this book.

Of course, I also want to thank everybody involved in the making of the Norwegian original. First and foremost, I want to thank Geir Mobakken, The Resident birdwatcher of Utsira *par excellence*, who scrutinized the Norwegian edition through his highly skilled eyes and filtered it all through his utmost critical sense. A special thanks also to Tor A. Olsen who has put a lot into improving parts of the book. Furthermore, the following 68 people have contributed by reading through and commenting upon one or more of the county chapters of the original book (county abbreviations in parenthesis): Arnfred Antonsen (Øs & He), Terje Axelsen (Ve), Georg Bangjord (SJ), Julian G. Bell (Ho), Nils Chr. Bjørgo (SF), Vegard Bunes (Tr, VA & Bu), Martin Dagsland (Ro), Espen Lie Dahl (MR), Even Dehli (Op), Martin Eggen (No), Stein Engebretsen (Øs), Frode Falkenberg (Ho), Michael Fredriksen (Ho), Olav Werner Grimsby (Op), John Grønning (Ro), Eirik Grønningsæter (SJ), Per Gylseth (OA), Øyvind Hagen (OA), Odd Hallaråker (Ho), Jon Ludvig Hals (Bu), Anders Hangård (Ve), Arild Hansen (Øs), Trond Haugskott (ST), Håkon Heggland (Fi & Ve), Oddvar Heggøy (Ho), Morten Helberg (Tr), Bjørn Ove Høyland (Ro & VA), Christer Jakobsen (Øs), Øivind W. Johannessen (Te), Henning A. Johansen (Ve), Runar Jåbekk (VA), Jan Helge Kjøstvedt (AA), Magne Klann (OA), Geir Klaveness (Ve), Terje Kolaas (NT & ST), Thor Edgar Kristiansen (No), Jonas Langbråten (He), Bjørn Harald Larsen (Op), Sebastian Ludvigsen (Øs), Jon Lurås (He), Kjetil Mork (SF), Magne Myklebust (MR, ST & NT), Jostein Myromslien (He & Te), Jan Kåre Ness (Ro & Ho), Gunnar Numme (Ve), Bård Nyberg (NT), Atle Ivar Olsen (No), Tor Audun Olsen (Ro, VA & Fi), Jon Opheim (Op), Svein Arne Orvik (MR and Selected species-section), Tore Reinsborg (NT), Kjetil Schjølberg (MR), Paul Shimmings (No), Rune Skåland (AA), Per Kristian Slagsvold (AA), John Stenersen (No), Steinar Stueflotten (Bu), Per Ole Syvertsen (No), Ståle Sætre (SF), Halvor Sørhuus (NT), Trond Sørhuus (NT), Jan Erik Tangen (Te), Morten Vang (ST), Rune Voie (Ho), Egil Ween (Ro), Ragnar Ødegård (Op), Ingar Jostein Øien (Fi) and Tomas Aarvak (Fi & No).

Many of these people have in addition contributed with pieces of information, small or large, regarding other counties. Such information has also been kindly handed to me, either beforehand or after the release of the first edition, by Geir S. Andersen, Berit Langdahl Andresen, Johannes Anonby, Steve Baines, Roald Bengtson, Gunnar Bergo, Kjell Inge Bjerga, Arne Kristian Borger, Ingar Støyle Bringsvor, Stian Edvardsen, Øivind Egeland, Stig Eide, Bjørn Fuldseth, Sverre Furre, Frank Grønningsæter, Andreas Gullberg, Jørn R. Gustad, Inge Hafstad, Ann Kristin Halvorsrud, Øystein Hauge, Maja Heger, Knut Hellandsjø, Olaf Henke, Svein Jan Hjelmeset, Dag Holtan, Arvid Johnsen, John Johnsen, Atle Karlstrøm, Geir Kristensen, Thomas Kvalnes, Terje Larsen, Tore Larsen, John Martin Mjelde, Kjell R. Mjølsnes, Øystein Mortensen, Grethe Gjerdingen Myrdahl, Ottar Osaland, Johnny Roger Pedersen, Peter Sjolte Ranke, Tom Resvoll-Holmsen, Knut Ring, Geir Ropstad, Jan Erik Røer, Jan Ove Sagerøy, Tom Schandy, Kjetil Aa. Solbakken, Paal-Magnus Solvang, Kjell Mork Soot, Frank Steinkjellå, Edvin Thesen, Thorleif Thorsen, Harald Totland, Knut Totland, Kjetil Lillebuan Vada, Bjørn Vikøyr, Per Inge Værnesbranden, Jim Wilson, Andreas Winnem, Roy Erling Wrånes and Olav Ås. Sincere thanks to all of you!

This book is embellished with pictures from several people, some of whom are among the best bird photographers in Norway. I am deeply grateful to all of them: Tormod Amundsen, Tommy Andre Andersen, Espen Lie Dahl, Bjørn Fuldseth, Bodil Gjevik, Eirik Grønningsæter, Per Gylseth, Arild Hansen, Christer Jakobsen, Magne Klann, Terje Kolaas, Jan Kåre Ness, Gunnar Numme, Svein Arne Orvik, Kjetil Schjølberg, John Stenersen, Christian Tiller, Ingar Jostein Øien and Tomas Aarvak.

Magne Klann has carefully edited all images for printing and, together with Katrine Mellingen Kaldal, has given invaluable feedback on my design and layout. Even though I have drawn most maps by hand in Adobe Illustrator, based on aerial photos, other maps and my own on-location measurements and notes, I wish to thank the map department at Cappelen Damm for letting me use the base files to their award-winning regional maps, which I have used in each chapter introduction.

Finally, I want to thank my beautiful and understanding wife Heidi and my wonderful daughters Matilde, Nora and Josefine, who have become a little bigger and even more adorable for every time I have looked up from behind the computer or come home from my many, but strictly needed, field expeditions.

Bjørn Olav Tveit
Bekkestua, April 2011

PREFACE TO 2ND EDITION

Fourteen years have passed since the first Norwegian edition of this book saw the light of day. The following year I translated the book into English and published it under the title *A Birdwatcher's Guide to Norway*, and since then I have regularly travelled to all corners of the country to update any changes. And you can rest assured that many changes have occurred. Most striking is the constant improvement in traffic. Linesøya in Trøndelag, Kobbevågen in Tromsø, and the islands in the Boknafjorden near Stavanger are all examples of bird locations that have been made easier to access by the fact that the ferry has now been replaced by a tunnel. Winding roads are constantly being straightened, making it faster and easier to visit bird locations. These changes are probably for the better for most people, but it creates an astonishing amount of extra work for us bird guidebook authors.

At the same time, there has been a dramatic increase in people's attitudes towards the use of environmentally friendly transport. Sales of electric cars when this book came out were marginal, but today 80 % of all new cars sold in Norway run on electricity. One can certainly discuss the environmental benefit of this, but I must say that being able to cruise around in a silent electric car, which glides through the landscape like a bicycle, contributes to increased enjoyment of the birding trips, particularly when driving around listening for birds. And, of course, because they are regarded as environmentally friendly, bird guide authors can drive around with a slightly better conscience when inspecting birding sites for updating book manuscripts. Another thing that has changed since the 1st edition is the growing popularity of tower hides and observation shelters for birdwatchers. In the 1st edition, I was concerned that the book should be a complete overview of such, which it was, with just over 100 listed towers and hides. But in recent years I have been so disappointed when visiting many of the new and fancy so-called "birdwatching tower hides" and "birdwatching shelters", both in terms of design, functionality and location, that in this edition I only mention the constructions that I believe add value to the birdwatching experience. Other practical changes for bird enthusiasts concern the steady improvement of both communication and registration of bird sightings.

The avifauna has also changed in these 14 years, but, unfortunately, almost without exception for the worse. The populations of many species are in sharp decline, particularly amongst seabirds, montane birds and farmland species. There has been a dramatic decline in the cliff-breeding birds along the coast. Many colonies of Kittiwake, Fulmar, Guillemot and several others, which used to breed by the thousands, have plummeted in numbers. Also Barred Warbler now seems to be gone from Stråholmen island, which was its last stronghold as a breeding bird in Norway. This dire situation actualizes more than ever the important work carried out by BirdLife Norway and other environmental organisations.

Fortunately, there are exceptions. It is particularly positive to note that the population of Lesser White-fronted Goose seems to be barely getting back on its feet, from being in danger of complete extinction in Scandinavia at the time when this book's 1st edition was published. Rustic Bunting was also on the brink of extinction in Norway, but during the last couple of years the population is showing signs of recovery. Great Grey Owl has become significantly easier to find as well, particularly in south-east and central Norway.

Acknowledgements

Also this time around I have received incredibly valuable help from a bunch of skilled and benevolent people, who have generously shared their local knowledge of the sites and the birds. However, I will first refer to the preface from the first edition. Although much text has changed in this 2nd edition, a great deal of the content is still based on the information and input I received during the making of the first edition. For this reason I have reprinted the preface from the first edition in its entirety in this edition. In addition, I would once more like to thank the photographers. They are all credited next to their photos. A special thanks to Alette Sandvik for granting permission to use the photos by the late Kjartan Trana.

Specifically in connection with the work with 2nd edition, I would like to pay tribute to the following people for their kind help in providing information: Stig Helge Basnes, Roald Bengtson, Lasse Blystad, Torgrim Breiehagen, Vegard Bunes, Håvard Eggen, Martin Eggen, Frode Falkenberg, Bjørn Frostad, Atle Grimsby, Eirik Grønningsæther, Jørn R. Gustad, Øyvind Hagen, Toril Hasle, John Haugen, Håkon Heggland, Oddvar Heggøy, Thorstein Holtskog, Bjørn Ove Høyland, Tarjei T. Jensen, Kjetil Johannessen, Per-Arne Johansen, Runar Jåbekk, Terje Kolaas, Rolf Terje Kroglund, Bård Kyrkjedelen, Jonas Langbråten, Jørn Helge Magnussen, Mariella Nora Isabella Filberg Memo, Geir Mobakken, Magne Myklebust, Grethe Gjerdingen Myrdahl, Ola Nordsteien, Gunnar Numme, Frode Omland, Tor Olsen, Svein Arne Orvik, Tore Reinsborg, Tom Resvoll-Holmsen, Simon Rix, Geir Ropstad, Kjetil Salomonsen, Alette Sandvik, Anders Braut Simonsen, Ole Skimmeland, Audun Brekke Skrindo, Rebecca Benedicte Solhaug, Paal-Magnus Solvang, Trude Starholm, Ole Knut Steinset, John Stenersen, Steinar Stueflotten, Ståle Sætre, Thorleif Thorsen, Klaus Maløya Torland, Heidi Tveit, Nora Tveit, Ulf Ullring, Per Inge Værnesbranden, Egil Ween, Ragnar Ødegård and Trond Øigarden.

Thanks also to Magne Klann and Katrine Mellingen Kaldal specifically for helping out with the design, and to Geir Mobakken for detailed review of the revised Norwegian 2nd edition, on which this book is based. A special shout-out to Tor A. Olsen, who has contributed as a specialist consultant on many chapters and has been incredibly useful to spar with during the process of revising these books.

Bjørn Olav Tveit
Hosle, April 2024

WELCOME TO NORWAY

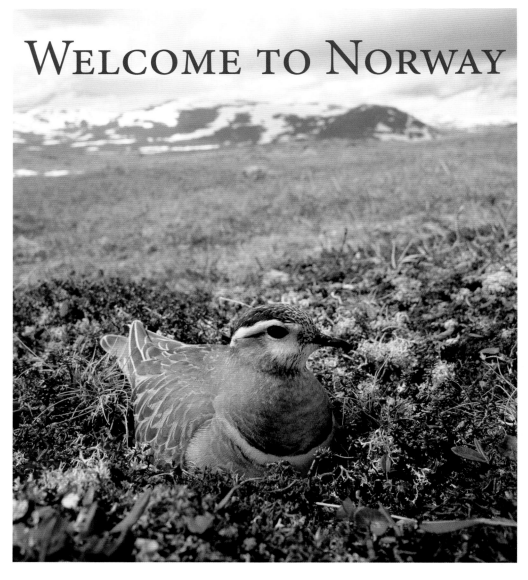

Dotterel on the nest. *Photo: Terje Kolaas.*

D ue to its location far north in Europe, Norway has a number of species that are difficult so see further south. This has resulted in Norway being a must-visit destination for many birdwatchers from abroad. This book describes in detail the best sites for finding the most sought-after birds, and for maximizing the avian outcome of your visit to the country.

Some key facts about Norway
The Kingdom of Norway is the northernmost country in Europe, comprising the western section of the Scandinavian Peninsula. It has a total area of 385,252 km², which makes it a little larger than Germany and a bit smaller than California. Mainland Norway is divided into 15 counties and two integral overseas Arctic territories, Jan Mayen and Svalbard. There are in addition three dependencies in and around the Antarctic, not covered in this book. Norway is hilly and mountainous, reaching an elevation of 2,469 metres at the summit of Mnt. Galdhøpiggen. The country is widely

recognized for having some of the deepest and most spectacular fjords on earth.

With only 5.5 million inhabitants, most of whom are concentrated around the capital city of Oslo, Norway is among the most sparsely populated countries in Europe. Norway is richly endowed with natural resources including oil, hydropower, fish, forests, and minerals. Large reserves of oil and natural gas were discovered in the 1960s, which led to a boom in the economy. Today, Norway has one of the highest income per capita in the world, and the largest capital reserve of any nation. The standard of living in Norway is also among the highest in the world. Although not a member of the EU, Norway is a highly integrated member of most sectors of the union's internal market. The local currency is Norwegian kroner (NOK). Norwegians speak Norwegian, but almost everybody can communicate well in English, and most people even in a third language. Sami and Kven are official languages in addition to Norwegian, for instance expressed on road signs in some areas.

Climate and seasons

Despite of the location far to the north, Norway has a temperate climate due to experiencing predominantly southwestern winds and ocean currents. Generally speaking, the coast is often rainy with temperatures varying little between summer and winter, as opposed to the drier interior with bitterly cold winters and rather warm summers. It is on average colder the higher up in the mountains or the further north you go, and the winter is progressively longer and the summer correspondingly shorter up north. There are four marked seasons in Norway:

Winter (December–February) being cold, with temperatures regularly below -25°C in the interior, although normally just below 0°C in Oslo. North of the Arctic Circle, the sun never rises above the horizon in mid-winter. This phenomenon is called the *arctic winter* and lasts for a longer period of time the farther north you get. Due to atmospheric conditions, however, the sun can still be seen north to around Bodø. In Tromsø, the arctic winter

lasts from November 27 to January 15, in Vardø from November 23 to January 19, and in Longyearbyen from October 26 to February 16. If skies are clear, there are still a few hours of daylight at midday, at least in mainland Norway. In the northernmost county of Finnmark, the light-conditions allow for serious birdwatching by early February. By March, in prime time for arctic gulls and eiders, there are about 15 hours of daylight per day. During the dark hours, you may be treated by the spectacular sight of the Northern Lights (Aurora Borealis).

Spring (March–May) often sees quite pleasant temperatures on sunny days, with ice and snow melting and a steady return of migrants. However, spells of cold temperatures and a new layer of snow may occur well into April in the south, even in May in the north and all summer on Svalbard.

Summer (June–August) is generally warmer, with temperatures ranging from 18–25°C in southern Norway, sometimes even exceeding 30°C inland in late summer. Cloud cover and winds vary from day to day, some days with showers and others with sun. The weather may change rapidly in the mountains, so don't be caught off guard wearing too little clothes. Even in Finnmark, temperatures in excess of 20°C in July and August are not uncommon, and spells of such conditions may occur in June and even as early as May. This climate dominates in areas such as in the Pasvik valley and on Finnmarksvidda, but in periods of southerly winds it may grace the outer coast as well. With northerly winds, however, an icy breeze from the Barents Sea enters the coastal areas, sending the summer temperatures down to a chilly 5°C. These are often the conditions you encounter at Slettnes and the Varanger Peninsula and is the reason why even in mid-summer you should bring your winter clothes to northern Norway and Svalbard. Just as the sun never rises above the horizon in mid-winter, it never sets in summer north of (or actually from a bit south of) the Arctic Circle.

Autumn (September–November) is generally windier and rainier than summer, but is equally variable. The dominating pattern is that of low-pressure weather systems entering from the Atlantic Ocean.

Infrastructure

Norway is tied together with railroads (north to Bodø) and national networks of airports and roads. The roads in the countryside are often winding, narrow and bumpy. Even though the number of charging stations for electric cars are steadily increasing, birdwatchers will still be better off using traditional cars when headed out in the countryside.

'The right to roam'

In Norway, the general public has a right to access on foot almost any public and privately owned uncultivated land. This freedom to roam is called *Allemannsretten* ("Everyman's right"). The right applies as long as the area is defined as wild nature or *utmark*, as opposed to cultivated *innmark*. To the latter are considered gardens, arable land, hay meadows, grazing pastures and

Visitor restrictions apply in some nature reserves at certain times of year, such as in this Shag colony on Røst archipelago. The sign reads that there is a traffic ban from April 15th to July 31st. *Photo: Tomas Aarvak.*

forest plantations, and similar areas where public access will be of undue disadvantage to the owner. *Innmark* can still be walked or skied upon when frozen or covered in snow in winter. The shoreline, even where inhabited, and roads through uncultivated areas are considered as *utmark*. You may camp freely in *utmark* for up to two nights at the same location without asking permission from the landowner, but no closer than 150 metres from a dwelling house or cabin. With *Allemannsretten* comes an obligation neither to harm or disturb wildlife, not to litter nor to damage vegetation.

Words of caution

Norway is considered to be a very safe country because of low crime rates and no really poisonous or otherwise dangerous animals, except for polar bears on Svalbard. Nature can still be very rough, though, and every year tourists are sadly killed and injured, often due to reckless conduct, such as by entering the wilderness with insufficient clothes, orientation equipment and other necessary hiking gear, or going out to sea in small vessels without life vests, or skiing in avalanche prone areas.

Sufficient fuel. Birdwatching in the Norwegian countryside often implies long driving distances in remote areas, so fill up your vehicle whenever possible when driving in remote areas.

Mosquitoes. These blood-thirsty insects are found all over Norway, and up north they are famous for their huge size and numbers. Do not enter the Norwegian wilderness in June, July and August without the best repellent available. In Finnmark, consider bringing mosquito gloves and a hat net as well. The numbers of mosquitoes vary from one year to the next, being lower in dry and windy summers.

Midnight sun and insomnia. If you are not used to the midnight sun it may cause difficulties sleeping and you end up exhausted. You may want to bring eye-blinds of the kind they hand out on long-distance flights, and make sure that your hotel room or tent is located in the shade.

Allemannsretten – 'the right to roam' – is a legal term in Norway, meaning that you may walk freely most places in the countryside and along the seashore. It is illegal for landowners to put up fences or signs restricting public access where this right applies. No wonder this House Sparrow is puzzled by the sign reading 'Private area' – such a sign is not a common sight in Norway! *Photo: Svein Arne Orvik.*

The birds of Norway

More than 530 species of wild birds have been recorded in Norway. Of these, about 220 breeds regularly and roughly 100 more are considered regular visitors either as migrants or winter visitors. Willow Warbler is considered to be the most numerous breeding bird in Norway, followed by Chaffinch. Birds that are common in many parts of the country, or that are so numerous and conspicuous that they are hard to miss, are not granted too much space in this guide book. Such easy-to-find species, and also a few which are widely distributed in the rest of Europe, are not systematically mentioned in the section *Notable species* in the site descriptions either. These common species are listed in the table on the next page.

Migrant species

Given the climate in Norway, many species leave the country and move south to warmer areas for the winter. In the mountains and the high north, the summer season for many species only lasts a few weeks before they head back to their winter quarters.

Most migrant land birds head south or southwest across the North Sea to Britain or the Continent, some continuing all the way to Sub-Saharan Africa. A very few species migrate southeast instead, such as Common Rosefinch and Rustic Bunting. A very few high-Arctic species consider Norway to be the warmer areas, and choose to migrate here to spend the winter, such as Purple Sandpiper, Glaucous Gull, King and Steller's Eiders.

Birds reluctant to cross bodies of water, such as soaring raptors and Cranes, migrate along the coast of Sweden, crossing over to Denmark where the strait is narrow. In order to follow this route in autumn, many raptors from southern Norway must follow the Skagerrak coast north – the wrong way, so to speak – crossing over at the narrow Oslofjord. In spring, on sunny days producing thermal winds, good numbers of such birds can be seen from anywhere with a good view in the Oslofjord area and along the Skagerrak coast. They are joined by a few waterbirds as well, notably Pink-footed Geese and Cormorants, migrating overland

Common birds in Norway

This is a list of species not always mentioned specifically at each site covered in this book, because they are considered to be common and/or easy to find. However, species marked * are usually mentioned where breeding.

Canada Goose *Branta canadensis*
Greylag Goose *Anser anser*
Mute Swan *Cygnus olor*
Common Shelduck *Tadorna tadorna*
Eurasian Wigeon *Mareca penelope*
Mallard *Anas platyrhynchos*
Eurasian Teal *Anas crecca*
Tufted Duck *Aythya fuligula*
Common Eider *Somateria mollissima*
Common Goldeneye *Bucephala clangula*
Common Merganser/Goosander *Mergus merganser*
Red-breasted Merganser *Mergus serrator*
Common Swift *Apus apus*
Common Cuckoo *Cuculus canorus*
Feral Pigeon *Columba livia* var. *domestica*
Common Woodpigeon *Columba palumbus*
Eurasian Collared Dove *Streptopelia decaocto*
Eurasian Oystercatcher *Haematopus ostralegus*
Grey Plover *Pluvialis squatarola*
European Golden Plover *Pluvialis apricaria*
Common Ringed Plover *Charadrius hiaticula*
Northern Lapwing *Vanellus vanellus*
*Eurasian Whimbrel *Numenius phaeopus*
Eurasian Curlew *Numenius arquata*
*Bar-tailed Godwit *Limosa lapponica*
Eurasian Woodcock *Scolopax rusticola*
Common Snipe *Gallinago gallinago*
Common Sandpiper *Actitis hypoleucos*
*Green Sandpiper *Tringa ochropus*
*Wood Sandpiper *Tringa glareola*
Common Redshank *Tringa totanus*
*Common Greenshank *Tringa nebularia*
*Turnstone *Arenaria interpres*
Red Knot *Calidris canutus*
*Ruff *Calidris pugnax*
Curlew Sandpiper *Calidris ferruginea*
*Sanderling *Calidris alba*
*Dunlin *Calidris alpina*
*Purple Sandpiper *Calidris maritima*
*Little Stint *Calidris minuta*
Arctic Tern *Sterna paradisaea*
Common Tern *Sterna hirundo*
Black-legged Kittiwake *Rissa tridactyla*
Black-headed Gull *Chroicocephalus ridibundus*
Common Gull *Larus canus*
Herring Gull *Larus argentatus*
Great Black-backed Gull *Larus marinus*
Lesser Black-backed Gull *Larus fuscus*
*Razorbill *Alca torda*
*Common Guillemot/Murre *Uria aalge*
Black Guillemot *Cepphus grylle*
Grey Heron *Ardea cinerea*
Eurasian Sparrowhawk *Accipiter nisus*
Tawny Owl *Strix aluco*
Great Spotted Woodpecker *Dendrocopos major*

European Green Woodpecker *Picus viridis*
Eurasian Jay *Garrulus glandarius*
Eurasian Magpie *Pica pica*
Western Jackdaw *Coloeus monedula*
Hooded Crow *Corvus cornix*
Northern Raven *Corvus corax*
Coal Tit *Periparus ater*
Crested Tit *Lophophanes cristatus*
Willow Tit *Poecile montanus*
Eurasian Blue Tit *Cyanistes caeruleus*
Great Tit *Parus major*
Eurasian Skylark *Alauda arvensis*
Sand Martin *Riparia riparia*
Barn Swallow *Hirundo rustica*
Western House Martin *Delichon urbicum*
Willow Warbler *Phylloscopus trochilus*
Chiffchaff *Phylloscopus collybita*
Eurasian Blackcap *Sylvia atricapilla*
Garden Warbler *Sylvia borin*
Lesser Whitethroat *Curruca curruca*
Common Whitethroat *Curruca communis*
Goldcrest *Regulus regulus*
Eurasian Wren *Troglodytes troglodytes*
Nuthatch *Sitta europaea*
Treecreeper *Certhia familiaris*
Starling *Sturnus vulgaris*
Song Thrush *Turdus philomelos*
Redwing *Turdus iliacus*
Common Blackbird *Turdus merula*
Fieldfare *Turdus pilaris*
Spotted Flycatcher *Muscicapa striata*
European Robin *Erithacus rubecula*
Pied Flycatcher *Ficedula hypoleuca*
Common Redstart *Phoenicurus phoenicurus*
Whinchat *Saxicola rubetra*
Northern Wheatear *Oenanthe oenanthe*
House Sparrow *Passer domesticus*
Eurasian Tree Sparrow *Passer montanus*
Dunnock *Prunella modularis*
Western Yellow Wagtail *Motacilla flava*
White Wagtail *Motacilla alba*
Meadow Pipit *Anthus pratensis*
Tree Pipit *Anthus trivialis*
Rock Pipit *Anthus petrosus*
Eurasian Chaffinch *Fringilla coelebs*
Brambling *Fringilla montifringilla*
Eurasian Bullfinch *Pyrrhula pyrrhula*
European Greenfinch *Chloris chloris*
Twite *Linaria flavirostris*
Common Linnet *Linaria cannabina*
Eurasian Siskin *Spinus spinus*
Common Redpoll *Acanthis flammea*
Red/Common Crossbill *Loxia curvirostra*
Yellowhammer *Emberiza citrinella*
Common Reed Bunting *Emberiza schoeniclus*

Little Bunting on the breeding grounds: one of many reasons to come to Norway. *Photo: Terje Kolaas.*

in large flocks from late March to early May. The geese are heading to their main staging grounds in the Trondheimsfjord and up in Lofoten and Vesterålen before crossing the Barents Sea to Svalbard. A few divers (mainly Black-throated) and ducks may be seen migrating through the Oslofjord and Skagerrak area as well, but most of these birds will rather follow the Norwegian west coast, all the way from Lista in the south to Vardø in the northeast. Any headland along this stretch of coast can produce good numbers of migrant waterbirds, particularly in May. Shorebirds are typically seen at migration sites from late spring through summer to early autumn. Pelagic birds are mostly encountered from late May through to November, often peaking in late summer and early autumn. Strong westerlies are typically best, but not too strong. A fresh breeze to near gale force (Beaufort scale 4–7) is usually considered to be good.

Report your sightings!

If you either by chance or by active searching come across a genuine local or national rarity, please document the record as best as you can and submit it to the local or national rarities committee. This is most easily done by reporting it directly to BirdLife Norway, (www.birdlife.no) or through the observations database www.artsobservasjoner.no. In this book, examples of vagrants that have been encountered at each site are sometimes briefly mentioned, just to show the potential of the site and perhaps to inspire you to search for new rarities.

Ready, set...

Before you go: remember that this book only tells you where, when and how to maximize your chances of encountering the wonderful birds of Norway. The birds, however, you have to search for and find yourself. Good luck!

THE OSLO AREA

O slo is the gateway for many people visiting Norway from abroad. It is the capital and by far the largest city in the country. Still, including suburbs, Oslo has only just above 1.5 million inhabitants, and so the city is quite small compared to most other capitals. The distances within the city are short, and public transportation and road networks are well developed. Compared to other capitals, Oslo is a green city with many gardens and parks. It is almost completely encircled by the Oslofjord on one side and

Common Rosefinch may be found in the Oslo area in summer, usually in open, sunny terrain with scattered bushes and trees. *Photo: Terje Kolaas.*

the Oslomarka forest on the other. Both are something the outdoor-minded inhabitants of Oslo know to take advantage from, and they are enthusiastic recreational users of these natural resources. Yet, it is still possible to find quiet spots both in the forest and along the fjord, particularly on weekdays and on days with less than perfect weather.

SITES IN THE OSLO AREA

The Oslofjord and the river systems and valleys in south-eastern Norway are important flyways for migrating ducks, geese, cormorants, cranes and birds of prey, giving plenty of opportunity to experience the migration in spring and autumn. In addition, there is a rich breeding fauna of birds in the Oslo area, including several attractive species typical of this part of Europe, including Whooper Swan, Capercaillie, Hazel Grouse, White-tailed Eagle, Crane, Woodcock, White-tailed Eagle, Tengmalm's Owl, Pygmy Owl, Wryneck, Black Woodpecker, Three-toed Woodpecker, Thrush Nightingale, Icterine Warbler, Red-backed Shrike, Nutcracker and Common Rosefinch. In this south-eastern part of Norway, you may also find a few species common in Continental Europe, which here are close to their northern breeding limits and hence are particularly popular with Norwegian birders, such as Hobby, Little Ringed Plover, Stock Dove, Nightjar, Black Redstart, Mistle Thrush and Goldfinch.

Although many of these species may be hard to find during a short visit to the area, you will surely have no difficulty in finding interesting birds and suitable sites for birdwatching in and around Oslo.

THE BEST SITES

For many birders visiting from abroad, the main ornithological attraction near Oslo is Oslomarka, the woodlands surrounding the city in three directions. Here you can find several species of bird typical of Scandinavian forests. However, the greatest diversity of birds near Oslo is found in association with wetlands and the fjord. The Oslo area has a variety of nutrient-rich lakes situated in areas with a mosaic of woodland and farmland. These may be teeming with birds, particularly in spring and summer. Lake Østensjøvann in central Oslo is popular with both birders and the general public alike, because of its convenient location and lots of confiding water birds, perfect for photographers and families with kids. And every now and then, the common species are accompanied by more scarce gems. The harbour and parks in Oslo are regularly scrutinized by birdwatchers in search of uncommon gulls and ducks. Fornebu peninsula in Bærum has some interesting breeding species and is also a prime migration site, sporting the most extensive species-list in the region. Another good place in the Oslo area in which to see migrating birds is the Nesodden headland just across the fjord from Oslo harbour. This is also a good spot from where to monitor birds in the inner Oslofjord basin. On the ferry trip from the docks in Oslo across to Nesodden,

Nutcracker nest in the woodlands surrounding Oslo, often appearing in the city's parks and residential areas in late summer and autumn. *Photo: Kjetil Schjølberg.*

you may have close encounters with auks, skuas, gulls and ducks, particularly after strong southerly winds in summer, or in any weather in late autumn and winter. If you want to make a serious attempt to see pelagic birds not too far from Oslo, you should head for the Drøbak sound, from where you gain views to the outer section of the Oslofjord and where seabirds are pushed near the shore and may pass at fairly close range. The Drøbak sound has the added bonus of breeding White-tailed Eagle.

The perhaps most notable of the general birding sites in the Oslo area, are the two major inland deltas in the lakes Øyeren and Tyrifjorden. These are wetlands of international importance, and host to a variety of bird species throughout the year. These are both generally the best sites for shorebirds, ducks, geese and other waterbirds near Oslo.

If you have a full day or more at your disposal, you should consider heading a bit further, into the deep forests and mountains of interior South Norway, such as Finnskogen or Trillemarka forests, or to Valdresflya mountain plateau – all of which are covered in the next chapter, *The Southern Interior*. Or you can head south towards the outer coast, covered in the chapter *Coastal Skagerrak*.

1. Urban Oslo

Oslo County
GPS: 59.90646° N 10.75416° E (the opera house)

Notable species: *Breeding*: Barnacle Goose (feral population), Lesser Black-backed Gull, Goshawk, Peregrine Falcon, Stock Dove, Collared Dove, Green Woodpecker, Lesser Spotted Woodpecker, Grey Wagtail, Black Redstart, Fieldfare, Nutcracker, Dipper, Goldfinch, Linnet and Hawfinch. *Migrants*: Wildfowl, gulls, auks (winter) and more.

Description: The parks and residential gardens in Oslo attract a variety of birds, and you may see many of the commoner species here. One of the finest birding areas in Oslo is at the Bygdøy peninsula. Here, mature deciduous forest is intermixed with a patchwork of lush pastures, where Tawny Owl, Stock Dove, Goldfinch and Hawfinch breed. A visit to Bygdøy can also be combined with visits to some of the most popular museums in Oslo. The same applies to the Vigeland Sculpture Park and the Natural History Museum. Most years you can see auks in the harbour during late autumn and winter, with Guillemot and Razorbill as the commoner species. Some years the fjord is invaded by Little Auks, and the odd Brünnich's Guillemot has spent the winter here. The many picturesque islands that freckle the inner Oslofjord, interconnected with a public ferry, can be well worth visiting, particularly the island Gressholmen.

Best season: All year, but late spring and early summer has the greatest diversity.

Directions: From the Oslo Central Station you can walk, hire a bike, or use the city's excellent public transport system, consisting of buses, subways, trams and intercity trains, and a multitude of taxi companies.

Arriving in Oslo by cruise ship

From the cruise ship landing site, you can walk along the docks and look for gulls, ducks and auks. On early spring mornings, Black Redstart may sing from rooftops behind the Akershus fortress. Take a cab, public transportation or a bicycle (electric bicycles are commonplace) to reach the rather bird rich sites - some even with cultural alibis - such as the Vigeland Sculpture Park, Bygdøy peninsula (several famous museums), the Botanical gardens (Natural History Museum), the beautiful Lake Østensjøvann, or the vast Oslomarka forest surrounding Oslo. Another option is to go to Aker Brygge and hop onto a ferry to the small, picturesque islands in the inner fjord or to Nesoddtangen headland. You can also rent a car and head further afield, or engage a local birding guide (see www.ornforlag.no).

Oslo · Interior · Skagerrak · Western · Central · Northern · Finnmark · Svalbard

Tactics: From **a.** The opera house in Bjørvika, by the mouth of the river Akerselva next to Oslo Central Station, you have a view to one of the best parts of the harbour for gulls, auks and ducks. Walk or drive along the docks and use your telescope often and thoroughly, preferably all the way from Sjursøya in the east to **b.** Frognerkilen inlet in the west. On

c. Bygdøy peninsula, you can follow small roads and foot paths through the woods and fields, particularly good in spring and early summer. The Hensenga fields on the west side can be particularly good for waterbirds in periods with flooding. The Huk beach in the south provides a good view of the inner Oslofjord, and the skerries here often produce

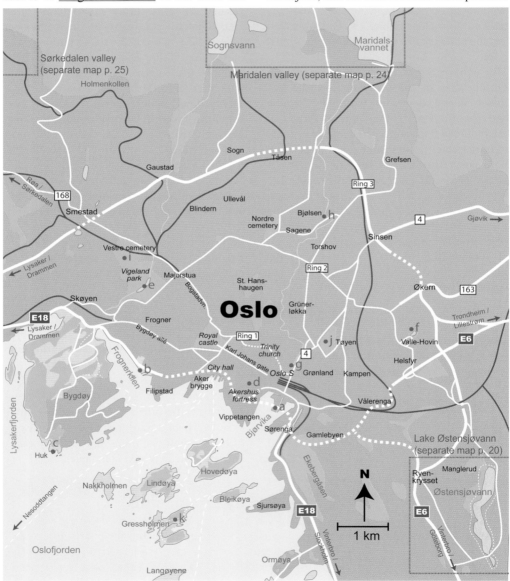

Urban Oslo: a. The opera house; **b.** Frognerkilen inlet (gulls in outer section, ducks and geese inner section); **c.** Huk, Bygdøy peninsula; **d.** core Black Redstart area; **e.** Vigeland sculpture park; **f.** Hovindammen pond; **g.** Vaterland riverside; **h.** Bjølsen riverside; **i.** Vestre gravlund cemetary; **j.** Botanical garden; **k.** Gressholmen island. See separate maps of Lake Østensjøvann and Maridalen and Sørkedalen valleys.

Purple Sandpiper in winter. It is not advisable, however, to walk with your binoculars among the naked sun-bathers occupying this beach in summer.

If you walk from the opera in Bjørvika west into **d.** the city centre, you may encounter the odd Black Redstart, particularly between Akershus fortress and Trefoldighetskirken church. In winter, bring some loaves of bread to attract gulls and ducks in **e-h.** the parks. **i.** Vestre Gravlund (*Vestre cemetery*) is particularly good for Nutcracker during late summer and early autumn, while **j.** Tøyenparken surrounding the Natural History Museum is good for Hawfinch and wintering passerines. **k.** Gressholmen and the other islands in the inner Oslofjord, are reached by ferry from the docs by Aker brygge.

2. LAKE ØSTENSJØVANN

Oslo County

GPS: 59.88124° N 10.83187° E (the south end)

Notable species: *Breeding*: Mute Swan, Great Crested Grebe, Moorhen, Coot, Goshawk, Peregrine Falcon, Green Woodpecker, Lesser Spotted Woodpecker, Marsh Warbler, Reed Warbler, Wood Warbler and Goldfinch. *Migration*: Wildfowl, gulls and more.

Description: Lake Østensjøvann is a nutrient-rich lake near downtown Oslo, hosting particularly good numbers of birds. The lake is surrounded by residential areas and lawns, some agricultural fields and a small coniferous forest. The lake is 1.8 km

Lake Østensjøvann in Oslo offers close encounters with birds such as Greylag Goose and Mallard (as shown), Barnacle Goose, Great Crested Grebe, Moorhen and Coot. *Photo: Bjørn Olav Tveit.*

long, and narrow, flanked by reedbeds and deciduous trees and bushes. Encircling the lake is a walkway that is a popular and busy hiking trail for the public. Birds at Lake Østensjøvann are fed by people on a regular basis, making the birds very confiding and popular amongst bird photographers.

Best season: Early summer has the most vibrant bird life, but Lake Østensjøvann offers exciting birding throughout the year,

at least as long as the lake is ice-free. Even during severe cold spells, a small section in the south end is usually free of ice.

Directions: *Public transport:* Take the metro towards Mortensrud. Leave the metro at either Skøyenåsen, Oppsal, Bøler or Bogerud stations, and walk westwards (see map). *By car from the city centre*, take main road E6 north towards Trondheim and make a turn inside the Ekeberg Tunnel towards

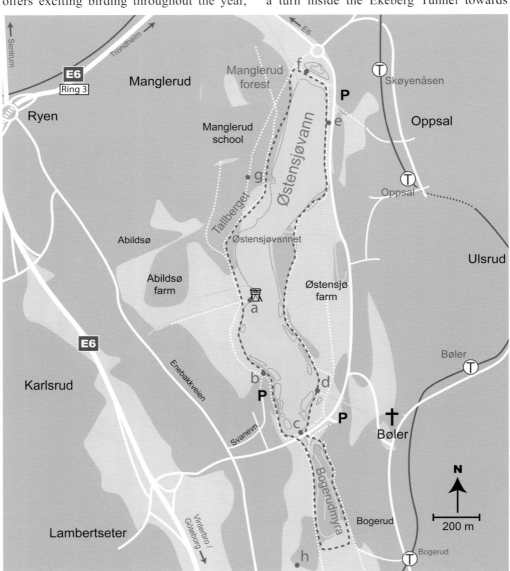

Lake Østensjøvann in Oslo: a–f. Places with a good view of the lake; g. Tallberget hill with nice mixed forest; h. Visitor centre.

Ryen. From the Ryen junction and onwards, see the map.

Tactics: Stroll along the footpath that leads around the lake. The birds are usually most approachable in the south end, where they are regularly fed. Migrants are often seen in the more undisturbed middle and northern sections of the lake.

3. OSLOMARKA FOREST

Oslo, Akershus and Buskerud counties

GPS: 59.99410° N 10.75277° E (Hammeren, Maridalen), 60.0191° N 10.5841° Ø (Skansebakken, Sørkedalen)

Notable species: *Breeding:* Whooper Swan, Hazel Grouse, Black Grouse, Capercaillie (scarce), Black-throated Diver, Goshawk, Honey Buzzard, Buzzard, Peregrine Falcon, Green Sandpiper, Snipe, Woodcock, Pygmy Owl, Tawny Owl, Tengmalm's Owl, Wryneck, Green Woodpecker, Black Woodpecker, Lesser Spotted Woodpecker, Three-toed Woodpecker (scarce), Mistle Thrush, Icterine Warbler, Wood Warbler, Long-tailed Tit, Willow Tit, Crested Tit, Coal Tit, Red-backed Shrike, Nutcracker, Goldfinch and Common Rosefinch. *Migration:* Pink-footed Goose, Cormorant, Crane, raptors and more.

Description: Oslo is surrounded by a vast forest collectively known as Oslomarka, with more specific names to each of its many parts, such as, from west to east, Vestmarka, Krokskogen, Nordmarka, Østmarka and Lillomarka, which are the names you find on local maps. Oslomarka is a popular recreational area, but you may still find tranquil places surprisingly close to densely inhabited areas. In such places you can experience many of the common woodland species of the region, and, with a little bit of patience and luck, also species one normally associates with deep taiga forest, such as Capercaillie and Three-toed Woodpecker. Virtually

any part of Oslomarka can prove to be interesting, and you might just as well try the parts most easily accessible from whatever part of Oslo you plan to visit. The rural, idyllic valleys of Maridalen and Sørkedalen are among the best and most easily accessible areas in which to go birdwatching. These valleys form good starting points for longer trips into Nordmarka. The farmland in the valleys creates a contrast with the conifer forest and contributes to making the area varied and particularly attractive to birds. At the bottom of both valleys are two larger freshwater bodies, Lake Maridalsvannet and Lake Bogstadvannet, respectively. The former is a source of drinking water for the city of Oslo, and is also the source of river Akerselva, which flows down to the opera house by the Oslofjord. There are often Black-throated Diver and a diversity of ducks to be found in these lakes as soon as the ice cover gradually disappears in April. Ospreys hunt here in late spring and summer. In recent years, Hobby has become quite regular. The surrounding fields and pastures may hold resting geese and other migrants, and in late spring and summer you can find a number of nesting birds associated with forests and farmland in this part of Scandinavia. Both valleys are

Hazel Grouse is amongst the many woodland species you may encounter in the Oslomarka forest. *Photo: Kjetil Schjølberg.*

Oslo
Interior
Skagerrak
Western
Central
Northern
Finnmark
Svalbard

At Mellomkollen in Maridalen valley, just outside of Oslo, even the most urban birdwatcher gets to enjoy several forest species without losing sight of the city. Lake Maridalsvann and the Oslofjord is also visible in the distance. *Photo: Bjørn Olav Tveit.*

great for nocturnal trips by car or bicycle. At Mellomkollen in Maridalen valley and at Heikampen in Sørkedalen valley, you find old, mature conifer forest with species typically associated with deep forest, such as Capercaillie, Black Grouse, Tengmalm's Owl and Black Woodpecker. Common Crossbill is found all year round, and you may some places find the sparser Parrot Crossbill, especially in areas of pine-dominated forest. From Sørkedalen you can also hike to Lake Triungsvann, one of the finest wetlands in Oslomarka, with nesting Black-throated Diver, Whooper Swan and more. Particularly during late fall and winter, you may be lucky and find Golden Eagle, Hawk Owl or perhaps a flock of Pine Grosbeak anywhere in Oslomarka.

Other wildlife: Despite the steady traffic of people, you may see quite a few mammals and other animals if you enter Maridalen or Sørkedalen valleys at dusk or dawn. Elk (moose), roe deer, red fox, badger and

mountain hare are sometimes seen roadside. Lynx is an inhabitant of Oslomarka as well, although they are notoriously shy and almost never seen. Beaver and several species of bat are to be found in the area, with northern bat as the commoner species of bat.

Best season: Late February to the end of March is best for calling owls, March–May for migrants and displaying Capercaillie and Black Grouse. The latter half of May through June holds the greatest variety of birds, a time when most of them are still active singing and displaying, making them easier to locate and identify.

Directions: *Public transportation from central Oslo to Maridalen valley:* Take the bus to Hammeren or – if you are headed for the mature forest at Mellomkollen – get off the bus at Nordre Vaggestein bus stop. *By car to Maridalen:* Take the exit from Ring 3 at Nydalen or Tåsen, and follow signs to Maridalen.

Public transportation from central Oslo to Sørkedalen valley: Take the subway to Røa station, from where you continue with bus into Sørkedalen. *By car to Sørkedalen:* From Ring 3 near Smestad, take the exit to Røa. Then turn towards Bogstad camping and keep straight until you enter the valley.

Besides these two valleys, Maridalen and Sørkedalen, the following can be recommended as particularly good starting points for birding trips into Oslomarka: Sognsvann and Skullerud in Oslo, Lommedalen and Kjaglidalen in Bærum, Semsvannet in Asker, and Losby estates in Lørenskog. It is also nice to ride a bicycle along the forest roads in Oslomarka, and in winter there is a widespread network of cross-country ski tracks. You may take a train to Stryken or a bus to Sollihøgda, and then make your way back through Oslomarka on bike or skis, mostly downhill, back to the city.

Tactics: Drive slowly along the roads in the valleys and stop to investigate wherever the habitat looks interesting. The woodlands are best explored on foot. The main tracks are well-marked with paint on tree-trunks and prominent rocks along the routes, in either blue (summer tracks) or red (ski tracks) paint. The same colours are repeated on signs and on hiking maps. It can be worthwhile to buy a map of the part of Oslomarka you plan to visit, before you set off.

In the following section, some particularly good birdwatching areas in the Maridalen and Sørkedalen valleys are described in detail.

Maridalen valley (see separate map): The fields at **a.** Skjerven farm are often moist and may host ducks and shorebirds. At this location, the shoreline of nearby Lake Maridalsvannet is swampy and good for many wetland species. At **b.** Hammeren, the main road passes close to the waters edge. Here, the river Skjersjøelva is responsible for making this where the ice cover forms the latest in autumn and opens up again earliest in spring, making it particularly popular with waterbirds. Dipper is often seen along the river all year. Near the ruins of the old church **c.** Margaretakirken, you get a nice overview of the surroundings. This is a good place to watch for migrating birds and birds resting in the surrounding fields. The best view of this part of lake Maridalsvannet is obtained

Goshawk is a fairly common breeder in woodland all over Norway, including the outskirts of Oslo. Although shy, it often enters suburban and even urban areas to hunt in winter. *Photo: Terje Kolaas.*

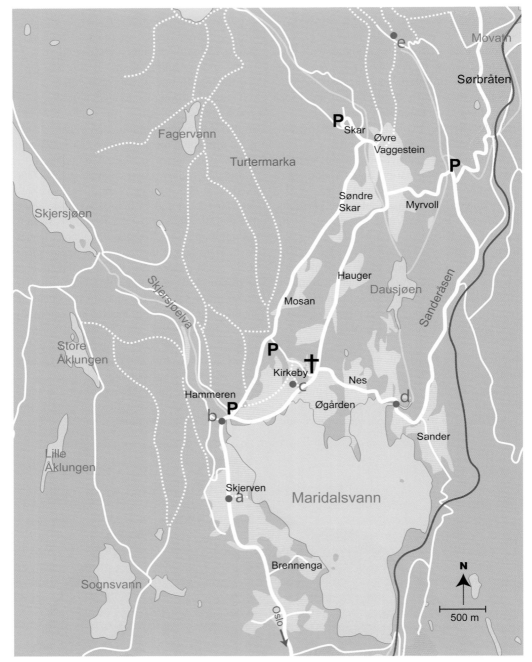

Maridalen valley: **a.** Skjerven farm; **b.** Hammeren; **c.** stone church ruins; **d.** mouth of river Dausjøelva; **e.** path to forested Mellomkollen hill.

roadside, just past the church. Where the creek **d.** Dausjøelva meets the lake, you often get birds close-up, and so this is a popular spot among photographers. Common Rosefinch is often seen here i late spring and summer.

e. Mellomkollen hill: Continue further into the valley and turn right onto the road signposted to Sørbråten. Continue 1 km and leave the car in the car park at Mobekken. From here, walk north along the forest road and turn left at the Y-junction up the hill. Soon there is a path leading to the right, signposted to Mellomkollen (you will arrive at this junction also if you instead walk from the Nordre Vaggestein bus stop). This path takes you through mixed forest with Tree Pipit, Wood Warbler and Treecreeper among the characteristic species, up to the lookout point at the top of Mellomkollen, 4 km from the car

park. Listen for singing Mistle Thrush in the distance. From the lookout point at the top, a path leads east past Lake Svarttjern and through magnificent, fir-dominated mature forest with the possibility of all three grouse, Three-toed Woodpecker, Pine Crossbill and more.

Sørkedalen valley (see separate map): You get the best view of Lake Bogstadvannet from

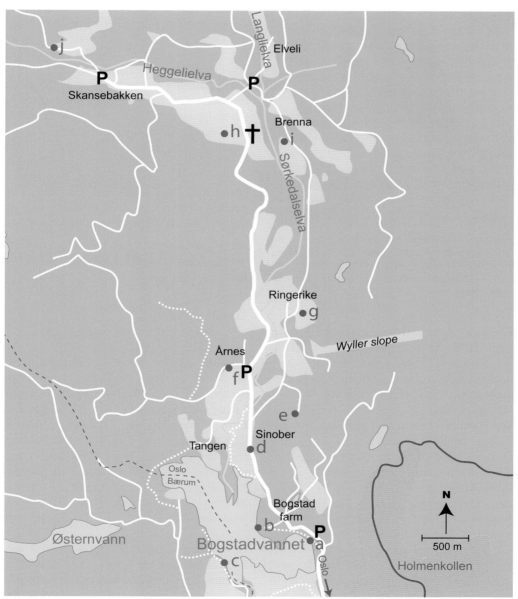

Sørkedalen valley: a. Badesvingen beach; **b.** Bogstad farm; **c.** Fossum; **d.** viewpoint to the fields north of the lake; **e.** Sinober forest; **f.** Årnes; **g.** Søndre Ringerike farm; **h.** Sørkedalen church; **i.** Brenna; **j.** Skansebakken.

Hawfinch has become increasingly common in the Oslo area over the last 20 years. It is usually found in association with mature deciduous forests, but also in subuarban areas and the city parks. *Photo: Terje Kolaas.*

and many of these animals may gather here during winters when the snow is particularly deep. **g.** Søndre Ringerike farm is another good spot for Common Rosefinch. The fields near **h.** Sørkedalen church may also be good for resting passerines in spring. **i.** Brenna still holds breeding Lapwing and sometimes feeding Cranes during migration. The lush birch woods to the east may produce Long-tailed Tit and Wryneck. **j.** Skansebakken is a good starting point for owling at night in early spring. Listen for e.g. Tengmalm's and Pygmy Owls along the forest roads leading into the woods.

When listening for owls at night, the usual tactic is to walk from starting points high up in the valley and walk as far in along the timber roads as you have time for, stopping to listen carefully every now and then. Pygmy Owl usually calls at dusk and dawn, while the other species in the area, including Tengmalm's Owl, usually start to call well after dark. The number of owls vary a lot from one year to the next, depending on the rodent populations.

4. LAKE DÆLIVANN AND STOVIVANN

Akershus County
GPS: 59.91040° N 10.53528° E (parking Gjettum)

Notable species: *Breeding:* Hazel Grouse, Moorhen, Green Sandpiper, Goshawk, Peregrine Falcon, Hobby, Honey Buzzard, Tawny Owl, Stock Dove, Green Woodpecker, Black Woodpecker, Lesser Spotted Woodpecker, Grey Wagtail, Reed Warbler, Icterine Warbler, Wood Warbler, Long-tailed Tit, Marsh Tit (scarce), Treecreeper, Nutcracker, Goldfinch, Linnet, Common Rosefinch and Hawfinch. *Winter:* Three-toed Woodpecker (scarce).

Description: Lake Dælivann is a small and nutrient-rich lake just outside Oslo, situated in an area of pastures and lush mixed forest just underneath the characteristically shaped Kolsås hill, a landmark visible from downtown Oslo. This is a good area in which to find many of the

a. Badesvingen and from stately **b.** Bogstad farm. From the latter you can see across to the outlet of river Sørkedalselva, where ducks, swans and geese often congregate in early spring, and where Stock Dove, Black Woodpecker, Green Woodpecker, Goldfinch and Hawfinch breed. You can also look out over Lake Bogstadvannet from **c.** Fossum in Bærum. From **d.** the farmland north of the lake you have a nice overview of the surroundings, good for scanning for migrant raptors or resting migrants in the fields. The mixed forest at **e.** Sinober may host a variety of woodpeckers, Wood Warbler and the slight possibility of Red-breasted Flycatcher. **f.** Årnes is usually the best spot in Sørkedalen valley for resting passerines in spring. Red-backed Shrike, Icterine Warbler and Common Rosefinch breeds in the area. There is also a feeding station for Elk here, bird species that inhabit the woodlands in this part of Scandinavia. After exploring the lake and its surrounding forest and fields, you can extend the trip with a walk east along the ancient road Ankerveien. This is a popular hiking trail along the often bird-rich border zone between mixed forest and farmland/golf course. A public feeder is situated along the trail at Øverland, active during winter. You can also hike along the foot of Kolsås or even follow the tracks up to the coniferous woods at its summit, where you may find species typical of this habitat.

In late spring and early summer, keep your ears and eyes peeled for the slight possibilities of Greenish Warbler and Red-breasted Flycatcher. Serin bred for the first time in Norway at Haslum cemetery in 2010. This cemetery is also a particularly good site for Hawfinch.

Best season: Spring and early summer.

Directions: *Public transport:* Take the metro to Gjettum. An alternative starting point is to enter the area from the east by taking bus destined for Bærums Verk from Oslo (or Bekkestua) to Haslum school or Øverland

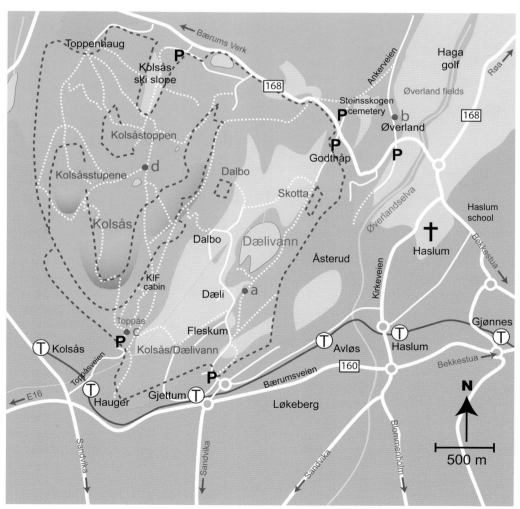

Lake Dælivann: a. Path through mature mixed forest, leading to a viewpoint of Lake Dælivann; **b.** BirdLife's public feeder; hiking trail; **c.** nice mixed forest at the foot of the Kolsås cliffs, at the top of Toppåsveien road; **d.** Kolsås hill, notable hill formation with forest dominated by spruce.

farm. The bus continues to Kolsås ski resort, which makes a good starting point for walks to the summit of Kolsås hill.

By car from Oslo, follow E18 west towards Drammen. In Sandvika, take the exit to E16 towards Hønefoss. After 2 km, exit again and follow signs towards Bekkestua. Then, see map.

Tactics: The area is best explored by foot in the early morning hours.

5. FORNEBU PENINSULA

Akershus County

GPS: 59.89407° N 10.59916° E (Lilløyplassen), 59.89064° N 10.63254° E (Rolvstangen)

Notable species: *Breeding:* Water Rail, Little Ringed Plover, Ringed Plover, Lesser and Great Black-backed Gulls, Goshawk, Long-eared Owl, Stock Dove, Marsh Warbler, Reed

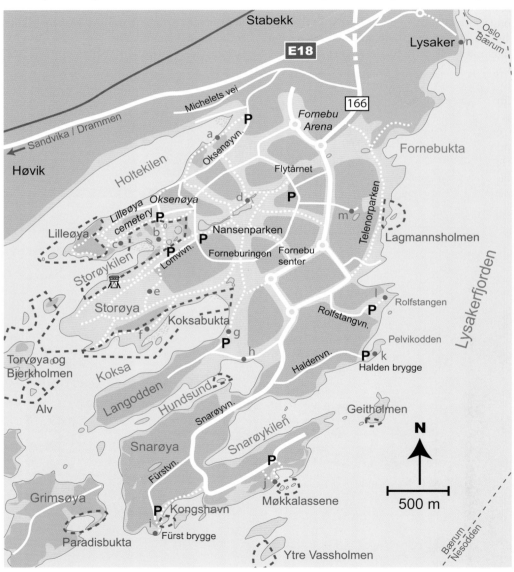

Fornebu: a. Holtekilen inlet; **b.** Storøykilen inlet; **c.** Lilløyplassen nature-house; **d.** Nansenparken; **e.** Storøya park and beach; **f.** and **g.** viewpoints Koksa inlet; **h.** Hundsund inlet; **i.** Fürst viewpoint; **j.** Lortbukta inlet; **k.** Halden viewpoint; **l.** Rolfstangen viewpoint; **m.** Telenorstranda park; **n.** Lysaker docks.

Warbler, Goldfinch, Linnet and Common Rosefinch. *Migration*: Pink-footed Goose, raptors, shorebirds, passerines and more. *Winter:* Water Rail and Bearded Tit (scarce).

Description: Fornebu is a rather flat peninsula in the Inner Oslofjord, indented by several narrow and shallow inlets which are fringed with marshland and lush reedbeds. Fornebu is one of the best sites for finding resting migrants close to Oslo. Especially on mornings with overcast weather in spring, significant numbers of pipits, thrushes and finches can migrate overhead or rest in the vegetation. On sunny days from late March to the end of May, it can be a good spot for watching migrating raptors and other soaring birds. Buzzards and large flocks of Cormorants may pass overhead early in this period, followed by Rough-legged Buzzard, Hen Harrier, Crane and Pink-footed Goose, the latter two usually in flocks. In May you may hope for Honey Buzzard, Hobby and

several species of diver and duck. Many species breed on Fornebu, and Common Rosefinch is frequently found singing in late spring and early summer. At low tide, exposed mud in the inlets may attract shorebirds, a group of birds that you seldom find in any good numbers in the inner Oslofjord.

In late autumn and winter, Fornebu may host a suite of wintering birds of the likes of Water Rail, Great Grey Shrike and Bearded Tit. Some years, the inner Oslofjord is invaded by substantial numbers of auks in this time of year, usually mainly Guillemot and Razorbill, sometimes Little Auk and even more exotic species, including the odd Brünnich's Guillemot.

A bird observatory was founded here in the early 1970s, now located together with a national wetland centre at *Lilløyplassen nature-house*. From the late 1930s to 1998, Fornebu hosted Norway's international airport, which is now located at Gardermoen

Fornebu peninsula on the outskirts of Oslo is a main migration hotspot, boasting the longest list of recorded bird species in the area. The prime bird-attracting features are the shallow inlets of Storøykilen (above) and Koksa, both surrounded by a variety of habitats that hold a multitude of birds at all times. *Foto: Bjørn Olav Tveit.*

Little Ringed Plover prefers open and dry areas near water, and is often to be seen on Fornebu peninsula. *Photo: Kjetil Schjølberg.*

north of Oslo. The airport, of course, had a great impact on the habitats and birds, and resulted in restricted access for birdwatchers. The old airfield is now subject to a major urban development project, but one that includes large green areas adjacent to the important wetlands, which are protected as nature reserves. The landscape has been modified to include small hills and lakes, and bushes and trees are planted. In some parts of the area, grazing livestock help keeping the terrain from overgrowing, simultanously restricting disturbance from people somewhat. The local chapter of BirdLife Norway is also conducting a wetland restauration project here. Hopefully, when this project is completed, it will result in suitable habitat for people and birds alike. Facilitated by the high level of ornithological activity in the course of the last four decades, Fornebu boasts the longest species list of the sites in the Oslo area.

Other wildlife: Roe deer, red fox and badger inhabit the area.

Best season: All year, but with the greatest diversity during migration (mid-March through early June, and from mid-July until November).

Directions: *Public transport from Oslo city centre:* Take the bus towards Snarøya and leave the bus at Fornebuparken bus stop. *From Oslo by car*, drive E18 westbound towards Drammen. Just after passing the Oslo city limit, exit to Fornebu. Then, see map.

Tactics: For general birding, the parks and wetlands surrounding the narrow inlets Storøykilen and Koksakilen are usually the best, particularly in the morning. Also check the small coniferous woods on Lilleøya and the open fields and bushes on Storøyodden. The artificial pond in Nansenparken is worth exploring, particularly on overcast and rainy days, often proving attractive for *Tringa*-waders, snipes, gulls, wagtails and pipits. The migration of raptors, cranes and geese is best watched from a hill, for instance at Storøya. The migration of buzzards and soaring birds does not commence before the sun warms up the landmasses and creates the essential thermal winds, from about 10 o'clock in the morning.

Head for the east coast of Fornebu peninsula for a view of the inner Oslofjord. Migrating ducks, gulls, divers and more may be seen in late spring mornings. Interesting gulls, seabirds and auks may show up, particularly after southerly winds from late summer through to winter.

6. SANDVIKA TOWN

Akershus County

GPS: 59.8935° N 10.5253° E (Lake Engervann parking)

Notable species: Shelduck, Tufted Duck, other ducks, Great Crested Grebe, Little Grebe (winter), gulls, Kingfisher, Grey Wagtail, Dipper, Icterine Warbler, Wood Warbler, Goldfinch and Hawfinch.

Description: Sandvika is a suburban town located where the Sandvikselva river meets the Oslofjord. The river mouth is largely open throughout the winter, which considerable numbers of ducks and swans appreciate. These are often fed by people, which in turn attracts quite a few gulls.

Occasionally, you can find auks or scarce ducks and gulls amongst the commoner species. Kingfisher has bred along the river and may be seen here or in Lake Engervannet, a 1 km long brackish body of water sandwiched between two ridges. At high tide, salty seawater flows into the lake. The salinity ensures that the water freezes later in the winter than other lakes in the area. In Lake Engervannet, you can occasionally find a few hundred ducks and other wetland birds, with Mallard and Tufted Duck being the dominant species. At the outlet, the part being closest to Sandvika town, the ducks and gulls are regularly fed by people. The inlet at the opposite end is more peaceful, with a nice swamp and a shallow lagoon where small numbers of Grey Herons, Cormorants and a few

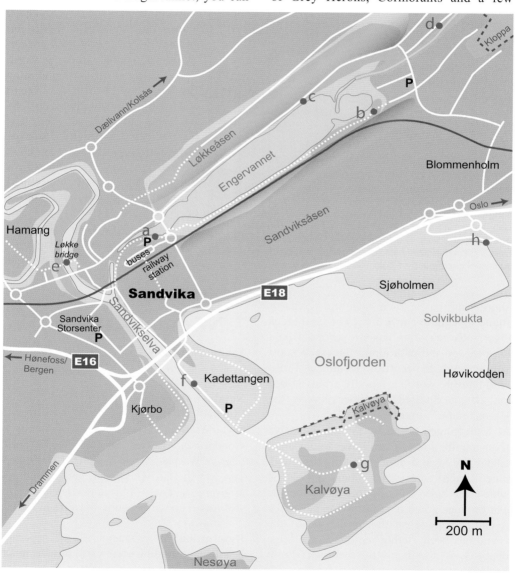

Sandvika town: a. Outlet of Lake Engervannet; **b.** BirdLife's public feeder; **c.** viewpoint Lake Engervannet; **d.** Øverlandselva river and Kloppa woodland; **e.** Løkke bridge (Kingfisher); **f.** outlet of river Sandvikselva and ferry landing to Ostøya island; **g.** Kalvøya island with mixed forest and pastures; **h.** Solvikbukta inlet.

Kingfisher occurs very sparingly along still-flowing waterways with clear water and overhanging branches from which it can fish. In winter it is also seen at estuaries along the coast. In Sandvikselva river, it nested regularly for several years, and became a popular subject of photography. *Photo: Kjetil Schjølberg.*

fewer people than Kalvøya island. The neighbouring Nesøya and Brønnøya islands, also have some fine bird habitats, including forests and wetland areas around Lake Nesøytjern and the Viernbukta inlet.

Best season: All year round.

Directions: *Public transport from Oslo city centre*: Several buses and trains stop in Sandvika. *From Oslo by car*, drive E18 westbound towards Drammen to Sandvika.

Tactics: Check the flocks of ducks and gulls thoroughly looking for more unusual species, and feel free to bring some bread to bring the birds closer to you. For Kingfisher, position yourself on Løkke bridge (made famous by French painter Claude Monet) or walk along the river. You can also see Kingfishers in Lake Engervannet and along Øverlandselva creek. It is also nice to walk the path encircling Lake Engervannet.

shorebirds may gather. A public feeder is found here, active in winter. The inlet creek Øverlandselva often has Grey Wagtail and Dipper, and the path along the creek takes you through a beautiful deciduous forest with Icterine Warbler, Wood Warbler, Green and Great Spotted Woodpeckers in the Kloppa nature reserve.

At Kadettangen beach by the Sandvikselva river-mouth, you can stroll over to the small island Kalvøya, a popular hiking area with nice mixed forest and open grasslands, which on days without too many people can produce some nice birding experiences. A nature reserve on the north side of the island has been established due to the large deposits of fossils in the bedrock. From Kadettangen, in summer, a ferry takes you to the islands even further out in the Oslofjord, including the beautiful Ostøya island, which has many exciting habitats and significantly

7. NESODDEN HEADLAND

Akershus County
GPS: 59.87068° N 10.65670° E (Nesoddtangen)

Notable species: *Breeding*: Stock Dove, Collared Dove, Tawny Owl, Long-eared Owl, Wryneck, Green Woodpecker, Black Woodpecker, Lesser Spotted Woodpecker, Icterine Warbler, Wood Warbler, Long-tailed Tit, Marsh Tit, Goldfinch and Hawfinch. *Migrants*: A variety of Scandinavian land- and seabirds are possible; gulls and auks in winter.

Description: Nesodden is a tapering headland protruding northwards into the inner Oslofjord basin and provides a flyway for migrating land-birds in spring. The northern tip of the peninsula, Nesoddtangen, is one of the best sites in Norway for diurnal migration of land birds in spring. The migration is at its most intense, as seen from the ground, in early mornings when there is a slight breeze from a northerly direction combined with overcast conditions and rising temperatures. Thousands of pipits, thrushes, finches and other common Scandinavian migrants may pass Nesoddtangen headland on the best days, crossing the Oslofjord in the direction of Bygdøy. On such days, birds often include a few surprises, such as short-distance migrants like woodpeckers, or generally scarce species such as Wood Lark. Raptors may also be seen from Nesodden in good numbers in spring. These are best watched on sunny days from a hill a bit south of the tip of the peninsula.

In autumn and winter, Nesoddtangen is a good headland from where to watch the migration of ducks, gulls, auks and other seabirds moving in and out of the Oslofjord. Many of these passes close to shore when they leave Bunnefjorden, the innermost part of the Oslofjord. Grey Phalarope and Sabine's Gull are seen occasionally, although they are not to be expected.

Nesodden headland has several areas of rich and mature deciduous woods. The best-known is Røerskogen forest. In these oak groves several pairs of Stock Dove breed. The doves can often be seen feeding in the surrounding fields. Røerskogen is also a pretty reliable place to find Wryneck, Lesser Spotted Woodpecker, Icterine Warbler, Wood Warbler, Long-tailed Tit, Marsh Tit and Hawfinch. The farm here is also the first place in Norway in which Canada Goose was released, back in 1936, resulting in a self-sustaining feral population.

Other wildlife: You may see harbour porpoise and sometimes other small cetaceans while sea-watching. Nesodden has a good population of roe deer.

Best season: From the end of March to mid-May for migrating land birds. Late autumn and winter for most seabirds, although seabirds may be seen all year after storms from a southerly direction. Røerskogen forest

Røerskogen forest has a bustling chorus of birdsong at sunrise in spring and early summer. The corral contains species associated with fine, old deciduous forest, such as Nuthatch, Spotted and Pied Flycatchers, Stock Dove and occasionally Golden Oriole. *Photo: Bjørn Olav Tveit.*

Oslo

Interior

Skagerrak

Western

Central

Northern

Finnmark

Svalbard

Wood Warbler breeds in small numbers in suitable habitat in this part of the country. *Photo: Terje Kolaas.*

has its greatest diversity from the end of May through June. However, the Stock Doves arrive already in early March.

Directions: The ferry from the docks at Aker brygge in Oslo traffics Nesoddtangen regularly all day. The trip takes about half an hour. Once at Nesodden, find a suitable spot from where you can stand and overlook the sea and the skies. To get from Nesoddtangen to Røerskogen forest, take the bus to Lovisenlund bus stop, close to the parking lot south of the entrance to Røer farm. Walk westwards and into the woodlands surrounding the farm. Also note the small pond Røertjernet about 200 metres south of the farm itself.

8. DRØBAK SOUND

Akershus County

GPS: 59.56363° N 10.65095° E (Krokstrand), 60.14923° N 11.45601° E (Skiphelle)

Notable species: *Breeding*: White-tailed Eagle, Goshawk, Peregrine Falcon, Eagle Owl, Stock Dove, Black Woodpecker, Green Woodpecker, Lesser Spotted Woodpecker, Nightjar, Mistle Thrush, Marsh Warbler, Icterine Warbler, Wood Warbler, Marsh Tit, Long-tailed Tit, Goldfinch and Linnet. *Migrants*: Ducks, divers, seabirds, raptors and more.

Description: Near the town of Drøbak, the Oslofjord narrows to just 1.5 km across. This offers an excellent opportunity to achieve close encounters with seabirds that sometimes move in and out of the inner basin. From the end of April through May, ducks, divers, gulls and sometimes skuas may use the fjord as a northbound flyway. In autumn the migration is less concentrated, but may still be evident, often including auks as well. In and after periods of strong winds from a south-westerly direction, pelagic species such as Fulmar, Gannet and Kittiwake may be pushed far into the Oslofjord. In such conditions, species like Manx Shearwater, Leach's Petrel, Storm Petrel, Grey Phalarope or Little Gull may be encountered as well, with the possibilities of several species of both divers and skuas. In fact, the Oslofjord has proved to be one of the best places in Norway for Sabine's Gull, even though it is not at all to be expected.

These birds may be seen from the docks at picturesque village Drøbak, but the closest encounters are usually achieved from Krokstrand further out. Hulvik even further south gives more extensive views of Breiangen, the wide middle section of the Oslofjord, and hence often produces greater numbers. The same birds may of course also be seen from Storsand, Filtvet or Tofte on the opposite side of the Drøbak sound. The light conditions are best on the east side of the sound in early morning, and best from the western shore in the afternoon.

In the general area along the Drøbak sound, you may see migration of passerines and raptors, amongst other things, as well, particularly in spring. These are birds following the coastline, crossing at the Drøbak sound or continuing up to Nesoddtangen. The habitat here is dominated by farmland and woodlands intermixed with pine. This makes up a nice selection of habitats, suitable for resting migrants and a few breeding birds, particularly woodpeckers, Nightjar (particularly on the west side) and Marsh Tit. The small inlet Sætrepollen near the village of Sætre may hold a nice selection of ducks and other waterbirds.

The island of Håøya just north of Drøbak, which is practically inaccessible, is well-known for hosting one nesting pair each of White-tailed Eagle and Eagle Owl. The latter may be heard calling from either side of the sound on calm evenings and mornings in February and March. The young may be heard begging at night in June. The White-tailed Eagles can sometimes be seen soaring above the island or along the sound. Try looking for them from the north side of Drøbak or from Storsand on the opposite side of the sound.

Other wildlife: Sea mammals, in particular harbour porpoise and common bottlenose dolphin, may be seen.

Best season: General migration from end of April through May. Seabirds after south-westerly winds summer and autumn.

Directions: *From Oslo to Drøbak*, drive E6 south towards Göteborg (Sweden). Exit onto Rv23 signposted to Drøbak to reach this town. You can continue along Rv23 in order to reach the west side of the sound via an undersea tunnel, and head north to Sætre or south to Filtvet/Tofte. From Drøbak on the east side of the sound, you can drive south towards Vestby, a road winding through mixed farmland and woodlands rich in birds. Side roads lead down to the shore here and there, for instance at Skiphelle by the parking area near the end of street Elleveien in Vestby, where there is very fine riparian habitat with woodpeckers and Marsh Tit.

From Oslo directly to Krokstrand or Hulvik, leave E6 further south, at Exit no 16, signposted to Hvitsten and Vestby syd. Follow signs to Hvitsten, for a while, but 7 km after leaving E6, make a right turn onto the road signposted to Hulvik and Krokstrand. At the T-junction, make a right to get to Krokstrand and a left to go to Hulvik.

9. LAKES OF FOLLO

Akershus County
GPS: 59.68508° N 10.82836° E (Østensjøvann south)

Notable species: *Breeding*: Great Crested Grebe, Moorhen, Coot, Honey Buzzard, Goshawk, Peregrine Falcon, Osprey, Hobby, Long-eared Owl, Stock Dove, Wryneck, Green Woodpecker, Black Woodpecker, Lesser Spotted Woodpecker, Thrush Nightingale, Reed Warbler, Marsh Warbler, Icterine Warbler, Long-tailed Tit, Marsh Tit, Goldfinch and Hawfinch. *Migrants*: Ducks and other wetland species.

Description: Just south-east of Oslo is the rural area Follo, with farmland interspersed with patches of mixed woodland and several nutrient-rich lakes lined with reeds, providing suitable habitat for a variety of birds. A nocturnal expedition to the area in late spring or early summer provides a fair chance of finding Quail, Corncrake or Grasshopper Warbler or even rarer species, such as River Warbler or Blyth's Reed Warbler.

Of these lakes, Lake Østensjøvann in Ås (not to be confused with Lake Østensjøvann in Oslo), is the most famous and generally the most bird-rich. In most years, Thrush Nightingale is heard singing here in late spring. Lake Pollevannet has a rather large reedbed in the south-end, where Bearded Tit have been seen on several occasions. Lake Årungen is marred by extensive rowing activities. But the lake has a nice swamp in the south end and wonderful mature deciduous forest around the mouth of the meandering Syverudbekken creek on the eastern shore. In the north end of Årungen, the outlet creek is particularly good for Grey Wagtail and

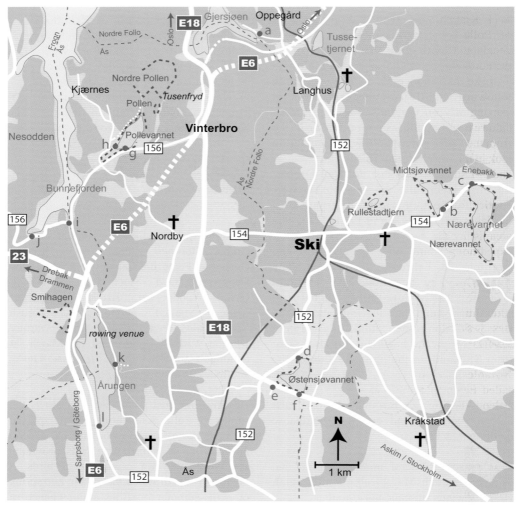

Lakes of Follo: a. Slorene wetland in Lake Gjersjøen; **b–h.** the most often used viewpoints to the lakes; **i.** and **j.** river mouth and mudflats in Bunnefjorden (part of the Oslofjord); **k.** mature deciduous forest and slow flowing river; **l.** lush marsh in Lake Årungen.

Dipper, and down-stream, where the water is more slow-flowing, Kingfisher may be found.

Best season: Late spring, summer and autumn.

Directions: *From Oslo by car*, take E18 or E6 southbound towards Göteborg/Stockholm about 20 km to Vinterbro, and then see the map.
Lake Østensjøvann in Ås: Highway E18 runs close to the lake. You may make a short stop by **e.** the side road to the farm Askjum for views to the western corner of the lake, and park briefly roadside 500 metres ahead, in the

curve on the south side in order to view the lake from **f.** the south corner. Watch out for heavy traffic! The lushest swamp and densest reedbed is in the northern end, which is best viewed from **d.** along Rv152.
Lake Pollevannet: The open water is best viewed from **g.** the rest area along Rv156. Drive a bit further south and make a right onto a road signposted to Kjærnes. This road takes you along the reedbed. Walk into **h.** the small industrial area with the address Kjærnesveien no. 19 for a view to the outer part of the reeds. If you drive a bit further and make a right onto Polleveien, you may park and walk the

track Kjærnesstien along the western shore of Pollevannet, through nice sections of deciduous forest to the swampy meadow in the north end.

k. Lake Årungen, east side is reached by following signs to Årungen Rostadion (rowing arena). A 2 km long nature and culture trail leads through the Syverud woods from the parking area. You should also have a look in the small but lush marsh in **l.** the south end of Lake Årungen.

Tactics: A telescope is essential for exploring the lakes and any distant raptors. The Follo area is highly recommended for nocturnal expeditions in late spring and early summer.

10. ØYEREN DELTA

Akershus County

GPS: 59.89350° N 11.11812° E (parking Årnestangen), 59.90123° N 11.16212° E (parking Furusand/Jørholmen)

Notable species: *Breeding*: Osprey, White-tailed Eagle (possibly), Goshawk, Peregrine Falcon, Hobby, Kestrel, Corncrake, Crane, Little Ringed Plover, Green Woodpecker, Thrush Nightingale, Sedge Warbler, Reed Warbler, Marsh Warbler, Icterine Warbler, Wood Warbler, Long-tailed Tit, Marsh Tit, Red-backed Shrike, Goldfinch, Linnet and Hawfinch.

Migration: A great diversity of species, particularly swans, geese, ducks, shorebirds and other waterbirds, but also raptors, passerines and others. The species list often include less common ones such as Pochard, Hen Harrier, Marsh Harrier, Little Gull or Great Grey Shrike, and sometimes stragglers like Pallid Harrier, Black or Caspian Tern and even greater rarities.

Description: Lake Øyeren is a more than 30 km long inland lake along the Glomma watercourse. In the north end, the three main rivers Nitelva, Leira and Glomma have created a large river delta. Sand, gravel and other sediments are transported with the water and deposited here, creating numerous sandbanks and islets. This is an ongoing process, shaping and reshaping the landscape over time.

The lakes of Follo are nutrient-rich and very attractive to a number of species. Depicted is the middle part of Slorene wetland in Lake Gjersjøen. *Photo: Bjørn Olav Tveit.*

Oslo · Interior · Skagerrak · Western · Central · Northern · Finnmark · Svalbard

Osprey is often seen hunting for fish in these lakes from April to September. *Photo: Tomas Aarvak.*

There are large areas of lush deciduous forest in the delta, and fertile soil here has formed the basis for extensive agriculture. The Øyeren delta is regarded as the largest inland delta in all of Scandinavia and is of great importance for the region's wetland birds, particularly as a resting place during migration. This has given the delta status as an internationally Important Bird Area (IBA).

The Øyeren delta can have exciting birds to offer in all seasons. It is of the finest places from which to follow the spring and autumn migration, it is the best site in the region for ducks and shorebirds, and it offers significant numbers of wintering birds. The most visited part of the Øyeren delta by birdwatchers, is the headland Årnestangen on the west side of the delta.

The water level in Lake Øyeren is highly variable through the seasons, and this has a major impact on bird occurrences. Because the lake is regulated artificially, variations in water levels are less than they would have been naturally. You don´t want a too high water level, which is often the case in late spring. Such conditions create flooding of the sandbanks, the lake-side meadows and mudflats completely, reducing the numbers and diversity of birds significantly. However, the intensity of the flood varies from year to year with snowfall, temperatures and rainfall. But even when the delta is flooded, you may still find gatherings of birds, particularly where the shoreline coincides with grazing pastures and ploughed farmland, for instance on Tuentangen. When the water level is reduced through the summer, large mudflats are exposed, creating suitable habitat for shorebirds – if not drying completely out for too long. When the exposure of moist mud coincides with the shorebird's migration periods, the Øyeren delta is the best site for such birds in the Oslo area. You may then find arctic species such as Curlew Sandpiper and Sanderling in addition to all of South Norway's common breeding shorebird species, often including the more uncommon species, such as Broad-billed Sandpiper and Great Snipe. In May–June and August–September, scarce species such as Garganey, Shoveler, Hobby, Marsh Harrier and Red-throated Pipit may occur. Little Gull is often seen as well, particularly in late summer.

Osprey breed in the surrounding forests, and several pairs hunt regularly in the delta. Hobby breeds nearby as well, and is seen occasionally by the lake. White-tailed Eagle can be seen all-year in the delta, and may become a regular breeder.

The Øyeren delta and surrounding agricultural and woodland areas are excellent for nocturnal expeditions in late spring and early summer, with Corncrake, Water and Spotted Rails, Nightjar, Thrush Nightingale, Marsh, Reed and Sedge Warblers amongst the regular species, and with the possibility of something rarer.

In late summer, autumn and winter, the Øyeren delta is an important staging area for ducks including up to one hundred Pochards and more than one thousand Whooper Swans. If you look carefully through the swans, you might find the odd Bewick's Swan. At this time of year, you might also see wintering raptors and Great Grey Shrike.

Other wildlife: Beaver has become quite common in the delta in recent years.

Best season: All year, but the Øyeren delta is perhaps most rewarding from late spring through summer to late autumn, partly depending on the water level not being too high, and not too low for too long.

Directions: *From Oslo by car*, drive E6 north towards Trondheim. At Furuset just out of town, keep right onto Rv159 signposted to Lillestrøm. Then see map.

Årnestangen headland. Just after the tunnel that leads traffic underneath the town Strømmen, exit onto Rv120 signposted to Enebakk and Fjerdingby. After almost 8 km, turn left signposted Årnes and the Leca factory, and then turn immediately left again, to the road signposted to Årnestangen. At this last intersection you may stop and take a quick overview with a telescope to get an idea of the situation on Årnestangen regarding water level, bird gatherings, location of other birdwatchers or fishermen etc., before driving downhill to the designated parking area near the end of the road. Scan **a.** Snekkervika inlet and the small meadow ponds before you walk or bicycle the 3 km

along the tractor track that leads to the tip of Årnestangen headland. Pipits and various shorebirds often feed in the fields so it can pay to stray off the tractor path (but, of course, do not walk on the farmland in the growing season). Ospreys are often seen hunting in the lagoons or resting on poles in the water or on the sandy banks. From **b.** the old tower hide close to the tip you still have a view east to Rossholmen island and the outer parts of the delta which are inaccessible without a boat. You also gain some height to get better views of the gatherings of ducks and other birds near the tip. However, to gain better views, continue along the path a few hundred metres past the tower hide to a signposted viewpoint platform. Here at the tip of the headland, you are not allowed to leave the path in order not to disturb the birds.

d. Gjellebekkvika. From Årnestangen, drive a little further south along Rv120. Some of the ducks are seen better from Gjellebekkvika than from Årnestangen, especially in the evening. You may find Lesser Spotted Woodpecker, Wood Warbler, Marsh Tit, Goldfinch and Common Rosefinch along the creek.

e. Svellet is a wide, slow-flowing section of

Oslo

Interior

Skagerrak

Western

Central

Northern

Finnmark

Svalbard

Flooded fields on Tuentangen in the Øyeren delta in early summer form favourable habitat for shorebirds. The wide river section of Svellet can be seen between the trees in the background. *Photo: Bjørn Olav Tveit.*

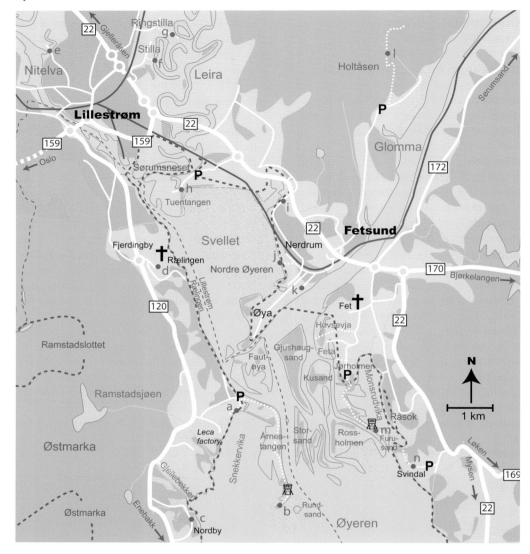

Øyeren delta: a. Viewpoint Snekkervika inlet; **b.** observation platform at tip of Årnestangen headland; **c.** Nordby, with a view to Årnestangen; **d.** Rælingen church, viewpoint to Svellet; **e.** Nebbursvollen; **f–g.** Stilla and Ringstilla oxbow lakes; **h.** Tuentangen headland; **i.** Merkja lagoon; **j.** Nerdrumstranda, Svellet; **k.** visitor centre; **l.** Holtåsen hill (Nightjar); **m.** Furusand tower hide; **n.** viewpoint Svindal farm.

river with vast, shorebird friendly mudflats at low water levels. You get an overview of Svellet from Rælingen church, but the birds will be distant and the view partially blocked by trees. If you find there are birds out there worth a closer study, you should enter the area from the opposite side, at **j.** Nerdrumstranda (which see).

f–g. Stilla and Ringstilla, two oxbow lakes that are obligatory stops on nocturnal expeditions in the area. Exit Rv22 east of Lillestrøm, on the road signposted to Skedsmohallen arena and Leiraveien. Follow these a few hundred metres and you get Stilla on the right-hand side, behind a warehouse and an archery training facility. If you continue a bit further on Rv22, pass under the railway and turn right at the next exit, you'll soon have Ringstilla across the field on your right. Follow the tractor track on foot in order to obtain closer views.

h. Tuentangen. Exit Rv22 on the road signposted to Tuen industrial area. Park behind the factory buildings and follow the road a bit further on foot (you need special permission to drive in). From this road you can see a stretch of the river Leira, and a small oxbow lake. When the water level in Svellet is high, the fields along this road may be partly flooded, providing favourable conditions for shorebirds, ducks and more.

i. Merkja is a fine little lagoon along Rv22. It is not advisable to stop along the highway, but if you exit towards Hovin and Nerdrum and pass under the highway, you will find suitable places to overlook the lagoon.

j. Nerdrumstranda (Svellet east). Continue past Merkja in the direction of Nerdrum until you cross the railway line. Keep hard to the right into Lundveien, and find suitable parking. Walk down to the beach between houses no. 45 and 47. You can also drive further on and turn onto Øyaveien and drive out to Øya, which can be good for nocturnal trips, and which in some places offers a view of the southern parts of Svellet. Feel free to stop by at **k.** Nordre Øyeren Naturinformasjonssenter (nature information centre), with a nature book shop and an exhibition dealing with the wildlife in the area.

l. Holtåsen is a traditional site for calling Nightjar in early summer evenings.

m. Furusand tower hide is reached by exiting from Rv22 just after crossing the Fetsund bridge over the river Glomma, following signs to Fet church. Pass the church and keep left along Jarenveien. Park in the parking lot at the end of the road, and continue 2 km on foot along the tractor road across Jørholmen peninsula to the tower hide.

n. Svindal farm provides the best overlook view of the eastern part of the Øyeren delta, and one of the best places from where to see the wintering Whooper Swans and ducks. Kestrel breeds here. To get there, follow Rv22 south towards Mysen. In Gan (signposted), turn right on Svindalveien. Park roadside just before the farm, but be careful not to block the road. You may walk through the farmyard and find suitable viewpoints.

Tactics: A telescope is essential at Lake Øyeren. The delta and the slow-flowing river Leira that meanders through the agricultural landscape are exciting places for nocturnal songbird trips. From June to August be prepared for significant quantities of mosquitoes and horseflies, especially at Årnestangen.

11. Oslo airport Gardermoen

Akershus County

GPS: 60.1908° N 11.1432° E (Gardermoen raceway), 60.2356° N 11.1652° E (Risebrutjernet pond)

Notable species: Slavonian Grebe, Little Grebe, Goshawk, Hobby, Honey Buzzard, Osprey, Little Ringed Plover, Long-eared Owl, Black Woodpecker, Green Woodpecker, Lesser Spotted Woodpecker, Woodlark, Marsh Warbler, Icterine Warbler, Wood Warbler, Great Grey Shrike, Red-backed Shrike, Goldfinch, Linnet and Common Rosefinch.

Description: Gardermoen is the main international airport in Norway, and many visitors from abroad use this as their gateway to the country. The airport hotels are also popular venues for business conferences, and many people may therefore have a few hours at their disposal near the airport. Fortunately, the area is so exciting in terms of birds that even the resident birdwatchers in Oslo city regularly make the trip up here. Gardermoen is particularly popular because it is the only place in the Oslo area where Woodlark is a regular breeder. The larks often sing enthusiastically just outside the airport fence. You may also find Hobby and Honey Buzzard in the forest areas here, especially up towards Hurdal valley. The alternating terrain with forest, agriculture and small water and streams in the hilly ravine landscape, makes Gardermoen and the surrounding area worth exploring, e.g. with nesting Slavonian Grebe. A number of birds of prey regularly migrate through. Gardermoen is also a good starting point in search of nocturnal songbirds. In

Oslo

Interior

Skagerrak

Western

Central

Northern

Finnmark

Svalbard

winter, the southern end of Lake Hurdalsjøen as well as the outlet in Andelva river and parts of the river Risa are often ice-free, which provides good conditions for swans, ducks and sometimes Little Grebe and other less numerous species.

Best season: Spring and early summer for general species diversity. The Woodlarks arrive in March, while Hobby and Honey Buzzard arrive in late April and early May, respectively. Winter is also a good time of year, as long as there is open water to be found.

Directions: From Oslo by car, follow the E6 for half an hour in the direction of Trondheim and exit to Gardermoen Airport. From there, see the map.

Tactics: You will need a car to explore the area effectively. A spotting scope is also helpful. **a.** Romerike landscape conservation area is in the immediate vicinity of the airport, with ravines and mixed forests, hosting Lesser

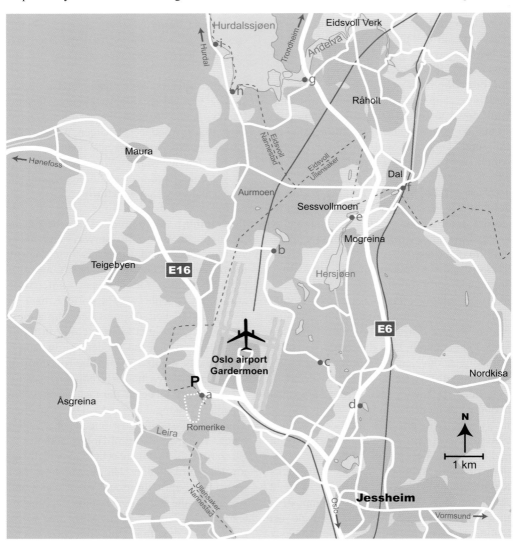

Gardermoen airport area: a. Romerike landscape conservation area; **b.** Aurmoen viewpoint; **c.** Gardermoen raceway (Woodlark); **d.** Lake Bonntjernet (Slavonian Grebe); **e.** Lake Risebrutjernet; **f.** Risa river by Dal; **g.** Lake Hurdalssjøen's outlet by Andelva; **h.** and **i.** viewpoint to Hona river mouth in Lake Hurdalssjøen.

Spotted Woodpecker and Wood Warbler, among others. Park by Scandic Hotel, and follow the signposted path. Woodlarks are usually found along the road along the eastern side of the airport, via **b.** the Aurmoen hill (which is also a nice vantage point to look for raptors) to **c.** Gardermoen raceway and Vilbergmoen. **d.** Bonntjernet pond is a regular site for Slavoinan Grebe. Park along the west side of the pond and follow the path down towards the bank. **e.** Risebrutjernet pond is a nice little wetland where both Slavonian and Little Grebe may possibly breed. The pond is best overlooked from the side of the road. **f.** Rise river near the village Dal often sports open water where ducks and Little Grebe regularly winter. Park 200 m north of the river and walk to the bridge for the best view, but watch out for the traffic along this narrow road. **g.** Lake Hurdal and Andelva river are best explored from Sundveien road and the lay-by along E6 southbound. The lake's south-western corner with the inlet from the Hona creek is best explored from **h.** Lima beach close to Stensgård church and from **i.** the lay-by 2 km further north. The northern part of Lake Hurdal also has a nice little delta, which is reached by driving 20 min. north. You may also drive E6 20 min. north to Minnesund, which also may host a variety of waterbirds, particularly in winter.

12. VORMA-GLOMMA CONFLUENCE

Akershus County

GPS: 60.14921° N 11.45615° E (Nestangen), 60.14923° N 11.45601° E (Grenimåsan, parking)

Notable species: *Breeding*: Hazel Grouse, Black Grouse, Capercaillie, Goshawk, Honey Buzzard, Osprey, Hobby, Crane, Green Sandpiper, Pygmy Owl, Long-eared Owl, Tengmalm's Owl, Green Woodpecker, Black Woodpecker, Lesser Spotted Woodpecker, Marsh Warbler, Icterine Warbler, Wood Warbler, Marsh Tit and Linnet. *Migrants*: Whooper Swan, Taiga Bean Goose, Pink-footed Goose, waterbirds and more.

Description: Glomma, the longest river in Norway, provides an important flyway for birds heading north in spring and south in autumn. Just east of Oslo's main international airport, Gardermoen, the river makes a sharp eastwards bend and intersects with the river Vorma, which empties Norway's largest lake Mjøsa. Here both rivers are shallow because of accumulation of sediments, and the under-water turbulence make the rivers stay ice-free much of the winter, providing favourable conditions for waterbirds. In the fork between the two rivers is Nestangen headland, providing a fine viewing point. The surrounding area consists of cultivated land, peat bogs and mixed woodland, with a variety of birds. Ortolan Bunting bred in the area until the turn of the millennium, and might show up again in the future. The area is now best known for being a regular stop-over site for geese, particularly flocks of Taiga Bean Goose in both spring and autumn, resting on their way between Scotland and the breeding grounds in Sweden. The general area is also good for Marsh Warbler and other nocturnal songbirds, sometimes including the likes of Quail, Corncrake or Blyth's Reed Warbler.

The boreal forest east of Glomma, for instance at Jansberg, can yield Capercaillie. Several leks are known in this general area, but these are usually kept secret, contrary to the leks of the louder Black Grouse, found more commonly here. This is also a good area for Tengmalm's Owl, particularly if you head southeast towards Rakeie and the Mangen forests (which see).

Other wildlife: Elk, roe deer, badger, beaver and red fox are commonly encountered in the area.

Best season: Most of the year, even in winter as long as the rivers are ice-free.

Directions: *From Gardermoen airport*, follow E6 south a few km towards Oslo and then head east on Rv2 towards Kongsvinger. From here, continue about 20 km (across

43

the river Vorma) and turn right onto road signposted to *Nes kirkeruiner* leading to the ruins of medieval Nes church at the tip of the Nestangen headland. *From Oslo city centre*, head north on E6 towards Trondheim until you reach the above mentioned Rv2 intersection.

Tactics: Follow the roads along and across the rivers and investigate both the rivers and the surrounding fields. The geese are usually seen feeding in the fields near the farm Horgen and resting on the river, near Udnes church. The road Nikevegen, providing

access to Grenimåsan peat bog, is signposted that entry is restricted to visitors to the land-owners, but birdwatching here is regarded as such visits, so you may drive here when looking for birds.

Jansberg forest is reached by continuing past Nes along Rv2 and turning right onto road signposted to Funnefoss. After the bridge crossing river Glomma, turn left towards Skarnes and turn right again towards Skøyenteiet. Follow this road 1 km and turn left onto Jansbergveien road. Continue 2 km, park roadside and walk the hill to your left (east of the road).

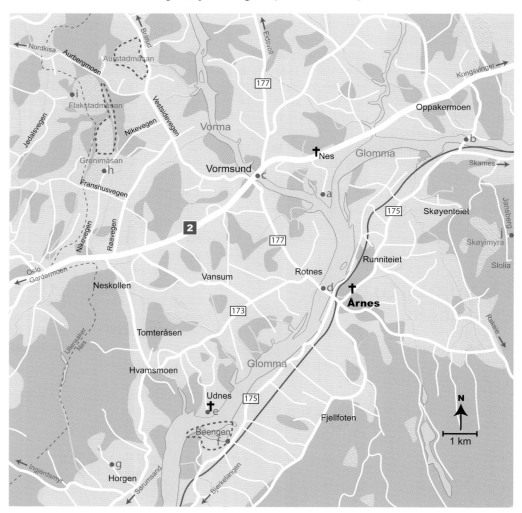

Vorma-Glomma confluence: a. Nestangen viewpoint; **b.** Funnefossen viewpoint; **c.** Vormsund viewpint; **d.** Årnes viewpoint from bridge; **e.** Udnes viewpoint; **f.** Beengen nature reserve; **g.** Horgen fields; **h.** Grenimåsan peat bog; **i.** Flakstadmåsan peat bog, **j.** Jansberg forest.

13. Bjørkelangen and Mangen

Akershus County

GPS: 59.8703° N 11.5640° E (Bjørkelangen), 59.8900° N 11.5803° E (Kjelle birdwatching hide)

Notable species: Geese and ducks, Red-throated Diver, Black-throated Diver, Great Crested Grebe, Osprey, Goshawk, Peregrine Falcon, Hobby, Capercaillie, Black Grouse, Hazel Grouse, owls, Mistle Thrush, Marsh Warbler and Parrot Crossbill.

Description: Bjørkelangen is the name of both a village and a nearby lake along the Halden watercourse. The northern end of the lake has a fine wetland area often holding a nice selection of both resting and nesting birds. Geese, ducks, swans and cranes can roost in the surrounding agricultural landscape. At certain times of year, especially in spring, these fields are flooded when the river Lierelva overflows its banks. The areas up along the river then become particularly attractive to birds and birders alike. At Kjelle, a small pond has been constructed to provide similarly favourable conditions for wetland birds even at lower water levels. The fields at Haugrim farm a little further north may also host resting and feeding wetland birds. Until the 1980s, several pairs of Ortolan Bunting nested at the Liermåsan peat bog just north of the centre of Bjørkelangen village, a species that may be worth looking out for here and in similar habitat in the region in May–June.

The deep conifer dominated Mangen forests, stretching north towards Lake Mangen between Bjørkelangen and the Vorma-Glomma confluence (which see), is host to several exciting forest birds, and is often a rewarding area to search for owls in March and April. Tengmalm's, Tawny and Pygmy Owls are most abundant, but also Long-eared, Hawk and even Ural and Great Grey Owls may be encountered here.

Best season: Spring and early summer.

Directions: *From Oslo by car*, follow the E6 north towards Trondheim. At Furuset, keep right on Rv159 signposted to Lillestrøm. Continue on Rv22 (and past the locations by the Øyeren delta) over Fet bridge and then follow Rv170 to

Kjelle wetland by Bjørkelangen village offers great habitat for a number of birds when river Lierelva is flooded. In periods of low water levels, the artificial pond in the middle of the picture ensures that there are always at least some birds to be seen here. *Photo: Bjørn Olav Tveit.*

Oslo

Interior

Skagerrak

Western

Central

Northern

Finnmark

Svalbard

Whooper Swan breeds sparingly in forest marshes and lakes, but can winter in large flocks in open water. *Photo: Bjørn Fuldseth.*

the centre of Bjørkelangen village.

Lake Bjørkelangen: Continue straight through the centre of Bjørkelangen village on Rv115 towards Askim. Just after you leave the village, turn left towards Ørje. You will soon see the northern end of Lake Bjørkelangen on your right.

Kjelle wetland: From the centre of Bjørkelangen village, follow Rv170 towards Rømskog/Setskog. Just before you exit Bjørkelangen village, turn left at the sign for Haugrim/S. Mangen and follow the road for 600 m before turning towards Kjelle farm on the left. Here you will find a neat shelter with a view of the artificial pond and the often flooded fields along the Lier river. You can also follow the road a few km further north and get a view of the Lier river and surrounding fields further up-stream.

Haugrim–Skjønhaug farmlands: From Kjelle, continue north. At the crossroads by Haugrim farm, signposted to Søndre Mangen, you get a view of the fields here, and this is a nice place to look for migrating raptors, cranes and others. Drive on and explore the fine agricultural areas north towards Haneborg and into Skjønhaug village.

Mangen forests: From Haugrim farm, turn right, signposted to Søndre Mangen. Stop at suitable places along the way – preferably with detours on the side roads – to watch and listen for owls and other birds. At lake Mangen, 12 km after Haugrim gård, turn left onto Engebret Soots vei, a forest road that takes you north to Rv236. If you turn left here towards Lierfoss, you will return just north of Bjørkelangen. If you turn right towards N. Mangen, you will reach Årnes (and the sites by the Vorma-Glomma confluence). Both directions lead through exciting forest terrain.

14. LAKE HELLESJØVANNET

Akershus County
GPS: 59.73725° N 11.45134° Ø (parking)

Notable species: Whooper Swan, Pochard, Red-throated Diver, Black-throated Diver, Great Crested Grebe, Slavonian Grebe, Marsh Harrier, Osprey, Goshawk, Peregrine Falcon, Hobby, Water Rail, Coot, Green Sandpiper, Wryneck, Green Woodpecker, Black Woodpecker, Lesser Spotted Woodpecker, Icterine Warbler, Sedge Warbler, Reed Warbler and Marsh Warbler.

Description: Lake Hellesjøvannet is a nutrient-rich body of water surrounded by farmlands and mixed forests. It is strategically located for migratory birds in the extension of the north-south-directed Halden watercourse. The lake is almost circular in shape and 900 m across at its widest, densely lined with reeds. An old tower hide is located in the forest on the south-western side. A number of wetland species regularly breed in Hellesjøvannet. Formerly breeders included Pochard and Slavonian Grebe, but both have disappeared in recent years. However, Slavonian Grebe is still often seen here and Pochard regularly appear in flocks, especially from July until the ice settles in November. Garganey and Shoveler breed sporadically and are seen regularly. Black-throated Diver, Osprey and Hobby nest in the area and use Lake Hellesjøvannet to feed. Wood Warbler and Tree Pipit are often heard singing in the forest surrounding the tower hide in spring. Honey Buzzard, Marsh Harrier and Little Gull are seen fairly regularly, and Black Tern and other unusual species occasionally appear. Lake

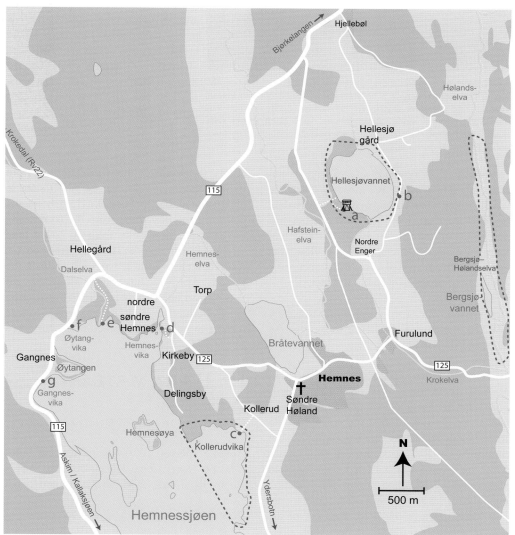

Lake Hellesjøvannet: a. Lake Hellesjøvannet tower hide; **b.** Lake Hellesjøvannet east; **c.** Kollerudvika bay in Lake Hemnessjøen. In the growing season, follow the edge of the field or a trodden path to the boat landing site; **d.** Hemnesvika inlet; **e.** Dalselv river mouth; **f.** Øytangvika inlet; **g.** Gangnesvika inlet.

Reed Warbler is often found in dense reedbeds in this south-eastern part of Norway. It is very similar in appearance to Marsh Warbler and the rarer Blyth's Reed Warbler, both of which are more often found along damp ditches and open areas with bushes and scrub. These species are best distinguished by their song. *Photo: Terje Kolaas.*

Hellesjøvannet is located in an area with several other interesting lakes and waterways that are worth visiting when you are in the area, including Hølandselva river. Lake Hemnessjøen lies just south-west of Lake Hellesjøvannet. It is 12 km long, and elongated in shape. It has shallow parts and several reed-covered coves that attract many of the same species as in Hellesjøvannet.

Best season: Mid-April to early July is usually best, but the area is of interest as long as the lakes and streams are free of ice.

Directions: *From Lillestrøm town to Lake Hellesjøvannet*, drive east on Rv22 towards Fetsund. After Fetsund, continue south on Rv22 signposted to Løken. In Gan, turn left onto Rv169, still signposted to Løken. In Løken, turn right on Rv115, signposted to Hemnes. After just over 5 km, turn left onto the road signposted to Furulund. From this junction, continue for 1.8 km and turn onto a small tractor path down towards the field. Park out of the way of agricultural machinery and walk along the forest edge down towards the water and into the forest 200 m to **a.** the tower hide. For a view from **b.** Lake Hellesjøvannet's east bank,

continue southwards along the main road and make a turn onto the gravel road by the Nordre Enger farm. Lake Hemnessjøen. Continue south 1 km along the main road and turn right at the sign for Rv125 Hemnes. In Hemnes village, you can choose to take the road signposted to Ydersbotn to the left, along a road that runs along the entire eastern shore of Hemnessjøen, with several roadside viewpoints, e.g. to the nature reserves in **c.** Kollerudvika and Kragtorpvika (the latter is partly included on the map of the Hæra nature reserve). You can alternatively continue straight ahead through Hemnes village for a view of **d–g.** the northern and western parts of Lake Hemnessjøen.

Tactics: The area is best explored by car. Find suitable vantage points to overlook the water bodies from the surrounding roads, but be careful not to obstruct other traffic. For Lake Hellesjøvannet, you have the best view from the road along the east side, as well as from the tower hide on the west side. In the Hellesjøvannet nature reserve, there is a traffic ban for visitors in the period 1 April–15. July, except along the path to the tower hide. There is a similar restriction in Kollerudvika and Kragtorpvika nature reserve in Lake Hemnessjøen in the period 15 April–15 July.

15. HÆRA WETLANDS

Østfold County
GPS: 59.66663° N 11.39373° Ø (Lake Kallaksjøen)

Notable species: A variety of wetland birds; Whooper Swan, Crane, Honey Buzzard, Osprey, Hobby, Reed Warbler, Marsh Warbler and Linnet.

Description: Hæra nature reserve consists of three small, nutrient-rich ponds: Kallaksjøen, Skottasjøen and Hærsetsjøen. On some maps, the latter is referred to as Dilleviksjøen. All are surrounded by bog, cultivated land and mixed forest. The ponds are primarily important as staging grounds for wetland birds during their migration in spring and autumn, and the area is exciting for nocturnal songbirds. Osprey and Hobby nest in the area and visit these ponds regularly. A few pairs of Whooper Swan nest, and several stage here in late autumn before the ice cover settles. These flocks may on rare occasions contain Bewick's Swan. Hæra nature reserve have been the subject of extensive improvement measures in the form of dredging, trenching etc. to ensure the birds thrive, and birdwatching is facilitated with parking and footpaths. A tower hide has been erected at Lake Hersetsjøen. When you are in this area, there are also other bodies of water that may be worth visiting, such as Lake Gyltetjernet.

Best season: Spring and early summer.

Directions: *By car from E18:* At Mysen town, take the exit and drive north on Rv22 signposted to Trøgstad. In Skjønhaug village, turn right from the main road and continue on Rv115 towards Bjørkelangen. Then, see the map. The entrance to the tower hide at Lake Hærsetsjøen is reached in just under 5 km on Rv115 and is signposted to Skjeringrud and *Hæra naturreservat.* Lake Kallaksjøen lies 2 km further on, signposted to Bingen.
By car from the north: Follow Rv115 south along the west side of Lake Hemnessjøen. 1.5 km after passing the border to Indre Østfold municipality, the road to Bingen turns left. Then, see the map.

Tactics: a. Lake Kallaksjøen is best viewed from the side of the road, at a point where the road passes over a small hill. **b.** Lake Hærsetsjøen is equipped with a tower hide. It is also possible to park at the southern end and walk the path down to the water's edge. **c.** Lake Skottasjøen can be viewed most easily from the farm road on the east side. Try to keep particularly quiet and move carefully in order not to flush the birds at these sensitive wetlands.

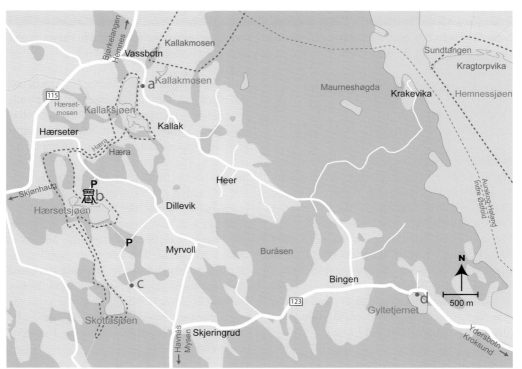

Hæra wetlands: a. Lake Kallaksjøen viewpoint; **b.** Lake Hærsetsjøen (Dilleviksjøen) tower hide; **c.** roadside viewpoint to Lake Skottasjøen; **d.** Lake Gyltetjernet.

Lake Gjølsjøen has two narrow sections, one north of the bridge (depicted) and one near the south end, which are completely overgrown with reeds and other aquatic plants. *Photo: Bjørn Olav Tveit.*

16. LAKE GJØLSJØEN

Østfold County

GPS: 59.43996 ° N 11.68611° Ø (southern tower hide)

Notable species: *Breeding*: Whooper Swan, Hazel Grouse, Black Grouse, Capercaillie, Black-throated Diver, Great Crested Grebe, Honey Buzzard, Marsh Harrier, Osprey, Hobby, Water Rail, Moorhen, Coot, Crane, Green Sandpiper, Wryneck, Green Woodpecker, Black Woodpecker, Lesser Spotted Woodpecker, Sedge Warbler, Reed Warbler, Marsh Warbler, Marsh Tit and Linnet. *Migrants*: Waterbirds, raptors and more, sometimes including Pochard, Red-throated Diver, Slavonian Grebe and Little Gull.

Description: Lake Gjølsjøen is a narrow and 5 km long, nutrient-rich lake near the Swedish border, surrounded by agriculture and spruce-dominated mixed forest. The banks and the narrowest parts of the lake are completely overgrown with reeds, water lilies and other vegetation. These are all fine habitats for a number of nesting species and many more staging during migration. The lake is strategically situated along the Halden watercourse, and in the extension of Østre Otteidvika, a tributary to the 65 km long Norwegian-Swedish lake Store Le. The lake is equipped with two tower hides. Marsh Harrier and Osprey can be seen regularly, and often quite a few geese. The forests surrounding the lake are well-known for its breeding woodland birds including Honey Buzzard and Hobby. Capercaillie, Black Woodpecker and even the odd Three-toed Woodpecker may be found a bit further into the forest, for instance at Tiurhøgda (*Capercaillie hill*).

When visiting the area, make sure also to stop by the smaller lakes Stikletjern and Solerudtjern a few km further south.

Other wildlife: Elk, roe deer, badger and red fox are commonly encountered in the area at dusk and dawn. There are also small populations of wild boar and grey wolf in the woodlands along the border with Sweden.

Best season: Mid-April to the beginning of July. During autumn migration as long as the lakes are ice-free, usually well into November.

Directions: *To Lake Gjølsjøen from Oslo:* Take the E18 towards Stockholm to Ørje. Exit from E18

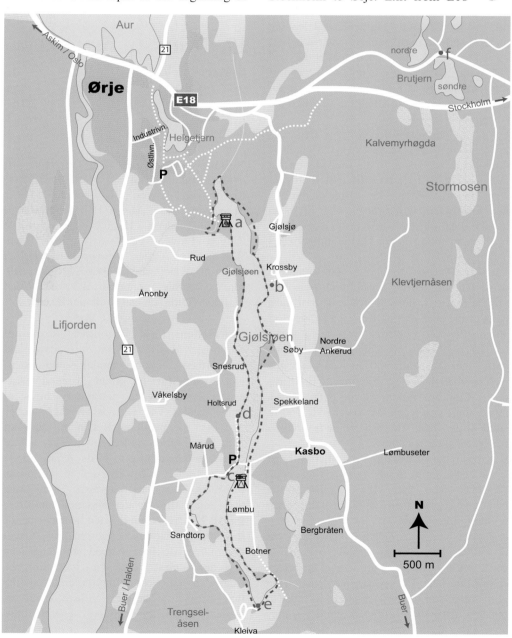

Lake Gjølsjøen: a. Northern tower hide; **b.** Krossby viewpoint; **c.** southern tower hide; **d.** path from bridge along the western bank; **e.** south end; **e.** Brutjerna lakes and starting point for hike north to Tiurhøgda hill.

either at the second exit signposted to Ørje, or pass them both and make a right onto the road signposted to Kasbo. Then, see the map of Lake Gjølsjøen.

To Tiurhøgda (*Capercaillie hill*): From Ørje, continue 3.5 km east on E18 and turn left (just after a lake to your left, and before a lake to your right, which both should be checked for e.g. divers) onto a gravel road leading north. The road may be closed, but it is best birded by foot anyway. Keep right at the road fork after 2.2 km and continue to the end of the road another 1.5 km ahead, looking for Capercaillie and more along the way and the surrounding woods.

Lakes Stikletjern and Solerudtjern: Drive 3.5 km south on Rv21 from the exit at the Sandtorp farms, and turn left onto the road signposted to Buer. You will see lake Stikletjern on your right after just over one km. Drive on through the farmyard and turn right and then left to get to lake Solerudtjern, which has a particularly nice swamp along the road at the southern end.

Tactics: Access to Lake Gjølsjøen is somewhat limited by agricultural areas and private property, but you get a view of most parts of the lake starting from the public road and the two tower hides.
a. The northern tower hide is reached by parking in the residential area and hike the path south to the bridge crossing the Bønselva creek. Continue further on the path along a fence, until you reach the tower hide 150 m south of the creek. This tower hide is rarely in use by local birdwatchers.
b. Krossby farm. Walk up on a small hill just south of the farm. Bring your spotting scope. From here, you obtain views to a part of the lake, and is also a nice vantage point when scanning for raptors.
c. The southern tower hide is more widely used by birdwatchers than the northern one, and is easily reached by parking just west of the bridge crossing the lake. The vegetation is dense on the northern side of the bridge, and so in order to obtain a better view in this direction you may walk along **d.** the path along the field on the western shoreline of the lake. You can also drive to the extreme **e.** south-end of the lake.

A suggestion for exploring some nice woodland in this area, is from the two small **e.** Brutjerna lakes, where you may see e.g. Black-throated Diver. From here you can walk along a forest road towards Tiurhøgda (*Capercaillie hill*) along which you might encounter grouses, woodpeckers and more.

17. TYRIFJORDEN DELTA

Buskerud County
GPS: 60.129347° N 10.154210° E (parking Karsrudtangen tower hide) 60.100848° N 10.287797° E (parking Steinsvika), 60.12247° N 10.18979° E (Røssholmstranda)

Notable species: Resting migrants of lakes, wetlands, farmland and forests; Quail, Red-throated Diver, Black-throated Diver, Slavonian Grebe, Marsh Harrier, Goshawk, Peregrine Falcon, Hobby, Osprey, Little Ringed Plover, Long-eared Owl, Black Woodpecker, Green Woodpecker, Lesser Spotted Woodpecker, Thrush Nightingale, Marsh Warbler, Icterine Warbler, Wood Warbler, Goldfinch and Common Rosefinch.

Description: Lake Tyrifjorden is one of the largest freshwater lakes in Norway, and the northern section is among south-eastern Norway's most important wetland areas, especially for resting migratory birds of many species. The lake has several branches, of which the two branches in the northern end, Nordfjorden and Steinsfjorden, are of particular importance to birds. A large delta area is formed in Nordfjorden, at the mouth of rivers Sogna and Storelva. Here, large areas of shallow water and mudbanks are found. The rivers run in wide meander bends through the flat landscape creating several favourable backwaters lined with meadows and agricultural fields. In river Storelva/Ådalselva these conditions continue all the way upstream to Hallingby village. The rivers have changed their course several times through the years. The old river bends have created nutrient-rich and gradually overgrown oxbow lakes. Although the delta for a large part is cultivated and developed, it is still a very important area for birds.

Little Ringed Plover, Stock Dove, Lesser Spotted Woodpecker and Wood Warbler are among the regular breeding birds in the delta. It is also an important hunting area for several pairs of Osprey nesting in the nearby woodlands. A substantial migration of Pink-footed Goose, Cormorant, Crane and raptors takes place through the area, particularly in spring. Many of these use the delta for feeding and resting.

Tyrifjorden is seldom completely frozen in winter. There is at least some open water by the mouth of Storelva and by the outlet near Vikersund. This makes Tyrifjorden a very important wintering site for Whooper Swan and ducks, often including Smew. Quite a few rare birds have been found in the delta through the years, most being waterbirds and nocturnal songbirds. The most notable rarity record is that of the Asian species Pallas's Fish-Eagle.

Other wildlife: Most of the mammals associated with mixed woods and cultivated land in this part of Norway, such as elk, roe deer, badger and red fox, are regularly seen while birdwatching in the area.

Best season: All year, but particularly in late autumn and when the ice starts to break up in April, a time when wildfowl are concentrated in the areas of open water.

Directions: From Oslo, follow E18 west towards Drammen. In Sandvika exit onto the E16 signposted to Hønefoss and continue 29 km to Sundvollen. Then, see the map.

Tactics: You need a car and preferably all day in order to explore this large area thoroughly. If you only have a few hours at your disposal, you should give priority to Steinsvika in Steinsfjorden, the oxbow lakes Juveren and Synneren, the Ask tower

Oslo

Interior

Skagerrak

Western

Central

Northern

Finnmark

Svalbard

Tyrifjorden delta is equipped with a tower hide near Ask village, overlooking the western and often very bird-rich corner of the delta, where river Sogna meets the lake. *Photo: Bjørn Olav Tveit.*

hide, and Røsholmstranda in Nordfjorden, which are where the largest concentrations of waterbirds are usually found. Along the roads to these sites, keep your eyes peeled, looking for birds flying overhead or feeding in the smaller wetlands, thickets and in the large agricultural fields.

The following is a suggestion for a more thorough tour of the delta area starting from Sundvollen:

a. Kroksund sound. Usually ice-free even in winter, often hosting a variety of ducks and other waterbirds. Continue along E16 towards Hønefoss, and stop where possible to scan the southern part of the lake Steinsfjorden.

b. Vikbukta bay is on your right 2 km past Kroksund. This shallow bay often holds dabbling ducks, Coot, Great-crested Grebe and swans. Continue E16 further.

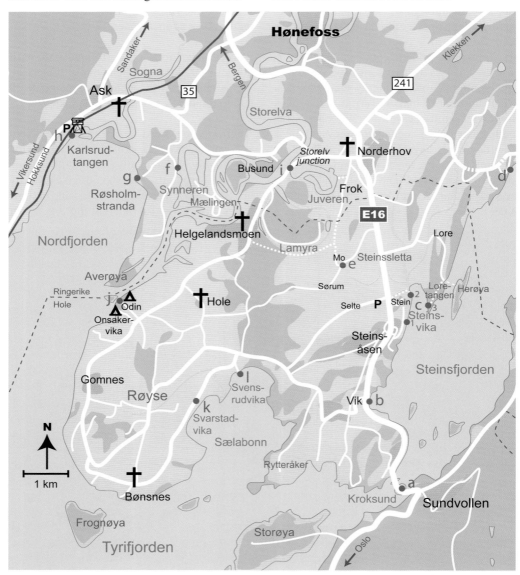

Tyrifjorden delta: a. Kroksund; **b.** Vikbukta; **c.** Steinsvika, three vantage points; **d.** Åsa, north end of Steinsfjorden; **e.** Steinssletta farmlands; **f.** Synneren oxbow lake; **g.** Røsholmstranda; **h.** Ask tower hide; **i.** Juveren oxbow lake; **j.** Storelva river mouth; **k.** Svarstadvika and **l.** Svensrudvika in Sælabonn. The boundaries of the nature reserves are not drawn on the map.

c. Steinsvika bay. Here you have three alternatives:

1) For just a quick scan from the outer part of the bay, leave the E16 on the road signposted to Steinsåsen. Then turn down towards the lake on the road signposted towards *Nedre Steinsåsen*. Continue 500 m to the car park at Høyenhall bathing area, and scan Steinsvika with a telescope from the cliffs by the shore.

2) For close-up studies of Steinsvika's inner part, follow the E16 for another km or so. Turn left on the road signposted Selte, and leave the car after 100 m in the designated road-side parking space. Walk back and cross the E16, and continue out along the field edge on the right side of the canal. After 500 m, the canal turns to the right (south) towards Steinsvika. Follow the tractor road up a hill with a good view of the cove's inner part.

3) For a view of Steinsvika from the eastern side, continue a further 1 km along the E16 at Steinssletta, turn right at the sign for Åsa. After 1 km, a farm road turns off to the right that leads past Lore farm and down to the parking lot for the ferry to Herøya. From the jetty you can go further out onto Loretangen, with a view of Steinsvika.

If you continue towards Åsa for 3 km and turn down an unmarked dirt road (*Grantoppsvingen*) to the right just before the second tunnel, you will come down to **d.** the reedbed at Åsa in the northern end of Lake Steinsfjorden.

Back at the E16 and the wide fields at **e.** Steinssletta, listen for nocturnal songbirds in late spring and summer. For the best view, leave E16 towards Hole church and stop by Mo farm. In autumn and winter, look for raptors, Great Grey Shrike and feeding swans and geese. Dotterel may be found here in May. Continue E16 towards Hønefoss for 3 km and turn south on the road signposted to Røyse and Gomnes. Follow the road towards Røyse, past Norderhov church, and at the Storelvkrysset junction, take a road signposted to Hokksund to the right. You immediately pass over the river Storelva, and on your left you will then find the small backwater Busund. Drive on and turn left into the road signposted towards Mælingen for a view further down the river.

f. Synneren oxbow lake. Drive up again and continue towards Hokksund town. You will soon see the lake Synneren on the left. Turn left, signposted to Røsholmstranda and Averøya. At the first crossroads, the road to the left leads to Synnern and the road to the right leads to Røsholmstranda. Turn left, park immediately and walk along the private road along the lake.

Drive back to the crossroads and further down to **g.** Røsholmstranda. From here you have a view over Averøya peninsula at the outlet of the river Storelva in the south, and Karlsrudtangen peninsula at the outlet of the Sogna river in the west, and the inner parts of the lake Nordfjord, where there are often good numbers of waterbirds. This is also one of the best places for shorebirds.

h. Ask tower hide. For a view of Karlsrudtangen from the west side, continue towards Hokksund, on a road that merges into Rv35 past the settlement Ask. 1.3 km after the exit signposted to Sandaker, leave the car at the parking on your left. Bring your telescope and walk the 200 m path signposted *Tårn* (tower), across the railroad tracks (be careful!), to the tower hide.

If you continue Rv35 further towards the south, after about 20 km you will come to Vikersund and the outlet of Tyrifjorden. Here, along the stretch from Vikersund to Bergsjøen at Geithus you always have open water with waterbirds in winter, often with Dipper by the Vikersund bridge.

Røyse peninsula. All the way back at the Storelvkrysset junction west of Norderhov, turn right towards Gomnes and Røyse. After a few hundred metres, you will see **i.** Juveren oxbow lake on your left. Here you can park at a suitable place on the side of the road and get a view of the lake from along the footpath. Be sure also to check Storelva river on the opposite side of the road. Driving a few hundred metres further on, you will see the heavily overgrown oxbow lake Lamyra on your left. It can be worth a stop, not least on nocturnal trips. Follow the main road for about 5 km and turn down towards **j.** Odin Camping. Park close to the Storelva river, where you have a nice view of Averøya peninsula in the outlet, where wetland birds often stop by during migration.

k–l. Sælabonn bay. Continue for 4 km along the Røyse peninsula and turn right towards Bønsnes. After a couple of km, you descend

Goldfinch is a common species in the lowlands around the Oslofjord, and is particularly fond of small seeds from thistles and other plants in open, sunny areas. Photo: Terje Kolaas.

into a lowland where the fields become of a particularly interesting calibre, and you soon have a view of **k. Svarstadvika inlet** in the larger bay Sælabonn. For a view of the inner part of this bay, continue 2 km further. Turn onto the slip road which turns abruptly to the right immediately after a yield sign (this is 150 m before the T-junction which meets the main road between Røyse and

Vik). This access road leads into the reed-covered bay **l. Svensrudvika** at the head of Sælabonn bay. If you turn right towards Vik at the T-junction, you will soon reach the E16 at Vikbukta back in Steinsfjorden again.

18. LINNESSTRANDA DELTA

Buskerud County
GPS: 59.75071° N 10.28394° Ø (tower hide parking)

Notable species: A variety of wetland species; Little Grebe, Water Rail, Little Ringed Plover, Peregrine Falcon, Kingfisher, Lesser Spotted Woodpecker, Thrush Nightingale, Reed Warbler, Marsh Warbler, Marsh Tit and Goldfinch.

Description: Linnesstranda nature reserve is a small, shallow delta at the outlet of the Lierelva river at the head of the Drammens fjord, located close to Drammen city. The shore is lined by reeds and a lush, almost impenetrable swamp forest. A number of ducks and other wetland birds can roost here during migration. Thrush Nightingale is regularly heard singing in May–June. Kingfisher is relatively often seen along the river and around the river mouth. A tower hide is erected here.

Best season: During migration spring and autumn migration. Also during winter, as long as the river mouth is free of ice.

Directions: From Oslo, take the E18 towards Kristiansand. Just before Drammen city, take the exit at junction 23 and follow the Rv23 signposted to Drøbak. Just over 3 km from the E18, after you have crossed the river Lierelva, turn right onto the Linnesstranda road. Find suitable parking just after crossing the bridge. To the left of the pump station opposite Gullaug school, continue on foot along the path that leads into the woodlands and along the river, ending at the tower hide.

Tactics: Start by taking a look along the river from the bridge, before following the path to the tower hide. Move carefully so that you do not flush the birds resting in the river mouth. From the tower hide, a spotting scope comes in handy. You can also drive a little further past the exit to Linnestranda and park at the sewage treatment plant on the east side of the Lierelva, and walk down the path past Dynodammen pond and a little further, along the river and the beach. Gilhusstranda beach, 300 m west of Linnestranda, has a nice oak grove. Be sure also to check the farmlands in the surrounding area of Lier. Lapwing breed and Golden Plovers and other open-country species are often found resting during migration.

19. MILETJERN POND

Buskerud County
GPS: 59.74992° N 10.03733° Ø (Miletjern, west-side)

Notable species: Ducks, gulls; Moorhen, Coot, Goldfinch and Common Rosefinch.

Description: Miletjern is a small pond, only 300 metres across, positioned adjacent to the Drammen River, just north of Drammen city. This small, nutrient-rich wetland is surrounded by the motorway on one side, and buildings, industrial areas and some cultivated land on the other. Miletjern pond appears as a small oasis in the middle of this "industrial desert". Along the edges, dense willow thickets and lush sedge and reedbeds grow, and the pond is partially covered with aquatic plants in summer. The pond attracts a variety of ducks and a number of other species, both as a nesting site and a resting place for birds migrating along the Drammen watercourse. Little Grebe is an annual visitor, and Shoveler almost so. Common Rosefinch is believed to breed in the area.

Just west of Miletjern pond is the Mile recycling facility. A fair number of gulls used to congregate here in winter, sometimes inter-spersed with the odd Glaucous or Iceland Gull, but in recent years the number of gulls has decreased at this site. The nearby Drammen River may still hold a few gulls, and you can find quite a few ducks and swans here in winter.

Best season: Spring, early summer and autumn in Miletjern pond, winter in Drammen River.

Description: From the E18 in Drammen city, take the exit at junction 25 and follow the E134 towards Haugesund and Kongsberg. Leave this highway on the off-ramp signposted to Mjøndalen/Krokstadelva. Exit towards Mjøndalen in both the first and second upcoming roundabouts. In both of the last two roundabouts, choose the exit signposted to Gulskogen. Then, just after passing the bridge across the highway, make a right turn onto the Ryghgata street. Drive along the industrial area and towards the residential area, before turning right into Eplegata street. Park opposite the kindergarten and walk 100 m back. Here, a footpath bridges over the stream. Follow the hiking trail 300 m to the west bank of Miletjern pond.

Tactics: Walk the footpath along the west side of Miletjern pond. You also get an

Little Grebe is most often seen in sheltered inlets along the coast in winter, but breed in small numbers in small nutrient-rich lakes a few places in Norway, such as in the Oslo area. *Photo: Terje Kolaas.*

Oslo

Interior

Skagerrak

Western

Central

Northern

Finnmark

Svalbard

overview of the pond from the residential area in the northeast. You get the best view of the Drammen River from the bus stops and the footpaths along the southern bank of the river.

20. LAKE FISKUMVANNET

Buskerud County

GPS: 59.71507° N 9.82177° Ø (tower hide parking)

Notable species: Ducks and other wetland species; Whooper Swan, Great Crested Grebe, Honey Buzzard, Marsh Harrier, Osprey, Goshawk, Hobby, Water Rail, Moorhen, Coot, Long-eared Owl, Nightjar, Wryneck, Green Woodpecker, Black Woodpecker, Lesser Spotted Woodpecker, Thrush Nightingale, Sedge Warbler, Reed Warbler, Marsh Warbler, Red-backed Shrike, Great Grey Shrike (winter), Goldfinch and Common Rosefinch.

Description: Lake Fiskumvannet is a 2.5 km long, shallow and nutrient-rich lake, surrounded by woods and farmland. It is

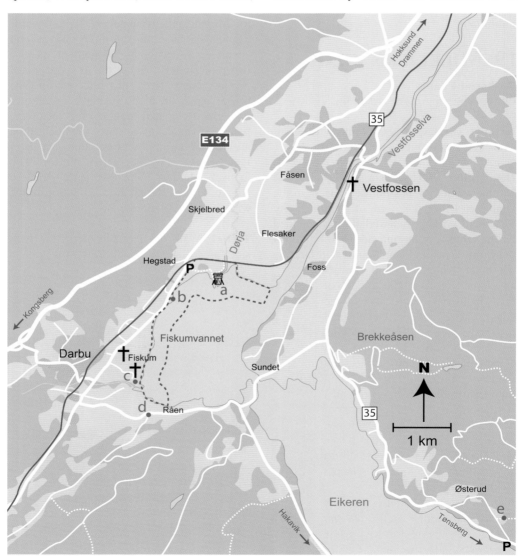

Lake Fiskumvannet: a. tower hide; **b.** lay-by viewpoint; **c.** Fiskum Old Church with access to the swamp in the SW corner of the lake; **d.** viewpoint by Råen farm; **e.** traditional Nightjar site by Lake Eikeren.

Oslo

Interior

Skagerrak

Western

Central

Northern

Finnmark

Svalbard

among the best and most popular birding lakes in south-eastern Norway. The northern and western shores are marshy and partly flooded at times, overgrown with reeds and other kinds of lush water vegetation. Lake Fiskumvannet is connected with the much larger and deeper lake Eikeren through a narrow strait. Lake Fiskumvannet has its outlet in the north-east through river Vestfosselva which runs into river Drammenselva a bit further downstream. Lake Fiskumvannet is home to several nesting birds, including Thrush Nightingale, Marsh Warbler, Red-backed Shrike and Common Rosefinch, but first and foremost it is an important staging site for wetland birds during migration. In addition to the commoner species, less numerous birds such as Black-throated Diver, Little Grebe, Slavonian Grebe, Pintail, Garganey, Shoveler and Scaup may be encountered.

There is a regular passage of Pink-footed Goose, Cormorant and raptors through the area during migration, usually most noticeable in April. Osprey and Hobby nest nearby and hunt frequently at the lake throughout spring and summer.

Lake Fiskumvannet forms the core of an extremely exciting area for nocturnal songbird expeditions. Often uncommon species like Quail, Spotted Crake, Grasshopper Warbler or even rarer species such as River Warbler and Blyth's Reed Warbler may be found. The pine-covered hills along the east side of Lake Eikeren is a traditional spot for Nightjar.

Other wildlife: Beaver is found in the river Dørja and is regularly seen from the tower hide. Moor frog is often heard in spring. Elk, roe deer, red fox, badger and stoat are common in the area.

Best season: Most of the year, from when the ice breaks up in April until it covers the lake again in early winter.

Directions: From E18 in Drammen city, turn at Exit 25 onto E134 towards Haugesund and Kongsberg. 10 km north of Mjøndalen town, exit onto Rv35 signposted to Tønsberg and Vestfossen. After 600 metres, make a right turn towards Darbu and follow this road almost 5 km until it passes under a railroad bridge. Park on the left side of the road 100 m after the bridge.

Tactics: From the parking spot by the railway bridge, walk through the gate and follow the footpath 700 metres through swampy woodland to the lake-side tower hide. Mosquito repellent may come in handy. Passerines including Common Rosefinch, Red-backed Shrike and – during migration – Bluethroat may be flushed along the track. In late autumn, Jack Snipe is sometimes encountered as well. The tower hide makes a perfect vantage point from where to scan the waterbirds out on the lake and to follow the overhead migration. A spotting scope is essential.

To explore other parts of the area, drive on towards Darbu. From **b.** you get nice views of the north-western corner of Lake Fiskumvannet. Drive on and make a left 200 m after Fiskum church. This road leads down to **c.** Fiskum Gamle Kirke (the old church), from where you can walk down to the lake and the river Fiskumelva. Out on the main road again, make a left onto the road signposted to Hakavik. Stop for another view of the lake at **d.** the intersection with Åssideveien. Continue on, and cross the strait that connects Lake Fiskumvannet with Lake Eikeren. Turn right onto Rv35 signposted to Tønsberg and continue exactly 5 km. Here at **e.** Østerud a closed gravel road takes off uphill to the left towards Slettfjell hill. This is a traditional site for Nightjar. Walk this gravel road after dusk between the end of May and early July and listen for the bird's characteristic, mechanical trill. Continue further down the main road along Lake Eikeren and stop to listen regularly at suitable places. Nightjars are sometimes seen perched on the road or flying across, with their eyes producing bright orange reflections in the car's headlight breams. At daytime these wooded hills are good for Black Grouse, Capercaillie, woodpeckers, Mistle Thrush and more.

INTERIOR SOUTH

The interior of southern Norway rises 2,469 metres above sea level at the summit of Galdhøpiggen in the mountain massif Jotunheimen, "the home of the giants". West of these mountains, the land plunges steeply down to sea level in the narrow fjords that cut deeply into the western coast. East of the mountains, however, the terrain descends more gently towards the valleys in the east, north and south. These valleys cut through the

Siberian Jay is fairly common in old-growth montane conifer forests, but can sometimes be difficult to find. As always, patience will eventually pay off. *Photo: Eirik Grønningsæter / WildNature.no*

south-eastern part of the country, generally in a north-south direction, with Østerdalen, Gudbrandsdalen, Valdres, Hallingdal and Setesdal as the most prominent valleys. In the bottom of each valley, you find rivers surrounded by farmland, particularly rich around Lake Mjøsa, the largest freshwater

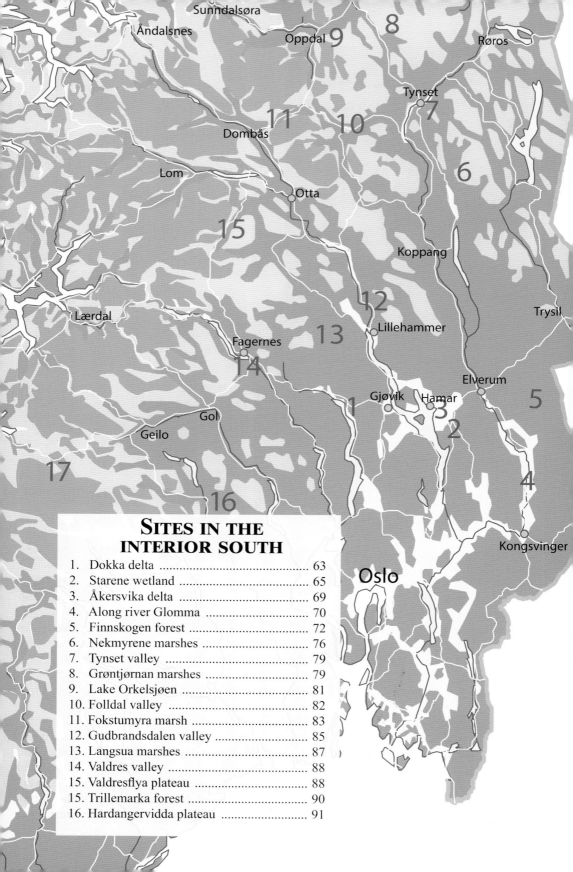

Sunndalsøra

Åndalsnes

Oppdal **9**

8

Røros

Tynset **7**

11

Dombås

10

Lom

Otta

6

15

Koppang

Lærdal

12

Trysil

13

Lillehammer

Fagernes

14

Elverum

Gjøvik

Hamar

5

1

3

Gol

2

Geilo

4

17

Kongsvinger

16

Oslo

Sites in the Interior South

Pygmy Owl can be quite common, particularly in mixed forest alternating with marshes and clearings. *Photo: Kjetil Schjølberg.*

Of course, some habitats in the southern interior of Norway are more diverse and bird-rich than others. Typically, marshy areas surrounded by a wide variety of vegetation will yield the most birds in the widest range of species. This holds true for wooded and montane areas alike. Foreigners thinking that birdlife is sparse this far north are often astounded by the shear number of birds encountered in certain habitats or locations if visited in prime time in late spring or early summer. On such occasions, the morning air may be filled with the sound of singing Tree Pipits, Redwings, Fieldfares, Bramblings, Willow and Wood Warblers, Dunnocks, Robins and several other Scandinavian songbirds, perhaps with ambient cries of displaying Black-throated Diver, Black Grouse, Black Woodpecker, Crane and others in the distance.

lake in Norway. The lowlands of south-eastern Norway are otherwise dominated by coniferous forest, actually the western fringe of the great Siberian taiga that stretches all the way across Russia to the Pacific coast in the east. In most areas the forest is dominated by Norway Spruce, in some areas with Scots Pine. The coniferous tree line is around 1000 m in this part of Scandinavia, with the hardier deciduous trees birch and willow covering the terrain up to about 1200 m.

A lot of specialized species of bird inhabit the southern interior of Norway, and many of them, such as Hazel Grouse, Capercaillie, Slavonian Grebe, Dotterel, Great Snipe, Red-necked Phalarope, Long-tailed Skua, Great Grey and Ural Owls, Black and Three-toed Woodpecker, Siberian Jay and Pine Grosbeak, are sparsely distributed in specific habitats. The key to finding them is to spend as much time as possible searching in the appropriate habitats, walking slowly and silently while listening and looking carefully. Other breeding specialities, such as Hen Harrier, Broad-billed Sandpiper, Ruff, Siberian Tit, Ortolan Bunting and Rustic Bunting are usually only found in a few specific locations.

THE BEST SITES

Generally, most of the Scandinavian forest species are to be found in any mature and continuous coniferous forest in southern Norway. Siberian Jay is more common in the northern parts of this region, particularly in high-elevation coniferous forest, and often near recreational cottages and lodges. The best areas for Ural Owl and Rustic Bunting (and brown bear and grey wolf) are along the border with Sweden in the east, in particular from Finnskogen forest and northwards. In the Folldal district and east of Lake Femunden, Siberian Tit may still be found, as the only remaining places in southern Norway. The best general areas for montane species are Fokstumyra, Valdresflya and Hardangervidda, all of them being areas which include rich marshlands above tree line. Nekmyrene marshes are the most reliable for breeding Ruff and Broad-billed Sandpiper, while Valdresflya is the most reliable south of Finnmark in northern Norway for Long-tailed Skua. In order to find many of the montane breeding species, virtually any location above tree line will do if you search thoroughly enough through the right habitats – see the species accounts in the back of the book for details on where, when and how to search. Wetlands in the cultivated lowlands are important staging areas for migrants waiting for the snow and ice to melt in the mountains in spring. These areas may provide suitable habitat for a number of breeding lowland birds as well. Of particular importance in this category are Åkersvika delta in Lake Mjøsa and the Dokka delta in Lake Randsfjorden.

1. DOKKA DELTA

Innlandet County
GPS: 60.79829° N 10.17308° E (Odnes tower hide), 60.79360° N 10.14007° E (Våten tower hide)

Notable species: Staging waterbirds; Great Crested Grebe, Slavonian Grebe, Black-necked Grebe (scarce), Osprey, Crane, Black Woodpecker, Grey Wagtail, Wood Warbler and Common Rosefinch.

Description: The delta where the river Dokka meets the north end of Lake Randsfjorden is one of the most important inland staging sites for wetland species in southern Norway.

It is situated in the wooded lowlands at an elevation of just 134 metres. Even though Lake Randsfjorden is artificially regulated, the natural condition of the delta itself is relatively intact.

Several species breed in the delta, including Great Crested Grebe. Uncommon species such as Shoveler, Garganey and Marsh Harrier have bred as well. Osprey hunts regularly in the delta. Hobby is seen with increasing regularity through summer, and it is suspected to nest in the nearby woodlands.

During migration, particularly in spring, substantial numbers and a relatively wide diversity of wetland species are found in the Dokka delta. Besides several species of duck, the delta is a regular staging area for Crane. Several hundred may gather in the delta and south to Fluberg bridge. Taiga Bean Goose is a regular visitor as well, with up to 150 individuals in the delta during the first half of May. The Dokka delta is a traditional site for Slavonian Grebe too, and the site gained national attention in the early 2000s when a small number of Black-necked Grebes, a very rare species in Norway, started to turn up on

Oslo · Interior · Skagerrak · Western · Central · Northern · Finnmark · Svalbard

The Dokka delta is well set up for birdwatching. Utsynsløa hide is a traditional farmer's building transformed into a birdwatching hide. *Photo: Trond Øigarden.*

a regular basis. The species was eventually found breeding in very small numbers in small ponds in the woodland between Randsfjorden and Mjøsa. However, this species seems to have disappeared recent years. In autumn and winter, ducks and Whooper Swan gather in the Dokka delta as long as the water is free of ice.

Further south, on the eastern side of Lake Randsfjorden, the shallow Røykenvika inlet may host a variety of ducks and grebes which may be studied from a roadside tower hide.

Best season: Late spring and early summer, but actually all year as long as there is open water to be found.

Directions: *By car from Oslo to Fluberg bridge*, drive route Rv4 signposted to Gjøvik until you reach Jaren, where you exit onto Rv34 signposted to Dokka. After 7 km on this road, you see Røykenvika inlet and the tower hide close to the road. 9 km north of Hov village make a left and drive out on Fluberg bridge.

Tactics: From the Fluberg bridge you have a good view of this part of Lake Randsfjorden, but a telescope is essential here and elsewhere in the delta. For a tour of the delta, drive back to Rv34 and turn north towards Fagernes. In the intersection with Rv33, turn left towards Dokka. From here it is 7 km to the exit on the left, signposted to *Dokkadeltaet*

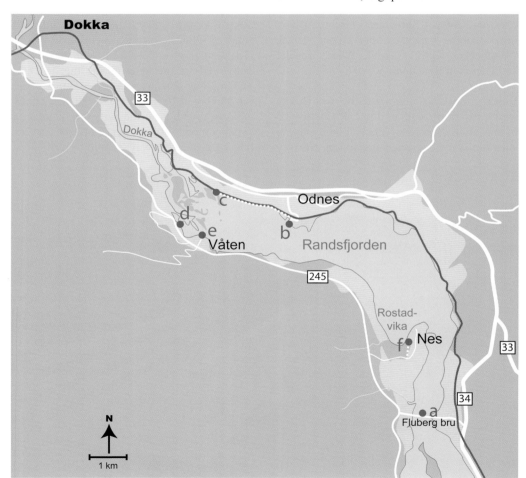

Dokka delta: a. Fluberg bridge viewpoint; **b.** Odnes tower hide, visitor centre; **c.** Utsynsløa hide; **d.** wall shelter hide; **e.** Våten tower hide; **f.** Rostadvika viewpoint.

Honey Buzzard is picky in terms of diet and the species is associated with deciduous and mixed forests alternating with open areas and lakes. It is widespread over large parts of south-eastern Norway south to Agder. The species does not have as much of a presence as Common Buzzard, which often sits on fence posts or circling for long periods of time over the nesting grounds. Honey Buzzard can therefore be more challenging to find. *Photo: Gunnar Numme.*

Våtmarkssenter (wetland visitor centre). Follow this road for 400 m and turn right down the hill, cross the railway track and park in the wetland centre's car park. You will find the bird tower on Odnes 50 m further ahead, where the Dokkadelta National Wetlands Centre is also located. *Utsynsløa hide*, a traditional farmer's building transformed into a birdwatching hide, can be reached by walking 1.8 km from Odnes along the old railroad track. Today, only railway biking is conducted on these tracks.

Back in the car, continue around the delta through agricultural areas with lush ditches and forest edges, where you must keep an eye out for roosting Cranes and geese, Great Grey Shrike, nocturnal songbirds and more, depending on the season. You first drive along Rv33 to Dokka village, and then head south along Rv245 signposted to Jevnaker. After just over 5 km on this road (or 200 m after you pass the entrance to the Nylinna side road) you will see the <u>wall shelter</u> on your left. Through peeking holes in this wall-like structure you can view up close one of the many fine, swampy ponds and coves in the delta, without disturbing the birds.

Continue another 600 m to Våten village, where there is a sign to the left to *Dokkadeltaet*, and where you will find the <u>Våten tower hide</u>. Continue south for 5 km and turn left into the road Nesgutua. Follow this for a few km, find suitable parking, and walk carefully on the edge of the field or in the forest to the north for about 400 m, where you reach the shallow <u>Rostadvika inlet</u>, which can sometimes hold ducks and other wetland birds.

2. STARENE WETLAND

Innlandet County
60.76866° N 11.22239° Ø (Horne farm parking)

Notable species: Staging geese, ducks, Crane and other wetland species in addition to passerines during migration periods; Slavonian Grebe, Honey Buzzard, Goshawk, Osprey, Hobby, Pygmy Owl, Long-eared Owl, Black Woodpecker, Lesser Spotted Woodpecker, Yellow Wagtail, Grey Wagtail, Marsh Warbler, Long-tailed Tit, Great Grey Shrike (winter), Red-backed Shrike, Rook, Goldfinch, Linnet and Common Rosefinch.

Description: Starene is a wetland area on a river plain located in an area of rich farmlands southeast of Hamar town. It is situated along the Svartelva and Starelva rivers, which are flooded every spring and during heavy rains so that large ponds are established. These used to be drained and cultivated, before the local BirdLife chapter took the initiative

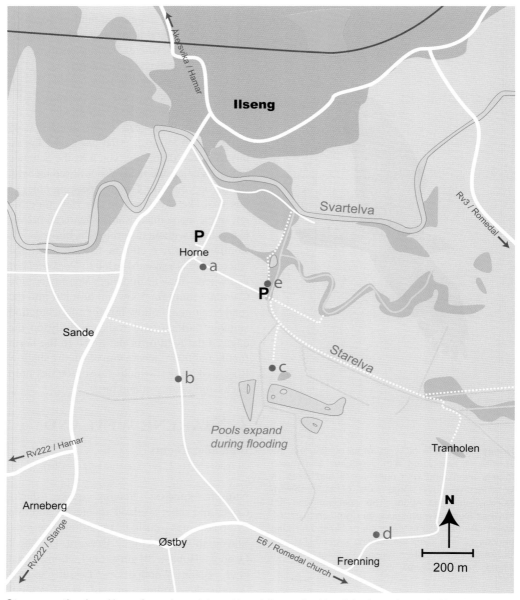

Starene wetland: a. Horne farm viewpoint, parking at the north side of the barn; **b.** road-side viewpoint to floodplains; **c.** nice and high vantage point on a pile, but should not be used when there are many birds nearby that might be flushed; **d.** alternative viewoint; **e.** Starene field station.

in 1989 to negotiate with the landowners. Several permanent dams have now also been created, to make conditions more stable and less dependent on floods. Thousands of ducks, geese, particularly Pink-footed Geese, Cranes and other wetland birds roost at Starene under favourable conditions during migration, especially in spring. Many of the birds commute between Starene and Åkersvika delta near Hamar (which see). The birds appear more concentrated at Starene than in Åkersvika, and are often seen at Starene at a better distance. The area also attracts a number of birds of prey. Quail, Water Rail and Corncrake are annual visitors. Slavonian Grebes nest on the islands in the largest pond. Ringing is carried out regularly at the Starene field station.

The surrounding landscape is also of interest, with large agricultural areas interspersed with smaller wetland areas and nutrient-rich ponds with the possibility of ducks and species such as Slavonian Grebe, Moorhen and Coot, as well as nocturnal songbirds, often including Water Rail and Spotted Crake.

Besides Starene wetland, the Frognertjernet pond in Hamar and the lakes Brynitjernet and Linderudsjøen further south are worth stopping by if in the area. Rotlia woods in Stange has deciduous forest with singing Icterine and Wood Warblers. Stein is a promontory that juts out into Lake Mjøsa in Ringsaker north of Hamar, where the large lake is at its narrowest. This is a good place to monitor movements on the lake during migration. Here, there are also shallow coves and a brook that becomes ice-free early in spring, and where considerable numbers of ducks may congregate. One of the county's largest resting places for Crane, with up to 200 individuals in autumn, is at Løten just east of Hamar town.

Best season: Starene wetland is at its best in spring for ducks and wetland species, early summer for nocturnal songbirds, and in autumn for migrating passerines.

Other wildlife: Beaver is found in the rivers and elk is a common sight.

Directions: *From Hamar to Starene*: Follow the side roads towards southeast, signposted to Ilseng. Park at Horne farm. See map.

Tactics: A telescope is essential on Starene wetland. Be especially careful not to flush the birds in this open countryside. Stick to roads and paths, and do not walk out into the fields. Other sites nearby: Lake Brynitjernet. Take the exit from E6

Oslo
Interior
Skagerrak
Western
Central
Northern
Finnmark
Svalbard

Starene wetland is basically floodplains, but a number of artificial ponds have been established to create more permanent conditions suitable for wetland birds. Pink-footed Geese often arrive in large flocks, as in this picture, and it is among the dominant species here during the spring migration. *Photo: Bjørn Olav Tveit.*

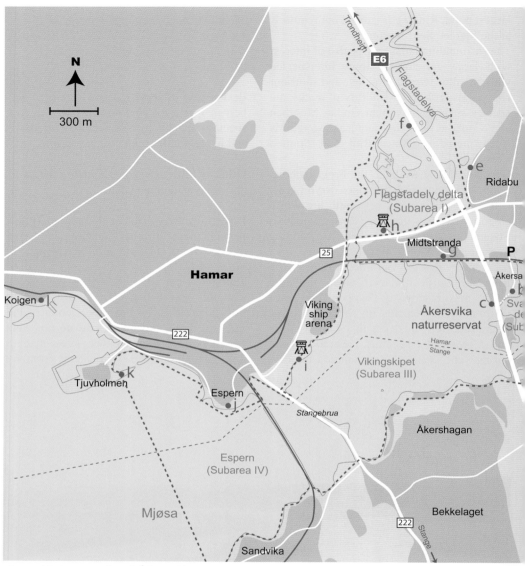

about 10 km south of Åkersvika delta to Rv24 signposted towards Kongsvinger. Follow Rv24 eastwards for 2.5 km and you will see the lake on your right.

Lake Linderudsjøen. Take the E6 exit at Tangen, 11 km further south. Follow Rv222 eastwards, signposted to Vallset. You immediately get the water on your left, and you will see a tower hide just over a km further ahead.

Rotlia woods. From Stange village centre, drive west following signposts towards Stange church. At the church, the road turns south (signposted to Sørum) along Lake Mjøsa. You must pay a small amount at a self-service tollbooth to continue all the way down to the car park just before the farm Rotlia, 7 km south of the church. Follow the path down to the bank of Mjøsa and follow this path 200 m southwards into the forest.

Stein. Follow the E6 just under 30 km north from Hamar town. At Moelv, just before the bridge crossing Lake Mjøsa, leave the E6 and follow Rv213 south towards Ringsaker church. You immediately get Korgerstuguvika inlet on your right, great for ducks. It is well explored from the

Interior

Skagerrak

Western

Central

Northern

Finnmark

Svalbard

Oslo

Åkersvika delta: a. Svartelv delta tower hide; **b.** viewpoint Svartelv delta north, and **c.** viewpoint Viking ship olympic arena east; **d.** Frognertjernet pond; **e.** pond near Ridabu; **f.** Flagstadelv delta north; **g.** Midtstranda viewpoint; **h.** Flagstadelv delta tower hide; **i.** Viking ship tower hide; **j.** and **k.** Espern viewpoints; **l.** Koigen beach.

3. ÅKERSVIKA DELTA

Innlandet County
GPS: 60.78930° N 11.13842° E (Svartelv delta tower hide)

Notable species: *Migration*: Large numbers of ducks, shorebirds and other wetland species. *Breeding*: Pheasant, Slavonian Grebe, Goshawk, Osprey, Kestrel, Hobby, Moorhen, Coot, Crane, Little Ringed Plover, Wood Pigeon, Collared Dove, Pygmy Owl, Long-eared Owl, Wryneck, Green Woodpecker, Black Woodpecker, Lesser Spotted Woodpecker, Yellow Wagtail, Grey Wagtail, Thrush Nightingale, Marsh Warbler, Long-tailed Tit, Rook, Nutcracker, Goldfinch, Linnet and Common Rosefinch.

Description: Åkersvika is a delta area along the eastern shore of Lake Mjøsa, the largest inland lake in Norway. It is one of the most important staging sites for ducks, shorebirds and other wetland species in the southern interior of Norway. The largest town in the region, Hamar, and its industrial areas are situated very close to Åkersvika. Furthermore, the main highway between Oslo and Trondheim, E6, actually crosses the delta. This makes the site susceptible to disturbance and in danger of further degradation. However, the site's convenient location makes it a popular and frequently visited site, both by local and travelling birdwatchers. Apart from the urban surroundings, the area also borders farmlands and patches of woodland. The heart of the area is the shallow bay at the mouth of the two rivers Svartelva and Flagstadelva, both of which have built up deltas of mudflats and small islands overgrown with vegetation beneficial to ducks and other waterbirds. Three tower hides are found at Åkersvika, but the one at the mouth of the Svartelv river is the only hide of any practical value.

bus lay-by after just a few kilometres on Rv213. If you want to walk out onto the headland, you can park at the water-skiing club just beyond. To scan the early ice-free section of Steinsvika inlet, continue 500 m on Rv213 and turn down the side road to the right.
You find <u>Løten</u> and the staging grounds for Cranes by following Rv25 from Hamar towards Løten and exit towards Løten after 11 km. The cranes roost in the fields along this 1.5 km long road between Løten church and Løten village centre.

The number of birds in the area is influenced by natural fluctuations in the water flow in the rivers and by the artificial raising and lowering of the water level in Lake Mjøsa. For ducks this is hardly an issue, but for shorebirds a high water level will limit the possibilities for feeding. Nevertheless, the water level during shorebird migration is often low enough even for short-legged stints to thrive. Osprey and Hobby hunt in the area on a regular basis throughout the summer. Rook, which is an uncommon breeding bird in Norway, nest in several colonies near Hamar and can be seen feeding in the fields.

Best season: Åkersvika is of interest from when the ice starts to break up at the creek and river inlets in the middle of March until the ice covers the wetlands again in November-December. The greatest numbers and diversity of ducks are often found in April. Shorebirds stop by from the first half of May and through to autumn, as long as the water level is low enough for the mudbanks to be exposed.

Directions: Highway E6 crosses Åkersvika just south of Hamar. The exit from E6 to **a.** the tower hide at the Svartelv delta is signposted to Hjellum, while the exit to **h.** the tower hide at the Flagstadelva delta is signposted to Olympiahallen. See the map for further directions.

Tactics: Åkersvika is best explored from **a.** the tower hide at the Svartelv delta and from vantage points along the roads in the area, as illustrated in the map.

4. ALONG RIVER GLOMMA

Innlandet County

GPS: 59.97063° N 12.14840° Ø (tower hide Lake Gaustadsjøen), 60.43060° N 12.04324° Ø (tower hide Lake Gardsjøen), 60.53481° N 12.02332° Ø (tower hiede Lake Strandsjøen)

Notable species: Ducks, shorebirds and Crane during migration; Black Grouse, Osprey, Honey Buzzard, Hobby, Coot, Little Ringed Plover, Collared Dove, Pygmy Owl, Long-eared Owl, Wryneck, Grey-headed Woodpecker, Green Woodpecker, Black Woodpecker, Lesser Spotted Woodpecker, Woodlark, Yellow Wagtail, Grey Wagtail, Redstart, Sedge Warbler, Marsh Warbler, Icterine Warbler, Wood Warbler, Long-tailed Tit, Marsh Tit, Red-backed Shrike and Ortolan Bunting.

Description: You will find many interesting oxbow lakes, shallows and marshes along Norway's longest river, the Glomma. Several of these, but especially Lake Vingersjøen and Lake Strandsjøen, function as staging grounds for ducks and other waterbirds during migration. At low water levels in river Glomma, there are favourable sandbanks for birds all the way from Kongsvinger and north to Elverum towns. The surrounding forests and farmlands can display species such as Crane, Osprey, Honey Buzzard and Hobby. Woodpeckers occur in suitable habitats along this stretch. Kingfisher can sometimes be encountered, and river Vrangselva in Eidskog and river Oppstadåa in Sør-Odal are among the best places to look for it. In the very south of the Vrangselva watercourse, Marsh Tit is found. On Melåsmoen and other peat bogs in Solør you may still find Ortolan Bunting. Starmoen is one of the few places in this region where Woodlark is found.

Other wildlife: Elk, roe deer, badger, American mink, weasel and red fox are often seen in these areas. Beavers are also common in several places. Along the river Vrangselva in Eidskog, e.g. at Lille Gaustadsjøen, there is a small population of harvest mouse. There are also unusual butterflies to be found here.

Best season: Migration seasons and early summer.

Directions: The sites along river Glomma are presented as they appear along E16/Rv2 between Skarnes to the south and Elverum in the north.

Lake Seimsjøen in Lake Storsjøen: In Skarnes, leave E16 onto Rv24 northwards, signposted to Odalen. After a few km, turn right onto Kvedalsvegen road, signposted to Ringås.

Keep left at the subsequent T-junction and continue past Lake Stortjennet and over the Vesleåa creek. Turn left, park in the parking lot here, and walk out onto the causeway crossing this most accessible part of the lake, called Nora. The farmlands around Oppstad often have nice flood ponds in spring. The small Lake Nusttjernet is also located here, which may be worth checking out. Look for Kingfisher in river Oppstadåa.

Lake Gaustadsjøen: From Kongsvinger, follow Rv2 along river Vrangselva south towards Magnor. At Skotterud, turn off onto Rv202 and follow this road for just under 3 km and turn down to the right after the agricultural centre. After 200 m, a tractor road exit to the left, crossing the railroad tracks. Here you can either head right and follow the path to the tower hide at Lake Gaustadsjøen, or whether you want to go left and follow the path to the swampy Lake Lille Gaustadsjøen.

Lake Vingersjøen: From Kongsvinger, follow Rv2 north towards Elverum. Just outside the town, take the E16 signposted to Austmarka. You will soon have a view of the lake on your right. After 2.5 km on the E16, stop in the lay-by below the farm Skansgården, and scan the lake with a telescope. In the meadow below, you can sometimes in May hear Great Snipe displaying on a migration stop-over.

Lake Gardsjøen: From Kongsvinger, follow Rv20 just over 3 km north. Turn right at the road sign for Gård. Lake Gardsjøen lies parallel to the road you are now on, on the left, hidden behind the deciduous trees on the opposite side of the fields.

Lake Silvatnet: In Kirkenær village, turn left onto Bruvegen road. Cross over river Glomma and turn north on Rv210 towards Elverum. Turn right after just over 1 km, and you will see the lake.

Lake Strandsjøen: Continue Rv210 north for another 6 km and take the exit to Hoff. You will soon see the tower hide by Lake Strandsjøen on your left. Continue on and take the first road to the left, and drive north along Lake Strandsjøen's eastern shore.

Kvislerdammen pond is the far northern section of Lake Strandsjøen, and this can be

Oslo

Interior

Skagerrak

Western

Central

Northern

Finnmark

Svalbard

Ortolan Bunting is a species in sharp decline in this country and may disappear from Norway within a few years. You can now only hope to find it in along river Glomma, associated with peat bogs, logging areas and burnt areas. Melåsmoen is the best-known place to encounter it nowadays. *Photo: Terje Kolaas.*

a superb place for ducks. Whooper Swan nest here.

Lake Gjesåssjøen: In Flisa, exit from Rv2 onto Rv206 signposted to Åsnes Finnskog. After a few km, turn left towards Braskereidfoss. You reach Lake Gjesåssjøen after just over 4 km. From the roads around the lake, you can explore it from all sides. Mute Swan, Great Crested Grebe, Marsh Harrier, Coot, Sedge and Reed Warbler are among the breeding species here.

Lake Tørråssjøen and Lake Tørråstjernet: From Rv2 in Braskereidfoss north of Flisa, turn east, signposted to Gravberget. After 3 km you will see Lake Tørråssjøen on the left and Lake Tørråstjernet on the right.

Melåsmoen: From Braskereidfoss, follow Rv2 north for 13 km and turn right onto the small Melåsmoenvegen road, signposted to Silkebækken. In recent years, this has been the most reliable site for Ortolan Bunting. The very few remaining individuals can be heard singing from the road. If you do not find Ortolan here, you can investigate Osmyra marsh just south of Jømna and Glesmyra marsh between Braskereidfoss and Våler. Please do not to use playback or otherwise disturb these birds.

Starmoen: Continue on the Melåsmoenvegen road and turn left on Melåsbergvegen. You reach Starmoen after 5 km. The go-kart track and the golf course are good places for singing Woodlarks.

Verken in Heradsbygd: This is a river plain along Glomma on the opposite side of Rv2 for Melåsmoen, close to Heradsbygd church. This can be a nice area during migration.

Tactics: Most of these localities can be explored from – or within a short walking distance from – the road. A telescope will come in handy, especially at Lake Gjesåssjøen.

Ural Owl in Norway breeds only regularly in Finnskogen forest. Its numbers are not as sensitive to fluctuations in the rodent populations as in Great Grey and most other owls, because the Ural Owl has a more varied diet. *Photo: Eirik Grønningsæter / WildNature.no*

5. FINNSKOGEN FOREST

Innlandet County

GPS: 60.75895° N 12.17108° E (Kynna bridge), 60.85999° N 11.69258° E (Starmoen)

Notable species: Black-throated Diver, Red-throated Diver, Hazel Grouse, Capercaillie, Black Grouse, Honey Buzzard, Buzzard, Goshawk, Golden Eagle, Osprey, Kestrel, Hobby, Crane, Whimbrel, Greenshank, Eagle Owl, Hawk Owl, Pygmy Owl, Ural Owl (uncommon), Great Grey Owl (scarce), Long-eared Owl, Tengmalm's Owl, Nightjar, Wryneck, Black Woodpecker, Three-toed Woodpecker, Woodlark, Yellow Wagtail, Redstart, Mistle Thrush, Siberian

Jay, Parrot Crossbill, Pine Grosbeak and Rustic Bunting (scarce).

Description: Kynndalen valley with its surroundings is one of the most exciting forest areas for birdwatchers in southern Norway. The valley is located in Finnskogen, a large conifer forest region close to the Swedish border. It is the most important area in the country for Ural Owl, Ortolan Bunting and Rustic Bunting. Ural Owl is possible in Kynndalen but more likely along the roads near the border, for instance in the low-lying marshy woodlands just south of Gravberget village, which is generally an owl hotspot in the area. Ural Owl is a critically endangered species in Norway, and so it is advisable not to go public with the precise location of any observations during breeding season. Do not expect to receive such information either. In

recent years, Great Grey Owl is found here as well. Eagle Owl can be listened for at dusk or dawn near cliffs in the valleys surrounding Finnskogen forest.

Rustic Bunting is an uncommon breeder along creeks and streams lined with deciduous trees in Kynndalen and Finnskogen in general, and they often need to be searched thoroughly for in flooded forestland. The population has decreased alarmingly in recent years.

An alternative place to search for large owls and Rustic Bunting is Ulvådalen valley east of Elverum and Skjeftkjølen marsh in Trysil. At Ulvådalen you may also find Crane, Whimbrel, Greenshank, Three-toed Woodpecker, Redstart, Mistle Thrush, Siberian Jay (scarce) and Parrot Crossbill. At Skjeftkjølen and nearby Lake Rysjøen, you may also see breeding Red-throated

Ural Owl prefers tall coniferous forest interspersed by lakes and open marshes. It may breed in large, preferably slightly open nest boxes hung up in the right place and in the right habitat. The picture shows precisely such terrain at an old nest box intended for Ural Owl in the Finnskogen forest. *Photo: Bjørn Olav Tveit.*

Black Woodpecker is the most important provider of natural, large cavities in the Scandinavian forests. It uses the hole only the year it makes it, leaving it for the enjoyment of owls and a variety of other species in subsequent seasons. *Photo: Kjartan Trana.*

Diver, Black-throated Diver, Whimbrel and Greenshank, and Crane and other wetland species may stage here during migration.

Look for Pine Grosbeak in upland coniferous forests with plenty of juniper undergrowth. Ortolan Bunting is almost extinct in Norway, but a very few pairs still breed in the Solør region in the western parts of Finnskogen forest, particularly at Meløyfloen near Starmoen, an area which is also a fairly reliable for Woodlark and Parrot Crossbill.

Other wildlife: Finnskogen forest, and adjoining areas on the Swedish side of the border, is an important area for the common Swedish–Norwegian population of grey wolf. Beaver and lynx inhabits the area as well, and brown bears occasionally wander through.

Best season: Generally spring and early summer. Rustic Bunting is easiest to find when the males sing actively in May, Ortolan Bunting in May and June. Owls may be encountered in Finnskogen year round, but they call most frequently from dusk til dawn in March and early April.

Directions: *From Oslo/Hamar*, aim for Kongsvinger, Elverum, Rena and/or Trysil.

Tactics: The most common way to explore these forests is by car – preferably at night for owling trips in spring. There is a well-developed network of forest roads in these areas. Driving on some of these roads requires you to pay a fee at an unattended toll booth upon entering. From the toll booth, it can be worthwhile to bring (or photograph with your mobile phone) a paper map of the road network, as mobile phone coverage and access to digital maps can be poor in some places. Note that the forest roads in spring can become very soft and difficult to access because of the snow melting, even with four-wheel drive. This can be dramatic if you drive into the forest onto frozen ground at night, and are not out again by the time the road thaws in late morning.

During the day, you should take your time to walk a bit in the terrain on foot. High impact boots and mosquito repellent are essential for a positive experience in the marshy areas.

Look for Rustic Bunting in heavily swampy deciduous and mixed forests along still-flowing rivers in the region, often in areas flooded by beaver dams. The species lives a reclusive life and usually performs its song from a hidden song post. It pays to have thoroughly rehearsed the song in advance. After mid-June they can be very difficult to locate.

You can find the owls by systematically driving around the forest roads at night and stepping out of the vehicle and listening at regular intervals. Ural Owl are usually found in areas with pine trees alternating with large bogs, preferably near open water. Great Grey Owl appear to be less picky but often appear in similar habitat or any old-growth coniferous forest.

At dusk and dawn, it is not unusual to come across grouses sitting on the dirt roads. Both the owls and the grouses can be found surprisingly close to farms or people's houses. Described below are a selection of well-known entrances to these forest areas, with directions from the nearest town – but it is even better, where possible, to drive between the described places via the forest roads.

Svullrya: From Kongsvinger town, follow Rv2 north towards Elverum and take the exit at Roverud onto Rv205/202 signposted to Svullrya. Here in the southern part of the Finnskogen forest, there have been found to be small populations of Nightjar and Woodlark, besides the possibility of many of the other mentioned species.

Kynndalen valley: From Kongsvinger, follow Rv2 north to Flisa. Here, take the Rv206 sign posted to Åsnes Finnskog. After 7.5 km on this road, turn left, signposted to Kynndalen. Follow the forest road further, and at the road forks, choose the road with the highest standard. At crossroads where there is any doubt as to where the main road goes, there are small signs to Kynndalen. Forest roads run along both sides of the river Kynna, which used to be a known area for Rustic Bunting.

Both Tengmalm's and Pygmy Owls can be found in the area. At Kynnmyrene marshes and the infield on the opposite side of the road, look carefully for large owls. Lake Rogsjøen can offer Black-throated Diver, ducks and more. This can be a good base for owling trips at night. Listen for Ural Owl here and in similar terrain elsewhere in Finnskogen. Hobby and Honey Buzzard are regularly seen in the area. From Kynndalen you can aim east towards Gravberget village. The areas around this village have potential for all the species mentioned at the beginning, especially Ural Owl.

Elverum forests: From Elverum town, drive Rv2 south towards Kongsvinger and just past Heradsbygd you turn left onto Melåsfloveien (an area with Ortolan Bunting and Woodlark). Drive this road to the end, cross over Nøtåsvegen road and enter the forest via Flotsvegen road. From here you aim towards Kvasstadsætra and then north towards Sørskogbygda, but preferably via as many detours as you have time for.

Ulvådalen and Ulvåkjølen marshes: From Elverum, follow Rv25 northeast towards Trysil. After about 21 km, in Ulvålia village, turn left signposted to Ulvådalen and turn left again just before Ulvåbrua bridge. After the toll booth, continue up the valley along the west bank of Ulvåa river to Ulvåkjølen nature reserve, which can be explored from the forest roads around the marsh area or on foot. The roads here are plowed in winter, but are closed for cars with gates. This limits access during the owls' most active calling period, except if you walk or ski the 7 km up to Ulvåkjølen.

Skjeftkjølen marsh and Lake Rysjøen. From Rv25 approximately midway between Elverum and Trysil, turn onto Rv208 south at the sign for Lutnes and follow signposts towards Rysjølia. There are two entrances to Skjeftkjølen marsh, and the first of these are reached after approx. 8 km, where there is a sign to the right towards Høljedalen. Keep left at the Y junction. A couple of kilometres further on, just as you get to the Søndagsmyr dam on your right (100 metres before the next crossroads), the road crosses a stream, Ørbekken. Park and follow this stream up and into the large marshy area in this nature reserve. In the past, Rustic Bunting was found along this stream. The second entrance to the area is reached by continuing Rv208 for another couple of km. You get a view over Lake Rysjøen 2.7 km after the sign that marks that you have arrived at Rysjølia, and a side road leads to the lake 3.6 km from the same spot. If you continue a further 200 m along Rv208, a path leads into the forest on the right-hand side, and there is also possible to park here. From here you can walk down to the water and follow the path halfway around until you reach Skjeftkjølbekken creek in the eastern part of Skjeftkjølen marsh. This was also a suitable place for Rustic Bunting. A large beaver dam lies across the stream half a kilometre below the water. A small tower hide provide a view over Lake Rysjøen.

Broad-billed Sandpiper is a threatened species in Scandinavia as a result of drainage of the wet breeding bogs, e.g. when driving on bare ground with large forestry machines or all-terrain vehicles. The Norwegian core area is in Finnmark County, but you can find the species in a few places in the south, such as in the Langsua or Nekmyrene marshes. *Photo: Kjetil Schjølberg.*

Åmot–Trysil. The Rv215 between Rena in Åmot and Jordet in Trysil takes you through 65 km of great terrain. Here you will find a number of forest species as well as Yellow Wagtail and Red-backed Shrike, and is one of the last places in the country where you might still be lucky to find Rustic Bunting. At Rena is the BT track (Rødsmoen "Basic Training" camp), an area set aside for military training. Here, Hen Harrier and other raptors are often seen during migration, and you can also find a variety of woodpeckers, grouse and sometimes Great Grey Owl. From Rv3 at Rena, take the Rv215 sign for Jordet. After 6 km, just after the exit to Julussdalen, turn left over the dam at the southern end of Lake Løpsjøen to enter the BT railway. Continue Rv215 to Jordet and take detours on side roads.

6. NEKMYRENE MARSHES

Innlandet County

GPS: 62.09173° N 11.21768° E (Nekmyrene), 62.03102° N 11.52021° E (Vesle Sølensjøen)

Notable species: Pintail, Scaup, Common Scoter, Rock Ptarmigan, Willow Ptarmigan, Black Grouse, Capercaillie, Red-throated Diver, Black-throated Diver, Goshawk, Rough-legged Buzzard, Golden Eagle, Kestrel, Crane, Dotterel, Broad-billed Sandpiper, Ruff, Red-necked Phalarope, Little Gull, Arctic Tern, Short-eared Owl, Tengmalm's Owl, Yellow Wagtail, Dipper, Bluethroat, Mistle Thrush, Siberian Jay, Siberian Tit (rare), Great Grey Shrike, Parrot Crossbill and Lapland Bunting.

Description: Nekmyrene marshes is one of the few remaining breeding sites for Broad-billed Sandpiper and Ruff in southern Norway. This wetland is situated above the tree line, just north of the prominent rock formation, Sølen. Human activities must be conducted with care at this important breeding site. However, the Broad-billed Sandpipers, in particular, are not very shy of people and tend to stay out in the softest and most inaccessible black bogs. The greatest threat to the population is actually not human disturbance, but drainage of the marshes they live in, for instance by the deposition of ruts by witless off-road driving of terrain vehicles and forestry machinery. Several more specialities breed at Nekmyrene marshes, including Dunlin, Red-necked Phalarope, Arctic Tern, Short-eared Owl, Bluethroat and Lapland Bunting. Rock Ptarmigan, Dotterel and Snow Bunting breed in the surrounding mountains.

The nearby Lake Vesle Sølensjøen is an extension of the slow flowing river Sølna, which flows through the valley east of Nekmyrene marshes. The soil in the area is low in nutrition and the vegetation thus rather sparse but Lake Vesle Sølensjøen is still one of the region's most important and bird-rich breeding grounds. Besides a number of common wetland species of the region, including Teal, Wigeon, Tufted Duck, Goldeneye, Greenshank, Wood Sandpiper, Ruff and Whimbrel, the site also offer favourable conditions for more scarce breeding species like Pintail, Scaup, Common Scoter, Red-breasted Merganser, Red-throated Diver, Black-throated Diver, Crane, Broad-billed Sandpiper, Red-necked Phalarope, Arctic Tern and Short-eared Owl. Little Gull also seem to be about to establish itself as a breeding bird along this watercourse. The pine dominated woodland in the valley is good for Siberian Jay. You might be lucky to find Pine Grosbeak or even Siberian Tit here as well.

On your way to these sites you pass the small islands Koppangsøyene in the river Glomma, and Lake Lomnessjøen in Rendalen valley, both of which are worth a stop to check for Crane, Little Gull and other waterbirds, especially during migration.

Nekmyrene marshes is one of the most interesting wetland areas in southern Norway and one of the very few places here where you can find both Ruff and Broad-billed Sandpiper. The latter feed in these very soft bogs. *Photo: Bjørn Olav Tveit.*

Capercaillie is an impressive sight during the males' display lek. The species is fairly common in old-growth coniferous forests of the interior south and can sometimes be seen displaying road-side. *Photo: Kjartan Trana.*

Other wildlife: Rendalen is one of the few areas in the country where breeding grey wolf has been confirmed in recent times. This is also part of the core area for brown bear in southern Norway. Lynx and wolverine roam the forests and mountains here as well, but you will be considered extremely lucky to spot any of these large and shy carnivores, even here in the heart of their range. Sølendalen valley has good populations of elk and beaver, and there are plenty of reindeer in the area.

Best season: Mid-May to mid-July.

Directions: From either village Koppang in the south or Tynset in the north, follow Rv30 through Rendalen valley. Koppangsøyene islets in the Glomma river are just by the exit to Koppang, while Lake Lomnessjøen is on the east side of Rv30 by the small town Otnes, where you might want to fuel up the car and acquire groceries before heading into the wilderness. The best part of Lake Lomnessjøen is seen from the road just north of Otnes. In Elvål (52 km north of Koppang

and 25 km north of Otnes), exit east on the road signposted to Unset. Pass through Unset village and make a right onto a gravel road signposted to Sølendalen. Pay a fee at the unmanned toll booth. Nekmyrene marshes is reached by driving another 14 km from the toll booth, following signs to Nekkjølen. You can walk along the dam to your left in a north-westerly direction and conveniently experience all the specialities in the area without having to plunge out into the wet marshland.

Lake Vesle Sølensjøen is reached from the toll booth by following signs to Haugsetvollen. After about 22 km the road crosses a wooden bridge (look for Dipper here) and passes through a cluster of traditional shielings or cottages. At this point the road network gets a bit complicated, but follow the road that is in best condition and look for the sign to Haugsetvollen, leading in a southerly direction (to the right). At 2.5 km after the shielings, a dirt road closed with a barrier exits to the left, immediately followed by a roadside parking area. Park here and follow the dirt road 1.5

km (a 20-minute walk) down to Lake Vesle Sølensjøen. Siberian Jay is sometimes seen along this road.

Tactics: Broad-billed Sandpiper is easiest to find during their characteristic aerial display over the marshes. You can also check the black bogs carefully with a telescope, a piece of equipment that also comes in handy from the bank of Lake Vesle Sølensjøen.

7. TYNSET VALLEY

Innlandet County
GPS: 62.2807° N 10.7716° Ø (Tynset bridge)

Notable species: Migrating and resting wetland species; Slavonian Grebe, Little Ringed Plover, Long-eared Owl, Black Woodpecker, Three-toed Woodpecker, Waxwing, Redstart, Mistle Thrush, Siberian Jay, Siberian Tit (rare), Two-barred Crossbill (scarce) and Parrot Crossbill.

Description: In the far north of Østerdalen valley, surrounded by mountains and higher-lying coniferous forest, the soft and lush agricultural landscape around Tynset town opens up in clear contrast to the surroundings. Here, in the very upper part of the Glomma watercourse, the river meanders calmly through the valley floor, and over time has left behind several beautiful lakes surrounded by marshland, birch forest and fields. This is called the Tjønn area and is a welcome place for ducks and shorebirds who need to rest while they wait for summer to arrive further up in the mountains. Slavonian Grebe, Pygmy Owl and Long-eared Owl are among the breeding birds here.
Ripan nature reserve just south of the centre of Tynset has a beautiful pine forest where you can find a number of forest species, including Three-toed Woodpecker and Siberian Jay, and where Two-barred Crossbill and Siberian Tit have been recorded. North-east of Tynset is Lake Nylandstjønna by Vingelen, a fine wetland equipped with a tower hide.

Directions: Tynset town is located along Rv3

at the head of Østerdalen valley. Ripan nature reserve is located 4 km south of Tynset, along Rv30 towards Rendalen.

Tactics: <u>Tynset bridge</u> cross the river in Tynset and make a good vantage point to the river banks. But look out for traffic! The wetlands in the <u>Tjønn area</u> down-river from the bridge, is the core birdwatching area, and should be explored on foot froom both sides of the river.
<u>Lake Nylandstjønna</u> is reached by following Rv30 north towards Røros and exit to Vingelen. Here you follow the signs to Vingelen church and then to Olaberget. After about 3 km after passing Vingelen, you will see the lake on your right. Walk down by the farm Røsli, where there is a signposted path to the Nylandstjønna tower hide.
<u>Ripan nature reserve</u> lies along Rv30 south from Tynset and can be explored on foot from several places along the road. A parking area with information signs 4 km from the centre of Tynset is the most used starting point for hikes into the forest. Feel free to walk all the way to the beautiful <u>Lake Riptjønna</u>.

8. GRØNTJØRNAN MARSHES

Innlandet County
GPS: 62.58007° N 10.56479° Ø (Grøntjørnan marshes), 62.39037° N 10.43769° Ø (Lake Stubsjøen tower hide)

Notable species: Scaup, Long-tailed Duck, Velvet Scoter, Rock Ptarmigan, Willow Ptarmigan, Red-throated Diver, White-tailed Eagle, Golden Eagle, Rough-legged Buzzard, Hen Harrier, Kestrel, Crane, Dotterel, Ruff, Temminck's Stint, Dunlin, Purple Sandpiper, Broad-billed Sandpiper, Great Snipe, Red-necked Phalarope, Arctic Tern, Short-eared Owl, Yellow Wagtail, Bluethroat, Ring Ouzel, Sedge Warbler, Great Grey Shrike and Lapland Bunting.

Description: Grøntjørnan (alternatively spelled Grøntjønnan) is a wetland in Forollhogna national park. It is located in the high mountains at 880 m above sea level and consists of very moist bogs with willow thickets and several smaller lakes and ponds along the river Ya, which meanders calmly through the landscape. You will find a number of mountain species nesting here, including Ruff, a species that has long been in decline in this part of the country. In years with a good supply of rodents, a variety of raptors and Short-eared Owl hunt in the area. White-tailed eagles have been seen regularly in recent years.

Down in the valley by Kvikneskogen, easily accessible along Rv3, you will find the two lakes Sørsjøen and Stubsjøen (also called Stugusjøen). Both are shallow and Lake Sørsjøen is almost overgrown and surrounded by large marshes, suitable as both resting and nesting places for a variety of wetland birds. A tower hide has been erected at the southern end of Lake Stubsjøen, along the northern border of Lake Sørsjøen nature reserve. From here you can see Arctic Tern and shorebirds up close.

Gyrfalcon breed scattered in the southern Norwegian highlands, and in northern Norway also along the coast. The species is vulnerable to human nest raiders who sell the young to falconers abroad, where they are trained for sport hunting. Therefore, be vigilant and report to the police if you see suspicious behavior near nesting sites for Gyrfalcons and other large birds of prey. Several known nesting sites are supervised by the local population. *Photo: Espen Lie Dahl.*

Other Wildlife: Forollhogna national park is known for having the country's most productive population of wild reindeer. You have a good chance of encountering a herd at Grøntjørnan marshes.

Best season: Late spring and early summer at Grøntjørnan marshes. Also late summer at Lake Sørsjøen/Stubsjøen.

Directions: *From Tynset*, follow Rv3 north towards Trondheim.

Tactics: Lake Sørsjøen is visible on the west side of the road, 20 km northwest of Tynset, one km south of Nytrøa. Park in the car park on the west side of the road and walk up a little to the outside of the guardrail for a view. A telescope is necessary. Drive 1.8 km further north and turn left to get to the tower hide at the south end of Lake Stubsjøen. You see the tower on your left after crossing the river.

You reach Grøntjørnan marshes by continuing north for 20 km to Yset. Here the road crosses the rivers Orkla and Ya. Immediately after the latter, turn right onto the road signposted Plasseterveien. Pay the toll at the booth and drive 14 km to the barrier at the national park border. Follow the closed road further into the Grøntjørnan nature reserve. It is approx. 4 km to the small shieling at the end of the road, at the centre of the nature reserve. Grøntjørnan marshes can be explored on foot in damp terrain, but you can also get a good view with a telescope from the shieling pastures.

9. LAKE ORKELSJØEN

Trøndelag County
GPS: 62.50101° N 9.86892° E (Lake Orkelsjøen)

Notable species: Scaup, Long-tailed Duck, Velvet Scoter, Rock Ptarmigan, Willow Ptarmigan, Golden Eagle, Rough-legged Buzzard, Hen Harrier, Kestrel, Gyrfalcon, Crane, Dotterel, Temminck's Stint, Great Snipe, Red-necked Phalarope, Long-tailed Skua, Short-eared Owl, Shore Lark, Yellow Wagtail, Bluethroat and Lapland Bunting.

Description: Lake Orkelsjøen and the nearby Vinstradalen valley are popular mountain sites, particularly with birdwatchers from the Trondheim area. All the typical alpine species are found here, just a two-hours drive from Trondheim city. These two sites are regarded as equally good or even better than the more famous site Fokstumyra, which is located an additional hour's drive further south. Especially at Lake Orkelsjøen, many of the birds can be conveniently studied and photographed at close range using your car as a mobile hide. The area has several known Great Snipe leks, and quite a few pairs of Hen Harrier breed in years with a good population of rodents. By the cottages in the southwestern and northern corners of Lake Orkelsjøen, Temminck's Stint and Lapland Bunting are usually found. The Olmflya upland on the south (right-hand) side of the road on your way up to Lake Orkelsjøen is good for Dotterel and Shore Lark.

Vinstradalen valley has many of the same species and characteristics as Lake Orkelsjøen, but at Vinstradalen Long-tailed Skua and Red-necked Phalarope are often easier to find. However, at this site you must be prepared to do quite a bit of hiking. The valleys in this area are among the most important breeding grounds for Gyrfalcon in Scandinavia, and several pairs of Golden Eagle nest here as well. The large raptors are exposed to nest looting by collectors and falconers. Do not hesitate to contact the police if you see any suspicious activity. Important resting grounds for Crane are found in the lower valleys of this region.

Best season: June and early July.

Directions: From the E6 1 km north of Oppdal town, the road to Lake Orkelsjøen is clearly signposted. A small fee is to be paid at the unmanned toll booth. The road is closed in winter and opens June 1. The road to Vinstradalen valley is signposted from E6 7.5 km south of Oppdal.

Tactics: Drive in to the sites and cover as much ground as possible on foot. At Lake Orkelsjøen you may well use the car as camouflage, but you will discover more birds outside of the vehicle. Listen for lekking Great Snipe on bright summer nights. Scan the mountain ridges with your binoculars regularly in search of raptors.

When en route to and from Trondheim, you pass through important staging areas for Crane, particularly if you make a detour

Red-necked Phalarope prefer to breed in small pools in marshy areas at higher elevations in the south, also along the coast in the north of the country. *Foto: Espen Lie Dahl.*

Hen Harrier is found sparingly on mountain moors, and in Norway some of its most important breeding areas are in the interior south. *Photo: Kjetil Schjølberg.*

along Rv700 between Berkåk and Orkanger, a road that leads through the agricultural fields in Meldal valley. The Cranes usually feed in the fields or roost along the river, particularly during migration, but normally a few individuals stay in these areas all summer.

10. FOLLDAL VALLEY

Innlandet County

GPS: 62.06577° N 9.91636° E (Grimsdalen valley), 62.28923° N 10.07594° E (Meløyfloen marsh)

Notable species: Pintail, Scaup, Willow Ptarmigan, Rock Ptarmigan, Black Grouse, Capercaillie, Hen Harrier, Goshawk, Rough-legged Buzzard, Kestrel, Great Snipe, Ruff, Short-eared Owl, Wryneck, Black Woodpecker, Yellow Wagtail, Bluethroat, Redstart, Mistle Thrush, Siberian Tit (rare) and Parrot Crossbill.

Description: Folldal valley with the distinctive, open pine forest on nutrient-poor moraine terrain may still hold some

of the very few individuals of Siberian Tit left in southern Norway. Even in this traditional core area, they are extremely hard to find. If you follow Folldal's side-valley Grimsdalen westwards, the forest gives way to an open valley with mountain pastures and a few marshes where Bluethroat and several breeding shorebirds thrive. The same applies in Meløyfloen marshes in Einunndalen valley just north of Folldal, where you also may find Hen Harrier, Pintail and Scaup. Grimsdalen and Einunndalen are considered to be two of the longest and lushest shieling valleys in Norway, to where dairy farmers from further down the valleys traditionally took their livestock up to graze during summer. You still find visible signs of old hunting and shieling traditions here. Areas like this, with mountain grazing pastures and clusters of willow scrub, make up favourable habitat for Great Snipe.

Best season: Early summer offers the greatest variety of birds and you can find lekking Great Snipe. Siberian Tit is present all year.

Directions: Along E6 there are exits signposted to Folldal at Hjerkinn (if you enter from the north) and at Ringebu (if you enter from the south), and from Rv3 in Alvdal.

The Grimsdalen valley exit is signposted from Rv27 8 km south of Folldal village centre and from E6 at Dovre. A small fee is paid at an unmanned toll booth.

Einunndalen valley. Along Rv29 18 km east of Folldal, and about 5 km west of the municipality border with Alvdal, a gravel road exits to the north, passing through a toll booth. From the toll booth continue 17 km. Here, the road to Klokkarhaugsætra in the south end of Meløyfloen nature reserve exits to the left. This road also gives you access to the west bank of Enunna river. From here, the main birdwatching area stretches 5 km up the valley, to the shieling Meløya.

Tactics: Look for Siberian Tits in areas of continuous pine forest. They are often accompanied by other species of tit, and are attracted to feeders. Even here in core Siberian Tit land, Willow Tits are far more numerous, so prepare to cover large areas and examine many dozens, perhaps hundreds of other tits before you eventually may succeed in finding a Siberian.

11. FOKSTUMYRA MARSH

Innlandet County
GPS: 62.11896° N 9.27980° E (parking)

Notable species: Pintail, Scaup, Long-tailed Duck, Common Scoter, Velvet Scoter, Rock Ptarmigan, Willow Ptarmigan, Black-throated Diver, Crane, Rough-legged Buzzard, Hen Harrier, Kestrel, Dotterel, Ruff, Whimbrel, Greenshank, Wood Sandpiper, Red-necked Phalarope, Short-eared Owl, Yellow Wagtail, Bluethroat, Redstart and Lapland Bunting (scarce).

Description: Fokstumyra marsh is one of the most famous bird sites in Norway, subject to ornithologists attention since the early 1800s. It was protected as a nature reserve in 1923 and is often referred to as Norway's first protected nature reserve. The area covers about 18 km² and is located at an elevation of about 950 metres above sea level. It is easily accessible along route E6 between Trondheim and Oslo. Fokstumyra (also spelled Fokstugumyrin) is situated in a wide valley surrounded by barren mountains, and with a mosaic of ponds and two tranquil meandering streams lined with willow scrub, juniper and dwarf birch. The marsh stretches about 10 km from Fokstugu lodge and north to Lake Vålåsjøen. When the railroad was built through the area early in the 1900s, the marsh's value as a breeding site deteriorated and species such as Broad-billed Sandpiper and Great Snipe gradually disappeared as breeding species. Nevertheless, the area is

Siberian Tit is extremely difficult to find outside of Finnmark County in the north, but by chance or systematic searching you can come across the species in old, glistening pine forest in Folldal valley, the northern part of Østerdalen valley, as well as in the Røros district. The chocolate brown – not black – cap is the safest character differentiating it from the far more numerous Willow Tit. *Photo: Terje Kolaas.*

Oslo

Interior

Skagerrak

Western

Central

Northern

Finnmark

Svalbard

Fokstumyra marsh. Here, during the breeding season, traffic is only permitted along this easy-to-walk boardwalk. Although this channelization of traffic reduces wear and tear on plant and bird life, the flow of tourists is so great that several species struggle to find sufficient peace to carry out their nesting. Additional traffic restrictions should be considered. Therefore, show extra consideration when you visit Fokstumyra marsh. The tower hide can be seen here in the distance. *Photo: Bjørn Olav Tveit.*

still one of southern Norway's best and most easily accessible montane birding sites, and is one of the most reliable places for species such as Hen Harrier, Ruff, Red-necked Phalarope and Short-eared Owl, although particularly Ruff is critically in danger of disappearing from the marsh. Fokstumyra marsh is also an important staging area for migratory birds, particularly for alpine species such as Long-tailed Duck, Common and Velvet Scooters.

The birds at Fokstumyra have been prone to serious disturbance due to the ease of access and the increasing influx of tourists. It is therefore very important that visitors comply with traffic rules and act cautiously.

The mountains surrounding Fokstumyra have a number of bird species associated with higher elevations and drier ground, in partiuclar Rock Ptarmigan, Dotterel, Shore Lark and Lapland Bunting, although the latter has been rather scarce here in recent years. Long-tailed Skua may also be encountered but is not to be expected. A particularly good area for montane species is along the road Snøheimvegen leading west from Hjerkinn, a former military firing range now undergoing rehabilitation.

Other wildlife: There are elk and wild reindeer in the area. Muskox was reintroduced to the mountain range of Dovrefjell in the early 1900s under the pretext that the species had lived naturally here in prehistoric times. The herd inhabiting Dovre and Sunndalsfjella count around 200 animals. If you stumble upon muskox, never approach closer than a few hundred metres, even if they seem calm and appear friendly. These powerful animals may charge at high speed out of the blue if you get too close for their liking. Muskox safaris are conducted in the area, but you can often find them on your own without too much effort, often along Snøheimvegen road. In spring the animals sometimes pull all the way down to E6.

Best season: Mid-May to the end of July. Be prepared to negotiate large build-ups of snow if you enter the area before mid-June.

Directions: *The train* from Oslo to Trondheim stops at Hjerkinn station. *By car:* Fokstumyra is easily accessible along route E6 11 km north of Dombås and 17 km south of Hjerkinn. Park at the designated parking area just north of Fokstugu lodge, signposted from the main road.

Tactics: Information boards are found by the parking lot with maps and signs pointing you to the 6 km long boardwalk leading you on a round trip through the marshland. The track leads past a tower hide, 1.7 km from the car park if you keep left in the track junction and follow the boardwalk clockwise. Keep strictly to the track and try not to flush any birds. The most famous place for lekking Ruff is at Sørre Horrtjønn, close to the track 500 m west of the bird tower. In Fokstumyra nature reserve, access is not permitted outside the marked track from May 1 to August 1.

Along E6 4.3 km north of the parking at Fokstumyra marsh, there is a roadside rest area. Here, the Storrhusranden observation tower gives a view to a different part of the Fokstumyra marsh. With a telescope, you may well see some of the more conspicuous species from here, such as Crane, Short-eared Owl, Hen Harrier – and elk.

Snøheimvegen road leads westwards from Hjerkinn train station, along E6 17 km north of Fokstugu. Several upland species of bird are found here, in addition to it being one of the most reliable sites for wild reindeer and

muskox. Stop regularly along the road and scan for birds and animals. When you reach the parking spot at the border of the national park, you can walk 1.5 km to the north until you get a view of Stroplsjødalen valley, where muskox gather in early summer.

12. GUDBRANDSDALEN VALLEY

Innlandet County

GPS: 61.53986° N 9.98731° Ø (Grafferdammen pond), 61.84238° N 8.58751° Ø (Åsjo pond), 61.82862° N 9.47822° Ø (Skottvatnet), 62.11081° N 8.73119° Ø (Lesjaleira delta)

Aktuelle arter: A variety of wetland species; Whooper Swan, Slavonian Grebe, Crane, Coot, Little Ringed Plover, Whimbrel, Great Snipe, Tengmalm's Owl, Yellow Wagtail, Grey Wagtail, Mistle Thrush, Wood Warbler, Siberian Jay, Common Rosefinch and Pine Grosbeak.

Description: The river Lågen flows through the Gudbrandsdalen valley before it empties into Lake Mjøsa near Lillehammer town. Along river Lågen – and in the Ottadalen side valley – there are a number of smaller wetland locations, of interest if you are staying in this area or passing through on your way between Oslo and Trondheim or Northwest Norway.
At these sites you can find a selection of wetland birds, especially during the migration periods. The most interesting areas relates to quiet parts of the rivers, where sediments have accumulated over the years and created shallows, mud embankments, ponds and lagoons. The areas are largely cultivated or developed, but are still well worth exploring. The conifer forests up here have species such as Tengmalm's Owl and Siberian Jay, besides a few scattered pairs of Pine Grosbeak.

Other wildlife: Elk is common. There are small populations of beaver and otter along the waterways.

Gudbrandsdalen valley has several birdwatching sites, well-suited for shorter stops if you travel along the E6 between Oslo and Trondheim. Hundorp nature reserve is one of them. A well-placed and neatly designed, wheelchair-friendly tower hide has been erected here. *Photo: Bjørn Olav Tveit.*

Best season: Spring, early summer and autumn.

Directions: The locations are located along the E6 from Lillehammer town and northwards to Dovre, and further on the E136 to Lesja, as well as along Rv15 which takes off westwards from the E6 in Otta village.

Tactics: From Lillehammer town in the south, driving the E6 north, you arrive at the sites in the following order:
Lågen delta: Leave the E6 just north of Lillehammer at exit no. 82, and keep right at the roundabout onto Korgvegen road. This road leads under the E6, after which you immediately turn left at the parking space signposted *Friområde*. From here you can walk 200 m along the footpath in the direction of Lillehammer to a tower hide with a view of the delta from the east. Here, where the river meets Lake Mjøsa, you can find gatherings of wetland birds during migration, especially at low water levels in spring. Little Ringed Plover nest on the gravel islands, and considerable numbers of gulls can gather in the delta in spring and autumn. For views from the west, there is a side road exiting from E6 on the south side of the Lillehammer bridge,

signposted towards Jørstadmoen, a road with several viewpoints.
You can see the Losna delta from the E6 by the exit to Fåvang church, but this tempting delta is difficult to investigate further without creating drama in the traffic.
You can see the Frya delta from the E6 4.2 km north of Ringebu village. Here you can drive out into the agricultural district by exiting to Frya airport. A little further north you can take the E6 at exit no. 91 signposted to Hundorp. After 1.5 km, turn west on Rv256 signposted to Lia, a road that crosses river Lågen. Just north of the bridge, along the railway line on the eastern bank, you will find Rykkhussumpene marshes. South of the bridge, on the west side of the river, you will find the shallow Grafferdammen pond and Olstadtjønna pond, surrounded by meadows and cultivated land. A wheelchair-friendly bird tower has been built here. You reach it by parking just after driving over the bridge and following the footpath.
At Otta village, you can turn west through Ottadalen valley on Rv15 towards Stryn. The most interesting areas in this side valley are Lake Lalmsvatnet just after Lalm village, the Åsjo pond 1 km before the centre of Lom village, and the river island Risheimøyi just over the border with Skjåk municipality, all of which can be viewed from the main road.
Back in Gudbrandsdalen valley, at Otta, you can follow the side road Selsveien along the west side of river Lågen (the road is signposted to *Skysstasjon*). Follow this road 8 km north until you get a good view of Lake Skottvatnet on your right. Further north along the E6 in Gudbrandsdalen, at the centre of Dovre village you can turn west on the road signposted to Jøndalen. Turn left again just after the bridge on the road Elvenær, and drive south through the housing estate to reach Lake Andgardstjern and Lake Vigerusttjern, both of which are

located along river Lågen's west bank 1 km below the bridge. Both here in Dovre and just south of Dombås village, there is a signposted exit eastwards to Grimsdalen valley (see the section about Folldal valley).

Lesjaleira delta, an agricultural area with several untied hook lakes, is located on the south side of the E136 between Lesja and Lordalen, about 10 km northwest of Dombås. In the centre of Lesja village, follow signs for Vestsida and Lesja church. Cross river Lågen, and immediately turn right onto a gravel road. After 1 km, turn right again on the road signposted as a bicycle path to Lora. Explore the areas here along the river further west. In the afternoon, you will have better light conditions if you explore the area in the opposite direction. If that is the case, continue from Lesja for just under 8 km along the E136, and exit onto a road signposted to Lordalen and drive along the small roads back east towards Lesja. The pine-dominated conifer forest areas here and west to Brøstdalen valley provide opportunities for finding Three-toed Woodpecker, Mistle Thrush and Siberian Jay.

13. LANGSUA MARSHES

Innlandet County

GPS: 61.17434° N 9.96201° E (Kittilbu lodge), 61.02855° N 9.68888° E (Lake Langtjedn)

Notable species: Willow Ptarmigan, Black-throated Diver, Slavonian Grebe, Rough-legged Buzzard, Golden Eagle, Hen Harrier, Kestrel, Crane, Dotterel, Broad-billed Sandpiper, Great Snipe, Whimbrel, Greenshank, Wood Sandpiper, Short-eared Owl, Yellow Wagtail, Dipper, Bluethroat, Great Grey Shrike and Siberian Jay.

Description: The mountains between Gausdal and Valdres is a remote wilderness dominated by marshes flanked by pine and birch woods. The only trace of human activity is a limited degree of traditional dairy farming. A number of birds may be encountered, including Great

Snipe and Hen Harrier, and this is one of southern Norway's most important nesting areas for Broad-billed Sandpiper. The marshes here act as important buffers that help prevent flooding during periods of high water flow in the river systems further south.

Other wildlife: Domestic reindeer roam the area.

Best season: From early June (depending on the snow melting) to the end of July.

Directions: From E6 north of Lillehammer, exit onto Rv255 signposted to Gausdal, later to Skåbu/Espedal. In Forset, exit left towards Fagernes, cross the bridge and turn right onto Vestfjellvegen (Fv204) signposted to Kittilbu/Fagernes.

Tactics: Vestfjellvegen road takes you up to about 900 metres. Stop and scan from any vantage point. The two roadside ponds Kittilbutjerna, just after the exit signposted to *Kittilbu Utmarksmuseum*, 13 km past Forset bridge, are particularly nice. Slavonian Grebe nests here. From the museum a 3 km long hiking track leads you on a nice round trip trough the area. In order to find all the birds up here, you should walk off the beaten track. You find Great Snipe in the willow thickets and Broad-billed Sandpiper in the wettest black bogs, such as Oppsjømyra marsh. The latter is reached by driving another 9 km west from Kittilbu to Oppsjøhytta lodge by the exit signposted to Dokka. Hike north from here along the shore of Oppsjøvatnet. The marsh is located at the north end of the lake. Broad-billed Sandpiper is found here and in similar terrain several places in this region, including in the north-western part of the complex Hynna marshlands north of Kittilbu. Use special care in these areas, both in the interests of the vulnerable species of birds here but also for your own safety. Closer to Fagernes town in the west (follow signs to Glenna motel), the Langtjedn wetland near Bakkebygdi in Etnedal is a nice roadside site with a few pairs of Slavonian Grebe, possibly Velvet Scoter, as well as migrants in season.

Oslo · Interior · Skagerrak · Western · Central · Northern · Finnmark · Svalbard

11. VALDRES VALLEY

Innlandet County

GPS: 61.13184° N 8.85171° E (Lomen delta), 60.87763° N 9.25161° E (Lake Ølsjøen)

Notable species: High-elevation forest species, resting ducks, shorebirds and Crane; Whooper Swan, Rock Ptarmigan, Willow Ptarmigan, Capercaillie, Black Grouse, Black-throated Diver, Hen Harrier, Rough-legged Buzzard, Golden Eagle, Kestrel, Gyrfalcon, Red-necked Phalarope, Hawk Owl, Pygmy Owl, Tengmalm's Owl, Wryneck, Black Woodpecker, Lesser Spotted Woodpecker, Three-toed Woodpecker, Yellow Wagtail, Bluethroat, Ring Ouzel, Mistle Thrush, Siberian Jay, Lapland Bunting and Snow Bunting.

Description: The Valdres area is a popular recreational region with many cottages and ski resorts situated in montane coniferous forest. Along the Begna watercourse that runs through Valdres south to Lake Tyrifjorden, you find several smaller wetland sites. The perhaps most notable of these is the Lomen delta in the western end of Lake Slidrefjorden. This is an important staging area for ducks, Black-throated Diver, Crane and shorebirds, in particular during spring migration. To a certain extent, this also applies to the area around the bridge crossing Lake Strondafjorden by Ulnes church. In late fall and winter, the slow-running part of river Begna near Bagn village in Sør-Aurdal is often host to one or two Little Grebes and a few Whooper Swans. At Lake Ølsjøen at Golsfjellet highlands, quite a few shorebirds may gather during migration when the water level is low. Whooper Swan and Bluethroat can be found here as well, and the feeders by the cabins here is one of the most reliable places for Siberian Jay. A host of other forest species may be found as well, and at higher elevations, you will find a variety of species associated with the Scandinavian mountains.

Best season: Late spring and early summer. Many of the forest species including the owls, woodpeckers and Siberian Jay are found all year.

Directions: You get a view of the Lomen delta from Lomen power station, albeit from a distance. To get a closer view, exit from the E16 approx. 400 m northwest of the exit to Lomen church. You can also take the E16 just west of Lake Lomen, on the road signposted to Riste bru, a road that crosses the delta.
If you drive 23 km closer to Oslo, 8 km west of Fagernes town, you can pull down to the watercourse at Ulnes and check the wetland on the west side of the bridge. In Fagernes town, you can turn north on the Rv51 signposted to Beitostølen and Vågå (the road to Valdresflya plateau, see next section) and after 25 km look for resting Crane, ducks and more at the north end of Lake Heggefjord.
You reach Lake Ølsjøen by exiting the E16 a couple of km southeast of Fagernes, on Rv51 signposted to Gol. The inlet delta is located in the north-eastern corner of the lake, reached by heading down the slip roads to the left approx. 13 km from the E16. You may drive around the entire lake.
Just west of Ølsjøen, you also have a nice entrance to Stølsvidda plateau, with the possibilities of Hen Harrier and many other montane species. Turn right at the grocery store in Tisleidalen (just SW of Tisleidalen church) on a road that runs parallel to Tisleifjorden, through Langestølen, further along Lake Flyvatnet to Nøsen. Here you can possibly detour further north before ending the round by driving east towards Vaset. There are several toll booths on this loop.
A further couple of miles to the south-east on the E16 you come to Bagn, where you may find wintering ducks, Whooper Swan and Little Grebe.

10. VALDRESFLYA PLATEAU

Innlandet County

GPS: 61.38963° N 8.81139° E (the hostel)

Notable species: Scaup, Long-tailed Duck, Rock Ptarmigan, Willow Ptarmigan, Black-throated Diver, Rough-legged Buzzard, Golden Eagle, Hen Harrier, Kestrel, Gyrfalcon, Crane, Dotterel, Temminck's Stint, Dunlin, Purple Sandpiper, Great Snipe, Whimbrel,

Red-necked Phalarope, Long-tailed Skua, Short-eared Owl, Shore Lark, Yellow Wagtail, Dipper, Bluethroat, Lapland Bunting and Snow Bunting.

Description: Valdresflya is a barren mountain plateau well above the tree line at an elevation of about 1400 metres. The moraine slopes and gently rolling hills of the plateau are in stark contrast to the magnificent peaks and precipitous cliffs in neighbouring Jotunheimen mountain range to the west. The plateau may seem quiet at first, but it is actually home to just about all the mountain species of southern Scandinavia. The highway Rv51 crosses Valdresflya plateau in a north–south direction and a nice rest area with a cafeteria is located near its highest point about midway across. This has given Valdresflya an accessibility that is quite unique. Long-tailed Skua is often found roadside, and a short walk around the cafeteria

will often be rewarded with many of the species of the area. In the willow scrub along the road to and from Valdresflya, Great Snipe may be found, Bluethroat is numerous in this area as well.

Other wildlife: Domestic reindeer.

Best season: June and (early) July.

Directions: *By car* from Fagernes town, en route along E16 between Oslo and Bergen, exit north onto Rv51 signposted to Beitostølen and Vågå. The road crosses Valdresflya 50–60 km from Fagernes. The road is closed during winter, but is usually open from April, long before the area is of interest to birdwatchers.

Tactics: The rest area with the cafeteria provides a good base for hiking in the easily walked terrain. Aim for the lakes to

Long-tailed Skua thrives on the easily accessible Valdresflya mountain plateau. A number of other desirable species are found up here, and many of them can be enjoyed from the road. *Photo: Bjørn Olav Tveit.*

the north-west and south-east of the hostel, where most of the ducks and shorebirds are found. On the drier hills in the area, such as Fisketjernnuten hill north-east of the cafeteria, a thorough search will usually produce Rock Ptarmigan, Dotterel, Shore Lark and Lapland Bunting. Drive 5 km north of the cafeteria and find parking to the left. From here you can walk along a willow-lined creek east towards Lake Øvre Heimdalsvatnet or north-west through the swampy Leirungsdalen valley, both extremely bird-rich areas with Bluethroat, Great Snipe and more.

Jotunheimvegen road: From the south side of Valdresflya plateau, you can exit Rv51 eastwards on a gravel road, signposted Jotunheimvegen, which takes you through a beautiful landscape with the possibility of many of the same species as on Valdresflya. This is also a well-known place to see Hen Harrier. Pallid Harrier has been seen here as well. It is also a good area for Short-eared Owl and sometimes Hawk Owl, as well as a good population of Great Snipe in some areas. Stop frequently along the road and scan the terrain, preferably with a telescope. You can also go hiking, e.g. into the beautiful Måbødalen valley. The Jotunheimvegen road leads through to Gudbrandsdalen valley and the E6.

16. TRILLEMARKA FOREST

Buskerud County

GPS: 60.03076° N 9.42653° E (Strandemyran marshes), 60.17476° N 9.32372° E (Lake Skodøltjenn)

Notable species: Hazel Grouse, Willow Ptarmigan, Black Grouse, Capercaillie, Black-throated Diver, Osprey, Golden Eagle, Rough-legged Buzzard, Buzzard, Goshawk, Kestrel, Peregrine Falcon, Gyrfalcon, Crane, Wood Sandpiper, Green Sandpiper, Greenshank, Hawk Owl, Tengmalm's Owl, Pygmy Owl, Wryneck, Grey-headed Woodpecker, Black Woodpecker, Lesser Spotted Woodpecker, Three-toed Woodpecker, Grey Wagtail, Dipper, Wood Warbler, Redstart, Mistle Thrush, Great Grey Shrike, Siberian Jay and Parrot Crossbill.

Description: Norway's largest continuous old-growth forest, Trillemarka, is located is only an hour and a half's drive from Oslo. The forest is dominated by spruce trees, with elements of pine, some of which are massive. In some areas the conifer forest is intermixed with deciduous trees. An important feature of Trillemarka forest is that there are trees of all generations including plenty of dead and dying trees, vital to woodpeckers and biodiversity in general. Trillemarka forest is situated in a broad valley centred around the watercourse Grønhovdsvassdraget, with a few steep cliffs and several small and medium-sized lakes, flanked by barren mountains in the north and south. The scenery is magnificent and almost completely unspoiled by power lines or roads – even hiking tracks are few.

In Trillemarka forest you will find a variety of species, including all the Norwegian breeding woodpeckers except White-backed. Three-toed Woodpecker and Black Woodpecker are among the commoner species. The area is one of the most reliable in southern Norway for Siberian Jay. The populations of Hazel Grouse, Willow Ptarmigan, Black Grouse and Capercaillie are variable from one year to the next, but are generally very good. You also have ample opportunities to encounter species such as Golden Eagle, Crane and Hawk Owl. Dotterel and Red-necked Phalarope breed in the mountains in the outskirts of Trillemarka forest.

Other wildlife: At least 26 species of mammals have been recorded in Trillemarka forest. Amongst the ones more commonly encountered are red fox, stoat, weasel, pine marten, elk, red deer and mountain hare. A small population of lynx live here, and wolverine and brown bear may pass through.

Best season: Spring and early summer.

Directions: From E18 in Drammen, exit at Exit 25 onto E134 towards Haugesund and Kongsberg. Just past Mjøndalen exit onto Rv35 signposted to Hokksund. Continue to Åmot and exit onto Rv287 signposted to Sigdal. You may alternatively enter

Three-toed Woodpecker is an uncommon and rather secretive species, linked to old spruce forests with a large proportion of dead trees. In spring, it hacks rings of small holes in the bark of spruce trees in order to obtain sap, an activity that leaves telltale markings on the trees. As with most woodpeckers: Listen for pecking in the forest and follow the sound to see what is making it. *Photo: Terje Kolaas.*

Trillemarka forest from other directions, such as from Rollag in Numedal valley or along Rv40 between Kongsberg and Geilo.

Tactics: As in most forest areas birds are often far between. In order to find them you need to have plenty of time and patience and move quietly about in the terrain.

<u>Strandemyran marshes</u>: This is one of Trillemarka's finest marshlands, a good spot for Crane and Lesser Spotted Woodpecker. In Sandsbråten just north of Sigdal, exit from Rv287 onto a road signposted to Lampeland. After another 8 km, turn north onto the road Gryteelvavegen (toll road). The marshland extends northwards from <u>Lake Grytevatn</u>.

<u>Lake Skodøltjenn</u>: This is a nice starting point for hiking trips into the woodlands in the core of Trillemarka forest. Continue further north on Rv287. After about 2.5 km past the Sigdal factory in Nedre Eggedal village, make a left up to Grønhovd, just after the bridge over the creek that runs out in Lake Solevatnet (you may want to check the delta in the north end of this lake before heading into the forest). Continue this gravel road past the unmanned toll booth all the way to Lake Skodøltjenn (also spelled Skoddølvatnet) and find suitable parking close to the road barrier. In winter, this road is not snow ploughed further up than to the toll booth. From the car park at Skodøltjenn you can follow marked tracks west to Vindolvatnet, south to Grunntjern/Grunntjernhytta, or north via Trillesetra and Finnevollane to Skårsetra.

17. HARDANGERVIDDA PLATEAU

Vestland, Buskerud and Telemark counties
GPS: 60.28713° N 7.56082° E (Lake Tinnhølen parking), 60.26880° N 7.48208° E (Langavassmyra marsh)

Notable species: Tundra Bean Goose (scarce breeder), Scaup, Velvet Scoter, Willow Ptarmigan, Rock Ptarmigan, Black-throated Diver, Rough-legged Buzzard, Hen Harrier, Kestrel, Gyrfalcon (scarce), Crane, Dotterel, Temminck's Stint, Dunlin, Purple Sandpiper, Great Snipe, Red-necked Phalarope, Whimbrel, Long-tailed Skua, Short-eared Owl, Shore Lark, Yellow Wagtail, Dipper, Bluethroat and Lapland Bunting.

Description: Hardangervidda is the largest mountain plateau in Europe, covering an area of about 6,500 km^2. With an average elevation

Oslo

Interior

Skagerrak

Western

Central

Northern

Finnmark

Svalbard

Dotterel nests in small numbers on barren, dry moraine hills in the mountains. It is known to be very trusting of humans, and may stand motionless and watch hikers pass by only a few metres away. This contributes to the fact that the species can be difficult to locate. In late spring, before the snow melts on the breeding grounds, the Dotterels often gather in small flocks on newly plowed fields in the lowlands. *Photo: Kjetil Schjølberg.*

of 1,100 metres, which is well above the tree limit, the climate is cold and alpine year-round. The barren landscape is dominated by rolling moraine hills, with marshes and lakes lined by willow and dwarf birch. In the north, the dark Hallingskarvet mountain massif rises in stark contrast to the gleaming white glacier Hardangerjøkulen. In the south-west the brick-shaped mountain Hårteigen makes a characteristic landmark.

Bird-wise, the central part of Hardangervidda is regarded as one of the richest mountain regions in southern Scandinavia. A popular entrance to the area is the gravel road that leads straight into the core, by Lake Tinnhølen at an elevation of 1,213 m. The ease of access has led to increased human traffic and a deterioration of wildlife. For instance, Snowy Owl is no longer a breeding bird here. The populations of Scaup, Ruff and Purple Sandpiper have declined in recent decades as well. Nevertheless, most montane bird species of southern Scandinavia are to be found, many of which still in good numbers and in comfortable walking distance from the car park at Lake Tinnhølen.

Willow Ptarmigan is common here, and in rocky areas you may find Rock Ptarmigan as well. Dunlin, Golden Plover, Common Snipe and Redshank are all common breeders, and with a bit of searching you should find all the notable species mentioned above. Dotterel is usually found on dry ridges. These birds are astonishingly unafraid of people and surprisingly well-camouflaged, so they can be hard to spot. Shore Lark and Lapland Bunting are often found in the same areas, although the populations have decreased markedly in recent years. Great Snipe tend to live on drier ground and more vegetated areas than the widespread Common Snipe, and is perhaps easiest to find in the willow thickets between Tråastøl tourist lodge and Lake Tinnhølen. Bluethroat abound in the same habitat. Crane try to breed in Bjoreidalen valley every year, but usually without success due to human disturbance. Temminck's Stint nest along the shore of the lakes, including Lake Tinnhølen, while Red-necked Phalarope prefer smaller marshland ponds, such as Sandtjørnane ponds and Langavassmyra marsh. Gyrfalcon breeds in the region but is only exceptionally seen during visits to these parts of the plateau. The same applies to Snow Bunting, which is more easily found at higher elevations and on more barren, rocky ground.

Several species, in particular skuas, owls and raptors, are totally dependent on the population of small rodents, which vary greatly from one year to the next. Thus, the population of these birds will fluctuate in line with the rodent populations. In fact, several other birds not directly dependent on rodents as a source of food, such as grouses, will indirectly fluctuate accordingly as well, because the raptors are busy catching mice. In years with good numbers of rodents, fair numbers of Rough-legged Buzzard, Kestrel, Long-tailed Skua and Short-eared Owl may be seen, with a chance of the odd Snowy Owl as well. In other years these species can be gone altogether.

An interesting feature at the Hardangervidda plateau are the distinct and slightly more luxuriantly vegetated raptor mounds that stand up to one metre tall on top of many of the moraine hills. They are caused by predatory birds that through the centuries have been fertilizing the highest vantage points with leftovers, regurgitated pellets and excrement while eating, resting, and scouting for new prey. The raptor mounds are still actively in use and should always be checked for fresh signs of raptor presence. Another great and easily accessible area in the Hardangervidda region is Lake Isdalsvatnet on the west side of the plateau. Here you find mountain bogs and steeper slopes in a confined area, with the possibility of seeing many of the above-mentioned species, including Great Snipe, Crane, Black-throated Diver, Golden Eagle, both ptarmigans, Yellow Wagtail and Bluethroat.

Other wildlife: Europe's largest population of wild reindeer roams Hardangervidda, the largest herds counting thousands of animals. Stoat is often seen in areas of loose rocks and boulders. Several species of rodent, including lemming and tundra vole, occur in quantities varying significantly from one year to another. Fluctuations in the rodent populations are one of the most important ecological factors in the mountains, and regulate the numbers of several bird and mammal

Rock Ptarmigan is associated with more barren terrain than the Willow Ptarmigan. In southern Norway you find Rock Ptarmigan high up in the mountains, often in hillsides with scree, large boulders and almost no vegetation, whereas Willow Ptarmigan thrives in areas with heather and willow thickets. Identification must be based on calls, bill thickness and/or plumage – the habitat alone gives no more than an indication of the species. *Photo: John Stenersen.*

Lake Isdalsvatnet on Hardangervidda plateau can display a fine selection of the region's mountain species. *Photo: Kjetil Salomonsen.*

species. Arctic fox used to be common on Hardangervidda. The species has been reintroduced to Finse in recent years, and there is hope for it to spread to other parts of the plateau.

Best season: Early summer is by far the best, particularly early June before all the snow has melted away and the birds are concentrated in the snow-bare areas. The gravel roads are closed during winter and does not open until early summer. Birding is still perfectly possible, however – see below. Few birds linger in the mountains past mid-August.

Directions: The Hardangervidda plateau can be accessed from many directions. The core of the area is reached along Rv7 which crosses the plateau between Eidfjord in the west and Geilo in the east. The 10 km long gravel road to Tinnhølen leads south from Tråastøl tourist lodge, which is located along Rv7 32 km east of Eidfjord and 30 km west of Haugastøl

lodge. A small fee is paid at the unmanned toll booth at the entrance. The gravel road is closed in winter but opens again from July 1 in most years. In June, before the road opens, Tråastøl and Dyranut (1 km further west from Tråastøl) tourist lodges makes practical bases for hiking trips in the area.

The road to Lake Isdalsvatnet is winter closed as well, but usually opens in late May, about one month earlier than the road to Lake Tinnhølen.

Tactics: Cover as much ground and as many different habitats as possible on foot. Scan the horizon regularly with binoculars or telescope for birds of prey, Long-tailed Skua, reindeer and more. Raptors and owls can often be found if you pay attention to what ravens devote their attention to. The telescope is useful for scanning marshes and lakes as well. A light tripod or monopod comes in really handy here.

Many of the species can be found just by walking the terrain along the road between Tråastøl and Tinnhølen. This is also the best area for Crane, Great Snipe and Short-eared Owl. Before the gravel road opens, you can hike the road on foot or skis. Alternatively, you can walk south 3 km from Dyranut tourist lodge along the path to Nybu in central Bjoreidalen valley, crossing Bjoreia stream at the wooden bridge. The morain hills just north of Dyranut is usually a reliable site for Dotterel and Rock Ptarmigan.

At Lake Tinnhølen, Temminck's Stint is usually easy to find along the shore, and on the hill Trondsbunuten north-east of Trondsbu tourist lodge, Rock Ptarmigan, Dotterel and Shore Lark may be found, in some years also Long-tailed Skua. At Byen south-east of Tinnhølen – a two-hours walk from the car park – you find Sandtjørnane and Bakkatjønnmyran marshlands where Scaup, Velvet Scoter, Black-throated Diver and Red-necked Phalarope nest. In order to get to the famous Langavassmyra delta, perhaps the most reliable spot for Red-necked Phalarope, Crane and Long-tailed Skua, you can hike from the car park at Lake Tinnhølen in a steady, off-piste route straight towards the prominent rock formation Hårteigen, or you can take out a compass course towards west-south-west. You reach Langavassmyra marsh after 4.5 km. You can alternatively follow the ancient track *Normannslepa* (connecting western and eastern Norway) from Lake Tinnhølen 3.5 km south-west to Hellehalsen tourist lodge by Lake Langavatnet (and where you may rent a small boat). From Hellehalsen walk to the west along Lake Langavatnet to the delta area in the west end. This route to Langavassmyra marsh takes about two hours from the car park. The upper delta is one of the most reliable places on Hardangervidda for Long-tailed Skua.

A word of caution: In fair weather you have several landmarks to help simplify navigation, but thick fog may appear very suddenly in these mountains making a map and compass vital accessories. The terrain in this part of the plateau is very easy to walk as long as you stay out of the wettest bogs. Wear sturdy hiking boots and plenty of clothes even in mid-summer, and always pack some extra warm clothes even on short trips. Mosquito repellant is a necessity in summer.

Lake Isdalsvatnet: From Rv7 18 km west of Dyranut lodge, take the road leading north, signposted to Vøringsfossen (a spectacular waterfall). Follow this road to the end, at the northern shore of Lake Isdalsvatnet. This road is winter closed and does usually not open until late May. In May and June, the Garen camping and golf course along Rv7 2 km east of the road signposted to Vøringsfossen, can be worth stopping by, looking for Great Snipe and other migrants waiting for the snow to melt in the mountains.

Lapland Bunting still breeds quite commonly on Hardangervidda plateau, although the population in some places has declined by as much as 85% in recent decades. On Hardangervidda you will find it in places with the most lush willow thickets, while especially in the north you will also often find it in more barren terrain. *Photo: Arild Hansen.*

Oslo

Interior

Skagerrak

Western

Central

Northern

Finnmark

Svalbard

COASTAL SKAGERRAK

S outhwards from Oslo, the Oslofjord widens before entering Skagerrak, the strait between Norway, Denmark and Sweden. Although seas may be rough at times, Skagerrak is a much more sheltered area of sea compared to the rest of the Norwegian coastal waters. Still there are pelagic birds to be seen here, particularly during and after periods of strong winds from the southwest. Some of the birds may be pushed far into the Oslofjord, often giving close views as the fjord narrows.

Thrush Nightingale is legendary for its powerful singing and correspondingly unostentatious appearance. It has become increasingly common in south-eastern Norway in recent decades, with a particularly dense population on the west side of the Outer Oslofjord. You can hear it singing both night and day, especially in swampy deciduous forest. *Photo: Bjørn Fuldseth.*

Several large rivers draining water from southeast Norway empty into Skagerrak, creating river valleys on their way stretching all the

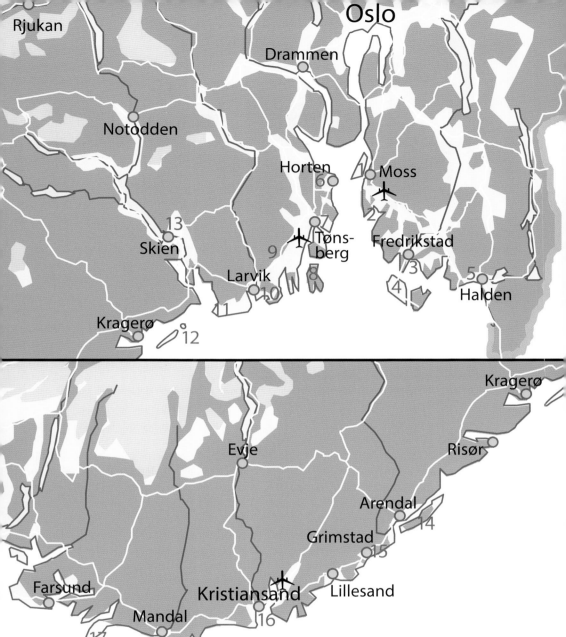

SITES ALONG COASTAL SKAGERRAK

Water Rail leads a hidden life within the reedbeds and other wetland vegetation but can sometimes be seen out in the open on the outskirts if you look closely, especially early in the morning and at dusk. The long bill is the most obvious difference from other rails breeding in Norway. *Photo: Terje Kolaas.*

way up into the mountains of interior southern Norway, acting as navigational landmarks for migrating birds, such as Pink-footed Geese, raptors, Cranes and Cormorants.

At the mouth of the Oslofjord, the landscape of the counties of Østfold and Vestfold is rather flat and fertile with large areas of cultivated fields and mixed woodlands, including areas of lush and tall deciduous forest, as opposed to the boreal coniferous forest that dominates the interior south. Characteristic for Østfold are the dry granite hills covered with pine and heather and patches of naked rock and gravel, providing habitat for Nightjar and Woodlark. Along the Skagerrak coast the landscape becomes slightly more hilly and rocky, still covered in mixed woods. The outer coast is peppered with islets and skerries, providing breeding grounds for colonies of gulls and terns as well as recreational opportunities for people in the densely populated Oslofjord area.

An important geological feature of this region is the large moraine ridge known as Raet, a relict from the last ice age 12,000 years ago, when large glaciers covered southern Norway. When the ice withdrew, this ridge of gravel

and sand was left behind. Raet is now evident several places along the coast, particularly in the outer Oslofjord. Some of the flat gravel islands further south, like Jomfruland and Stråholmen islands, are also a part of it. Many migrant birds use this ridge as a navigational landmark which they follow during their spring and autumn movements. The Mølen headland at the southern tip of Vestfold County benefits from this, making it the prime site in the region for migration in autumn. Raet is also responsible for blocking waterways resulting in several bird-rich lakes, such as Lake Borrevannet. Unsurprisingly, given the proximity to mainland Europe, the Skagerrak coast is the part of Norway that has most in common with the European mainland, with breeding birds including Quail, Thrush Nightingale, Reed Warbler, Goldfinch and Hawfinch, which are uncommon in most other parts of Norway. This is also where new colonists from the southeast first try to establish a foothold, such as Marsh Harrier, Woodlark, Bearded Tit and Cormorant of the continental subspecies *sinensis*. Through the 20th century, birds of prey, such as Peregrine Falcon and Eagle Owl has come close to extinction due to hunting and agricultural poisons, but thanks to human effort, the populations of both are showing clear signs of recovering again. Particularly Peregrine is nowadays a rather common sight along the Skagerrak coast.

THE BEST SITES

For general migration of Scandinavian land birds, and for wintering waterbirds, Mølen headland is traditionally considered the best site in the region, although the quantity and quality of birds encountered here are largely dependent on the weather conditions. Kurefjorden bay, Lake Borrevannet, the Tønsberg inlets and Klåstadkilen inlet all combine visible migration with a variety of favourable habitats, providing somewhat more stable conditions. For seabird migration, Ravn headland on the tip of Brunlanes peninsula can be good even in light winds, whereas Brentetangen seawatch near Kurefjorden bay and Møringa headland near Borrevannet situated at the eastern and western shores of the Oslofjord respectively, often provides better views of the pelagic birds when the conditions are ideal in the Skagerrak, meaning strong winds from a southwesterly direction. For nocturnal songbirds the wetlands and agricultural areas of both Østfold and Vestfold counties are equally promising, although Nightjar and Woodlark are easier to find in Østfold and Thrush Nightingale is more numerous in Vestfold. There are several sites along the Skagerrak coast which are known for hosting species considered rare or uncommon in Norway, such as Hobby (Lake Gjennestadvannet), Kingfisher (Halden area), staging Broad-billed Sandpiper (Kurefjorden bay), Bearded Tit (Øra delta and Tønsberg inlets) and Golden Oriole (Jomfruland island).

1. JELØYA ISLAND

Østfold County

GPS: 59.42401° N 10.60819° E (Alby farm)

Notable species: Migrant shorebirds and seabirds; Pheasant, Stock Dove, Green Woodpecker, Black Woodpecker, Lesser Spotted Woodpecker, Thrush Nightingale, Marsh Warbler, Wood Warbler, Marsh Tit, Goldfinch, Linnet, Common Rosefinch and Hawfinch.

Description: Jeløya island is situated in the Oslofjord, close to shore just outside the town of Moss. Originally a peninsula, it is now separated from the mainland by an artificial channel. Jeløya island has a mixture of cultivated land and mixed forest with batches of lush, mature deciduous trees with a number of breeding birds. In recent years, Three-toed Woodpecker has been seen regularly as well. Jeløya island offers decent habitats for ducks and shorebirds, particularly near its southern tip. It is a very interesting area to explore on nocturnal songbird trips. You might also see migrating passerines, raptors and seabirds here in season.

Best season: All year.

Directions: From the E6 between Oslo and Sarpsborg, exit onto the road signposted to Horten and Moss. Then, see the map. Gulls congregate at the roadside lake in a. Nesparken park along the main road through town, in and around b. Moss harbour and at c. Sjøbadet beach by the canal. Continue through Moss town centre, and follow signposts to Jeløya. On Jeløya island, the road to Alby farm (and *Galleri F15*) is clearly signposted. Besides art exhibitions, Alby farm can offer a cafe and *Jeløy Naturhus*, providing information concerning the nature and culture of the area.

Tactics: f. Jeløy Radio is surrounded by an open area with scattered bushes and hedges, good for resting migrant passerines and a few breeding ones, including Common Rosefinch. You can follow movements of seabirds on the fjord from here as well. From Alby farm, walk the paths leading down to the seaside shorebird friendly Albystranda beach and g. Reierbukta beach. This is a very popular recreational area for the general public and is often crowded on weekends with

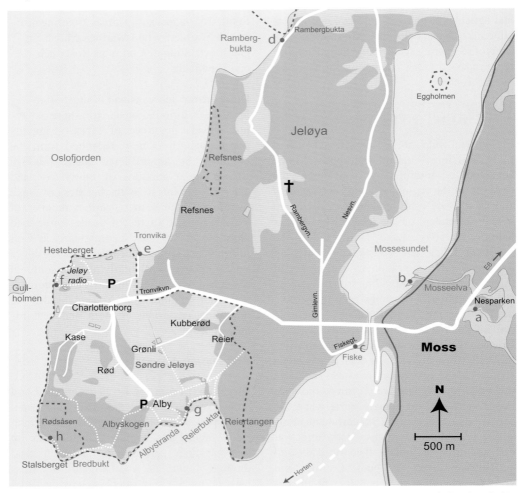

Jeløya island, southern half: a. Nesparken park; **b.** Moss harbour; **c.** Sjøbadet beach; **d.** Rambergbukta bay; **e.** Tronvika bay; **f.** Jeløy Radio; **g.** Reierbukta bay; **h.** Stalsberget seawatch. All these are natural stopping points when visiting Jeløya island. Also scan the agricultural areas north of Alby farm.

immaculate weather. To watch seabirds when conditions are suitable, follow the broad path from the southern corner of the parking area through very interesting woods and out to **h.** Stalsberget headland. Alternatively, drive to Brentetangen seawatch, described under the Kurefjorden bay section below.

2. KUREFJORDEN BAY

Østfold County
GPS: 59.34899° N 10.74838° E (Rosnesbukta bay), 59.31916° N 10.75142° E (Ovenbukta lagoon), 59.34518° N 10.66050° E (Brentetangen seawatch)

Notable species: Staging ducks, shorebirds and other waterbirds; seabirds; Quail, Pheasant, Great Crested Grebe, Slavonian Grebe, Buzzard, Little Ringed Plover, Broad-billed Sandpiper, Wryneck, Green Woodpecker, Lesser Spotted Woodpecker, Thrush Nightingale, Marsh Warbler, Reed Warbler, Red-backed Shrike, Marsh Tit, Goldfinch, Linnet and Common Rosefinch.

Description: Kurefjorden bay is a 6 km deep side-arm of the Oslofjord and is perhaps the best all-round site in the Østfold region. The inlet is rather shallow and thereby attractive to a substantial number of ducks,

low tide, providing excellent shorebird habitat. Most regular migrant species can be found resting here and this is the most reliable site in southern Norway for Broad-billed Sandpiper during migration. The Kurefjorden bay area is also strategically positioned for migration of terrestrial species, raptors in particular. During nocturnal songbird trips in the agricultural areas between Kurefjorden and Moss town in early summer, Quail, Thrush Nightingale and Marsh Warbler are all usually encountered. Staging Dotterel sometimes occur in fallow or newly ploughed fields in this area as well, particular during periods of rain in May. The nearby Brentetangen headland is one of the best and most easily accessible seawatching viewpoints on this side of the Oslofjord. Several rare birds have been found in the Kurefjorden bay area through the years.

Leach's Storm Petrel and other true seabirds can be pushed far inland into the Oslofjord by strong southerly winds, especially in summer and autumn. Under such conditions, you can e.g. at Brentetangen seawatch often encounter such species at closer quarters and in flatter seas than at locations facing the open sea. *Photo: Tomas Aarvak.*

divers, Great Crested Grebe, Slavonian Grebe (autumn and winter), gulls and terns. Rosnesbukta bay at the head of the fjord and Ovenbukta lagoon a bit further out have both extensive mudflats exposed at

Best season: All year, but particularly in the shorebird migration periods. Broad-billed Sandpiper is uncommon but regular in the last part of May to early June, and in August as well.

Ovenbukta lagoon in Kurefjorden bay provides nice habitat for wetland species. *Photo: Bjørn Olav Tveit.*

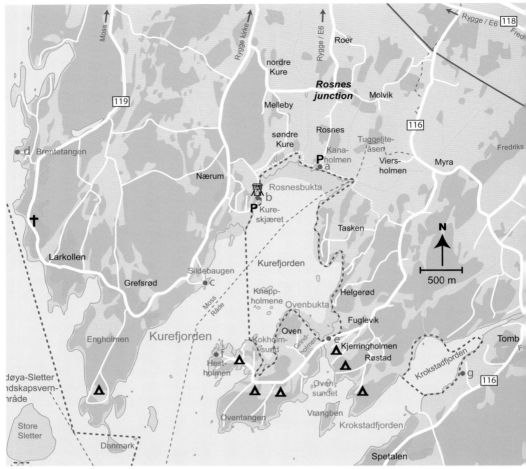

Kurefjorden bay: a. Kanaholmen viewpoint to Rosnesbukta bay; **b.** Rosnes tower hide; **c.** Sildebaugen viewpoint; **d.** Brentetangen seawatch; **e.** Ovenbukta viewpoint; **f.** Hestholmen viewpoint; **g.** Krokstadfjorden viewpoint.

At Brentetangen seawatch, strong winds from the southwest from the end of April to the beginning of November are usually the most rewarding.

Directions: *From E6 to the Rosnes junction*: Kurefjorden is situated near Moss airport Rygge. Exit from the E6 at Exit 12, signposted to the airport and to Larkollen. Continue a few hundred metres on Rv118 south and then follow signs to Larkollen through several junctions, before heading towards Rosnes. Follow the Rosnes road 2 km until it makes a 90-degrees right bend. Here, gravel roads exit both straight ahead (signposted Rosnes) and to the left (signposted Molvik). This is the Rosnes junction, from where further directions are described.

From the Rosnes junction to **a.** *Rosnesbukta bay*, continue straight ahead on the gravel road towards Rosnes, past two farms, down a steep hill and across the fields. Park at the designated parking. Bring your telescope and walk the path through the wooded hill between the fields and the bay. This the best general viewpoint.

From the Rosnes junction to **b.** *Rosnes tower hide*, continue along the paved road towards Kure 2 km to a T-junction and turn left on the main road to Larkollen. After another 2.6 km, pull left again onto a road signposted *Fugletårn* (tower hide). Park at the end of this road and walk 400 m to the tower hide.

Continue the main road towards Larkollen almost 1 km and turn left onto the side road **c.** Sildebaugen and get a view of the outer section of Kurefjorden bay from the marina by the sea. This is a good viewpoint to the deeper parts of Kurefjorden bay, often good for *Melanitta* ducks, divers and more.

d. Brentetangen seawatch. Back on the main road towards Larkollen, continue for about 5 km and turn left onto the side road Evjesundveien. Follow this road and keep left in the upcoming junction, signposted to Fristranden. Park at the end of the road and walk out onto the rock by the sea.

From the Rosnes junction to **e.** *Ovenbukta lagoon,* turn left towards Molvik and follow this road until it meets Rv118, where you turn right towards Tomb. Check the fields for feeding passerines and shorebirds, and hunting harriers. After 2 km, make a right signposted to Oven and after 3.5 km you will see Ovenbukta lagoon on your right. It is possible to park one or two cars roadside on the hilltop. From this hill you have a nice view of the lagoon and salt meadows. You can explore the area further ahead all the way to **f.** Hestholmen Camping, from

where you can scan the outer part of Kurefjorden bay from the opposite side from Sildebaugen. You should also stop by **g.** Krokstadfjorden inlet, of which you get a nice view from a cliff at Hestevold sewage treatment plant (*Renseanlegg*).

3. ØRA DELTA

Østfold County

GPS: 59.17758° N 10.96790° E (tower hide), 59.17018° N 11.01583° E (Marikova)

Notable species: Ducks, shorebirds, gulls and other waterbirds; Whooper Swan (winter), Shoveler, Little Grebe, Great Crested Grebe, Cormorant, Marsh Harrier, Osprey, Little Ringed Plover, Green Woodpecker, Lesser Spotted Woodpecker, Woodlark, Reed Warbler, Marsh Warbler, Bearded Tit, Marsh Tit, Goldfinch and Linnet.

Description: Øra delta is a large wetland at the mouth of the longest river in Norway, Glomma river, just outside the town Fredrikstad. It is a strategically positioned staging area for

Øra tower hide provides good overview of the large delta and to the gulls on the landfill just behind it.
Photo: Bjørn Olav Tveit.

Oslo

Interior

Skagerrak

Western

Central

Northern

Finnmark

Svalbard

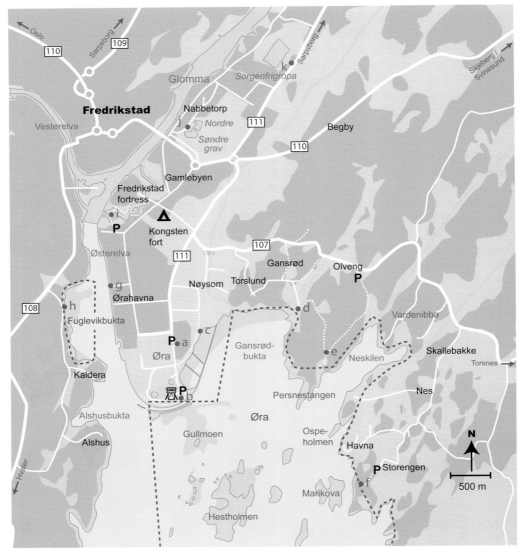

Øra delta: a. Entrance to the landfill and the **b.** tower hide; **c.** and **d.** overview Gansrødbukta; **e.** Persnestangen (Woodlark); **f.** viewpoint to Marikova and outer part of the Øra delta; **g.** and **h.** viewpoint to Glomma river mouth and to Fuglevikbukta wetland; **i.** fortress moats; **j.** Nordre and Søndre grav ponds; **k.** Sorgenfrigropa pond.

migrant waterbirds, and a very important breeding area for several species, including Cormorant of the Continental subspecies *sinensis*, which has bred here since 1997. The population here is now in excess of 2,500 pairs. Shoveler, Marsh Harrier, Little Ringed Plover, Woodlark, Reed Warbler and Bearded Tit are also among the regular breeders in and around the delta. The latter is also a relatively new nesting species, regularly seen along the channel by the tower hide. The Øra delta is frequented by several nearby breeding pairs of Osprey through the summer. A seaside waste landfill attracts gulls by the thousands, sometimes including the odd rarity such as Glaucous, Iceland, Caspian or Yellow-legged Gull. Red-throated Pipit is often found at the landfill in September, and Shore Lark, Great Grey Shrike and Arctic Redpoll in winter. Eagle Owl is sometimes seen here as well, or may be heard calling at dusk or dawn nearby in late winter and spring. White-tailed Eagle is a regular winter visitor to the delta. At least 273 species have been recorded in the delta, about 90 of them breeding.

Best season: All year, but particularly in April and May, and from August to October.

Directions: *From Fredrikstad,* see the map. You can drive in to the landfill area during opening hours, although driving is restricted to driving directly to the tower hide. When the landfill is closed, park at the gate and walk.

Tactics: From **a.** the landfill gate, head south along the sea. Just before you get to the channel, a road exits to the right up onto the hill where the **b.** tower hide is located, and where you will be in great need of your telescope. This area is at its best at low tide, when shorebird-friendly areas of exposed mud emerge. Also walk the path along the channel, and north along the mouth of Glomma river. Check the landfill and its ponds. Back in the car, follow the map to **c.** and **d.** to view Gansrødbukta inlet. **e.** Persnestangen peninsula is an area with breeding Woodlark and Black Woodpecker. At **f.** Marikova you can find Great Crested Grebe, divers and diving ducks. This is also the best viewpoint to the colony of Cormorant on Hestholmen further out in the bay. Also check the

outer section of **g.** Glomma river with **h.** Fuglevikbukta wetland. Further up the river, a few nutrient-rich ponds may be worth checking for ducks, swallows and more, such as **i.** the moats at Fredrikstad fortress, **j.** Nordre and Søndre grav, and **k.** Sorgenfrigropa ponds.

4. HVALER ARCHIPELAGO

Østfold County
GPS: 59.03892° N 11.01124° E (tower hide Arekilen)

Notable species: Seabirds; migrants; nocturnal songbirds; Black Guillemot, Nightjar, Green Woodpecker, Lesser Spotted Woodpecker, Woodlark, Thrush Nightingale, Redstart, Reed Warbler, Marsh Warbler, Icterine Warbler, Wood Warbler, Marsh Tit and Linnet.

Description: Hvaler archipelago stretches southwards from Fredrikstad towards the border with Sweden. Some of the larger islands are interconnected with each other and the mainland through a series of bridges and an underseas tunnel. The myriad of small islands, low rocks, narrow straits and an underseas choral reef make excellent habitats for nesting Common Eider, Black Guillemot, gulls and terns. Most of the area is protected as Ytre Hvaler National Park. The main islands of Hvaler archipelago have a varied selection of habitats, from lush agricultural land, mixed forests, coastal meadows and dry, rocky hills with heather and pine woods. The latter is particularly good for Woodlark and Nightjar. Lake Arekilen on Kirkeøy island, is almost completely overgrown with reeds and surrounded by black alder-dominated mixed forest and some agricultural fields providing good conditions for species like Bearded Tit, Little Grebe and Water Rail. In autumn, the archipelago can sometimes experience morning rushes of finches, pipits, thrushes

Woodlark has a fairly strong population in open and dry pine tree habitat in this district. It is an early migrant. *Photo: Terje Kolaas.*

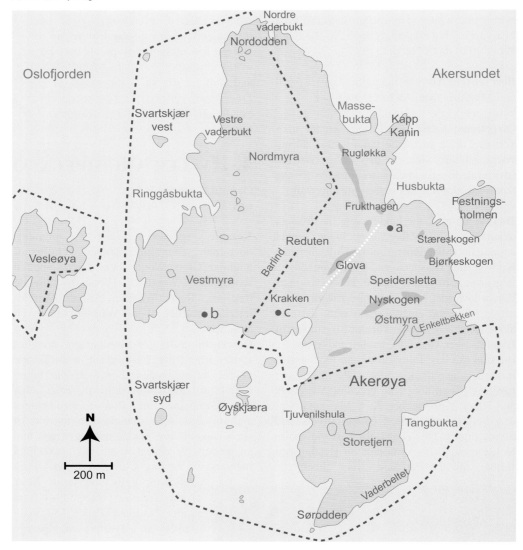

Hvaler arhipelago, Akerøya island: The area dotted in green is closed to traffic April 15th to July 15th.
a. Akerøya Bird Observatory; **b.** preferred seawatch; **c.** seawatch in period of traffic ban.

and others. Here are also nice viewpoints for overlooking migration at sea. One of the westernmost islands, Akerøya island, has a bird observatory and is something of a rarity hotspot in late spring and autumn, famous for records of e.g. Song Sparrow and Rock Thrush. With its mild climate, Hvaler may support interesting wintering birds such as Water Rail, Great Grey Shrike and many Rock Pipits worth checking for rarer options.

Best season: Year-round, peaking during migration. On sunny summer days, the archipelago is crowded with tourists.

Directions: From Rv110 leading through Fredrikstad, exit south onto Rv108 signposted to Hvaler.

Tactics: The best places in Hvaler archipelago, from north to south, are as follows:
Bastangen seawatch on Vesterøy island is the best mainland-connected viewpoint for seabird migration. Once on Vesterøy island,

exit onto the first road to the right, and then keep left signposted to Utgårdskilen. Just before the outermost parking area, walk up a steep gravel road. Follow markings on the rocks past the outer cottage area, 1.1 km from the car park.

On Spjærøy island along Rv108 further south you pass through areas of exposed rock and scattered pine trees, a perfect habitat for Woodlark.

Skipstadkilen lagoon on Asmaløy island is shallow and situated along the outer coastline. On Asmaløy island, exit right from Rv108 signposted to Brattestø, and after 500 m pull down to the left by a stable. You may well walk from this site to the next. Thrush Nightingale and Woodlark can often be found in the woodlands of Asmaløy island.

Vikerkilen inlet is reached either by walking south along the coast from Skipstakilen lagoon, or you can drive further south on Rv108 and make a left just before the tunnel to Kirkeøy, signposted to Viker. Follow this road to Vikerhavn harbour and turn right towards the parking area by Vikerkilen inlet.

From here, you can walk a bit further out on the Vikertangen headland, from where you may observe migration both at land and at sea.

Fugletangen headland at Håbu, 1 km east of Vikertangen headland, is considered to be a better vantage point for migrating land birds than Vikertangen. This place is reached by leaving Rv108 as described for getting to Vikertangen, but then exit onto the second road to the left, signposted to Håbu, before you make a right turn onto the road signposted to Ekevika. Follow this road as far as possible towards the little marina and walk a short distance along the rocks to your left, along the east side of the small Ekevika inlet.

Lake Arekilen. Continue Rv108 through the underseas Hvaler tunnel to Kirkeøy island, and turn right into the Ørekroken parking lot. Follow the path on the opposite side of the road. After passing the last cottage, the path leads to a T-junction. Go left to the tower hide, or go right to a viewing platform. The lake is lined with nice and varied swamp forest.

Akerøya island is somewhat isolated from the rest of Hvaler archipelago, positioned relatively far off shore. It is known as a rarity magnet of national calibre. *Photo: Bjørn Frostad.*

Hvaler archipelago is strategically positioned geographically for migrants. It also sports several interesting bird habitats and sites. Lake Arekilen on Kirkeøy island is one of them. This wetland used to be almost completely inaccessible to people, but has in recent years been equipped with both an observation platform (as shown) and a tower hide, providing great views of this exciting locality. *Photo: Bjørn Olav Tveit.*

In the north there is a colony of Grey Heron which should not be disturbed.

Kirkøy island is generally well worth exploring. The beaches and coast line on the west side can hold ducks, shorebirds and passerines. The woodlands have breeding Nightjar, Woodlark and Redstart.

There is a ferry from Skjærhalden at the far end of Kirkeøy to Herføl island and other islands in the archipelago.

Akerøya island is reached by private or taxi boat. You may want to contact the bird observatory for details if you plan on visiting the island.

5. HALDEN AREA

Østfold County

GPS: 59.19688° N 11.34361° E (Lake Rokkevannet), 58.96079° N 11.49177° E (Berby)

Notable species: Whooper Swan, Hazel Grouse, Black Grouse, Capercaillie, Red-throated Diver, Black-throated Diver, Great Crested Grebe, Honey Buzzard, Buzzard, Goshawk, Osprey, Crane, Bittern (irregular), Little Ringed Plover, Stock Dove, Nightjar, Eagle Owl, Pygmy Owl, Tengmalm's Owl, Kingfisher, Wryneck, Green Woodpecker, Black Woodpecker, Lesser Spotted Woodpecker, Woodlark, Grey Wagtail, Dipper, Redstart, Mistle Thrush, Reed Warbler, Marsh Warbler, Icterine Warbler, Wood Warbler, Red-backed Shrike, Long-tailed Tit, Marsh Tit and Hawfinch.

Description: The extreme southeast corner of Norway, in the area around Halden town, has several good birdwatching qualities. For one, it is perhaps the easiest accessible site in this part of the country for finding Eagle Owl, with a few pairs breeding along the Iddefjord. In spring they can actually sometimes be heard calling from Halden town centre just before dawn. The dense reedbed at Lake Rokkevannet is in some years a site for booming Bittern. This site also supports Crane, Marsh Harrier and – during migration in late May – often displaying Great Snipe. The Rokkesletta farmlands south of the lake can be good for raptors, Dotterel and other species that seek to farmlands during migration. The woodlands surrounding the lake are good for Nightjar. Obviously, the Halden area is a rewarding district for nocturnal songbird trips. The Berbyelva river is a traditional all-year site for Kingfisher, and in winter you may find it along Tisla river as well. In Remmendalen valley just outside Halden town centre, Hawfinch is reliable and Red-breasted Flycatcher has been found singing a few times in late spring.

The Halden area is also a stronghold for breeding Osprey, and in winter it is a regular site for both White-tailed and Golden Eagles. The coniferous woodlands east of Halden support several forest species, including Hazel Grouse, here often found in association with batches of alder along rivers and streams.

Other wildlife: A small population of grey wolf roams the interior of Østfold County, particularly in the forests of Lundsneset nature reserve and Tresticklan national park on the Swedish side of the border.

Best season: Late spring and early summer, winter as well for Kingfisher and eagles.

Directions: Follow signs to Halden from E6 just before the border crossing at Svinesund. From Halden town centre, the sites mentioned above are found as follows:

Remmendalen valley: Follow Rv21 west out of town and exit right to *Høgskolen i Østfold* (Østfold University College). Park at the college and follow the creek behind it downstreams towards the Iddefjord.

Tista river: Follow Rv21 east signposted to Ørje, which is a road running alongside the river. The parallel road on the north side of the river has more suitable places to stop. Lake Femsjøen, from which the river empties, is also worth checking. Continuing Rv21 east will take you to areas mainly covered in coniferous woodlands.

The Iddefjord and Berby: Drive south along

Nightjar has its strongest population in the south-eastern parts of the country. Like Woodlark, it is usually found in association with dry, open pine forest. Nightjars are seldom seen, and their presence is usually noticed by the buzzing song, which is performed on quiet early summer nights. *Photo: Bjørn Olav Tveit.*

Lake Rokkevannet has been upgraded with a suitable path, picnic benches and a tower hide that gives a nice view of the lake and reedbeds. *Photo: Per-Arne Johansen.*

Rv22 signposted to Holtet and pull down sideroads to the right towards the fjord. About 20 km south of Halden, turn right signposted to Strømstad and Berby and follow this road until it crosses Berbyelva river.

Lake Rokkevannet: *From Oslo/Sarpsborg*, leave the E6 at junction 7 and follow Rv118 through Sarpsborg. After Rv118 has crossed Glomma river, you follow it for 5 km before turning left on the road signposted to Rokke church. You follow this road past the exit to Buertjernet (which should be checked) and continue for about 11 km. You are now crossing Rokkesletta farmlands, which must be checked carefully. At the end of Rokkesletta farmlands, before Rokke church, turn left onto Rokkevannsveien road, signposted to the tower hide. Continue through the yard at Strand farm and leave the car in the designated parking area by the barn (a small parking fee must be expected). Follow the footpath across the fields and down to the tower hide by Lake Rokkevannet. For an overview of the western cove, drive 200 m back on the main road, and turn right onto Fjerdingveien road.

6. LAKE BORREVANNET

Vestfold County

GPS: 59.39782° N 10.45286° E (parking tower hide), 59.38941° N 10.42902° E (parking west side), 59.42900° N 10.49514° E (Møringa seawatch)

Notable species: Migrants; passing seabirds; Great Crested Grebe, Osprey, Marsh Harrier, Honey Buzzard, Buzzard, Kestrel, Peregrine Falcon, Hobby, Crane (spring), Moorhen, Coot, Water Rail, shorebirds, Lesser Black-backed Gull, Stock Dove, Tawny Owl, Long-eared Owl, Green Woodpecker, Black Woodpecker, Lesser Spotted Woodpecker, Dipper (winter), Thrush Nightingale, Sedge Warbler, Marsh Warbler, Reed Warbler, Icterine Warbler, Wood Warbler, Bearded Tit (late autumn), Long-tailed Tit, Marsh Tit, Red-backed Shrike, Great Grey Shrike (winter), Nutcracker, Rook, Goldfinch, Linnet, Common Rosefinch and Hawfinch.

Description: Lake Borrevannet is nutrient-rich and 4 km long, situated near the fjord by Horten town. It is bordered with cultivated fields and mixed woods, with spruce dominating in the north and tall, mature deciduous trees in the south. In the south end of the lake is Vassbånn wetland, a marshy area with birch and willow scrub and reedbeds, framed by dykes in order to keep the lake from flooding the surrounding farmland. Vassbånn holds the country's densest population of Thrush Nightingale, providing a choir consisting of up to 40 singing males from the second week of May until mid-June. In Vassbånn wetland and the adjacent open water you can find several species of duck, Great Crested Grebe, Coot and more. The area is best viewed from the tower hide. A few pairs of Moorhen, Water Rail and Sedge Warbler can also be found, besides many pairs of Reed Warbler. The surrounding farmland support several pairs of Marsh Warbler. Quail is usually heard calling in early summer nights as well. Marsh Harrier has bred in recent years, and the vicinity holds Osprey, Honey Buzzard, Buzzard, Hobby and Peregrine Falcon, all of which may be seen down by the lake through summer. The mature

deciduous forest support Long-eared Owl, Icterine Warbler, Wood Warbler, Goldfinch and Hawfinch. Nutcracker is often seen here, particularly during late summer and autumn. Common Rosefinch used to be rather common in Vassbånn wetland, but seems now only to be reliable at Kongsodden peninsula, where Red-backed Shrike breed as well. Rooks from the small colony in Horten town centre can often be seen among the Hooded Crows and Jackdaws which frequent the farmland.

Lake Borrevannet is also famous for being a very good migration site, particularly in spring. At that time of year, it is among the best sites for raptor watching in the country. The raptors migrate north along the west coast of Sweden before entering Østfold on the opposite side of the Oslofjord. Here, they gain height before crossing the fjord. Often the same birds can be seen at both Kurefjorden bay in Østfold and Lake Borrevannet in Vestfold. Common Buzzard is the dominating species of raptor, usually arriving in mid-March, migrating

When the fields at the southern end of Lake Borrevannet are flooded, an eldorado is formed for wetland birds. The picture is taken from point **e.** in the direction of point **d.** on the map on the next page. This is the best raptor viewpoint. You get a glimpse of the reedbed in Vassbånn wetland to the left. *Photo: Bjørn Olav Tveit.*

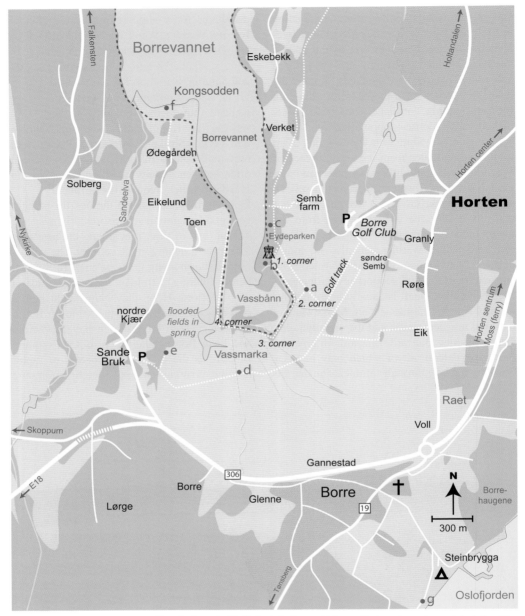

Lake Borrevannet, south: a. The Golf track (*Golfstien*); **b.** tower hide; **c.** Eydeparken viewpoint; **d.** tracktor track; **e.** raptor-watch; **f.** Kongsodden headland; **g.** Steinbrygga jetty and coastal beach.

through to early May. On sunny days, 20–30 or so Buzzards are usually seen, with a maximum day count of 140 birds. Rough-legged Buzzard is less numerous and usually arrives two-three weeks later than Buzzard. On good days in April you will also often get to see flocks of Pink-footed Goose, Cormorant and Crane. Honey Buzzard and Hobby usually arrive in

early May.

In most years, melting snow results in large pools of water on the farmlands surrounding Vassbånn wetland in spring, creating perfect conditions for ducks, geese, swans, gulls and – if the pools are retained until late spring – also fresh-water shorebirds. Scarce birds such as Garganey, Pintail, Shoveler, Gadwall, Little

Ringed Plover and Temminck's Stint are often encountered under such conditions. Great Snipe are regularly encountered in areas of tall grass in May, and Jack Snipe is often found in the wetter areas in October. In late autumn and winter, a few migrants may try to winter here if the climatic conditions allow it. In the cold season, Dipper, Great Grey Shrike, Bearded Tit and Arctic Redpoll may be found as well. Many rare birds have been encountered at Lake Borrevannet, incl. Blue-winged Teal, Purple Heron and Short-toed Snake Eagle. The nearby coastline is always worth exploring, and one of the best seawatching viewpoints on this side of the fjord is Møringa just outside Horten town centre.

Other wildlife: Roe deer, red fox, badger, beaver and bats are all regularly seen in the area.

Best season: Spring and early summer is particularly good, but Lake Borrevannet can be rewarding all year as long as there are ice-free sections of the lake.

Directions: Leave route E18 between Tønsberg and Holmestrand at Exit 34 and continue Rv306 signposted to Horten and Moss. Then, see the map. You may also reach Borrevannet from the east by crossing the Oslofjord with a car ferry from Moss to Horten.

Tactics: Park at Borre Golf Club. Walk past the club house and follow **a.** the golf track, signposted *Golfstien*. This track takes you around the golf court alongside the farmers fields and down to **b.** the tower hide. From here you overlook Vassbånn wetland and the south end of Lake Borrevannet. An alternative view is obtained by walking further along the track through the mature deciduous forest to **c.** the Eydeparken cliff. When the fields are flooded, the opposite side of Vassbånn wetland i usually better. You can walk along the dike or follow **d.** a path through the fields. But please, do not enter the farmland itself nor the golf course. **e.** The west side is also best for watching migration in spring. Here you can overlook

both the flooded fields and scan the horizon for raptors and other migrants, many of which follow along the moraine ridge Raet separating Lake Borrevannet from the Oslofjord. You may also birdwatch along the roads and paths in the area all the way to **f.** Kongsodden peninsula. Drive down to the fjord at **g.** Steinbrygga jetty and work your way east and northwards along the coast through Horten to Falkensten, at the outlet stream of Borrevannet. This stream is a pretty reliable site for Dipper in winter. Møringa seawatch is reached from Horten town centre by following signs to Karljohansvern fortress. Continue past the church and find parking at Vollen beach. Walk the seaside path north towards the military gate. Here is a good vantage point for seawatching, and you may also find staging shorebirds, gulls and ducks here in season.

7. TØNSBERG INLETS

Vestfold County

GPS: 59.27939° N 10.37748° E (parking Ilene inlet), 59.27132° N 10.43748° E (parking Presterødkilen inlet)

Notable species: Staging geese, ducks, shorebirds and gulls; migrating raptors; Pheasant, Buzzard, Stock Dove, Tawny Owl, Green Woodpecker, Lesser Spotted Woodpecker, Thrush Nightingale, Reed Warbler, Marsh Warbler, Icterine Warbler, Wood Warbler, Marsh Tit, Goldfinch and Hawfinch.

Description: Two of the best birding sites in the outer Oslofjord area are the shallow Ilene and Presterødkilen inlets, situated on either side of Tønsberg town. Ilene inlet on the west side contains the delta of Aulielva river, surrounded by farmland, and is sheltered from the sea by Nøtterøy island. Presterødkilen inlet has large areas of exposed mud at low tide and is lined by dense reedbeds. The surroundings, however, are urban-industrial and not very bird friendly. Both inlets can support large numbers of wetland species. Presterødkilen inlet is considered reliable for Bearded Tit and Water Rail. For this part of the country, the numbers of *Calidris*

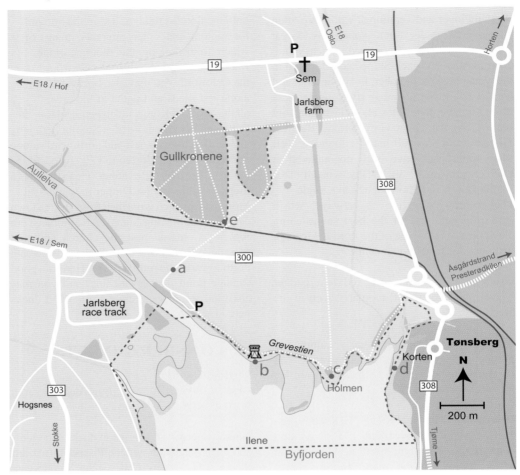

Ilene inlet: a. Raptorwatch; **b.** tower hide; **c.** visitor centre; **d.** eastern viewpoint; **e.** Gullkronene forest.

The Tønsberg inlets are well organized for birdwatching. Here is the tower hide marked as point **f.** on the maps. *Photo: Bjørn Olav Tveit.*

shorebirds can be quite good, often including Temminck's Stint. Raptors migrate through the area in spring and autumn. Early summer can be very rewarding in terms of nocturnal songbirds. Close to Ilene inlet, a small forest with mature deciduous trees surrounded by farmland can hold a number of woodland species.

Best season: All year, but greatest diversity in April through to October.

Directions: *By car from Oslo*, leave E18 at Exit 35 signposted to Tønsberg and Tjøme.

Follow Rv308 to Tønsberg. Just before entering town, follow signs to Kr.sand/ Sem in order to get to Ilene, or to Tjøme and then Åsgårdstrand to get to Presterødkilen. See the maps for further instructions.

Tactics: A telescope is essential at both Ilene and Prestrødkilen inlets. There are nicely prepared paths along both inlets, adapted to wheelchair use. Ilene has one large tower hide, while Presterødkilen is equipped with three smaller tower hides and one viewing platform.

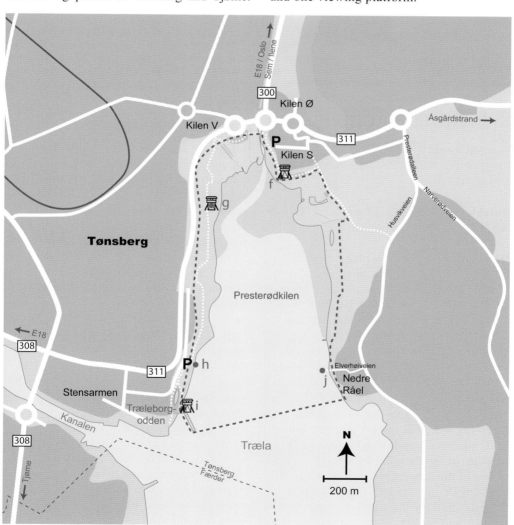

Presterødkilen inlet: f. innermost tower hide; **g.** birdwatching hide; **h.** wheel-chair adapted observation platform; **i.** Træleborgodden tower hide; **j.** eastern viewpoint.

Hooded Merganser is one of the characteristic birds along the Norwegian coast all year round, and the species nests at several places along the Oslofjord. *Photo: Christian Tiller.*

8. FÆRDER ARCHIPELAGO

Vestfold County
GPS: 59.07182° N 10.39731° E (Moutmarka), 59.06546° N 10.44271° E (Pirane)

Notable species: Migrants, Thrush Nightingale; Little Grebe (winter).

Description: The Færder archipelago consists of the 12 km long and narrow Tjøme island, surrounded by several smaller islands, including Hvasser island. The archipelago stretches south, serving as a navigational landmark for birds during autumn migration. The south end is called Verdens Ende (*The end of the World*) and is the last bit of dry land before the Skagerrak sea crossing towards Sweden and Denmark. Substantial numbers of Buzzards and other raptors migrate through in favourable conditions, particularly in autumn. In summer, the archipelago is host to one of the densest populations of Thrush Nightingale in Norway. It is also a popular summer holiday destination, but in late autumn and winter, few people other than the few residents are around. At this time of year, Rock Pipts abound and among these

a few Meadow Pipits and the odd Water Pipit may be found. It is also a regular wintering site for Little Grebe, Water Rail and Great Grey Shrike. A large portion of the outer archipelago is incorporated into Færder National Park.

The easternmost island in the Færder archipelago is Store Færder island. It is rather isolated situated in the mouth of the Oslofjord and is well-known for producing records of rare birds during the migration periods, including several firsts for Norway, incl. Alpine Accentor.

Directions: From E18 through Vestfold, exit to Rv300 signposted to Tjøme and Tønsberg. In Tønsberg, continue on Rv308, signposted to Tjøme.

Tactics: One of the archipelago's best sites for resting migrants is the coastal heathland Moutmarka, with its ponds (both natural and dug out) and rocky beaches with washed up seaweed and kelp. It is reached by driving about 10 km past the Vrengen bridge connecting Tjøme island to the mainland. Exit right onto Moveien road signposted to Mostranda Camping and turn left again onto one of the side roads before the campsite. You may alternatively enter Moutmarka from further south by driving all the way south to the parking area at Verdens Ende and walk northwest from there. The gardens and paddock by the parking should be checked for resting migrants. Færder National Park Centre is located in this area. The rocky hill northeast of the parking is a nice lookout for raptors and other migrants.

Hvasser island is reached by following signs to Hvasser further north, just south of Tjøme village centre. This road crosses Røssesund sound, a narrow sound between Hvasser and Tjøme islands, often holding Little Grebe in winter. 1 km after passing this bridge, turn right onto the side road Oppegårdsveien and continue past a nice farmland valley down to the parking area by the sea. From here, walk across the rocky hill to the west, leading to the area Pirane, which can be described as a smaller version of Moutmarka.

Store Færder island is reached by private boat or by contacting Store Færder Bird Observatory.

9. Lake Gjennestadvannet

Vestfold County

GPS: 59.23533° N 10.24024° E (viewpoint)

Notable species: Ducks and other waterbirds, migrating raptors; Honey Buzzard, Marsh Harrier, Hobby, Crane, Wryneck, Woodlark (irregular), Icterine Warbler and Wood Warbler.

Description: Lake Gjennestadvannet is 700 m long, nutrient-rich and lined with reeds, surrounded by farmland and spruce-dominated mixed woods. It is probably the most reliable and easiest available site for

finding Hobby in Norway. The birds breed in the vicinity and can regularly be seen hunting insects and small birds by the lake. Honey Buzzard and Wryneck breed here as well, and in some years, Woodlark has been found singing in the area of barren pine hills to the west of the lake. Migrating raptors can often be seen passing through.

Best season: Late spring and early summer. To a somewhat lesser extent, autumn as well.

Directions: From E18 south of Tønsberg, exit to Stokke and Borgeskogen industrial area, and continue towards Stokke. Then make a right onto a road signposted to Løke and Gjein. After 2 km on this road, you see Lake Gjennestadvannet on your left.

Tactics: Scan the lake and its surroundings with a telescope from the roadside viewpoint by the western end of the lake.

Lake Gjennestadvannet, seen from the viewpoint by the western end of the lake. Honey Buzzard, Hobby and other raptors can often be seen circling above the hill in the background. *Photo: Bjørn Olav Tveit.*

Klåstadkilen inlet and the rest of the inner part of Viksfjord is a pearl of a bird locality, with large shallow water areas, several small coves and lush edge vegetation, adjacent to cultivated land and mixed forest. The inlet itself is the jewel in the crown of this wetland system. Unfortunately, it has been partially filled and deserves to be restored to its original state. The picture was taken towards the north, just north of Tangen. *Photo: Bjørn Olav Tveit.*

10. KLÅSTADKILEN INLET

Vestfold County
GPS: 59.05974° N 10.17503° E (parking)

Notable species: Staging wetland birds, passing raptors and other migrants; Quail, Little Grebe, Honey Buzzard, Water Rail, Jack Snipe (autumn), Lesser Spotted Woodpecker, Grey Wagtail, Thrush Nightingale, Marsh Warbler, Reed Warbler, Icterine Warbler, Wood Warbler, Marsh Tit, Goldfinch and Hawfinch.

Description: Klåstadkilen inlet is the innermost corner of Viksfjord, which is a 7 km long side-arm to Larviksfjord. It was the heart of an important wetland area until the 1970s, when it was partially filled to give way to farmland. Klåstadkilen inlet is shallow, lined with reeds and surrounded by farmland on all sides. Mudflats are exposed during low tide, making suitable conditions for a variety of shorebirds, ducks, rails and more. The inner part of Viksfjord remains a very attractive site for a variety birds.

Best season: Late spring to late autumn.

Directions: *From E18* exit to Larvik and continue on Rv303 east towards Sandefjord. About 7 km after crossing the river Lågen, make a right onto the road signposted to Ula and Eftang. Park behind the bus turning area in this junction.

Tactics: Klåstadkilen inlet and the inner part of Viksfjord is best at low tide. A telescope will come in very handy. From the parking, walk along the reed covered ditch which crosses the fields and continue down to the marshy area by the fjord, the remnants of Klåstadkilen inlet. Do not walk on growing farmland, only on stubble fields from late summer through winter. You may cross the ditch via a small bridge just underneath the power lines. Look for birds along the shore

at suitable places, such as <u>viewpoints a–d</u>. The best place to overlook the general area is at **b**. <u>Tangen</u>. You can walk along the shore all the way to **d**. <u>Skisakerkilen inlet</u> (along the way there is a small bridge crossing the creek Ivjua), or you can walk back to the car and drive

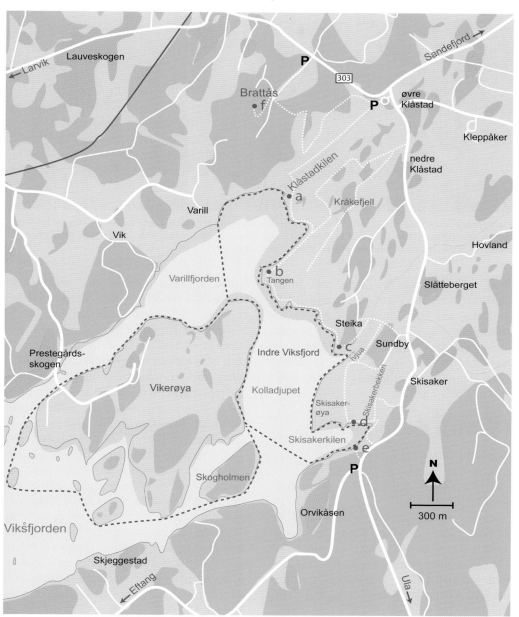

Klåstadkilen inlet and inner Viksfjord: **a**. Klåstadkilen inlet; **b**. Tangen viewpoint; **c**. Ivjua creek outlet; **d**. and **e**. Skisakerkilen inlet; **f**. Brattåsen hill raptor-watch.

It's a fantastic experience to watch the migration from dawn and onwards, and a great challenge to identify the passing birds. Here is a flock of Pink-footed Goose, the most numerous species of goose regularly passing through the interior of Norway. *Photo: Kjetil Schjølberg.*

there. If you choose the latter, park in the junction where the roads to Ula and Eftang separates, and walk down to **e**. viewpoint. To watch migrating raptors, it is best to position oneself on top of **f**. Brattåsen hill. Many of the raptors seen here, can also be seen at Mølen headland, although at Brattås hill they often pass at closer quarters.

11. MØLEN HEADLAND

Vestfold County
GPS: 58.97207° N 9.83348° E (Mølen), 58.96528° N 9.96120° E (Ravn), 58.98185° N 10.02562° E (Rakke)

Notable species: Migrating birds in all categories; staging ducks and other waterbirds, often including Common Scoter, Velvet Scoter, Scaup, Black-throated and Red-throated Divers, Slavonian and Red-necked Grebes; *summer:* Nightjar and Thrush Nightingale.

Description: Mølen headland is the southwestern tip of Brunlanes peninsula,

which is the southern end of Vestfold County, stretching out into Skagerrak. This is reckoned to be the number one site in Norway for migrating land birds in autumn. When conditions are right, the early morning rush of birds can be breathtaking. The air is filled with large flocks of thrushes, finches and pigeons, often stacked in several layers in the sky, some migrating close to the ground, even some flicking from one bush to the next, and others at very high altitudes, only barely visible through binoculars. The impressions can be overwhelming, and really put your identification skills to the test. Day-counts can reach figures in the hundred thousands, with the most numerous species being Woodpigeon (maximum day-count 105,000 individuals), Chaffinch/Brambling (230,000) and Siskin (75,000). Other species can in some autumns occur in large influxes also resulting in impressive day-counts, such as Redpoll (40,000) and Pine Grosbeak (639). The composition of species varies through the autumn, with higher proportions of warblers, Yellow Wagtail and Tree Pipit in

August–September, and more pigeons, tits, finches and short-distance migrant influxes in September–October. On good days, you can also expect a few less common species, such as Mistle Thrush, Grey Wagtail, Nutcracker, Hawfinch and Woodlark, and even bigger surprises from the interior, such as a Three-toed Woodpecker or Hawk Owl. Raptors can also migrate through in good numbers, with Sparrowhawk, Buzzard and Rough-legged Buzzard the dominating species. On the best raptor day ever, 1. October 2007, a total of 1233 individuals were counted, including all-time high day-figures of Sparrowhawk (215) and Buzzard (820). Ringing has revealed that a number of owls pass through in autumn, including good numbers of Pygmy and Tengmalm's Owls. A maximum of 115 species has been recorded at Mølen in one day (30. September 1988), and a total

exceeding 320 species has been recorded at the site through the years, including several extreme rarities such as Willet and Blyth's Pipit.

The landscape on the Mølen headland is evidently a part of the large ancient moraine ridge Raet. The shoreline on the sea-side of the headland is littered with gravel and rocks, battered, rounded and smoothened by the sea for aeons. On the opposite side, finer-grained moraine particles dominate, making way to agricultural fields and sandy beaches. Low and dense scrub vegetation characterize the otherwise dry and barren peninsular tip. Several large iron-age burial mounds, made from the rounded rocks from the outer coast, is an eye-catching physical feature of Mølen headland.

Thrush Nightingale dominate the soundscape during late spring and early summer. Until a

Mølen headland seen from the northwest. The characteristic pebble beach has been a landmark since the dawn of time. Mølen is south-eastern Norway's best location for direct migration of land birds in autumn. Here you can witness that tens of thousands of Woodpigeons and passerines escape the country before winter sets in. The varied range of habitats, the strategic location and the sheltered shallow water areas on the leeward side of the peninsula contribute to making Mølen headland a year-round birdwatching location. *Photo: Magne Klann.*

Nevlungstranda lagoons east of Mølen headland is an area of nice salt meadows alternating with small brackish water lagoons where you can find resting migrants and wintering Shag, Rock Pipit and more. *Photo: Bjørn Olav Tveit.*

few years ago, this was a regular breeding site for Barred Warbler as well, but the increasing vegetation is probably the main reason for its decline. It is still recorded here occasionally, though, but mostly juveniles in early autumn. The bay on the north side of the peninsula is a sheltered haven for a variety of seabirds throughout the year. In summer, substantial numbers of ducks stage here. The numbers and diversity increase through autumn, when it eventually also includes divers, grebes and auks. Throughout the winter, this is also a favoured spot for a number of passerine species, sometimes including Great Grey Shrike and Shore Lark.

The Brunlanes coast has plenty of headlands, inlets and sheltered beaches that are worth exploring for passing seabirds, resting migrants or wintering birds. In early summer, Nightjar can be heard at night in the interior of Brunlanes peninsula.

Best season: Mølen is a typical year-round site, although bird highlights intensify during April through early June and even more so in September-October.

Directions: *From E18 near Larvik,* exit onto the southern of the two exits signposted to Larvik, and then follow Rv302 signposted to Helgeroa and follow this road out onto Brunlanes peninsula. Just before Nevlunghavn village, turn right onto a road signposted to Mølen. Follow this road to the parking area at the end. Several paths lead from here out to the headland (see the map).

Tactics: *Weather:* For migration in spring, and for shorebirds in summer and early autumn, you should hope for overcast weather and high temperatures, preferably with a slight breeze from the southeast. A few drops of rain are just fine. In autumn, the best conditions are caused when a period of bad weather with strong wind and heavy rain suddenly changes to blue skies and calm winds, preferably blowing from between west and north. For passerine migration, slightly overcast skies will do as well, but rain usually stops them. If these autumn conditions are combined with a significant drop in temperatures, preferably with nightly frost inland, it can cause a tremendous rush of migrants at Mølen headland, commencing at dawn. On sunny mornings, raptors can begin migrating from about 9 o'clock. In winds from the north, the raptors may pass straight overhead, whereas on most days they pass at a distance over the quarries to the north and are often seen at closer range at Brattås hill by Klåstadkilen inlet (which see) or at sites even further north. In strong winds from the southwest, you may look for seabirds from g. Saltstein. However, if you plan to indulge into some serious seawatching you are probably better off travelling a bit further east along the Brunlanes peninsula, e.g. to Ravn seawatch.

Mølen headland: During migration periods in autumn, you should position yourself at a. the ridge near the tip at dawn (which would be about one hour before the officially announced sunrise), and keep your eyes and ears open. Some days the morning rush continues for several hours. The closer to

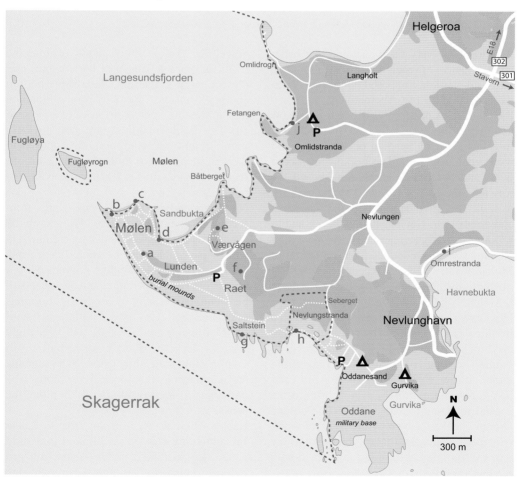

Mølen headland: a. Traditional viewpoint; b. Odden viewpoint; c. viewpoint; d. Sandbukta inlet; e. cabin area; f. Mølen Bird Observatory; g. Saltstein seawatch; h. Nevlungstranda lagoons; i. Omrestranda beach; j. Omlidstranda beach.

Velvet Scoter is common around Mølen headland all year. *Photo: Kjetil Schjølberg.*

b. Odden (the very tip) you stand, the better are the chances of seeing some of the grounded birds up-close, as opposed to having them passing quickly by. Alternatively, or when the number of passing birds begin to tail off, you can walk around looking for grounded migrants. Start by heading out to the tip and work your way inwards, checking the scrub, fields and beaches **(b–d)** thoroughly, all the way to **e.** the cottage area and by the car park. A walk (or drive around) to the saltmarsh and ponds at **h.** Nevlungstranda beach and to **j.** Omlidstranda beach can also prove rewarding. Ringing has been conducted out on the tip of Mølen peninsula, but in recent years most of this activity has been going on just outside **f.** Mølen Bird Observatory's cabin.

Brunlanes peninsula apart from Mølen headland can be explored by continuing out to Nevlunghavn town and **i.** Omrestranda beach, before heading back to Helgeroa. From there, follow Rv301 towards Stavern, and just before Berg church, turn right onto the coastal road signposted to Naverfjorden and Hummerbakken. Down by the creek-crossing, side roads to the right takes you down to the narrow Hummerbakkfjorden inlet lined with reedbeds and with mudflats at low tide.

Ravn seawatch. Continue 1.5 km and pull off by Anvikstranda and Donavall. Turn immediately right and then right again onto a gravel road. Follow this road 800 m along the fields and turn left up past the cottages and down a steep gravel hill. Park at the bottom and walk out past the outermost cabins and find a suitable spot with an ocean view to sit. The light conditions are best in early morning and late afternoon. As opposed to the seawatching sites further up the Oslofjord, there can be seabirds at Ravn even on days of rather calm winds.

Naverfjorden bay. The coast eastwards from Ravn has many nice areas for shorebirds, waterbirds and passerines during autumn and winter, particularly in the Naverfjorden bay. There are numerous camping grounds packed with people during summer, but birds can still be found, particularly in very early mornings of overcast and rainy days. Continue east until the road again meets Rv301 and head right towards Stavern. Shortly after, pull down onto a side road signposted Nalumstranda and continue to the shore. From here, it is particularly nice to walk the 4 km along the coast east to Rakke.

Rakke. Alternatively you can drive to Rakke by continuing east on Rv301 and pull right, signposted to Rakke. In the Y-junction just before Solplassen Camping, keep right and continue to the parking lot. From here, you can explore the cottage area, the fields and salt-marshes at Rakke.

Tanum. Follow signs from Rv301 or Rv302 to Tanum church. The nearby woodland is the best place to hear displaying Nightjar in early summer nights.

12. JOMFRULAND AND STRÅHOLMEN ISLANDS

Telemark County
GPS: 58.88151° N 9.61058° E (bird observatory), 58.86968° N 9.41581° E (Kragerø harbour)

Notable species: Migrating birds in most categories; Shag, Tawny Owl, Lesser Spotted Woodpecker, Thrush Nightingale, Marsh Warbler, Icterine Warbler, Wood Warbler, Golden Oriole (uncommon), Red-backed Shrike and Common Rosefinch.

Description: Jomfruland island is narrow and 7 km long, running parallel with the Telemark coastline. The northern tip of Jomfruland, together with Mølen headland, is regarded as the best migration site along the Skagerrak coast, good for land birds and seabirds alike. The island's oblong shape and strategic position along the outer coast makes it a natural navigational mark for birds moving along the coast in both spring and autumn. The island itself is rather flat and made up of gravel, boulders and sand, obviously a part of the afore mentioned moraine ridge Raet. The island is covered in tall, mature deciduous woodland, with small pastures and paddocks here and there. Vast blankets of flowering white wood anemones dominate the forest undergrowth in summer,

Oslo · Interior · **Skagerrak** · Western · Central · Northern · Finnmark · Svalbard

Jomfruland island's northern tip in a birds-eye perspective. *Photo: Magne Klann.*

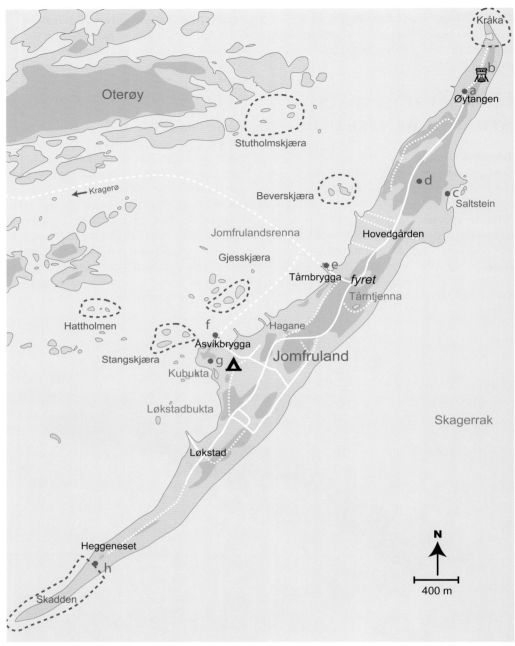

Jomfruland island: a. Bird Observatory; **b.** tower hide; **c.** Saltstein rock; **d.** oak forest; **e.** Tårnbrygga and **f.** Åsvikbrygga ferry landings; **g.** Kubukta wetland; **h.** Skadden headland. Most of the area is part of the Jomfruland National Park; the protected areas marked in green on this map only show the parts that are closed to traffic in the period April 15th to July 15th (referred to as zone B in the protection regulations).

adding to the picturesque and somewhat continental atmosphere of the island. The continental feel also affects the avifauna, most notably by this being the only regular site in Norway for Golden Oriole, although it is not to be expected on a short visit. Jomfruland Bird Observatory is positioned near the northern tip of the island. Here is also a tower hide overlooking the salt-marsh and beaches at the tip.

The much smaller Stråholmen island is positioned north of Jomfruland like "the dot on top of the i". Stråholmen island used to be the only reliable breeding site in Norway for Barred Warbler, but this is now history. Stråholmen has a higher proportion of scrub, and fewer trees compared to Jomfruland, and it has nice shorebird-friendly areas of rocky beach and tidal pools, well worth investigating. Sandwitch Tern is fairly regular. The sea surrounding Stråholmen island is particularly good for wintering ducks, divers and cormorants, including Shag.

Other wildlife: The population of red squirrels at Jomfruland island is particularly good.

Best season: Spring and autumn. First half of June is best for Golden Oriole.

Directions: <u>Jomfruland island</u>: Leave E18 at Exit 54 and continue on Rv38 signposted to Kragerø. You need to leave your vehicle here, as there is no ordinary car traffic on Jomfruland. A passenger ferry takes you from Kragerø harbour to Jomfruland. There are two landing sites at the island, Åsvikbrygga to the south and Tårnbrygga further north. The latter is the closest (3 km) from the northern tip of the island.

<u>Stråholmen island</u>: Reached either by private or taxi boat from Valle marina (alternatively from Kragerø town or Jomfruland island). In order to drive to Valle marina, which is where the taxi boat terminal is located, leave E18 further north (if arriving from Oslo), signposted to Valle. Outside the tourist season, you should call and make an appointment with the skipper beforehand. The trip is rather expensive, and you get the best deal pr. person if you travel as a little group. It can be a good idea to negotiate a pick-up time at Stråholmen before waving good-bye to the taxi boat.

Tactics: <u>Jomfruland island</u>: Bicycles can be rented at the campsite and is a practical means of transportation on the island. Most of the ornithological activities are conducted on the north end of the island where a. <u>Jomfruland Bird Observatory</u> is located. The b. <u>tower hide</u> makes a good viewpoint for overlooking the migration at land and sea. A

Oslo

Interior

Skagerrak

Western

Central

Northern

Finmark

Svalbard

Barred Warbler used to breed regularly along the Skagerrak coast, although in very small numbers. It was particularly regular at Mølen headland, Jomfruland island and – as the last regular breeding site – Stråholmen island. Even if the species no longer breeds regularly, you should still keep a special lookout for it. *Photo: Gunnar Numme.*

Stråholmen island seen from the southeast. The mosaic of pebble beach, mudflats and brackish water spits in the foreground is a particularly important nesting and resting place for wetland birds. *Photo: Magne Klann.*

telescope is essential. Raptors often pass on the landside of the island. Shorebirds stage in the marshy areas around the tip, and you can find pipits and such in the grassy fields. c. Saltstein is one of the few places on the island with solid rock. Here, stranded seaweed often attracts shorebirds, gulls and a few passerines, particularly in autumn. Golden Oriole is usually found in the canopies of the tallest and densest part of d. the oak wood in the north. Listen for it in late spring and early summer.

At e. Tårnbrygga there are nice areas of shallow saltwater and muddy shores attracting a few shorebirds. Check the sound between Jomfruland and the islands further towards the mainland for ducks, grebes and such during migration and winter. The sound is best overlooked from the road at Hagane and from f. Åsvikbrygga, where you also find a grocery store. g. Kubukta has areas of shallow water, small tidal flats and salt marsh with grazing cows and a little pond perfect for dabbling ducks, wagtails and

more. Further south, in Løkstadbukta is also a salt marsh, pastures and scattered bushes that may host resting migrants. h. Skadden nature reserve at the south end of Jomfruland is a pebbled headland formerly hosting a colony of seabirds, now mainly noted for being surrounded by waters suitable for resting and wintering ducks, grebes and divers. Access is not permitted inside the reserve from April 15 to July 15.

Stråholmen island (separate map): Follow the footpath from i. the taxi boat landing over to the group of houses. Check the fields on your right along the way. Continue down to Rabbestranda and j. Rabberumpa, from where you can scan the tidal area inside the nature reserve, in which access is not permitted from April 15 to July 15. At low tide it is possible to wade the shallow strait to Mostein islet. At times when the visitor restrictions do not apply, you can walk south along the shore past the tidal ponds. But walk carefully so you do not spook the birds. The best point to overlook this area is from the **n.** Søndre Huet rock. You can also walk the **l.** path down to the reedbed to listen for birds, as vision is restricted due to the tall vegetation. The best area for Barred Warbler used to be **m.** near the cottages east of Søndre Huet, and this is still a good place for warblers in general and other bush-dwelling passerines. From Søndre Huet, continue walking west along Vestrestrand to **o.** Nordre Huet, and check the small pools Doplene a few metres in from the shore, although at times they may be all dried out. Nordrestrand

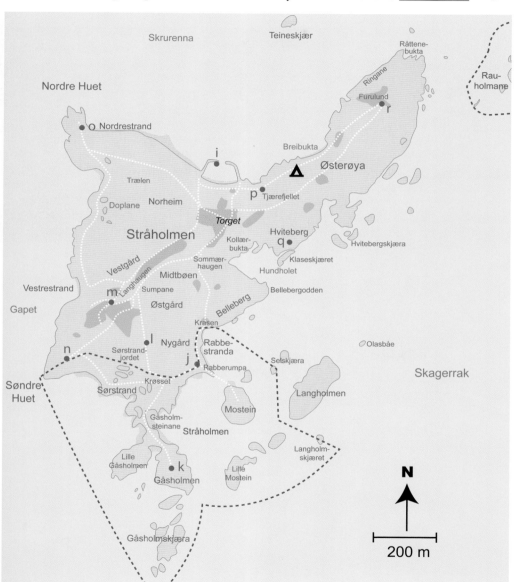

Stråholmen island: i. Taxi boat landing; **j.** viewpoint to the nature reserve when the traffic ban applies; **k.** Gåsholmen islet; **l.** path to reedbed; **m.** lush vegetation (Barred Warbler habitat); **n.** Søndre Huet, viewpoint to the nature reserve from the west; **o.** Nordre Huet viewpoint; **p.** public restroom; **q.** Hviteberg seawatch; **r.** Østerøya viewpoint. Most of the area is part of the Jomfruland National Park; the protected areas marked in green on this map only show the parts that are closed to traffic in the period April 15th to July 15th.

too may be good for shorebirds, gulls and terns, and in winter pipits and more as well. This beach should be viewed from **i.** the the taxi boat landing jetty as well. From **r.** Østerøya you have a nice view to the sea northeast of Stråholmen. Look for migrating seabirds from **q.** Hviteberg or **k.** Gåsholmen in the south.

13. LAKE BØRSESJØ

Telemark County

GPS: 59.23094° N 9.60759° E (Børsesjø Wetland Centre), 59.20495° N 9.64169° E (Storemyr viewpoint)

Notable species: Staging and breeding waterbirds; Quail, Great Crested Grebe,

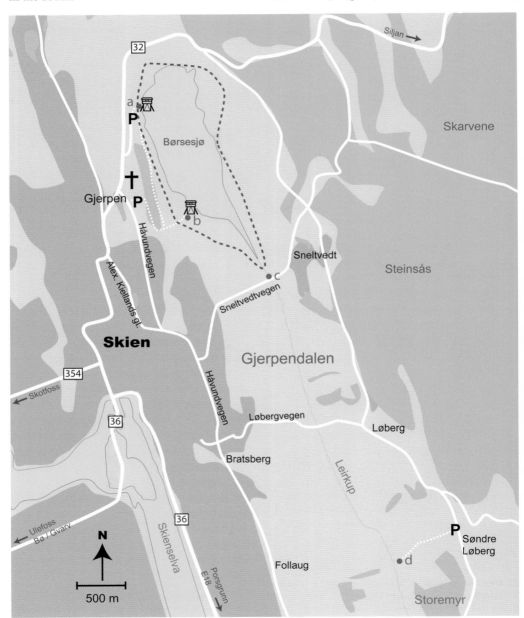

Lake Børsesjø: a. Northern tower hide; **b.** southern tower hide; **c.** Sneltvedtveien road, viewpoint to the outlet creek; **d.** Storemyr viewpoint.

Osprey, Water Rail, Corncrake, Thrush Nightingale, Sedge Warbler, Reed Warbler, Marsh Warbler, Marsh Tit, Goldfinch and Common Rosefinch.

Description: Lake Børsesjø, just east of Skien town, is shallow and nutrient-rich, providing a sanctuary for waterbirds in the otherwise industrialised region of Grenland. Parts of the lake are completely overgrown with reeds and other water plants. The lake is equipped with two tower hides. Several species breed, and Osprey and Hobby is found feeding here in summer. In late spring, Lake Børsesjø and the surrounding agricultural fields of Gjerpendalen valley make a nice area for nocturnal songbird outings. Particularly good is the Storemyr farmlands area, about 2 km southeast of Lake Børsesjø. Here is also a good viewpoint for migrating raptors.

Best season: Late spring, summer and autumn. In winter, Lake Børsesjø is completely frozen over and relatively quiet.

Directions: From E18 in Telemark, exit onto Rv36 signposted to Porsgrunn and Skien. Pass through Porsgrunn and continue to Skien, and exit onto Rv32 signposted to Siljan. Continue about 2 km on this road.

Tactics: To get to the **a.** northern tower hide continue straight past Gjerpen church, and after 400 m, turn right onto the road Gamle Siljanveg and park near the end of the road. Here, an old barn is converted into Børsesjø Wetland Centre. From here, walk the path 200 m to the tower hide, signposted to *Fugleamfi*. To reach **b.** the southern tower hide, you can either walk 15 minutes south along a path, or you can drive back to Gjerpen church and turn onto the Håvundvegen road, and make a left into the parking area signposted to *Fugletårn*. By the kindergarten, follow a path downhill through the woods and via the boardwalk through the reedbed to **a.** the southern tower hide.

Osprey breeds at several places along the Skagerrak coast and also inland. They can often fly far from the nest in order to hunt for prey. *Photo: Tomas Aarvak.*

Bjellandstranda beach at Tromøya island is not a large area, but is perhaps the site in the southern part of this region that best combines the potential for great species diversity with easy access. *Photo: Bjørn Olav Tveit.*

To explore the farmlands of Gjerpendalen valley south of Lake Børsesjø, continue along Håvundvegen road, which runs along the west side of the valley. Several side roads lead across the valley. The first one is Sneltvedtvegen which passes **c.** the southern part of the reeds surrounding Børsesjø, while the next road is Løbergvegen which leads to the fields and the raptor-viewpoint at **d.** Storemyr farmlands.

14. TROMØYA ISLAND

Agder County

GPS: 58.45673° N 8.87946° E (Bjellandstrand beach)

Notable species: Migrants; shorebirds; Little Grebe (winter), Wryneck, Black Woodpecker, Thrush Nightingale, Reed Warbler, Marsh Warbler and Red-backed Shrike.

Description: Tromøya island, including the adjacent and much smaller Tromlingene islet, is the best migration site along the outer

Skagerrak coast of the eastern part of Agder County. Most of the sites mentioned here are part of Raet National Park. Tromøya island is 12 km long, situated just outside the town of Arendal, connected to the mainland with a bridge. This populated island is covered with mixed forest, with some agricultural areas. Its outer coastline is littered with rounded rocks, just like at Mølen headland and Jomfruland island, clearly showing that Tromøya too is a part of the moraine ridge Raet. It is particularly the southeastern part of the island that stands out as interesting for birdwatchers. Bjellandstrand beach is a nice little spot, good for resting shorebirds and passing migrants over both land and sea.

Tromlingene islet is a 2 km long moraine island, separated from Tromøya by a narrow, shallow strait. Tromlingene has several areas of fine-grained beaches and mudflats suitable for resting shorebirds and other wetland species. Tromlingene is sparsely vegetated with grass and juniper scrub and, like Mølen headland, it has several iron-age burial mounds made out of rounded rocks. Whereas the areas of southern Tromøya are relatively large, the habitats and birds of Tromlingene are more concentrated.

Best season: All year, but particularly during migration. The area is usually crowded during summer holidays.

Directions: Leave the E18 at Exit 71 onto Rv410 signposted to Arendal. In Arendal, continue on Rv409 signposted to Tromøy. The southern part of Tromøya island is reached by going straight ahead after the Tromøy bridge and then making a left after another 1.5 km onto the road signposted to Hove. Thereafter, follow signs to Hoveodden and park by the barrier. Tromlingene is reached by private or taxi boat from Arendal.

Tactics: On Tromøya island, the area around Hove has nice farmland and paddocks. Outside the tourist season you can walk the sea-side lawns and beaches on Hoveodden peninsula. The woods surrounding the former campsite can also host interesting migrants. If you follow the path along the ocean, you will eventually reach Såda, the local bird association's club house. It is situated upon a rock with a fantastic view to the sea – the perfect viewpoint for watching seabird migration.

If you drive north towards Tromøy church you get access to the sea several places, such as at Spornesstranda beach just north of Hove. Here, Thrush Nightingale and Red-backed Shrike breed. Also check the fields and the little Skottjenn pond opposite the church. Just north of the church, exit onto the side road towards Bjelland and park after 700 m near *Bjellandstrand gård* (farm). Just beyond this farm, a gravel road leads between the houses on your right and down to Bjellandstrand beach. Walk carefully the last stretch before the seashore, so you do not flush the gulls and Cormorants often resting here. Continue to the end of

this side road and walk out onto Botstangen headland, from where you can overlook Botsfjord, the bay south of Tromlingene islet. Ducks, divers and grebes may rest here.

Out on Tromlingene islet, check every inlet, marsh and scrub for migrants. The middle and southwestern parts are usually the best. The summer tourists fancy the northern part of the island.

15. Grimstad area

Agder County
GPS: 58.40505° N 8.70394° Ø (Saulekilen inlet), 58.39255° N 8.72275° Ø (Hasseltangen headland), 58.38139° N 8.64085° Ø (Lake Temse tower hide), 58.32701° N 8.47436° Ø (Lake Reddalsvannet)

Notable species: Wetland birds, migrants; Wryneck, Black Woodpecker, Thrush Nightingale, Reed Warbler, Marsh Warbler, Icterine Warbler, Wood Warbler, Red-backed Shrike, Common Rosefinch and Hawfinch.

Description: The area surrounding Grimstad town has several interesting wetlands in a

Red-backed Shrike thrives in dry bushland and sunny slopes. It has a habit of sitting well exposed on top of bushes. During the breeding season, however, it takes cover and can be difficult to find. It spends the winter south of the Sahara, and like many other Norwegian migrants to Africa, the population has declined sharply in recent years. *Photo: Gunnar Numme.*

Oslo

Interior

Skagerrak

Western

Central

Northern

Finnmark

Svalbard

Lesser Black-backed Gull is common along the Skagerrak coast in summer, from March to September, but it migrates south in winter. The breeding subspecies is *intermedius. Photo: Terje Kolaas.*

generally fertile and bird-rich farmland district, good for geese, ducks and a variety of passerines. This whole area is good for nocturnal songbird trips. Central to the area are Sømskilen and Ruakerkilen inlets, consisting of several shallow coves shielded from the sea. Here there are good conditions for shorebirds, geese and ducks. The beech forests are good for woodpeckers and warblers, as well as for resting passerines during migration. The Hasseltangen headland is a good place to look for salt-water species. Little Grebe regularly winter in the area. Gjervoldsøy island also sport nice, moist fields, a place that became famous in 2019 for turning up Norway's first record of Grey-headed Lapwing.

Lake Temse is a 1.3 km long lake, surrounded by deciduous forest and agricultural land. The lake is shallow and lined with reedbeds, making it one of the region's best places for resting ducks and other wetland birds. The lake is equipped with a tower hide.

Lake Reddalsvannet is another good option for ducks and wetland species. It is located in a hilly area with alternating forest, cultivated land and a number of other lakes. One of these, Lake Landvikvannet, has a particularly

interesting narrow and reed-covered section, Inntjorkilen inlet. Kingfisher is sometimes seen here, and Red-backed Shrike are among the breeding birds.

Best season: All year, particularly during migration. In summer, leisure boat traffic makes life challenging for both birds and birdwatchers along the coast.

Directions: The area described is located around Grimstad, between Arendal and Lillesand towns, along the E18.

Tactics: <u>Søm farmlands</u>: Leave the E18 at either exit 76 (southbound) or exit 77 (northbound) and follow Rv420 signposted to Fevik. Just north of Fevik, turn east and follow the road signs to Søm. Scan the fields and reed-covered ditches on the way out to Søm. The fields along <u>Vessøyveien road</u> are good for shorebirds. In several places there are detours to the left that take you out to places with a view of <u>Sømskilen inlet</u>. In the left turn 400 m before you reach the harbour at Hasseltangen headland, a path goes into the forest on the right. This path takes you the 300 m away to the west side of Ruakerkilen inlet. Drive on and park as far out on Hasseltangen headland as possible and walk the last bit out. On nocturnal songbird trips, the wetlands along the banks of the Nidelva river in Arendal and out towards Hisøy may be worth checking. Much the same applies to Gjervoldsøy island. In order to get here, follow Rv420 instead north towards Arendal, and turn onto the road signposted to Gjervoldsøy.

Lake Temse is located along Rv407 between Arendal and Grimstad. Take the exit from the E18 at junction 75 (from the north) or 77 (from the south), in both cases signposted to Rv407 and Rykene. The tower hide is located along this road, at the entrance to Kleppekjærveien road. At Lake Temse, the tower hide is a good starting point, scanning the lake with your telescope. You may then walk the path that runs along the entire south-eastern bank. There is also a nice viewpoint from the slip road along the northern side of the lake.

Lake Reddalsvannet is reached by leaving the E18 just south of Grimstad, at exit 81 (southbound) or 82 (northbound) to Rv420 and follow the signs for Reddal. After the crossing at Landvik church, you will immediately see Lake Landvikvannet on your left, then an exit to Inntjore – which takes you the 2 km to Inntjorkilen inlet. If you instead continue straight ahead on Reddalsveien road, you will soon see Lake Reddalsvannet on your left. 1.3 km after the exit to Inntjore you can stop and view the northern and often most bird-rich corner of the lake. Continue on and check out the fine Reddal farmlands.

16. KRISTIANSAND AREA

Agder County

GPS: 58.07140° N 8.00228° E (Flekkerøya island), 58.10956° N 8.15158° E (Sodefjed seawatch), 58.19364° N 8.07901° E (Hamresanden beach)

Notable species: Migrants; Capercaillie, Nightjar, Wryneck, White-backed Woodpecker, Icterine Warbler, Wood Warbler and Hawfinch.

Description: Kristiansand, with its 117,000 inhabitants, is by far the largest town along the Norwegian Skagerrak coast. It is also a popular holiday destination. For these reasons, it is quite likely that you will end up in or near Kristiansand during your stay in Norway.

The woodlands on the outskirts of the city are particularly good for Wryneck and Hawfinch, and the surrounding forests have strong populations of Capercaillie. Although White-backed Woodpecker has become almost extinct in south-eastern Norway, it is still possible to find it in the woodlands from Kristiansand and southwards. The coast outside Kristiansand is dotted with small islands, making the coastline untidy with little or no concentrating effect on the migrating land birds that undoubtedly pass through the area in substantial numbers. There are also few possibilities for reaching the outer coast without a private boat, so seawatching is restricted. Flekkerøya island is your best bet for both grounded migrants and passing seabirds. The latter can also be appreciated from Sodefjed. In winter and during migration, the mouth of Topdalselva river in Topdalsfjorden can hold some staging shorebirds and ducks, and so can Kjosbukta inlet along the road to Flekkerøya island. The agricultural fields in Søgne village just south of Kristiansand can also be a good spot for migrants during spring and autumn. In early summer the same district can host a diversity of nocturnal songbirds, including Nightjar.

Other wildlife: Elk and beaver are very often seen at dusk and dawn outside the city, particularly along Topdalselva river.

Best season: Migration periods spring and autumn. Summer can be very crowded with holidaymakers.

Directions: E18 leads south to Kristiansand, and E39 leads to the city from the south. Kristiansand airport Kjevik is situated near the mouth of Topdalselva river.

Tactics: The Kristiansand area is best explored by car.

Sodefjed seawatch: From Kristiansand,

White-backed Woodpecker is an uncommon inhabitant of the woodlands along the southern Skagerrak coast. *Photo: Kjetil Schjølberg.*

follow route E18 east towards Oslo. Just after Varoddbroa bridge, exit onto Rv401 signposted to Høvåg. After 3 km on Rv401, exit towards Kongshavn. The woods you pass through here are good for Wryneck and Icterine Warbler. In Kongshavn, make a left towards Stangenes and follow this road 2.3 km before turning right to *Sodefjed hytteområde (cabins)*. Park by the road barrier and walk on for 500 m. Make a right just after the tennis court, keep walking for 20 m and leave the road for a indistinct path leading up to a rock which provides a good seaview.

Topdalselva river: Back on E18 heading towards Oslo, exit onto Rv41 signposted to Kjevik and Birkeland. Do not despair if your travel mates insist on stopping at the nearby Sørlandssenteret shopping mall, because the woodlands to the south and east of the

mall is a favoured site for White-backed Woodpecker. Back onto Rv41, after 3 km, exit to Hamresanden Camping and walk the footpaths along the river downstream to the river mouth. If you drive further along Rv42 towards Birkeland you can stop several places to overlook the river. The woodlands in this area is great for Capercaillie, Wryneck, White-backed Woodpecker and other woodland species.

Flekkerøya island: From Kristiansand, follow route E39 towards Stavanger. Just outside of town, exit onto Rv456 signposted to Vågsbygd. After 2.5 km you see Kjosbukta inlet on your left. You can scan it from behind the first bus stop and from opposite the petrol station. Continue straight ahead onto Rv457 signposted to Flekkerøya. At the roundabout on Flekkerøya island, just after the tunnel, you may turn right and then immediately to the left onto Mæbøveien road. Keep on to the end and walk *Kjærlighetsstien* (The love path!) along the coast.

Søgne farmlands: Back at Kjosbukta inlet, you could have chosen to make a turn onto Rv456 signposted to Søgne. This road takes you through Nightjar territory, to Søgne village, where you bird roadside along the fields down towards Høllen harbour. A quicker route directly to Søgne from Kristiansand, is by following E39 a bit further towards Stavanger, and then exit to Søgne.

17. LINDESNES PENINSULA

Agder County
GPS: 57.98249° N 7.04667° E (Lindesnes lighthouse)

Notable species: Migrants; Grey-headed Woodpecker, White-backed Woodpecker, Lesser Spotted Woodpecker, Marsh Tit and Hawfinch.

Description: Lindesnes peninsula is the southernmost point of mainland Norway. Although, after the Spangereid channel opened in 2007, the peninsula was technically transformed into an island. Nevertheless, Lindesnes is still an easily accessible site positioned where birds from both the

Skagerrak and the North Sea coasts meet before heading out across the sea en route to Denmark and the European continent. The concentration of birds is not as notable as the location might indicate, however, probably due to the somewhat untidy coastline with a myriad of small islands, unlike, for instance, Falsterbo in Sweden and Skagen in Denmark. Lindesnes lighthouse on the outermost tip is the strategic vantage point for observing the migration of land birds as well as seabirds. The vegetation by the lighthouse is sparse, with only a few scattered bushes. A bit further inland, migrants are offered more shelter. The remainder of Lindesnes peninsula is dominated by rather barren heathland. Spangereid village 8 km north of the lighthouse, however, has a variety of attractive habitats for resting migrants, including agricultural fields, woodland, gardens and sandy beaches with washed-up seaweed.

Other wildlife: Harbour porpoise and harbour seal are often seen in this area.

Best season: Migration periods, particularly in autumn. In summer, seawatching can be good during strong winds from between south and west.

Directions: Exit from E39 in Vigeland, midway between Mandal and Lyngdal, onto Rv460 signposted to Lindesnes.

Tactics: There is usually no need to rush to get all the way out to the lighthouse. Instead, use your time checking the mouth of the Audna river and the agricultural fields and other interesting habitat along the way. Kjerkevågen bay just southwest of the Spangereid channel often supports Slavonian Grebe, divers and ducks in winter. The beach here is also worth checking thoroughly, particularly in late autumn. The slopes covered with deciduous woodland are home to Grey-headed, White-backed and Lesser Spotted Woodpeckers and more. Look for resting migrants near the tip before finally entering the lighthouse area. The bunkers in front of the lighthouse make good viewpoints.

Oslo Interior **Skagerrak** Western Central Northern Finnmark Svalbard

Kjerkevågen bay at the base of the Lindesnes peninsula has exciting habitats to offer, not least in late autumn. Behind the conifer plantation are nice agricultural areas. *Photo: Bjørn Olav Tveit.*

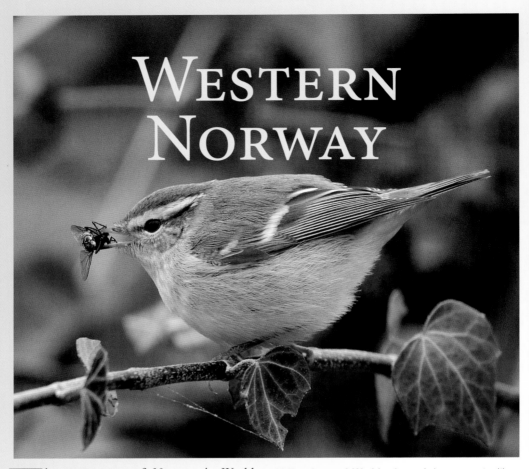

WESTERN NORWAY

The west coast of Norway is World-famous for its narrow fjords, spectacular mountains and gleaming glaciers. These areas are home to a few highly specialized species including Golden Eagle, Gyrfalcon, White-backed and Grey-headed Woodpeckers, Icterine Warbler and Nutcracker. To many birdwatchers, the west coast is particularly interesting also because of its qualities of giving the opportunity to encounter a variety of migratory species. Many of these, both seabirds and land birds, migrate along the western coast of Norway. Particularly good bird areas are the two contrasting fringes of flat and fertile land along the coast, Lista and Jæren in the south. Here you find a combination of coastal wetlands, cultivated land and other attractive habitats supporting a large number of resting birds in the migration seasons. Due to the proximity to the sea, these areas are relatively mild in winter and the

Yellow-browed Warbler is an Asian vagrant with a special status in Norway. Not only is it seen here every autumn in increasing numbers, but the species was also the driving force that in its time led to the discovery of Utsira island, one of the country's most fabled birdwatching sites. *Photo: Bjørn Fuldseth.*

lakes and rivers seldom freeze completely. This enables these areas to support substantial numbers of wintering birds as well. Also, these are the parts of Norway where summer migrants arrive earliest in spring, often two–tree weeks before they show up around Oslo. Lista and Jæren are also important for breeding birds, with species such as Gadwall, Shoveler, Pochard, Spotted Crake, Corncrake, Black-tailed Godwit and Grasshopper Warbler all having their Norwegian strongholds here. The west coast in general is important for breeding seabirds as well, with many colonies of gulls, terns, auks, Shag and Common Eider. Most famous is the easily accessible Runde

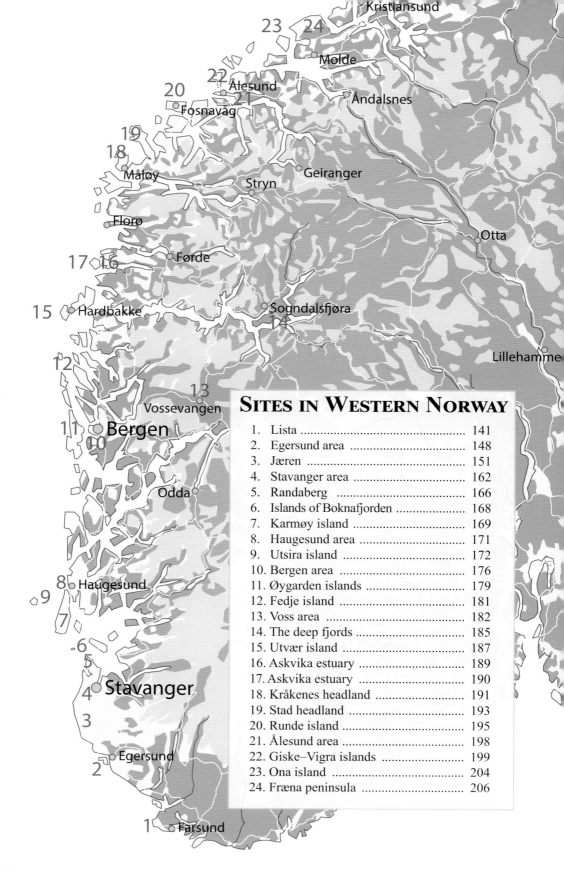

Kristiansund

23 24

Molde

22 Ålesund
20
Fosnavåg 21   Åndalsnes

19
18
Måløy   Stryn   Geiranger

Florø

Otta

17 16   Førde

15 Hardbakke   Sogndalsfjøra
14

Lillehamme

12

13
Vossevangen

11 Bergen
10

Odda

8 Haugesund
9

7

6
5

4 Stavanger

3

Egersund
2

1 Farsund

Lista lighthouse as it must appear for a migrant bird arriving from the sea. The "lebelte" forest plantation behind the lighthouse, the Voien inlet to the right, the shoreline, pastures and marshland, all combine to make one of the best birdwatching sites in Norway. *Photo: Magne Klann.*

island, with its precipitous cliffs containing thousands of seabirds including Puffins, Guillemots, Razorbills, Kittiwakes, Fulmars, Gannets and Great Skuas.

Off-shore islands are notoriously unpredictable during migration but at the same time having the potential of turning up vagrants from North America, Asia or Southern Europe. In Norway, Utsira is the most famous of these islands, but any island along the outer coast can exhibit much of the same qualities, and can even surpass Utsira in terms of avian results if the weather is favourable and the season is right. If choosing other islands than Utsira, you will more often have the advantage of finding your own birds.

Rogaland County, containing both Jæren and Utsira, two of the prime sites in Norway, is by far the richest bird county in the country, with more than 430 species recorded over the years.

THE BEST SITES

Two of the best birdwatching areas in the entire country are Jæren and Lista, both being worth visiting at all times of year. They are both vast, Jæren in particular, and contain a number of sub-sites. Runde island is an easily accessible and truly spectacular site for breeding seabirds. The remote island of Utsira is an international hotspot for rare vagrants, but the outcome during a short visit is highly unpredictable. Islands like Utsira, Kvitsøy, Fedje, Bulandet & Værlandet and Ona can only be expected to be rewarding in late spring, late summer (passing seabirds) and in autumn. The areas Giske/Vigra and Fræna at Møre and Herdla near Bergen are also sites of national interest, well worth visiting year-round. The breeding specialities of the famous Norwegian fjords can be found with some effort many places, but the terrain is steep and access often difficult. In this guide, some of the most reliable sites for woodpeckers and others near or along the way to the most popular tourist destinations are described.

1. LISTA

Agder County

GPS: 58.10905° N 6.56914° E (Lista lighthouse), 58.07154° N 6.64734° E (Fuglevika inlet), 58.07065° N 6.69004° E (Kviljoodden headland parking)

Notable species: *All year:* Shag, Peregrine Falcon, Water Rail, Tawny Owl, Collared Dove, Grey-headed Woodpecker, White-backed Woodpecker, Lesser Spotted Woodpecker, Marsh Tit, Bearded Tit, Carrion Crow, Goldfinch, Twite, (Lesser) Redpoll and Hawfinch.

Spring and summer: Brent Goose (end of May), Gadwall, Garganey and Shoveler (Apr–Jun); Quail (May–Aug), White-billed Diver (Apr–May), Manx Shearwater (May–Sep), Marsh Harrier (Apr–Sep), Spotted Crake (May–Jun), Crane (Mar–May), Dotterel (May), Pomarine Skua (May), Great Skua (May–Oct), Sandwich Tern (Apr–Aug), Turtle Dove (May–Jun), Short-eared Owl (Apr–May), Wryneck (Apr/May–Jul), Yellow Wagtail *flava* and *flavissima* (Apr/May–Aug), Black Redstart (Mar–May), Stonechat (Mar–Apr), Grasshopper Warbler (Apr/May–Aug), Marsh Warbler (May/

Jun–Aug); Sedge Warbler, Reed Warbler and Icterine Warbler (May–Aug); Wood Warbler (Apr/May–Aug), Red-backed Shrike (May–Aug), Linnet (Mar/Apr–Nov) and Common Rosefinch (May–Aug).

Autumn and winter: resting geese incl. Tundra and Taiga Bean Geese and White-fronted Goose (Nov–Mar), migrating Brent Goose (*hrota* in Sep and *bernicla* in Oct); Shoveler (Sep–Oct), Scaup (Oct–Mar), Pochard (Sep–Mar), Smew (Oct–Apr), Great Northern Diver (Oct–May), White-billed Diver (Oct); Slavonian Grebe, Red-necked Grebe and Little Grebe (Sep–Apr); Sooty Shearwater (Aug–Oct), Honey Buzzard (Aug–Sept), Hen Harrier (Sep–Apr), Gyrfalcon (Oct–Mar), Coot, Crane (Aug–Sep), Black-tailed Godwit (Aug–Sep), Jack Snipe (Sep–Nov), Glaucous Gull, Greenland Gull (Dec–Mar), Turtle Dove (Sep–Oct), Tengmalm's Owl (Sep–Oct), Short-eared Owl (Sep–Nov), Shore Lark (Sep–Apr), Richard's Pipit (Sep–Nov), Red-throated Pipit (Sep), Grey Wagtail, Black Redstart (Oct–Nov), Stonechat (Sep–Nov), Barred Warbler (Aug–Sep), Yellow-browed Warbler (Sep–Oct), Great Grey Shrike (Sep–Apr) and Little Bunting (Oct).

Description: Lista is a flat moraine plain on the outer coast just west of the southernmost point of Norway, providing a multitude of bird-friendly habitats in contrast to the hilly, forested coastline and interior surrounding

Great Northern Diver winters scattered along the Lista coast. When the sea is calm and the birds are easy to spot, it is not unusual to see dozens of individuals on the stretch between Lista lighthouse and Lomsesanden beach. *Photo: Terje Kolaas.*

141

it. At Lista you find long sandy beaches lined with beach-grass covered dunes, fertile farmland and large wetlands, together creating what is one of the country's very best birding areas at all times of year. The physical similarities to Jæren in Rogaland a bit further north are obvious, although Lista, covering 4x16 km, is just 1/10 of Jæren. This means that you get many of the same species in these two areas, but at Lista they are more concentrated. About 380 species of bird has been recorded in the area, most of the within a one-kilometres radius from the lighthouse.

Lista comprises a number of smaller sites, one of the most important being Lake Slevdalsvannet, a 4.5 km² waterbody almost completely overgrown with reeds. A wetland restoring program has been conducted in recent years, opening up segments of the lake and erecting a couple of tower hides. The lakes Nesheimvannet, Hanangervannet and Kråkenesvannet have more open water naturally, and are surrounded by meadows and farmland providing excellent conditions for a variety of waterbirds. The position of Lista near the southern tip of the country also yields diurnal migration both over land and sea, often best appreciated at the lighthouse, where the bird observatory (founded in 1989) is located, or at Kviljoodden headland.

The list of breeding birds at Lista includes Marsh Harrier, Wryneck, Wood Warbler, Red-backed Shrike, Twite, (Lesser) Redpoll, Common Rosefinch and Bearded Tit. It is also the most important breeding area in the country for the few remaining pairs of Yellow Wagtail of the two subspecies *flava* and *flavissima*. Grey-headed Yellow Wagtail (*thunbergi*) is the commoner subspecies on migration, particularly in May. The combination of farmland and wetlands make perfect conditions also for nocturnal songbirds, and you can be pretty sure to hear Quail, three species of rail, Grasshopper Warbler and all three common *Acrocephalus* warblers during a calm and warm early summer night. On the wooden slopes inland from Lista, Grey-headed and White-backed Woodpeckers, Wood Warbler and Hawfinch are all regular breeders.

Many species winter at Lista, including several species of geese, ducks, divers, grebes, gulls and passerines, and a few raptors and shorebirds. Uncommon species like Scaup, Pochard, Smew, Great Northern Diver, Little Grebe, Hen Harrier, Gyrfalcon, Peregrine Falcon and Great Grey Shrike are regular from late autumn through winter.

Being a prime site for regular Scandinavian migrants, Lista is also a real rarity magnet with many first's for Norway. Several of them are eyebrow-raising even in a Western European context, like Greater Sand Plover, Rufous Turtle Dove, White-throated Needletail, Dusky Thrush, Masked Shrike and White Wagtail of the Asian subspecies *personata* and *leucopsis*. These, of course, come in addition to an array of more obvious ones. Annual, or nearly so, among the national rarities are (main seasons in parentheses): Little Egret (May–Jun), Great Egret (Apr–May), Black Kite (May and Sep), Red Kite (Mar–May), Pallid Harrier (Apr/May and Sep), Montagu's Harrier (May), Red-footed Falcon (May/Jun and Sep), Pectoral Sandpiper (Jun–Sep), European Bee-eater (May–Jun), Short-toed Lark (May), Red-rumped Swalow (Apr–May), Water Pipit (Nov–Dec), Citrine Wagtail (May and Sep–Oct), Blyth's Reed Warbler (Aug–Sep), Rose-coloured Starling (Aug–Nov) and Corn Bunting (May–Jul).

Other wildlife: Harbour porpoise is often seen during seawatching, and sometimes bottlenose dolphin, killer whale and other whales. Flocks of long-finned pilot whales can be seen in late summer. Both harbour and grey seals are regular along the shores, and so is weasel and introduced American mink. Red fox, badger, mountain hare, beaver, roe deer and a few species of bat are among the animals commonly encountered at Lista. Inland you can add marten, elk and red deer to the list.

Best season: All year, but with the greatest diversity in late spring and from late summer throughout autumn. Migration at Lista seems to be continuous from the first signs of spring in early February, through the summer season

Corn Bunting had a small breeding population on Lista and Jæren until the end of the 19th century. It has made nesting attempts again here in recent years. Look and listen for it in the agricultural landscape of these areas. *Photo: Terje Kolaas.*

Lista, both at land and sea. Here is also a seawatching wind shelter. A telescope is a must. To look for resting migrants, check the lighthouse gardens and adjacent meadows and bushes, but be aware that the bird observatory conducts ringing here. Particularly interesting is Voien, the lagoon just below the lighthouse, often holding roosting ducks, shorebirds and gulls. Do not approach the lagoon too closely in order not to spook the birds away, and particularly not in late spring and summer because of the breeding waterbirds and Yellow Wagtails. Instead, scan Voien either from the slope beneath the lighthouse, from the main road or from **d.** the tractor track on the east side of the lagoon. From the latter, you can also scan the bay Vågsvollvika. To explore the area further, you can walk the dirt road through the conifer plantation and out onto **b.** Steinodden headland to check the bays Sevika and Verevågen (the latter alternatively reached by car, see below), and along the fields and the wet, reed-covered ditches at **c.** Gunnarsmyra. Note: walking on the farmland or driving the dirt roads here is prohibited.

You can then take the car and drive to the other sites as follows:

e. <u>Lake Slevdalsvannet</u>. Drive into Borhaug village and make a left onto Rv463, signposted to Ore. Then turn right into Villaveien. After 400 m, turn left between the houses and park the car in the car park here. Bring your telescope and walk the tractor road down-hill towards Slevdalsvannet. After 800 m you reach a large tower hide with a view of the recently rehabilitated part of Slevdalsvannet, with an open water surface and salt meadows, a fine spot for ducks, *Tringa*-waders and other waterbirds. Marsh Harrier is a common sight here in summer, and Hen Harrier – and perhaps other harrier species – often come in to roost in the reed forest during the migration periods. Listen and look for rails, warblers and Bearded Tit. Follow the tractor road a bit further, and turn left through a gate, to get a view of the open water surface from a different angle. You can also walk the tractor road along the reeds in the opposite direction

with its shorebirds and seabirds, until autumn migration ends in late November or even early December. Late spring and early summer are the best time for nocturnal songbirds. A significant movement of raptors takes place from mid-August to mid-October. Finches, tits and woodpeckers migrate through in September and October. Substantial numbers of waterbirds, raptors and passerines winter here.

Directions: *From the east (Oslo/Kristiansand)*, leave route E39 by Lyngdal and continue on Rv43 signposted to Farsund, and later, to *Lista fyr* (lighthouse). This road leads through Farsund town and Vanse village, all the way to the lighthouse. On the way out, from Farsund and on, you pass several premium sites – see the maps.
From the west (Stavanger), leave E39 at the junction 2 km past the bridge crossing Fedafjorden, and continue on Rv465 signposted to Farsund.

Tactics: a. Lista lighthouse is a perfect vantage point for experiencing the migration past

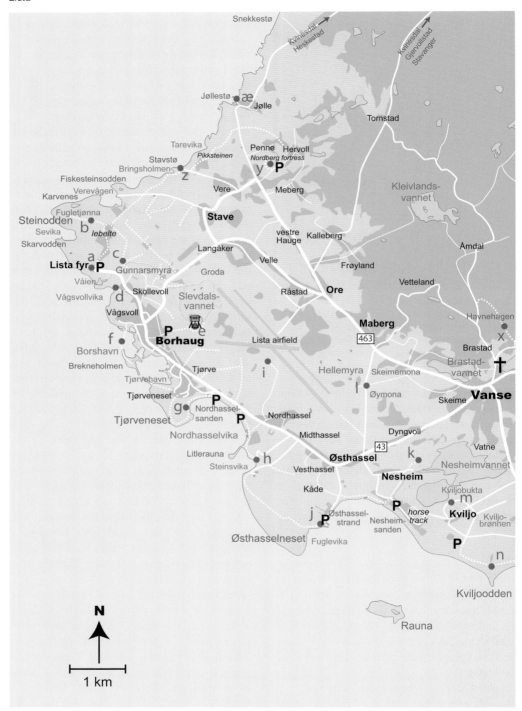

Lista, western half: a. Lista lighthouse and Bird Observatory; **b.** Steinodden headland; **c.** Gunnarsmyra fields; **d.** Våien inlet; **e.** Lake Slevdalsvannet; **f.** Borshavn harbour; **g.** Tjørveneset headland; **h.** Steinsvika inlet; **i.** Nordhassel fields; **j.** Fuglevika inlet; **k.** Lake Nesheimvannet north; **l.** Hellemyra fields; **m.** Lake Nesheimvannet south (a.k.a. Kviljobukta); **n.** Kviljo headland; **x.** Havnehagen pastures/forest; **y.** Nordberg fortress; **z.** Stavstø harbour; **æ.** Jøllestø harbour. See next page for map of eastern half for the other reference points.

from the tower hide, in order to view Lake Slevdalsvannet from the north, also providing some elevation, making it a good look-out point for raptor migration.

f. <u>Borshavn harbour</u>. Drive back into Borhaug and down to Borshavn harbour. Look for gulls, Little Grebe, ducks and more from the breakwater in winter. You can do the same from the docks in <u>Tjørvehavn harbour</u>, the eastern part of the same harbour, where you in addition may encounter a few shorebirds. In Borhaug village you find *Natur og Fritid* (<u>www.naturogfritid.no</u>), which is the only dedicated birdwatcher's shop in the country, owned by BirdLife Norway, selling optical equipment, books and more.

g. <u>Tjørveneset headland</u>. Follow Rv43 just out of Borhaug and park in the road-side parking area by the beach. Walk out onto Tjørveneset headland, an interesting shorebird site. Also scan <u>Nordhasselvika bay</u> for ducks, divers and grebes. You may walk along the beach, or you can drive to the far end, where you should also check **h.** <u>Steinsvika inlet</u>. Scan thoroughly all the fields you pass for shorebirds, gulls and passerines, hunting harriers and more.

i. <u>Nordhassel fields</u>. Walk the dirt track from the small Nordhassel village towards the old airfield. Here, Dotterel is often found in May, Crane during summer, and raptors, swans and geese in winter. The area is also good for passerines, often with substantial flocks

of Skylark and Yellowhammer in the cold season. Among them, you may find uncommon species like Woodlark and Lapland Bunting.

j. <u>Fuglevika inlet</u>. Leave Rv43 onto the road signposted to Østhasselstrand. The fields around this junction may hold interesting birds. Collared Doves can be found near the houses and single Turtle Doves are sometimes found among them. Drive down to the coast to scan <u>Nesheimsanden beach</u> and <u>Østhasselvika bay</u> with the telescope. Drive all the way to the end of this road, to the gate by the small marina, park and walk through the gate to explore Fuglevika inlet and surrounding fields, clusters of trees and the shoreline. From the parking, you can see the low gravel island Rauna in the distance. This is an important breeding site for Common Eider, Cormorant and Lesser Black-backed Gull, and Sandwich Tern has bred here as well.

k. <u>Lake Nesheimvannet</u>. Continue on Rv43 a bit further east, exit onto the road signposted to Nesheim. Park where possible, and walk a bit further to obtain views of Lake Nesheimvannet from the west. You should also continue down to Nesheimsanden beach by making a right before the farm, signposted to *Nesheim travbane* (horse track). The area around the mouth of <u>Nesheimbekken</u> creek is a good site for shorebirds, gulls and pipits,

Lake Slevdalsvannet is large and almost completely covered in reeds. In recent years, a restauration project has been conducted, opening up the lake in a few places, such as here in the southern corner. The main tower hide is visible in the distance. *Photo: Bjørn Olav Tveit.*

Oslo Interior Skagerrak **Western** Central Northern Finnmark Svalbard

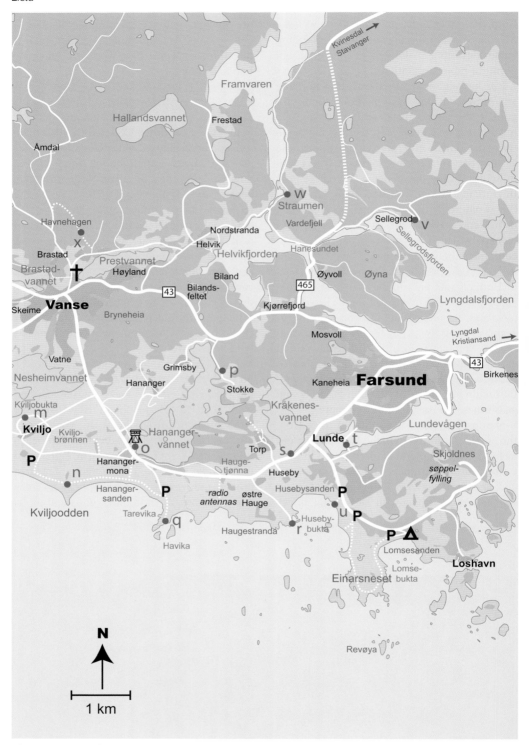

Lista, eastern half: m. Kviljobukta (Lake Nesheimvatnet); **n.** Kviljo headland; **o.** Lake Hanangervannet south;
p. Lake Hanangervannet north; **q.** Havika; **r.** Haugestranda harbour; **s.** Lake Kråkenesvannet; **t.** Lundevågen inlet;
u. Husebysanden beach; **v.** Sellegrod forest; **w.** Straumen sound; **x.** Havnehagen. See also map on previous page.

Bewick's Swan is traditionally a rare winter visitor in Norway, mainly on Lista and Jæren, often singly or in small groups mixed in with flocks of Whooper Swan. In recent years, however, Bewick's Swan has become increasingly rare as a winter visitor. Your best bet to find it nowadays is during the migration periods, in March and November. *Photo: Terje Kolaas.*

and the seaside cluster of trees may hold interesting passerines during migration and in late autumn and winter.

l. Hellemyra fields has many of the same qualities as mentioned for the Nordhassel fields above. In addition, *flava* Yellow Wagtail and Red-backed Shrike breed here, and it is a good viewpoint for raptor migration in autumn.

m. Lake Kviljobukta. Continue on Rv43 into Vanse village centre and turn right onto the road towards Lunde. After almost 2 km, exit to the right, signposted to Kviljo. This gravel road, with side roads, leads through interesting fields, often yielding Stonechat. Make a right again after 1 km, and you will soon get a view to Lake Nesheimvannet's southern section, called Kviljobukta. Be particularly careful not to block traffic in this area.

n. Kviljo area. Follow the gravel road signposted as a bicycle route east from Kviljo farm. The potato fields on the south side of the farmhouses here often holding interesting birds, sometimes including Dotterel. Collared Dove is common here. Drive down towards the beach and park by the information poster. From here, a path leads through the dunes east to Kviljoodden headland, one of the best shorebird sites at Lista. There is no need to

walk all the way out onto the headland – you only risk flushing the birds. Instead, sit down upon the dunes and scan with the telescope from there. The shorebirds are replaced more frequently than it may seem at a glance, as small groups come and leave throughout the day. On weekends, hikers and dog-walkers often spook the birds, making very early morning the best time of day for visiting here. From the dunes, scan the sea and the fields in-land as well, as this is a good general migration watch point. From Kviljooodden you may walk further east along to Hanangersanden beach.

o. Lake Hanangervannet. This lake is best viewed from various points along the roads surrounding it. The tower hide in the south end is of minimal value. The gravel road along the west side of the lake is signposted to Grimsby. There is a nice little reedy swamp in the north-end of the lake, from where you can make a right turn and continue to **p.** a viewpoint overlooking the northern section of Lake Hanangervannet.

From the south end of Hanangervannet, continue east on the main road towards Lunde. Side roads head off towards the sea to your right, both at **q.** Havika inlet (with Tarevika inlet) and **r.** Haugestranda harbour. These are good for sea ducks, divers and grebes, with a potential for resting shorebirds, passerines and more. Stop by the small, nutrient-rich **r.** Haugetjønna pond to your left. Shoveler and Garganey have bred here in the past, but the pond seem to have lost much of its value to birds in recent years. You may still have a look, and park here in order to walk down past the tall radio antennas to the central section of Haugestranda.

s. Lake Kråkenesvannet is one of the best lakes of Lista, usually holding a good selection of ducks in winter, often including Smew. The best viewpoint is from a designated, lake-side lay-by along the main road.

t. Lundevågen inlet is the easternmost

shorebird site at Lista, particularly good for *Tringa*-species, and for ducks and herons. Leave the main road and turn onto the road signposted to Loshavn. Turn left two times to view the inlet from the causeway crossing the bay.

The road towards Loshavn provides access to the beach several places, such as at **u.** Husebysanden, from where you have a nice view to the many sea ducks and divers that exploit the area all year, being particularely numerous in winter. As opposed to the coast further west at Lista, this eastern coastline is speckled with small islands and rocks providing sheltered conditions for the seabirds. The area around Lomsesanden Camping is crowded by people during summer but can attract birds at other times of year. You may park here and walk the footpaths south onto the headland, in order to look for migrants and to get a view of the ducks, grebes and divers often is found along the coast.

v. Sellegrod and the wooded slopes inland. Drive to Kjørrefjord, along Rv43 west of Farsund. Turn north onto Rv465 signposted to Åpta and Kvinesdal, and follow this road 2 km. Make a right just before the Ravnehei tunnel, onto a road signposted to Bjørnestad. Follow this road uphill and stop at suitable places. These wooded slopes are home to a healthy population of White-backed Woodpecker. The birds are around all year but are easiest to find when drumming on sunny mornings in early spring. The area is also good for Wryneck (late spring and summer), Grey-headed Woodpecker, Green Woodpecker, Lesser Spotted Woodpecker, Wood Warbler (late spring and summer), Long-tailed Tit, Marsh Tit and Hawfinch.

w. Straumen tidal current. The narrow sound leading in to the Framvaren fjord produce a strong tidal current preventing it from freezing during winter. If the lakes at Lista are frozen during spells of extreme cold, Straumen can prove very attractive to the wintering waterbirds. The species mentioned for Sellegrod can also be found in the wooded slopes surrounding Straumen.

x. Havnehagen woodland. The mixed woods here support a range of breeding species in late spring and early summer, including Tawny Owl, Wryneck, Green Woodpecker, Grey-headed Woodpecker, Lesser Spotted Woodpecker, Tree Pipit, Icterine Warbler, Wood Warbler, Nuthatch, Treecreeper, Bullfinch and Hawfinch. In Vanse village centre, pull off from Rv43 signposted to Tomstad. Pass Vanse church and keep right towards Helvik in the upcoming T-junction. Pass Vanse school on your left, and continue 250 m around the bend. Here, turn left onto a gravel road and continue to the barrier. Walk along the gravel road and paths into the woods. Back in Vanse, follow Rv463, the inner road across Lista, west towards Ore. In Ore village, keep right towards Kvinesdal and Gjervollstad along interesting fields and wooded slopes. From this route, you also reach **y.** the raptor viewpoint at the hill-top bunkers at Nordberg fort, to Verevågen inlet (mentioned above) and to the small settlements **æ.** Stavstø and **æ.** Jøllestø by the Lista fjord. Stavstø, particularly the Bringsholmen inlet to the south, is very good for *Tringa*-species in late summer and for passerines in late autumn and winter. The Jølle-area is among the best for Stonechat, Red-backed Shrike and Common Rosefinch at Lista in summer. On the rocky slopes at Snekkestø north of Jølle, Ring Ouzel and Twite breed.

2. EGERSUND AREA

Rogaland County

GPS: 58.44150° N 5.88668° E (parking Eigerøy), 58.52045° N 6.01206° E (Gådå, Tengesdal)

Notable species: Migrating seabirds, resting and migrating land birds; nocturnal songbirds; Tawny Owl, Eagle Owl, Wryneck, Grey-headed Woodpecker, Green Woodpecker, White-backed Woodpecker, Lesser Spotted Woodpecker, Grey Wagtail, Dipper, Stonechat, Icterine Warbler, Wood Warbler, Long-tailed Tit, Nutcracker, Common Rosefinch and Hawfinch; wintering White-tailed Eagle, Golden Eagle, Iceland Gull and Glaucous Gull.

Description: Eigerøy lighthouse is situated on the outer coast, strategically positioned for observing migration on both land and sea at the southwestern corner of Norway. The terrain out here is rocky and rather barren but there is a small coniferous forest by the lighthouse, attracting passerines during migration, particularly on overcast days. A bird observatory is located here, ringing and counting the migrants.

Several sites in the vicinity are also worth exploring. Nordre Eigerøya has a variety of bird-friendly habitat, such as beaches, farmland and gardens. Particularly nice is Skadbergsanden beach in Lundarviga bay. Interesting terrain is also found at Svanes and Stapnes just south of Eigerøya, especially during migration. Raptor passage can be noticeable through the area, particularly in autumn.

The harbour of Egersund town has landing facilities for fishing vessels attracting a number of gulls, particularly in winter. Usually, at least one Glaucous Gull is found among them. After periods of westerly winds, Iceland Gull may turn up as well.

Several lakes, wetlands and slow-running rivers are found inland from Egersund. Slettebøvatnet near Egersund holds a selection of swans and ducks in winter, and the gulls of Egersund harbour come here to rest and clean-up after feeding. Bjerkreimselva river and Fotlandsvatnet lake in Tengesdal support waterbirds as well, the latter being particularly good also for White-tailed Eagle in winter. Kingfisher is sometimes seen in these areas. In early summer, nocturnal songbird trips can prove rewarding. The reedbeds at Gådå between Eikesvatnet and Fotlandsvatnet in Tengesdal is particularly interesting, and so is the areas around Hegrestad and Hellvik. The valleys of the area hold Wryneck, Grey-headed Woodpecker, Icterine Warbler and Wood Warbler. Further inland you may also encounter White-backed Woodpecker. A very few pairs of Stonechat are thought to breed annually in the coastal moorland of the region. Look for them along route Rv44 from Egersund to Brusand and along the old highway between Hegrestad and Ogna.

Eigerøy island near Egersund mostly consists of bare heather and pastures. Many resting passerines are concentrated in the island's few clusters of trees and bushes. *Photo: Bjørn Olav Tveit.*

Jæren is predominantly flat, with a mosaic of farmland, water and wetlands, a veritable eldorado for birds and birdwatchers. This picture shows one of Norway's perhaps most species-rich and fabled areas. In the foreground is Horpestadvatnet, which is connected to Ergavatnet on the left of the picture. Behind is the southern end of the large Orrevatnet, with Orreosen. Out by the coast, to the right of the centre of the picture, you can make out the Reef Tangen. *Photo: Bjørn Olav Tveit.*

Other wildlife: Marine mammals can be seen at Eigerøy lighthouse while seawatching. Red deer and roe deer are common, some places beaver as well.

Best season: Spring and autumn, as well as summer for seabirds and nocturnal songbirds.

Directions: From E39, exit onto Rv42 to Egersund. From Jæren you may alternatively take Rv44 south to Egersund, passing Hellvik and Hegrestad on the way.

Tactics: Eigerøy lighthouse. From Egersund follow Rv502 west onto Eigerøya island, from where you immediately turn right, signposted to *Eigerøy fyr* (lighthouse). You will soon see Storesanden beach at the head of Lundarviga inlet to your left, well worth checking. Drive on and turn right again, still signposted to Eigerøy fyr. Park at the end of the road and continue on foot the 2 km to the lighthouse. Signs ask you to please stay on the path.

Svanes and Stapnes peninsulas. Drive Rv44 south from Egersund and turn right after 6 km onto Nordre Svanesvei. The road to Stapnes (Stapnesveien) turns right 1 km further down Rv44.

Hegrestad and Hellvik. Explore the farmlands and remnants of wetlands along Rv44 about 9 and 11 km north of Egersund.

Tengesdal. Drive Rv44 north to Tengs village, 2 km from Egersund town. Exit north onto the road signposted to Tengesdal. The road leads along Lake Fotlandsvatnet, and crosses Gådå reedbeds almost 5 km after you left Rv44. Turn left just before the bridge and keep immediately to the right to enter the parking area. In addition to nocturnal songbirds, this is a good spot for migrants, such as Swallow and White Wagtail, roosting in the reeds at night. The fields here can also hold quite a few birds during migration and raptors may pass overhead.

3. JÆREN

Rogaland County

GPS: 58.83033° N 5.57507° E (Lake Harvalandsvatnet), 58.79679° N 5.62582° E (Lake Grudavatnet *east*), 58.73592° N 5.55192° E (Orreosen), 58.75472° N 5.51160° E (parking Revtangen), 58.67736° N 5.54653° E (Nærlandsstranden beach), 58.54461° N 5.68102° E (Kvassheim lighthouse), 58.58344° N 5.79503° Ø (Lassaskaret raptorwatch parking)

Notable species: Resting and migrating birds of all categories, but waterbirds in particular; vagrants.

Regular breeding birds: Gadwall, Garganey, Shoveler, Pochard, Quail, Little Grebe, Great Crested Grebe, Marsh Harrier, Water Rail, Spotted Crake, Corncrake, Moorhen, Coot, Black-tailed Godwit *limosa*, Long-eared Owl, Collared Dove, Yellow Wagtail *flava*, Pied Wagtail, Grasshopper Warbler, Sedge Warbler, Reed Warbler, Bearded Tit, Rook, Carrion Crow and (Lesser) Redpoll.

Regular on migration: Manx Shearwater, Sooty Shearwater, Brent Goose, White-billed Diver, Dotterel, Curlew Sandpiper, Pectoral and Broad-billed Sandpipers (scarce), Great Skua, Caspian Gull (scarce), Little Gull, Sandwich Tern, Black Tern, Richard's Pipit, Stonechat and Yellow-browed Warbler.

Regular wintering birds: Bewick's Swan (scarce), Tundra and Taiga Bean Geese, White-fronted Goose, Pochard, Scaup, Smew, Red-throated Diver, Black-throated Diver, Great Northern Diver, Slavonian Grebe, Red-necked Grebe, Little Grebe, Great Crested Grebe, Hen Harrier, Rough-legged Buzzard, Gyrfalcon, Glaucous Gull, Grey Wagtail, Black Redstart, Carrion Crow and Rook.

Additionally, breeding in the hills of Høg-Jæren: 'southern' Golden Plover, Dunlin, Eagle Owl, Grey-headed Woodpecker, Green Woodpecker, Lesser Spotted Woodpecker, Dipper, Stonechat, Willow Tit, Long-tailed Tit and Red-backed Shrike. *Also:* Migrating raptors; Golden and White-tailed Eagles year-round.

Description: The flat coastal lowland area Jæren is the *El Dorado* for birds and birdwatchers in Norway. It is safe to say that

Lake Grudavatnet on Jæren is one of Norway's most famous shorebird sites, responsible for a long list of rare bird records. The picture shows the viewpoint on the south side of the lake. *Photo: Bjørn Olav Tveit.*

151

Oslo

Interior

Skagerrak

Western

Central

Northern

Finnmark

Svalbard

this is the single most important and most varied area for birds in the country at all times of year. The area is vast, about 700 km², and contains a number of legendary sites such as Revtangen headland, Lake Grudavatnet, Lake Orrevatnet and Nærlandsstranden beach. But actually, any little field, ditch or pool of water at Jæren can be regarded as a birdwatching site of national significance, and both birds and birdwatchers seem to move freely from one site to the next. Hence, drawing meaningful lines between the sites at Jæren is difficult and not very fruitful. The sites near and north of the cities of Stavanger and Sandnes are regarded as parts of Jæren as well, but in this book they are treated separately for practical reasons.

Similar to Lista, Jæren is a coastal moraine plain contrasting with its hilly surroundings. Jæren is, however, ten times the size of Lista. Sandy beaches dominate the 86 km coastline from Tungenes lighthouse in the north to Brusand village in the south. The sea is shallow even at a distance from the shore, and except in the north, hardly any islands give shelter from the North Sea. Unfortunately, this feature attracts a lot of kiters and surfers as well, flushing birds resting on the sea and beaches. Luckily, this activity is now banned along most of the coastline, although this disruptive activity is still allowed in a few places.

The terrain and fine-grained soil of Jæren is very fertile, resulting in highly productive agriculture. The farmlands are marbled with ditches, streams, lakes and marshes, making terrific bird habitats, even though many marshes and coastal meadows have been drained and cultivated. Still, the combination of large areas of bird friendly habitat and the the location in the southwestern corner of Norway, is what makes Jæren important to breeding, resting and wintering birds alike. The numbers and variety of birds found here surpasses that of anywhere else in Norway. Of course, commoner species dominate the picture, and these often come in large numbers, also attracting raptors. Desirable species such as Hen Harrier, Gyrfalcon and Peregrine Falcon regularly swoop in and create panic

Corncrake is about to return to the cultural landscape, after having been almost extinct in Norway as a result of industrialized agriculture. Farmers with calling Corncrake on their fields are encouraged to wait to harvest the crops until the breeding season is over. *Photo: Espen Lie Dahl.*

and disorder in the bird flocks. There are few or no places in the country where you have a greater chance of finding the uncommon species, such as the ones mentioned in the *Notable species* section above. Jæren is the only regular breeding site in Norway for Black-tailed Godwit of the southern, nominate form *limosa*, and the most important breeding area for Corncrake, Grasshopper Warbler and several species of ducks.

Høg-Jæren, the rolling highlands bordering the flat coastal plain of Jæren, still holds a few pairs of 'southern' Golden Plover, having summer plumages with restricted black underneath. Also, a very few pairs of Dunlin resides here, as well as supporting one of the highest densities in Europe of Eagle Owl. Høg-Jæren also provides one the most exciting raptor viewpoints in the country. Unfortunately, a grand wind farm is constructed in this area.

Jæren is the number one rarity hotspot in Norway with a number of extreme records to its credit over the years, like Baikal Teal, Pied-billed Grebe (successfully interbreeding with Little Grebe!), Swinhoe's Petrel, Egyptian Vulture, Red-necked Stint, White's Thrush, Sykes's Warbler, Eastern Crowned Warbler, Blackpoll Warbler and Black-faced Bunting, to name a few. Many of the species mentioned as regularly occurring rarities at Lista (see that section), are regular in Jæren as well. Revtangen Bird Observatory was founded in 1937, making it the oldest still active bird observatory in Scandinavia.

Other wildlife: Harbour and grey seals are regular at Revtangen headland. Hedgehog, red fox, badger, roe deer and several species of bat are common.

Best season: Jæren offers high-quality birdwatching at all times of year:
Winter (November–March): Substantial numbers of wintering swans, geese, ducks, divers, grebes, raptors and gulls. Species like Bewick's Swan, both Bean Geese, White-fronted Goose, Scaup, Smew, Great Northern Diver, Hen Harrier, White-tailed Eagle and Glaucous Gull are all usually seen during a day in the field. Every year a few rarities in the likes of American Wigeon, Stellers Eider or Surf Scoter winter here.
Spring (February–May): Common spring migrants gradually show up in the fields, lakes and beaches of Jæren from the end of February, typically starting with Starlings, Lapwings, Oystercatchers and Skylarks, followed by other common Scandinavian breeding species. The chance of encountering a rarity increases in late spring. The best weather for resting migrants is overcast skies with light winds from the southeast. Under such conditions, big falls of passerines may occur in May.
Summer (May–August): The seasons overlap at Jæren, and in summer, many birds breed while some still are migrating north. Some, particularly the Arctic shorebirds, have even started on the journey south. In the main shorebird season, from May to October, rarities are regularly found, annually with more then 10 records of Pectoral Sandpiper. Jæren is an exciting area for nocturnal songbird trips and, in westerly winds, seawatching.

Autumn (August–November): In addition to shorebirds and seabirds, autumn is a time for migrating raptors at Jæren. At the lakes birds change often, and the numbers increases towards winter. In late autumn, beach walking is popular amongst birdwatchers, and a number of rare pipits, wheatears and other vagrants have been found over the years.

Directions: Jæren is just south of Stavanger city. Stavanger can be reached by plane to Sola airport, or by train from Oslo.
By car from Oslo: Follow E18 to Kristiansand and continue on E39 towards Stavanger. Exit at one of four junctions (see the maps p. 156 and p. 158):
1) to Rv42 signposted to Egersund, continuing on Rv44 towards Stavanger, entering Jæren from Brusand in the south,
2) exit from E39 to Rv504 towards Varhaug,
3) from E39 by Ålgård to Rv506 towards Bryne,
4) continue E39 to Sandnes/Stavanger and enter Jæren from the north.
From Bergen Jæren is reached by bus or boat, or by driving E39 south to Stavanger.

Tactics: Due to the distances involved, Jæren is best explored by car. Fortunately, many of the sites can be birded from the roads, a great advantage if encountering wind and rain, commonly occurring in this part of the country. The usual approach is to take a round-trip to the best sites, including some of the sites mentioned under the *Stavanger and Northern Jæren* section. Stop wherever it looks promising and scan thoroughy through any congregation of birds. In order to ease tension between farmers and birdwatchers, please respect private property and never park in a way that blocks other traffic. A thorough round-trip takes all day, but if you are short of time, you might choose to visit only the very best areas close to Stavanger, for instance by starting at Lake Harvalandsvatnet and continue via Lakde Grudavatnet, Nese and Orreosen in Lake Orrevatnet, Lake Horpestadvatnet/Ergavatnet, Nærlandsstranden beach to Lake Søylandsvatnet. You should also visit Revtangen headland.

A suggested route to the best sites of Jæren is described here, starting in the north (see maps p. 156 and p. 158):

a. Lake Harvalandsvatnet. From Sola airport follow route Rv510 south, signposted towards Kleppe. You see the lake on your right soon after passing through Tjelta village. You can scan the lake from the roadside *Byberg bus stop,* and from the side roads along the northern and southern shores. The best viewpoint, however, is from the top of Bybergsnuten hill, which requires a bit of off-road walking. The hill is reached by turning onto Bybergveien (signposted to Byberg), park the car, and walk along the tractor tracks leading up to your right (700 m from Rv510), until you reach the top of the hill. Most ducks and other wetland species of Jæren are found in Lake Harvalandsvatnet, and it is a favoured spot for nocturnal songbird trips.

Sele farmlands. Continue 1 km south along Rv510 and make a right towards Sele. This road makes a 5 km loop through farmland, leading back onto Rv510 further south. Check the fields for resting geese and shorebirds, depending on season, and check the harbour **b.** Sele havn at Lyratangen headland for shorebirds. The Sele loop encircles the small, reed-covered pond **c.** Alvevatnet, which supports breeding Marsh Harrier and interesting nocturnal songbirds such as Spotted Crake and Grasshopper Warbler. Walk the dirt roads from the south to scan the lake.

d. Lake Vasshusvatnet and Lake Grudavatnet north. Further south on Rv510 make a left at the bend by the petrol station and drive through Voll village. Make a right signposted to Skjæveland and continue 0.6 km before parking the car outside *Vasshus barnehage* (kindergarten) on the left side of the road. Bring your telescope, cross the main road, and walk the gravel road discretely marked with a white sign reading the number *101.* Walk this gravel road and continue on the footpath along the pastures, all the way to the lake and to the tower hide overlooking Vasshusvatnet. To overlook the legendary Grudavatnet with its ducks and shorebirds, you need to walk a bit closer along the edge of the farmers field to

a large standing stone in the water. Both lakes Vasshusvatnet and Grudavatnet harbour large numbers of ducks all year round. During cold spells in winter, Lake Grudavatnet is often the last of the waters on Jæren to freeze over. You can then experience several thousand ducks here, including up to 25 Smew. From May to September, mud banks in Lake Grudavatnet are favourable for shorebirds. The water level in the lake and the extent of the mud banks vary a great deal, coinciding with the water flow in the river Figgjo.

e. Lake Lonavatnet. Back in the car, continue on the main road. When you meet Rv44, turn left (north) and after 400 m turn into Skjæveland industrial area and further into a housing estate, where you turn down past the address *Vagleskogveien 59* and find suitable parking. From here you can walk down to Lonavatnet and follow the footpath/roads around the water. Lake Lonavatnet is partly overgrown with patches of reeds and difficult to overview. It is a hidden gem, however, and particularly good for Reed Warbler.

Lake Øksnavadtjørna. Continue south along Rv44 and make a right signposted to *Øksnavad Videregående Skole* and park at the first parking area, by the horticulture outlet. This is also a designated parking for the nature hiking trail (*Natursti*). Walk underneath Rv44 and follow signs to *Natursti* to the left. Follow this path along Rv44 150 m to the tower hide overlooking the small and nutrient-rich Lake Øksnavadtjørna, surrounded by marshy woodland and pastures with grazing cattle.

f. Lake Grudavatnet south. Continue south on Rv44. After almost 3 km, make a right in the roundabout, signposted to Gruda, and follow along this side road (Grudevegen) 1.8 km. Make a right, marked with signs reading the numbers *155, 159,* and drive down the hill. Park so that you do not block traffic to the properties, and walk along the tractor tracks through a cluster of spruce trees. Walk silently here, in order not to spook the birds in the lake just behind the trees. After checking the closest birds, you can climb carefully onto a stone wall, providing you with some height. **g.** The Bore farmlands. Continue Rv44 south. In Kleppe, exit onto Rv510 signposted to Sola, and then turn left onto Rv507 towards Orre. A bit up this first slope you can stop and scan the Bore farmlands roadside for resting swans, geese, raptors and more, especially in autumn and winter.

h. Nese in Lake Orrevatnet. From Rv507, make a left signposted to Horpestad. You will soon see Orrevatnet on your right. Make a right, again signposted to Horpestad. From here, continue 1.5 km to the top of the hill. At the top, pull right onto the side road and find suitable parking not blocking other traffic. Just past a sharp bend to the left, you can overlook the northeastern corner of Orrevatnet. This part of the lake usually holds good numbers of ducks, often including 20–30 Smew in winter. You need a telescope from here, but it is also possible to walk dirt roads in order to obtain closer views.

i. Orreosen in Lake Orrevatnet. Continue past the hilltop and make a right towards Orre in the next junction. Stop roadside to scan Lake Orrevatnet wherever possible. 600 m after you cross the Horpestad channel connecting Lake Orrevatnet with Lake Horpestadvatnet, park by the farm on the right. Bring a telescope and follow the path down towards the lake to a tower hide 300 m from the parking. You have to walk past the hide to get a view of the mud banks at Orreosen. Follow the hedge past the hide and turn right along the edge of the field and down along a new hedge that leads down to the lake. Be very careful when you approach the bank, and do not walk all the way out to the water's edge, it is neither necessary nor advisable as you will spook the birds. Orreosen is one of best sites in all of Jæren. Gulls, terns, ducks, geese and shorebirds often rest here, and is the best place

Pectoral Sandpiper breeds in both North America and Siberia, and is the most regular of the "rare" shorebirds on Jæren. You have the best chance of finding it at freshwater locations such as Lake Grudavatnet and Orreosen, but the species is also seen along the beaches. *Photo: Bjørn Fuldseth.*

for Yellow Wagtail *flava*. Over the years, the place has produced an incredible number of rarities. The species selection in Orreosen depends a lot on the water level, i.e. whether there are favourable visible sandbanks. In winter, the areas around Orreosen are very good for raptors, and you can often see 5-6 species within just a few minutes. Please respect private property and do not walk out into the fields.

j. Lake Horpestadvatnet/Ergavatnet. Lake Horpestadvatnet on the opposite side of the road from Lake Orrevatnet is connected to the smaller lake Ergavatnet through a narrow sound, and the two appear to be one single lake. They are among the most bird-rich lakes in the entire country, and species such as Gadwall, Shoveler, Garganey and Pochard are regular in season. These lakes are best overlooked from the farm roads on the east side. To get there, continue to the T-junction with Rv507 and turn left (south), and then left signposted Erga, and finally left again onto the farm road before the hilltop.

Orre. Drive back to Rv507 and turn north (right) across the fields between Lake

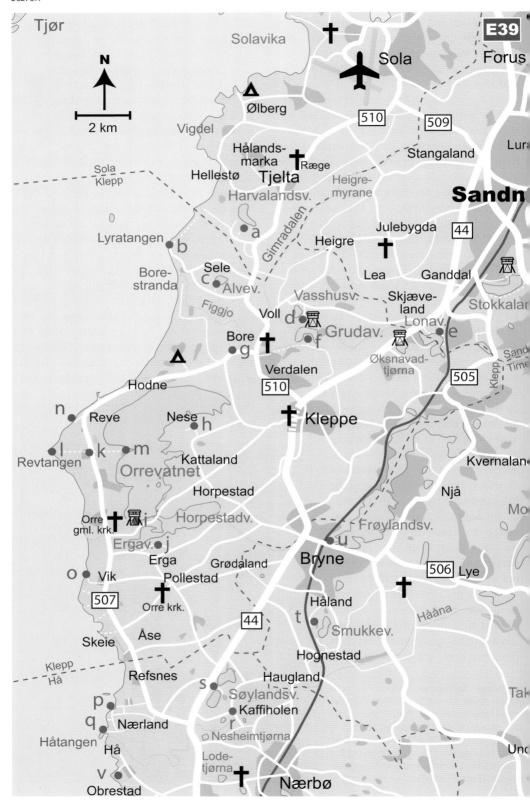

Orrevatnet and the coast, supporting breeding Black-tailed Godwit in June and July. Look for passerines and other migrants in the coniferous woods during spring and autumn.

k. Revtangen headland. Follow Rv507 north almost 3 km from the junction with the road from Orreosen. Just south of a go-kart racetrack, by a tall rectangular electrical transformer, a gravel road crosses the main road. Park where possible. Check the coniferous wood **l.** Reveparken for migrants. Bring your telescope and walk the gravel road 1.2 km across the fields west of Reveparken, down to the dunes by the coast. Alternatively, you can pay a fee to the landowner for permission to drive almost all the way down – see instructions on signs along this road. Check the fields thoroughly for resting passerines, shorebirds, gulls and more. Where the field on the right-hand side is replaced by salt meadow, you can step over to the opposite side of the fence and follow a footpath all the way through the dunes to a small brick bunker by the coast. Move particularly calmly here, because all of a sudden you are standing close to one of the country's finest shorebird beaches, often with a large number of birds in mixed-species flocks. Do not walk out onto the headland itself, only to scare the birds away, but instead walk up onto the sand dunes and study the birds through your telescope. The gradual replacement of feeding birds is considerably larger than it may appear at first glance, so take your time, and see birds steadily come and go. Revtangen headland is Jæren's most westerly point and the shallow seas just outside is good for resting divers, ducks and auks. The headland itself often attracts resting gulls and terns, and from the dunes you can sit down and look for migrating seabirds. You may find some shelter from the wind behind the brick bunker. At the far end of the pebble beach, grey and harbour seals reside.

m. Maleneset headland in Lake Orrevatnet. Back at the car, cross the main road and walk the gravel road on the opposite side, leading to Maleneset headland in Orrevatnet. The fields and spruce forest at Maleneset can support resting migrants and is a good viewpoint to the muddy shores and sandy islets of Lake Orrevatnet, often holding a good selection of ducks, shorebirds, gulls and terns.

n. Revesvingen and Revekaia jetty. Continue 1.2 km north on Rv507 (passing Revtangen Bird Observatory along the way, on your left). Park roadside and walk the private road to the left to get access to the outer coast at Revesvingen. If you drive 700 m further on, you reach Revekaia marina. This is a good seawatching viewpoint where you may use your car as a shelter, as opposed to Revtangen headland. Just like at Revtangen, you must pay a fee to the landowner to drive all the way; see instructions on the roadside signpost. The breakwater at Revekaia is where most Storm Petrels are ringed during calm, dark late-summer nights. A few Leach's Petrels may be caught as well, and even Swinhoe's Petrel has turned up a few times. During winter, passerines may be found feeding on insects in the dead seaweed that has been washed ashore, including scarce species such as Black Redstart and Stonechat. On the opposite side of Rv507 you get a view to the northwestern section of Lake Orrevatnet, often holding good numbers of ducks.

o. Vik–Skeie coast. Drive all the way back south along Rv507, past the junction with the road to Orreosen. About 2 km south of this junction, you will see the football pitch at Vik on your right. Here, you may drive down to the seashore. The 3 km stretch of coastline from here south to Skeie alternates between sandy and rocky beaches, often good for shorebirds and with a potential for rare migrants in late

Jæren, middle section: a. Lake Harvalandsvatnet; **b.** Sele harbour; **c.** Sele fields and Lake Alvevatnet; **d.** Lake Vasshusvatnet and Lake Grudavatnet north; **e.** Lake Lonavatnet; **f.** Lake Grudavatnet south; **g.** Bore fields; **h.** Nese viewpoint; **i.** Orreosen tower hide; **j.** Lake Horpestadvatnet and Lake Ergavatnet; **k.** Reveparken forest; **l.** Revtangen headland; **m.** Maleneset headland; **n.** Revesvingen seawatch; **o.** Vik–Skeie beaches; **p.** Nærlandsstranden beach; **q.** Håtangen river mouth; **r.** Kaffiholen farmlands; **s.** Lake Søylandsvatnet; **t.** Lake Smukkevatnet; **u.** Lake Frøylandsvatnet; **v.** Obrestad lighthouse. See also map of the southern section on the next page.

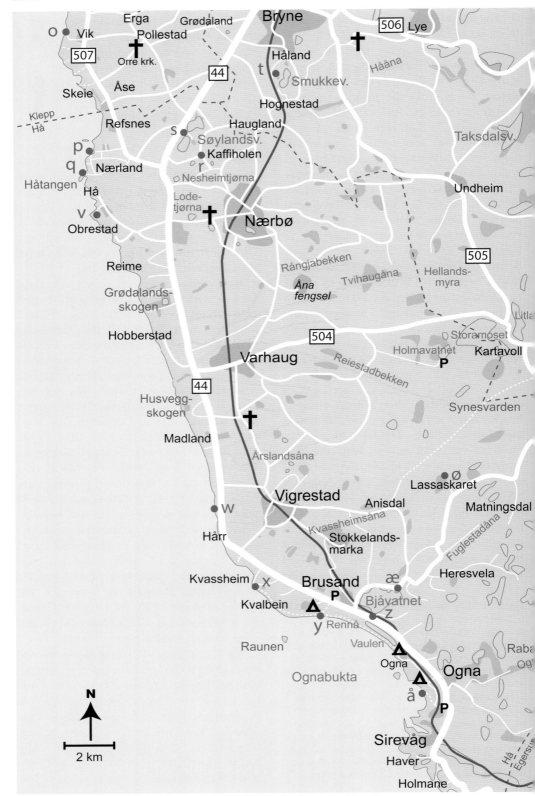

autumn. Also check the airfield at Skeie for resting gulls, plovers and more.

p. Nærlandsstranden beach. Continue south on Rv507 and make a right onto a road signposted to Nærlandsparken, an area with vegetation suitable for resting passerines close to the seashore. This also provides the only designated parking in this area. Continue towards the sea, make a left signposted to Nordsjøruta bicycle track, and then a right turn again. Down by the beach you may study the ducks and shorebirds up close from inside the car without flushing them. It is prohibited, however, to drive on the beach itself, and you may not park and leave the car here. Nærlandsstranden is one of Jæren's most favourable shorebird beaches, but it is subject to a lot of disturbance by hikers, so it pays to be there early in the morning and on days with unfavourable hiking weather.

q. Håtangen is a rocky headland by the outlet of the river Hååna. It represents the end of the pristine sandy beaches stretching almost 30 km continuously from Solavika. Southwards from here a more rugged shoreline predominates. At the river mouth, the sheltered lagoon is a popular stop-over for ducks and gulls, often including Glaucous Gull in winter. Shorebirds thrive among the boulders on the seashore. The pastures and fields along river Hååna are good for Curlews, Golden Plovers and other farmland species. In spring, this is a favoured spot for Oystercatchers and other early migrants. You reach Håtangen from Nærlandsstranden by driving south along the gravel road and walk through the gate blocking the road turning right, leading 300 m down to the headland. Parking is a challenge in this area, so you may want to leave your car at Nærlandsparken, mentioned above, and walk the stretch of beach from Nærlandsstranden to Håtangen, a stretch that often proves rewarding.

Jæren, southern section: o. Vik–Skeie beaches; **p.** Nærlandsstranden beach; **q.** Håtangen river mouth; **r.** Kaffiholen farmlands; **s.** Lake Søylandsvatnet; **v.** Obrestad lighthouse; **w.** Hårr seawatch; **x.** Kvassheim harbour; **y.** Brusand campsite; **z.** railway bridge viewpoint; **æ.** viewpoint Lake Bjåvatnet north; **ø.** Lassaskaret raptorwatch; **å.** Ognasanden beach. See also map of the middle section on the previous page.

r. Kaffiholen farmlands. Continue south on Rv507 and make a left towards Stavanger along Rv44. Then make a right, signposted to Kaffiholen. This road leads past fields bordering the Hååna river. In periods of heavy rain or snow melting, the river may flood these fields, providing excellent habitat for gulls, shorebirds, ducks, swans and more.

s. Lake Søylandsvatnet is one of Jærens most bird-rich lakes. The reedbeds here are the largest on Jæren, supporting breeding Marsh Harrier, Water Rail and Bearded Tit. A substantial number of ducks are constantly present. The lake is divided in two separate sections. The southern section can be viewed from along the road to Kaffiholen (above), by parking almost 1 km from Rv44 and walking a short distance along the edge of the farmer's field. You may also scan the lake from bus stops along Rv44, and from the gravel road that exits to your right 1 km north of the Kaffiholen junction, along a stone wall. This gravel road leads down between the two sections and gives close access to the reedbeds, essential for looking and listening for reed-bound species, such as Bearded Tit. To see the northern section of Søylandsvatnet, continue north on Rv44 and make a right onto the road signposted to Haugland and find road-side viewpoints.

t. Lake Smukkevatnet is first and foremost a ringing site, made famous by quite a few records of rare birds, including Aquatic Warbler on several occasions. The ringing activities are conducted on private land. The lake itself is best viewed from the south end.

u. Lake Frøylandsvatnet is a 6 km long lake and a popular recreation area. The south end has a particularly lush reedbed. It is reached from Bryne town centre, signposted Tursti. Drivable roads encircle the entire lake. The lake is particularly good for diving ducks and species associated with reedbeds. A colony of Rook is found by Bryne secondary school.

v. Obrestad lighthouse. Drive back towards Lake Søylandsvatnet and continue south on Rv44 a couple of kilometres past the Rv507 junction. Turn right towards Obrestad fyr (lighthouse) and follow the signs. On the way

Oslo

Interior

Skagerrak

Western

Central

Northern

Finnmark

Svalbard

Brusand in the far south of Jæren has many exciting sites, including Vaulen lagoon surrounded by rugged sand dunes. When the water level is low, large, shorebird-friendly mudflats are exposed. *Photo: Bjørn Olav Tveit.*

to the lighthouse, you pass *Obrestad hamn* (harbour) which is worth checking for ducks and divers, and the stranded seaweed may hold interesting passerines in winter, and gulls and shorebirds during migration. Continue to Obrestad lighthouse, a good seawatching point. You may also want to investigate the spruce forest here for resting migrants. If you drive further north along the coast, you can drive down to view Håtangen headland from this side. A footbridge crosses the river Hååna. Reime–Madland coastline. Southwards from Håtangen headland the coast is more rocky and harder to walk than the sandy beaches to the north. It can be worth the struggle, though, because this is a favoured area for gulls, Gyrfalcon, wheatears, pipits and more, depending on the season. Ducks, shorebirds and gulls often gather at the mouth of rivers and creeks, the largest of which are marked on the maps. These carry water even in dry periods. Additional creeks, originating from flooded fields, may appear after heavy rainfall. Any flooded field should be checked as well. Along this stretch of coast several conifer plantations provide shelter for passerines, especially Grødalandsskogen and Husveggskogen forests.

w. Hårr seawatch. 12 km south of the junction with Obrestad, in a bend where Rv44 almost touches the shoreline, is a lay-by on the seaside of the road. This is the best and most accessible seawatching point in southern Jæren. You can watch from inside your car or walk a couple of hundred metres further north to sit next to a stone wall, providing some shelter. If you continue a kilometre further south, just over a hundred metres after the exit to Vigrestad, a side road flanked by a hedge leads down to the sea at Øyrtangen, a fine locality for shorebirds.

x. Kvassheim lighthouse. The side road to Kvassheim harbour, signposted from Rv44 south of Hårr and Øyrtangen, crosses the creek Kvassheimsåa which empties in Øyrvika. This inlet is often the best shorebird site in southern Jæren (but see Vaulen, below). It is also a good site for ducks, gulls and passerines associated with open country, often including Carrion Crow. From Kvassheim lighthouse you can walk south along the coast through the very exciting area Kvalbein, made famous by several records of rare late autumn passerines. Again, check all conifer plantations including the one surrounding **y.** <u>Brusand Camping</u>. Just south of this campsite, just before arriving in Brusand village, a parking area signposted *Friområde* is found on the east side of Rv44. Park here and walk underneath the highway along a footpath leading to the beach at the mouth of river Rennå. Many of the afore mentioned birds may be found here. In addition, it is a favoured site for divers, often including Black-throated, a rather uncommon species along the west coast of Norway. You may cross the river using the steppingstones and climb up onto the dunes in order to gain some height to overlook the 1.2 km long <u>Vaulen lagoon</u> behind the dunes. This shallow lagoon is particularly interesting in summer and when the water level is low, exposing shorebird-friendly mudflats. Under such circumstances, Vaulen lagoon is the best shorebird site in southern Jæren. You may also overlook the lagoon, at least in part, from **z.** <u>the railway bridge in Brusand</u>, reached by exiting from Rv44 onto a road signposted to *Rom/Hytter*. From this bridge you also get a view to a section of <u>Lake Bjåvatnet</u>, which can be good for ducks including Smew (Nov–Apr), as well as Slavonian Grebe and Coot in winter. Drive back down from the bridge and turn right on the side-road just before Rv44 to check the outlet and south end of Bjåvatnet. From here, you can explore the areas further south, including **å.** <u>Ognasanden beach,</u> or you can drive north along Rv44 into Brusand village and turn right (north) a few hundred metres, and turn right again onto the road signposted to Kartavoll and Stokkelandsmarka. Along this road, stop to overlook Bjåvatnet wherever possible, such as **æ.** to overlook the northern section.

<u>Høg-Jæren highlands</u>. The road from Brusand past Bjåvatnet continues up through one of the wooded valleys in the rolling highlands of Høg-Jæren. This valley supports breeding Grey-headed Woodpecker, Green Woodpecker, Lesser Spotted Woodpecker, Willow Tit, Long-tailed Tit and Dipper. Eagle Owl, Stonechat and Red-backed Shrike may be found in this area as well. It is also a year-round stronghold for Golden and White-tailed Eagles, often soaring above the ridges on calm, sunny days. To look for raptors, any peak might do but the most commonly used is the peak just west of Lassaskaret, marked as point **ø.** on the map. In order to reach it, turn in onto the side road by the bus stop Engjane 3.2 km north of the junction signposted to Herredsvela. Park out of the way and walk past the houses and make a right onto the walkway after the farm <u>Lassaskaret</u>. Take the second dirt track to your left, leading up to the hill on the north side of the settlement. Do not forget to close the gates behind you. At the top you

Eagle Owl is an endangered species in Scandinavia, but it has a fairly good population in the hills of the Høg-Jæren highlands. These birds are very difficult to spot amongst the cliffs and heather, but their characteristic calls can be heard at dawn and dusk in early spring. *Photo: Kjetil Schjølberg.*

161

can sit down and scan the horizon for raptors, which even in autumn usually arrive from the southeast. Day-counts of more than 50 each of Sparrowhawk and Kestrel is not uncommon in autumn, and Buzzard, Rough-legged Buzzard, Goshawk, Hen Harrier, Golden Eagle, White-tailed Eagle, Peregrine Falcon, Gyrfalcon and Merlin are all regular migrants. Big surprises, such as Egyptian Vulture, Eastern Imperial Eagle and Black Stork have been encountered, not only passing by but also lingering for weeks and even months in the area.

To look for breeding 'southern' Golden Plover and Dunlin, walk the moorland in Synesvarden nature reserve, reached via Rv504 east of Varhaug village, following signposts to Synesvarden and Steinkjerringa. Park in the designated parking near Holmavatnet and follow the trails leading east. Be careful not to disturb the breeding birds.

4. STAVANGER AREA

Rogaland County

GPS: 58.91131° N 5.59248° E (Kolnes parking), 58.82562° N 5.73500° E (tower hide Lake Stokkalandsvatnet)

Notable species: Ducks, divers, grebes, shorebirds and other wetland species; Scaup (winter), Pheasant, Water Rail, Spotted Crake, Sandwich Tern (summer and autumn), Tawny Owl, Eagle Owl, Wryneck, Grey-headed Woodpecker, White-backed Woodpecker, Grasshopper Warbler, Reed Warbler and Icterine Warbler.

Description: The northern section of Jæren is strongly urbanized, hosting the twin cities of Stavanger and Sandnes, as well as suburbs, a major airport (Sola) and vast industrial areas. Still, there are many interesting birdwatching sites here. Several lakes in the area attracts ducks, gulls and other wetland species all year. In some places the birds are fed and become quite confiding, popular with photographers and families with kids. The most bird-rich lakes are Store Stokkavatnet, Litla Stokkavatnet, Mosvatnet and Breiavatnet in Stavanger, and Stokkalandsvatnet in Sandnes. The latter is equipped with a tower hide. Hafrsfjord, the

bay between Stavanger and Sola airport is an important resting and wintering area for ducks, supporting the largest gathering of Scaup in the country, often counting more than 100 individuals in winter. This is also the most reliable site in Norway for Sandwich Tern in late summer and early autumn. This species probably breeds just off the coast. The outer coastline has some very interesting sites as well, particularly Kolnes peninsula and Solasvika beach. The former can support substantial numbers of shorebirds in season and Wigeons and other ducks in the thousands during winter. Great Northern Diver is a common sight both here and at Solavika inlet in the cold season, as is Slavonian and Red-necked Grebes. Solastranda beach is reliable for Sanderling and Wheatear in late autumn.

Just east of Sandnes are several interesting lakes and slow-flowing rivers and streams, worth checking for a variety of wetland species, sometimes including Kingfisher and nocturnal songbirds. Particularly good are the lakes Grunningen and Kyllesvatnet. The woodlands here support Eagle Owl, Tawny Owl, Wryneck, Grey-headed Woodpecker and White-backed Woodpecker with the possibilities of Buzzard, Black Woodpecker and Wood Warbler. A long list of vagrants has been recorded in these areas over the years.

Best season: All year.

Directions: From Sola, the main airport in the Stavanger-area, you may well walk to **k.** Sømmevågen inlet and **m.** Solastranda beach. **i.** Kolnes peninsula is but a short taxi trip away. The railway station in Stavanger is positioned next to **a.** Lake Breiavatnet, and you may reach the other lakes using public transportation.

Tactics: To visit these sites by car you can follow these instructions:

b. Lake Mosvatnet is reached by drive E39 through Stavanger, exiting to Mosvangen Camping. Gulls and ducks congregate where people feed the birds, such as lake-side by the camping site. A telescope comes in handy to identify birds out on the lake.

Sandwich Tern is an irregular breeding bird in Norway. The species is mostly seen roaming along the coast from Østfold to Rogaland counties in April–September. It is quite regular along the Stavanger and Jæren coast. *Photo: Espen Lie Dahl.*

c. Lake Litla Stokkavatnet is reached from E39 2 km further north. Exit onto the road signposted to Stokka and continue westwards along Gustav Vigelandsvei. You reach the lake after 200 m on this road. Afterwards, drive back onto E39 and exit to Rv509 signposted to Tananger. Just after the roundabout at Sanddal, park by the industrial buildings and walk the footpath along the creek north to **d.** Lake Stora Stokkavatnet. Also check **e.** Lake Hålandsvatnet, reached by continuing south on Rv509 and follow the road signposted to Krossbergveien. Turn left into side road Krossbergkroken, park by the barrier and walk a bit further to overlook the south end of the lake. To view **f.** the north end, drive

back to Rv509 and turn right in the next roundbabout, signposted to Kvernevik. Further along Rv509, in the roundabout just before the Hafrsfjord bridge, make a right and then a left onto the road leading underneath Rv509. From the footpath under the bridge, you have a nice view to **g.** the narrow sound into Hafrsfjord, often holding a nice selection of waterbirds including divers, grebes, ducks and auks. This is also a favoured spot for terns, often including Sandwich Tern.

Back in the car, cross the bridge and continue 1.5 km south along Rv509. At the roundabout you have the marshes Kvitemyr to your left and Storemyr 500 m to your right, along the side road signposted to Tananger. Both these marshes are interesting during nocturnal songbird trips. 1 km further south along Rv509, by Tananger church, turn left onto Hagavegen and continue down to the shore to overlook **h.** Hagavågen inlet, often holding a nice collection of ducks. Shorebirds may gather here at low tide. Further south along Rv509, make your way to the road Søndre Hogstadveg. From along this road, you obtain views to **i.** Strandsnesvågen inlet, another good place for ducks and shorebirds. You may also walk down to the shore.

j. Kolnes peninsula. Across from the Rv509, drive to the end of the road Raffinerivegen and find suitable parking by the electrical transformer building. From here, starting at the often bird-rich Brunnavika inlet, you can walk south along this very exciting stretch of coast. Keep to the tide line; do not walk across the fields.

k. Sømmevågen inlet. Continue south along Rv509 and leave at the exit just past the Sola airport runway. Then follow Rv510 towards Røyneberg, crossing the highway, and exit from Rv510 towards the Joa industrial area. Then take the first road to the left, and park near the end of this road. Bring your telescope and walk further out to the water's edge. More inlets and beaches are found along the eastern shore of Hafrsfjord, such as **l.** Grannesbukta inlet. These are reached by continuing the Rv510 eastwards, signposted to Stavanger, exploring side-roads to the left.

Håsteinsfjorden
Bø
Mekjarvik
Randaberg Stavanger

Randaberg

Randaberg (see separate map)

Hundvåg

Visteviga
f
Randaberg Sola
Kvernevik
Hålandsv.
Litla Stokkav.
e
Stora Stokkav.
c
Breiav.
a

g
Store-myr
509
Sanddal
Mosv.
b

Tananger
Haga
Madla
Stavanger
h

Rott
u
509
Hafrsfjord
l
44

i
Kolnes
j
Sømme-vågen
510

k
Gausel

Solavika
E39

m
Sola
Forus

Ølberg
Sola Klepp

Vigdel
n
510
509
Lura

Hålands-marka
Ræge
Stangaland

Hellestø
Tjelta
Heigre-myrane

Harvalandsv.
Sandnes
Grunn
Hana

Lyra-tangen
Julebygda
44
Austrått

Sele
Heigre
p
Høyland

Alvev.
Lea
Ganddal

Figgjo
Vasshusv.
Skjæveland
Stokkalandsv.

Voll
Gandsfjorden
Storemyr
Stavang
Kristiansand

m. <u>Solastranda beach</u>. Back on the Rv509 by the airport runway, exit south towards Sola airport. Make a right just before the airport area on the road signposted to *Sola Strandhotell (Sola beach hotel)*. This road leads past Solastranda beach lining Solavika bay, famous for supporting ducks, divers and grebes – if the kiters and surfers haven't spooked them away, that is!

n. <u>Ølberg peninsula</u>. Past Solastranda beach, make a right signposted to Ølberg, and then follow signs to Tjelta. This leads you through a loop around the Ølberg peninsula and past Ølberg harbour, which may hold gulls and provides a different view to outer Solavika. This loop also leads through an area of gardens lining the seashore, and past fields and beaches, all worth exploring. From here, you can continue south to the sites on Jæren (above), or you can drive east toward Sandnes town and check out the sites in this area.

The innermost section of Gandsfjorden, called **o.** <u>Sandnesvågen inlet</u>, should be checked for gulls and ducks in winter. **p.** <u>Lake Stokkalandsvatnet</u> is reached by following signs from Sandnes to Ganddal, and continuing along Rv505 towards Kværnaland. From Kværnalandsveien road in Ganddal, just after passing underneath the railway bridge, make a left and then a right onto Elgveien. This street borders the lake. In order to reach the tower hide, park near the end of Elgveien and continue 300 m on foot along the bicycle/footpath which encircles the lake.

If you continue 2 km along Rv505 you reach <u>Fosseikeland</u>, a calm little section of the river Figgjo with potential for Grey Wagtail and

Stavanger area (Jæren, northern section):

a. Lake Breiavatnet; **b.** Lake Mosvatnet; **c.** Lake Litla Stokkavatnet; **d.** Lake Stora Stokkavatnet; **e.** and **f.** Lake Hålandsvatnet south and north; **g.** Hafrsfjord sound; **h.** Hagavågen inlet; **i.** Brunnavika inlet, parking Kolnes peninsula; **j.** Strandsnesvågen inlet; **k.** Sømmevågen inlet; **l.** Grannesbukta inlet; **m.** Solastranda beach; **n.** Ølberg–Vigdel peninsula; **o.** Sandnesvågen inlet; **p.** Lake Stokkalandsvatnet; **q.** Skjørestadmyra and Grunningen wetlands; **r.** Lake Kyllesvatnet; **s.** Horve woodlands; **t.** Lake Kleivadalsvatnet; **u.** Rott island.

Kingfisher. From E39 south of Sandnes, exit onto Rv13 signposted to Røldal. After 3 km, make a right in the roundabout, signposted to Sviland, and from here continue 1.8 km and turn left into Skjørestadveien. This road, although not drivable all the way, leads to Skjørestadmyra marsh with the nutrient-rich and partly overgrown pond q. Grunningen, a very exciting and seldom visited site with good populations of Water Rail, Woodcock, Tawny Owl, Grasshopper Warbler, Reed Warbler and Icterine Warbler. Spotted Crake is almost annual.

Continue the main road towards Sviland a couple more kilometres, and you will see the south end of r. Lake Kyllesvatnet on your left. The stream Svilandsåna should be checked as well, for Kingfisher and more.

If you drive back and continue Rv13 further east and follow signs to Vårli, you will reach s. Horve, with some of the best woodlands in this part of the country, with breeding Lesser Spotted, Green, Grey-headed and White-backed Woodpeckers, Wood Warbler and more. The woodlands surrounding t. Lake Kleivadalsvatnet east of Hommersåk village is also a good and easily accessible area for woodland species. From Hommersåk, take the road signposted to Bersagel. After 500

m, exit towards Øvre Hetland and follow the road 2 km to the parking area by the end of the road.

u. Rott is a small island with few facilities, but with great potential for rarities during migration. It can only be reached by private boat.

5. RANDABERG

Rogaland County

GPS: 59.01535° N 5.57925° E (Børaunen), 59.03386° N 5.58663° E (Tungenes lighthouse)

Notable species: Diurnal migration of passerines and more; resting ducks, shorebirds etc.

Description: The northern section of Jæren is an oasis of beaches, wetlands and cultivated land in contrast to the urban areas of the Stavanger region. The bird life here is very similar to what is described for Jæren, but at the tip of this peninsula you see more of the diurnal migration of passerines and such. Because the headland is pointing north, it would be natural to think that it leads and concentrates migration in spring only, but the visible migration is significant in autumn too, moving north just as in spring. Heavy overcast and calm winds are usually best. On such days at Tungenes lighthouse in autumn, you can experience a nice selection of northwards moving pipits, thrushes, tits, finches and many more. Look for resting migrants in the surrounding terrain. The wetland Børaunen just southwest of Tungneset consists of a unique saltmarsh attractive to

Scaup breeds in small numbers in the mountains of southern Norway, but has also been found to nest in the lowlands, including on Jæren. In northern Norway, it also breeds spread out to the coast. The species overwinters in coastal freshwater as well as in sheltered fjords and brackish pools. One of its main wintering sites are in the Hafrsfjord bay near Stavanger, where it often can be encountered in two-digit numbers. *Photo: John Stenersen.*

Tringa-species, Ruff and other fresh-water species. Several rarities has been recorded over the years.

Best season: All year, but migration periods in particular.

Directions: From Stavanger, follow E39 north towards Bergen. After about 5 km, make a left signposted to Randaberg and Tungenes fyr (lighthouse). Drive through Randaberg village and follow signs towards Tungenes. Then, see map.

Tactics: Find suitable parking along the road at Grøde village. From here you can walk down to **a.** Sundet with a view of the shorebirds and ducks out at Børaunen peninsula. Pay particular attention to the vulnerable birdlife in this entire area, especially during the breeding season. To minimize disturbance of the birds, do not walk the fields, but go back instead and walk the roads to **b.** Bøvika inlet. The jetty here provides the best view of the sea and shoreline. Do not walk out on Todden peninsula either, you will only flush the birds. To scan Bøvika inlet, you can also drive to the **b.** carpark. Back at Sundet, you can also walk south along the coast and look for resting birds along Bøstranda beach.

Out at **d.** Tungenes lighthouse you can watch

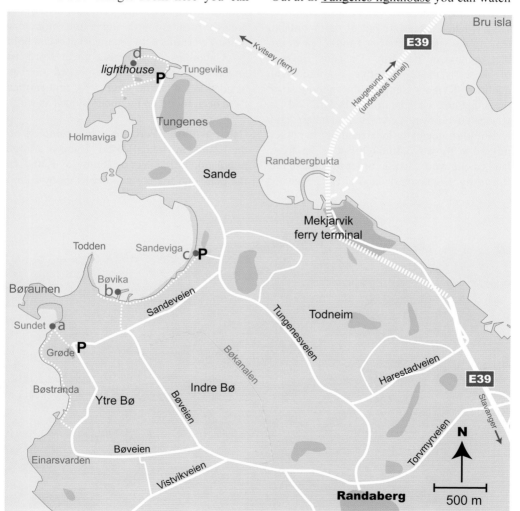

Randaberg: a. Børaunen peninsula, Sundet; **b.** jetty viewpoint and **c.** carpark viewpoint to Bøvika inlet; **d.** Tungenes lighthouse.

migrating birds pass overhead and search for resting birds in the surrounding terrain. Look for waders on the rocks west of the lighthouse. The pastures on the way down towards Holmaviga inlet may be productive as well.

6. ISLANDS OF BOKNAFJORDEN

Rogaland County
GPS: 59.06131° N 5.40121° Ø (Kvitsøy island)

Notable species: Migrants, rarities; wintering divers and grebes.

Desription: North of Stavanger city, the sea has taken a big bite out of the mainland, leaving the wide Boknafjord and a few crumbles of land in the form of islands. Several of these islands have fine ornithological qualities. Kvitsøy island is the outermost of the inhabited islands in the Boknafjord. It is located midway between Randaberg and Karmøy, strategically located along the outer migration fly-way, although not as far from land as one would like in order to achieve the full rarity-producing "island effect". Kvitsøy

is connected to the mainland with a ferry and a higher-speed passenger boat. The island has nice pastures and lush gardens, and is small enough to be explored in full during one long day, especially if you have a car or bike at your disposal. It is therefore an easily accessible alternative to Utsira island for birdwatchers in the Stavanger region. Some of the rarities encountered here in recent years have not at all fallen short of those at Utsira, and includes American passerines like Red-eyed Vireo and Alder Flycatcher. The shallow seas and straits between skerries, especially on the south side of Kvitsøy, hold fairly good numbers of ducks, grebes and divers in winter.

The inner islands of the Boknafjord have areas with lush farmlands and wetlands, favourable both for breeding birds and resting migratory birds. The straits between the islands can hold good numbers of wintering Great Northern Diver, Slavonian and Red-necked Grebes, besides a few Little Grebes. A few rarities have been found on the inner islands as well.

Other wildlife: Porcupines are common on the islands in Boknafjorden. There is a large population of red deer on Rennesøy island. In the ponds on the north side of Finnøy island,

Kvitsøy island has a nice combination of habitat for resting migrant passerines. *Photo: Bjørn Olav Tveit.*

pool frogs have been released, with their special croaking call that can be heard on nocturnal songbird trips in spring and early summer.

Best season: Late spring and autumn for migrants; winter for grebes and divers.

Directions: Kvitsøy island can be reached by car via a 40-minute car ferry ride from Mekjarvik harbour, 15 minutes' drive north on the E39 from central Stavanger. Without a car, you can take a faster passenger ferry from the main pier in central Stavanger (35 min). Several of the islands in the Boknafjord have a road connection along the E39 from Stavanger, including Åmøy, Mosterøy, Rennesøy, Talgje and Finnøy islands.

Taktics: Kvitsøy island is explored by fine-combing the terrain, like you do on most rarity-producing islands; for more on this, see the section about Utsira island. It is an advantage to have a car or bicycle to cover the entire island in one day. Do not enter gardens without asking permission.

The larger, inner islands should preferably be explored by car. On Åmøy island, Torsteinvika inlet is particularly good – you will see the cove on your left 2 km after you have crossed the bridge to the island. Aksjesundet sound between Sokn and Mosterøy islands often has a good stock of divers and grebes. On Mosterøy island, it is particularly favourable in Dysjalandsvågen inlet in the south and around Klosterøya and Fjøløy islands in the north. The latter island has nice gardens and fields, which provide good conditions for resting migratory birds. In Fjøløysundet sound west of the monastery, there are often Little Grebe in winter. On Rennesøy island, Peregrine Falcon, Eagle Owl and Goshawk breed, and the many small lakes out here often offer calling Water Rail and Spotted Crake on bright summer nights; in particular, Lake Bjergavatnet and Førsvollvatna lakes can have exciting experiences to offer. On Finnøy island, it is particularly nice in the farmlands at Sevheim on the west side and at Lake Hauskjevatnet on the east side.

7. KARMØY ISLAND

Rogaland County

GPS: 59.21709° N 5.18155° E (Taravika inlet)

Notable species: Resting shorebirds, migrating seabirds; Black Grouse, Great Northern Diver (winter), Eagle Owl, Water Rail, Spotted Crake, Corncrake, Grasshopper Warbler, Sedge Warbler and Stonechat.

Description: Karmøy is a 30 km long island along the outer coast, separated from the mainland with narrow sounds to the north and east. However, it appears like a peninsula protruding out into the Boknafjord, concentrating birds which follow the coast towards the south. The terrain on Karmøy is predominantly rocky heathland with scattered marshes and grazing pastures, as is typical of the outer coast in this part of Norway. Gardens and a few conifer plantations provide shelter for passerines. A few sandy beaches can be found along the otherwise rocky coast, and these beaches are all worth walking at any time of the year. The most famous of these, at Taravika inlet, is a small but favoured resting site for shorebirds. Actually, Taravika inlet is one of a very few good resting grounds for shorebirds along the Norwegian west coast between Giske and Jæren. The number of birds is seldom high here, but in favourable conditions – heavy overcast and a soft breeze from the east – a surprisingly great variety of species may show up. A birdwatching hide here provides good opportunities to study the waders up close. The adjacent fields attract plovers, larks, wagtails and pipits, and on the sea, ducks, divers and grebes rest. Several rare birds have been found at Taravika inlet and along the other beaches on Karmøy island, including Red-necked Stint and Asian Buff-bellied Pipit, and with up to three Water Pipits simultaneously.

The town Skudeneshavn is situated at the southern tip of Karmøy. The gardens here are well worth checking for resting migrants. Rare passerines, including Ovenbird, has been recorded here.

The interior of Karmøy island supports a small population of Eagle Owl and – probably annually – breeding Stonechat. The latter is fairly regular near Lake Mjåvatnet in the south. There are many wetlands on the island, such as the two almost overgrown lakes Tjøsvollvatnet and Heiavatnet by Åkrehamn village on western Karmøy, where Moorhen, Coot and perhaps also Little Grebe breed. Lake Hilleslandsvatnet in the south is interesting as well, particularly for wintering ducks. Karmøy island provides many good areas for nocturnal songbirds, particularly along the west coast and around Torvastad village in the north. Grasshopper Warbler, Sedge Warbler, Corncrake, Water Rail and Spotted Crake can often be heard on quiet early summer nights.

On islets off the coast of Karmøy important seabird colonies are found, with Fulmar, Shag, Black Guillemot, gulls and terns, and small numbers of Puffin and Razorbill found breeding.

Other wildlife: Red deer is very common.

Best season: All year, but particularly in the migration periods.

Directions: Haugesund city airport is situated on Karmøy island. By car from Haugesund, Karmøy is reached by driving E134 south (signposted to the airport) across the Karmsund bridge. From Stavanger city, drive north along the E39, and after crossing the Boknafjord via the Mortavika–Arsvågen ferry, head west and follow Rv47 towards Haugesund airport Karmøy.

Tactics: Karmøy island is best explored by car. If you arrive via E134 from Haugesund city, you may take a roundtrip to the best sites following these directions:

Start by checking the little roadside Bøvatnet pond on your left, just south of the Karmsund bridge. Shoveler has bred here. Continue south and turn right towards Viken/Torvastad to explore the agricultural areas in the north. Further south along E134 you soon get the entrance to Haugesund airport on your right. Here you make a left towards Husøy, along a road leading passing the sheltered Gylsethøra inlet on your left.

Drive back to the airport junction and head south along Rv547 signposted to Skudeneshavn. Continue 3 km past Kopervik town, and you get Lake Heiavatnet on your left. The lake is visible from Rv547, but is better explored from the narrow side road Aureivegen, exiting to the left. Further on along Rv547, when entering Åkrehavn town, you see Lake Tjøsvollvatnet on your left as well. It is best explored by exiting onto the side road Tostemvegen, signposted to Tjøsvold Ø.

Further south along Rv547 you soon see signpost to Åkrasanden beach on your right, followed further south by Stavasanden beach reached by turning right towards *Ferkingstad havn* (harbour). 1.4 km south of this junction, a small side road discretely signposted *Søre Langåkerveg* leads down to the famous Taravika inlet on your right. This is 22 km south

Brent Goose occurs in Norway in two subspecies, the light-bellied *hrota* which passes by in concentrated periods at the end of May and in September, and the dark-bellied nominate form (depicted) which is more sparse and which is mainly encountered in October. *Photo: Kjetil Schjølberg.*

of the Karmsund bridge and 11 km north of Skudeneshavn town. Follow this side road through the small village and continue along the fields to the coast. Park in the designated area just before the birdwatching hide.

Continue Rv547 further south and, just before Skudenes, exit to <u>Syre</u>, and follow a road-loop that crosses through nice fields favourable to a variety of migrants. In Skudeneshavn village centre, exit left (east) onto Rv511 signposted to Falnes church. Continue through the village and exit left towards Dale. This road leads to <u>Lake Hilleslandsvatnet</u>, which may be viewed from the roads. Keep going a bit further east along Rv511 to Falnes church. Here, turn right onto a road signposted to <u>Beiningen</u>, a hotspot for resting migrants.

8. HAUGESUND AREA

Rogaland and Vestland counties

GPS: 59.44195° N 5.25148° E (Tornesvatnet west), 59.54883° N 5.37879° E (Bjellandsvatnet)

Notable species: Wetland species; Little Grebe, Water Rail, Glaucous Gull (winter), Eagle Owl, Collared Dove, White-backed Woodpecker, Grasshopper Warbler and Sedge Warbler.

Description: Haugesund is the largest town between Stavanger and Bergen and is the base for trips to Utsira. Several interesting sites are found nearby, many of which are described under the section about Karmøy island. Haugesund harbour, the parks, lawns and lakes, such as Lake Skeisvatnet, attract gulls and ducks, particularly in late fall and winter. Among the gulls, Glaucous is annual and Iceland Gull nearly so. The overgrown Lake Røyrvatnet supports Little Grebe in winter (as does the harbour) and Grasshopper Warbler in early summer. Lake Tornesvatnet north of Haugesund is 1 km long lined with extensive reedbeds. It is an important resting site for waterbirds. The nearby seashore is good for sea ducks, divers and grebes, and the beach is lined with scrub,

Lake Tornesvatnet is probably the most exciting birdwatching site near Haugesund town. It is an important wetland for local birds, and a number of rarities have showed up here. *Photo: Bjørn Olav Tveit.*

providing suitable habitat for migrant passerines and more. You can watch passing seabirds from here, as well as from Ryvarden lighthouse in Sveio further north. Several interesting lakes are found in the heathland close to Sveio village centre, producing wintering waterbirds in the cold season and nocturnal songbirds in the summer. The inlet Øyr in Viksefjord can have rather extensive mudflats at low tide, and may support a few ducks and shorebirds. White-backed Woodpecker and Capercaillie can be encountered by Naustvika at the mouth of Ålfjord. Eagle Owl breed several places in Sveio.

Other wildlife: Great crested newt is common in ponds between Mannavatnet and Sveio centre. There are good populations of red deer and roe deer in the area.

Best season: All year.

Directions: Haugesund town can be explored on foot or by public transportation. This can be particularly useful if your visit to Haugesund is restricted to spending a couple of hours waiting for the ferry to Utsira island. A car will come in handy if you have more time and want to visit sites further afield.

Tactics: *Without a car:* The town lakes in Haugesund are situated in the northern section of the city, within walking distance from the harbour. There is a nice footpath from the Killingøy breakwater at the north end of the harbour leading north to Kvalsvik. At the headland here, an old bunker makes a suitable seawatching shelter. You can alternatively take the bus to Kvalsvik and walk westwards. To get to Lake Tornesvatnet, walk east from Kvalsvik until you reach the street Årabrotsvegen, and follow this road northwards until you see the western section of Lake Tornesvatnet to your right. This road eventually leads to the seashore.
By car: Drive north along Rv47 through Haugesund town centre. Kvalsvik is reached by turning left onto Kvalsvikveien, signposted to the Haraldshaugen monument. Continue past the campsite and the monument and turn left 1.1 km after you left Rv47. Follow Kvalsvikveien to the end of the road and walk 150 m to the bunker.

Lake Tornesvatnet and Årabrot, continue on Rv47 a bit further north and turn left onto Årabrotsveien, leading past Tornesvatnet to the seashore at Årabrot.
Viksefjord and Ryvarden lighthouse are reached by turning left from Rv47 onto Rv541 signposted to Ryvarden, just after crossing the county border from Rogaland to Vestland. Øyr in Viksefjord is reached after about 1 km, while Ryvarden lighthouse is 8 km further out.
Lake Mannavatnet and Bjellandsvatnet in Sveio are both situated along Rv47 just north of the second junction to Sveio village centre (which is signposted Sentrum/Sveioåsen), 18 km north of Haugesund. A side road turns right just after this junction, leading to a good viewpoint to Mannavatnet. 1 km further north, on your left, is Bjellandsvatnet. Here you can pull left, following signs for *Fuglereservat*, leading to a designated parking spot. From here you can walk along a path to a tower hide.
The other lakes in Sveio can be reached by going back to Sveio centre and from there head north towards Tjernagel. You will soon get Åsevatnet and then Nordskogvatnet on your left.
Naustvika is reached by following Rv47 north to the E39 and continuing on the latter towards Bergen. Exit the E39 just south of the Bømlafjord tunnel, signposted to Tittelsnes/Valevåg. From Tittelsnes, continue south through the sheltered and lush wooden slopes along the fjord and stop wherever it looks fine.

9. UTSIRA ISLAND
Rogaland County
GPS: 59.30905° N 4.87524° E (ringing site)

Notable species: Resting migrants; rarities; passing and breeding seabirds. *Regular unusual autumn migrants:* Sooty Shearwater, Storm Petrel (late summer), Jack Snipe, Turtle Dove, Richard's Pipit, Black Redstart, Barred Warbler, Yellow-browed Warbler, Red-breasted Flycatcher, Common Rosefinch, Ortolan Bunting and Little Bunting. *Almost annual are* Sabine's Gull, Short-toed Lark, Olive-backed Pipit, Subalpine Warbler (spring), Pallas's Warbler and Rustic Bunting.

Breeding birds: Fulmar, Shag, Guillemot, Razorbill, Black Guillemot and Puffin (many of these on remote Spannholmane islets).

Description: Utsira is a small, weather-beaten island in the North Sea, 16 km off the mainland. Being a former vibrant fishing community, it now only supports about 200 inhabitants. Utsira has won international reputation as a rarity magnet in the European top-league together with Fair Isle in Scotland and Heligoland in Germany. Its potential for producing vagrants was discovered in the 1930s when, among other exceptional records, Europe's first Olive-backed Pipit was documented. Utsira was for a long time regarded as a one-of-a-kind in Norway. In recent years, however, it has gained competition from several other islands along the Norwegian coast, such as Ona and Røst. Utsira still reigns as number one, however, with the central and most productive section of the island having the highest density of national rarity records anywhere in Norway.

Utsira is first and foremost a site for resting passerines and other land-based migrants, although seabirds can pass by in impressive numbers. The 6.2 km² island is hilly, reaching 68 m above sea level. It was originally sparsely vegetated and dominated by heathland, grasses and marshes. Clusters of hardy, exotic conifer species was planted during the 1960s and 1970s, however, in order to provide some shelter from the wind. These have spread and turned into rather dense and, in some places, almost impenetrable woods, some of which have been cut down in recent years. A central valley divides the island in two halves, providing a fertile haven with agricultural fields and several lush gardens amid the otherwise barren surroundings. The coastline is rocky and rugged, leaving the short-grass fields and a few freshwater reservoirs as the only acceptable resting habitats for most species of shorebird.

When weather conditions are favourable from a migrating bird's perspective, most of them pass over Utsira at great altitudes and hence go unnoticed on the island. In such conditions, birdwatching at Utsira can be slow for days and even weeks on

Utsira island with its Herberg gardens has probably Norway's highest accumulated concentration of vagrants. Blue skies and "nice weather" like this is good migration conditions, but for the birds to go in for landing on offshore islands like Utsira, heavy clouds and a light easterly wind is hoped for. *Photo: Egil Ween.*

Oslo

Interior

Skagerrak

Western

Central

Northern

Finnmark

Svalbard

end. But when the weather deteriorates and becomes favourable from a birdwatcher's point of view, with dark clouds and a bit of precipitation accompanied by a light breeze, birds fly at lower altitudes and may be forced to land at the first possible spot, often proving to be Utsira. In such conditions great falls of thrushes, pipits, finches and many other migrants may occur. Days like that may produce a few rare birds as well, but the really exotic ones are often discovered in the aftermath, often a day or two later, when numbers of the commoner migrants start to tail off and the weather has changed again.

Utsira can boast of some exceptional season totals of species normally considered uncommon in Norway, such as Manx Shearwater (446), Richard's Pipit (22), Barred Warbler (25), Yellow-browed Warbler (102), Red-breasted Flycatcher (25), Ortolan Bunting (16) and Little Bunting (20) – all time high year totals in parentheses. More than 335 species have been recorded on the island, several of which are extreme rarities such as Black-browed Albatross, Spotted Sandpiper, Siberian Rubythroat, White's Thrush, Siberian Thrush, Eye-browed Thrush, Swainson's Thrush, Grey-cheeked Thrush, Pallas's Grasshopper Warbler, Syke's Warbler, Thick-billed Warbler, Brown Shrike, Red-eyed Vireo, Yellow-rumped Warbler, Cape May Warbler, Chestnut Bunting, Pallas's Reed Bunting, Pine Bunting and Rose-breasted Grosbeak.

Some rare species are at Utsira encountered with unsurpassed regularity, like Sabine's Gull (Sep–Oct), Short-toed Lark (late spring and Sep–Oct), Olive-backed Pipit (Oct), Citrine Wagtail (Aug–Oct), Nightingale (Apr/May–Jun), Siberian/Amur Stonechat (Sep–Oct), Eastern/Western Subalpine Warbler (Apr/May–Jun), Greenish Warbler (late spring and early autumn), Pallas's Warbler (Oct) and Rose-coloured Starling (Aug–Oct) – all close to annual (main seasons in parentheses).

In addition to being a migration hot-spot, Utsira is one of the most important breeding areas for seabirds in this part of Norway.

Most of the classic cliff-breeding species are represented. On the main island you find colonies of Common Eider, Fulmar and Black Guillemot. Additionally, the small neighbouring archipelago Spannholmane supports Shag, Guillemot, Razorbill and Puffin. On late summer nights, Storm Petrels

Strategy on "rarity islands"

1. Keep on your toes! Check each and every bird you find thoroughly and do not leave it until it is identified with certainty. Furthermore, don't hesitate to "walk the extra mile" in order to get to bushes or bogs that seem interesting.

2. Work systematically through the terrain. Don't bother yourself with thoughts that you should rather be somewhere else on the island. Dense, lush gardens can hide far more than they seem. Approach quietly and sit down and wait. Newly arrived migrants often move around restlessly and can appear in front of you when you least expect it. Walk along ditches and zigzag through reedbeds and tall grass fields. This is how many Jack Snipe and unusual passerines have been found. Look for shorebirds, ducks and the like on open bodies of water, but avoid flushing the birds by walking all the way down to the water's edge, primarily for the sake of the resting birds, but also keep in mind that ducks and waders act as effective decoys that will help to get even more birds to settle. Geese, Crane and other larger, shy birds seek peace on the more remote sections of the island.

3. Focus on a selection of sites, rather than stressing about trying to cover the entire island. This is especially true if there are several other birdwatchers present.

4. Plan your tactics according to the weather. If a fresh wind is blowing from the western sector, you should consider looking for seabirds with a telescope from a suitable vantage point, which on Utsira will be the *Stormhytta* shelter on Pedleneset, or possibly from the bunkers on Jupevikshoien on the east side of the island. In strong winds, the passerines often disappear from the island, although some will still linger in the deepest and least disturbed gardens.

5. Fill up the theoretical ballast in advance. The more you know about identification and the more experience you have from home and abroad, the greater chance you have of finding and determining the identity of difficult-to-identify species.

6. Be critical of species identification, both your own and others birdwatcher's conclusions.

7. Familiarize yourself with local faunistics, i.e. the local status of the various species. Many species that are common elsewhere in Norway are very rare or have never been recorded on Utsira, such as Crested Tit.

are caught for ringing, sometimes also a few Leach's Petrels, indicating that at least the former might breed here as well.

Utsira Bird Observatory conducts observations and ringing, primarily in the conifer plantation **c.** Søre merkeskog. Several hundred birdwatchers visit the island in late spring and in autumn, particularly in the first week of October, when the bird observatory arranges social events and lectures.

Other wildlife: Marine mammals are often encountered during seawatching.

Best season: Migration periods, particularly from late April to mid-June and from mid-September to mid-October. The highest number of birdwatchers gather on Utsira in the first week of October. In autumn, the two–three weeks before and after this core period can

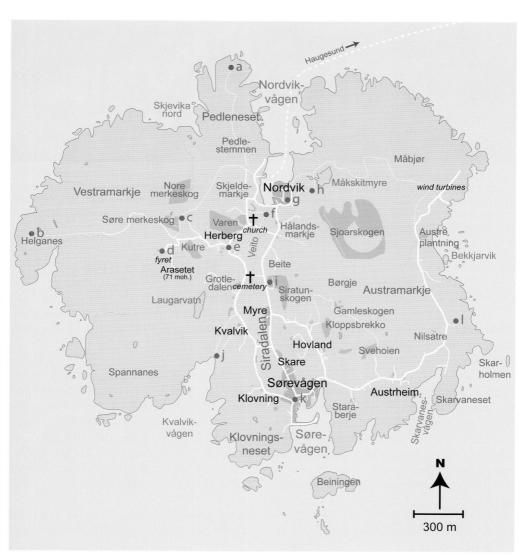

Utsira island: a. *Stormhytta* seawatch shelter; **b.** Helganes, alternative seawatch; **c.** ringing site; **d.** Utsira lighthouse; **e.** Herberg gardens; **f.** Veito, reed-covered creek; **g.** Dalanaustet; **h.** Lake Måkskitmyre (ducks etc.); **i.** Siratun municipal centre; **j.** Kvalvikvågen inlet with a small reedbed; **k.** grocery store; **l.** Jupevikshoien bunkers, viewpoint.

Utsira island

Utsira island has been a key destination for bird enthusiasts from all over Norway since the early 1970s. During the first week of October, there can be hundreds of birdwatchers on this small island. It's a nice social event, but makes it difficult to spot the birds on your own. This is probably part of the reason why more and more new rarity islands have been explored along the Norwegian coast. The picture is from the football field at Hovland. *Photo: Egil Ween.*

be good as well, depending on the weather conditions. For migrating seabirds, May and from July to October are the best periods.

Directions: Utsira is reached by a 70-minute car ferry from Haugesund harbour. It runs 3–4 times per day.

Tactics: See the *Strategy on "rarity islands"* section, p. 174. Even though many rare birds have been here found over the years, rare birds are rare on Utsira as well. The residents of Utsira are used to birdwatchers, but please respect private property. Check with the local birders which properties may be entered. Keep strictly to the edges of growing cultivated fields. Be particularly careful when crossing fences so that you do not bend them or kick down any stones in stone walls. Leave all gates as you found them, whether open or closed. Please report all your bird sightings to the bird observatory.

10. BERGEN AREA

Vestland County
GPS: 60.39090° N 5.32658° E (Lake Lille Lungegårdsvann), 60.27527° N 5.41787° E (tower hide Kalandsvika)

Notable species: Gulls, many winters including Ring-billed Gull; Whooper Swan and Smew (winter), Rock Ptarmigan, Little Grebe, Golden Eagle (winter), Rough-legged Buzzard, Water Rail, Moorhen (winter), Coot, Dotterel, Grey-headed Woodpecker, White-backed Woodpecker, Grasshopper Warbler, Sedge Warbler and Snow Bunting.

Description: Bergen is Norway's second largest city, with about 270 000 inhabitants. It is also one of the most popular tourist destinations in the country. The city is located in a sheltered spot between what is known as *The seven mountains,* and behind Øygarden, a row of islands protecting the

176

city from the often severe winds from the North Sea. The moist air is pushed up to higher altitudes by the mountains, leaving Bergen with an average of 242 rainy days per year. The city has a few lakes attracting interesting gulls and ducks, mainly in winter when up to 2200 Common Gulls have been counted at the most often visited Lake Lille Lungegårdsvann in the central city park. This small lake, and the larger Lake Tveitevannet at Slettebakken, are particularly famous for having produced Ring-billed Gulls almost annually for long periods of time, since 1983. Of course, some of these are returning individuals, but at least 16 different birds have been involved. Two ringing recoveries indicate that they might actually perform biannual migration back and forth to North America. Glaucous and Iceland Gulls are annual in Bergen too, and Caspian, Ivory and even Thayer's Gull have been recorded here.

It is fairly easy to find a few inland breeding specialities near Bergen, such as Grey-headed and White-backed Woodpeckers, for instance near Lysekloster monastery. At Gullfjell mountain not far from the city you reach high elevations with Rock Ptarmigan, Golden Eagle (winter), Rough-legged Buzzard, Dotterel (May), Ring Ouzel, Snow Bunting and more. A few good conventional birdwatching sites are found south of the city, particularly around Flesland airport. Lake Kalandsvatnet at Fana is the largest lake in the area. Here, Sedge Warbler is common, and Grasshopper Warbler usually found in early summer. The lake is also an important resting and wintering area for waterbirds, annually including Smew.

Other wildlife: Red deer and hedgehog are often seen in the Bergen area.

Best season: Autumn and winter is best in Bergen city centre, while the migration periods and early summer is best at Flesland and Lake Kalandsvatnet.

Directions: The network of public transportation in and around Bergen is well developed.

Lake Kalandsvatnet at Fana is one of the most important wetlands in the Bergen area. A tower hide gives a view to the Kalandsvika cove in the south-eastern corner of the lake. *Photo: Bjørn Olav Tveit.*

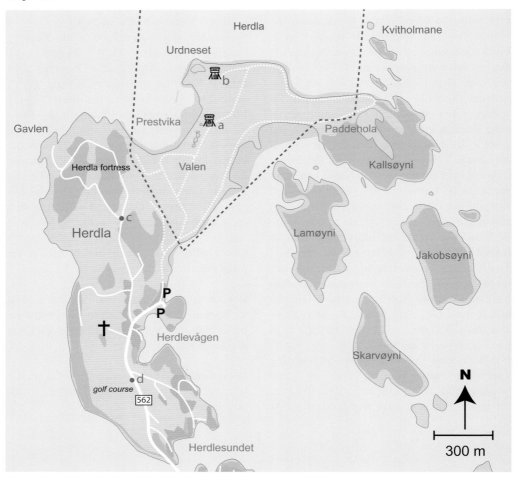

Øygarden islands, Herdla island: a. Prestvika tower hide; **b.** Urdneset birdwatching hide; **c.** Herdla fortress (museum and bird observatory); **d.** golf course.

By car: Lake Lille Lungegårdsvann is situated in the city centre, close to the railway station. Lake Tveitevannet is close to Slettebakken school in Årstad, in the southern part of the city. The sites around Flesland airport can be found by following signs along E39 south of the city. Explore the surrounding fields, small lakes and coastline. Grey-headed Woodpecker breeds in this area. Lake Kalandsvatnet is reached by following E39 further south towards Stavanger. About 4 km south of the Flesland junction, just after an exit signposted to Fana church, you see Kalandsvatnet on your right. Stop at the bus stop just opposite the exit to Bontveit and find suitable parking. From the bus stop, walk along the tractor tracks along the fields 100 m to the tower hide. Do not cross the fields during the growing season. Lysekloster monastery is reached by continuing past Lake Kalandsvatnet and turning right onto the road signposted to Lysekloster. Park by the monastery ruins and follow the path behind them and into the valley. Gullfjell mountain is reached by leaving E39 at Nesttun, just south of Bergen, heading north onto Rv580. After about 10 km, turn right towards Unneland and follow this road to the parking area at the top. From here you can take walks for instance along the path signposted to *Redningshytten* (rescue cabin).

11. ØYGARDEN ISLANDS

Vestland County

GPS: 60.57014° N 4.95565° E (Herdla), 60.56850° N 4.81749° E (Herdlevær), 60.59856° N 4.82730° E (Tjeldstø)

Notable species: Migrating land- and seabirds, often including White-billed Diver, Manx Shearwater and Pomarine Skua; Eagle Owl, Grey-headed Woodpecker (autumn) and Stonechat.

Description: West of Bergen is a neat row of islands called Øygarden, *the guard of islands*, protecting the inner coast from the rough North Sea and providing a guideline for land- and seabirds migrating up and down the Norwegian west coast. The landscape here is quite barren and typical of the region, with knolls, heathers and a number of conifer plantations. Any moist pasture, garden or collection of shrubs and trees has the potential for resting migrants. Especially the northern part of Øygarden, with Herdlevær, Hjelme and Hellesøy, are particularly good for passerines. Here you will find shrubs, clusters of trees, scattered gardens and pastures. The small, northernmost island Hernar have proved to be particularly attractive to passerines. Quite a few rarities have been found in Øygarden, including Dark-sided Flycatcher. The marshes and lakes at Tjeldstø are the best resting grounds for wetland birds in the area, ducks and fresh-water shorebirds in particular. At Ona, geese sometimes stage. The rocky and rough parts of Øygarden, is home to a few pairs of Eagle Owl and Stonechat.

Skogsøy island protrudes a bit further west than the other islands in Øygarden. A seawatching shelter is offered here. White-billed Diver is regular in fairly good numbers, particularly on days in May with a gentle breeze from the north. Pomarine Skua and Manx Shearwater can pass by in good numbers in late spring as well, particularly in westerly winds. From late summer, Sooty

Herdla island has beautiful beaches and grazing fields with water flashes that make an eldorado for shorebirds, ducks and more. The island is equipped with two hides, both visible in the picture. *Photo: Kjetil Salomonsen.*

Øygarden islands has many interesting sites. One of them is Skogsøy island, where you can sit dry and warm in an observation shelter on top of the cliff and observe the migration of seabirds. You would normally hope for more stormy weather than in this picture. *Photo: Kjetil Salomonsen.*

Shearwater is regular in strong westerlies. The outer coast can be good for sea ducks, divers and such during migration and in winter.

As a contrast to the islands of Øygarden, the flat and fertile Herdla island is located on the lee side of Øygarden. The island is just 2 km², with grazing pastures and long stretches of muddy shoreline, particularly on the north side, perfect for shorebirds, geese, ducks, pipits and more. Herdla island is arguably the best all-round site in the vicinity of Bergen. In overcast conditions, and when the pastures are partially flooded with rainwater, Herdla is at its best. Under such conditions up to 1000 shorebirds have gathered here. Several rarities have been encountered at Herdla, and Richard's Pipit is a regular visitor in September/October. The sea surrounding the island is shallow and a very important feeding ground for sea ducks, divers and grebes all year. White-tailed Eagle and Gyrfalcon are often seen in winter.

Other wildlife: Otter and harbour porpoise are rather common. Other whales are sometimes seen while seawatching. Red deer is common on the islands.

Best season: All year, but migration periods in particular.

Directions: *Public transportation to Øyøgarden:* Øygarden may be reached by bus from Bergen city centre, see local timetables. Aim for the last bus stop at *Hellesøy* if taking the express boat to Hernar island.
Øygarden by car: From Bergen, follow Rv555 west towards Sotra. In the T-junction at Sotra, turn right onto Rv561 signposted to Hellesøy.
Herdla by bus: from Bergen, take bus no. 499 Herdla/Askøy to *Herdla*.
Herdla by car: from Bergen, follow Rv555 towards Sotra and exit north after 5–6 km onto Rv562 signposted to Askøy. Continue about 30 km to the end of the road at Herdla.

Tactics: At Øygarden, the best sites may be explored along Rv561 towards Hellesøy as follows:

Turøy island: Turn left, signposted to Turøy, 14.5 km north of the T-junction at Sotra. This is primarily a ringing site, most actively in use during the 1990s, but it can be explored conventionally as well by searching the dense vegetation along the roads thoroughly.

Skogsøy seawatch: 15 km further north along Rv561, turn left onto the road signposted to Herdlevær. After 2.7 km, stop roadside just before the second bridge. Bring your telescope and walk along the marked path the 2 km out to the seawatching shelter on the headland.

Herdlevær: To search for resting migrants at Herdlevær, drive a few hundred metres further. You explore Herdlevær like you do a rarity island (see Utsira island). Do not cross the pastures and be especially considerate when exploring the gardens, which are preferably viewed from the roads.

Tjeldstø wetland: This wetland is visible on the right-hand side of Rv561, 4 km north of the Herdlevær junction. A bus shelter with an observation hatch is located just after the entrance to a side road signposted to Tjeldstø, adnd another observation shelter is visible from the main road a little further ahead. However, you are probably better off viewing the wetland from the roads on both sides of the wetland.

Hellesøy and Hernar islands: Continue to Hellesøy at the end of Rv561. Hernar is reached by express boat from here.

Herdla island: You will definitely need your telescope here. Follow the paths and tractor tracks at Herdlevalen fields – do not stray out onto the grazing pastures or in the nature reserve in the period of traffic restrictions. Observations can be made from the **a.** tower hide and the **b.** shelter hide. Check the **d.** golf course as well, often attracting pipits and the like when the course is not in use. The planted area around **c.** Herdla Fortress can host resting passerines.

At Herdla, access onto the beach on the west side, between Prestvika and Urdneset inlets, is not permitted from April 15 to September 30, except along the marked tracks.

12. FEDJE ISLAND

Vestland County

GPS: 60.78857° N 4.83044° E (ferry-landing Sævrøy), 60.78233° N 4.71435° E (ferry-landing Fedje)

Notable species: Resting passerines; vagrants; seabird migration.

Description: Fedje (pronounced: *feh-yeh*) is the most classic of the rarity islands in the proximity of Bergen city. The island is situated as an extension of Øygarden to the north, 4 km west of the mainland. The island is 7,2 km², a bit larger than Utsira island. The terrain on Fedje is much the same as in Øygarden, dominated by rocky heathland, and similarly, the most important bird areas are the gardens, conifer plantations and grazing pastures. These kinds of habitats are mainly found around the ferry-landing in the north, where most of the about 450 inhabitants live, and to a lesser extent around the village Stormark on the south side of the island. Fedje is also an excellent place for seawatching. Several rarities have been found over the years, including Oriental Turtle Dove, Calandra Lark, Lanceolated Warbler and Hume's Warbler. Yellow-browed Warbler and Richard's Pipit are annual in autumn.

Best season: Late spring and autumn.

Directions: The bus from Bergen to Sævrøy takes 2.5 hrs and corresponds with the ferry to Fedje.

By car from Bergen, drive north along E39 across the Osterfjord (which is good for resting and wintering sea ducks). Next, exit onto Rv57 and after another 1.5 km turn left onto Rv565. Follow this road 33 km, and just after Austrheim church, exit onto Rv568 signposted to Fedje. This road leads to the ferry-landing at Sævrøy.

Tactics: Fedje is explored like any other rarities island, such as described for Utsira island. Generally speaking, the lushest gardens are found near the ferry-landing

At Fedje island you will find the most lush gardens and conifer plantations on the north side of the island. The best pastures are on the south side. *Photo: Bjørn Olav Tveit.*

in the north and the best pastures are found in the south. You might want to bring a bicycle or even a car in order to work both the north and the south of the island, situated 3 km apart with rather barren heathland in between. Sewatching is best conducted from the bunkers at Søre Vidnappen in the far northwest. To get there from the ferry, follow signs to Mulen, then make a left and follow the road signposted to Langedal. Keep left by the maritime pilot station and follow the road to its end. Walk through the gate and a path will lead you to the bunkers. This route also leads you past many of the best gardens on the island.

If you must wait a while for the ferry at Sævrøy on your way to Fedje island, drive back a bit and turn north towards Krossøy. This road leads past some nice areas with gardens and pastures. Stonechat probably breeds in this area. There are several nutrient-rich lakes towards Mongstad worth visiting, such as Lake Purkebolsvatnet, good for nocturnal songbirds and waterbirds in general.

13. VOSS AREA

Vestland County

GPS: 60.69847° N 6.48957° E (Lønaøyane delta), 60.62689° N 6.30946° E (Rekvesøyane delta), 60.62827° N 6.44639° E (Haugo farm)

Notable species: Resting waterbirds; Quail, Green Woodpecker, Bluethroat, Nutcracker and Marsh Tit.

Description: The landscape of the inner Hardanger area is famous for its tall mountains and narrow fjords and valleys, but at Voss the valley widens and give way to fertile agriculture. Along the Vosso watercourse running through the valley are some of the best resting areas for migrating waterbirds in the inner fjord system of the Norwegian west coast. The sites are at the mercy of the changing water levels, however, and are at their best when the water level is low so that mud is exposed, leaving small pools of water. One of the best sites here is Lønaøyane delta, also known as Reppen, by the mouth of the river Strandaelvi in the north end of Lake Lønavatnet. This delta area is partly cultivated and equipped with a tower hide. An interesting delta has also formed a bit further north, by the mouth of river Myrkdalselvi in the north end of Lake Myrkdalsvatnet. Here at the Myrkdal delta, Bluethroat and in some years Sedge Warbler breed, and both Marsh Warbler and Common Rosefinch have bred. Quail is usually represented with at least a couple of calling individuals annually. Rekvesøyane at the mouth of the river Dyrvo along the northwestern shore of Lake Vangsvatnet is worth checking. This site becomes free of snow early in spring, which is a quality migrants appreciate. At the mouth of river Vosso in the east-end of Lake Vangsvatnet is Grandane, an area of long gravel beaches and a few mud banks, ideal for shorebirds when the water level is low. At all times it is a favoured area for ducks. This area becomes ice-free early, in March/April. Little Ringed Plover has bred several times. This site is conveniently situated near Vossevangen, the village centre in Voss. Unfortunately, it is also a favoured spot for dog-walkers, so early mornings and days with inclement weather are usually preferred for birdwatching. Another site near Vossevangen village is Lundarosen, a calm stretch of the Vosso river. When the fields, particularly the moist ones, are recently ploughed in spring they can be very attractive to birds. One such place often proving to be good is at the farm Haugo in Vossevangen village.

Other wildlife: Red deer is very common.

Best season: April–May and from August to early October.

Directions: These sites are located near Vossevangen, the village centre in Voss, which is situated along E16 90 km east of Bergen.

Tactics: a. <u>Rekvesøyane</u> is along E16 4 km west of Vossevangen, and just west of the junction signposted to Dyrvedalen. It is visible from several places along the highway, but be

Lønaøyane delta in Voss in spring, with a high water level due to meltwater from the surrounding snow-covered mountains. Notice the tower hide behind the pine trees out in the delta. *Photo: Gunnar Bergo.*

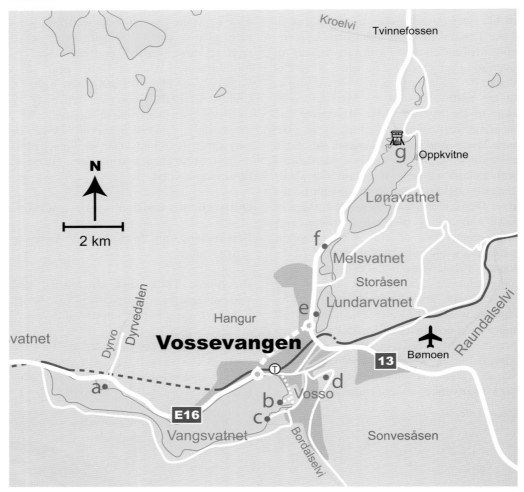

Voss area: a. Rekvesøyane; **b.** Grandane; **c.** Gjernesmoen; **d.** Haugo farm; **e.** Lundarosen; **f.** viewpoint Lake Melsvatnet; **g.** Lønaøyane delta.

aware of traffic. You may also park just west of a bridge crossing E16 and walk along the edge of the field along the river out to the headland. **b.** Grandane can be approached from several directions, but the most common entry is from the sports grounds by Voss school and Voss Camping on the west side of the river Vosso. On nice-weather days and in weekends, this area is often crowded with people. Then go there in very early morning, or try **c.** Gjernesmoen on the other side of river Vosso instead: drive in from the Voss water treatment plant, and walk out on the shoreline. The **d.** Haugo farmland is reached by crossing river Vosso along Rv13 at Langebrua bridge and then exit onto Bordalsvegen road towards

Rong. Make a stop after about 1 km, near the sharp right bend before the farm Haugo, which is the first farm you see. **e.** Lundarosen is along E16, 300 m north of the northernmost of the two exits to Rv13. Park along one of the side roads at the opposite side of E16. In order to get to **g.** Lønaøyane delta, continue along E16 towards Oslo (make a stop at the **f.** viewpoint to Lake Melsvatnet) and then exit onto the first side road signposted to Nedkvitne. This road leads along the east side of Lake Lønavatnet, and the site is located at the north end of this lake. Park at *Nesheimstunet* and walk 200 m further along the road. Head down along the tractor tracks to the left, signposted to the tower hide (*Fugletårn*) 600 m further on. If you arrive along E16 from the east, you can reach this site by exiting to Norheim/Nedkvitne just after the spectacular Tvinnefossen waterfall. The Myrkdal delta is found by following the E16 towards Oslo 8 km past Tvinnefossen and exiting onto Rv13 signposted to Vik. You will soon see Lake Myrkdalsvatnet on your left, but continue to the north end of the lake. Here, several side roads lead through the delta. Rv13 continues uphill to the Vikafjell mountain area, good for birds of higher elevations.

14. THE DEEP FJORDS

Vestland County

GPS: 61.13420° N 6.99704° E (Fimreiteåsen), 61.42005° N 6.75396° E (Bøyaøyri), 61.77113° N 6.19745° E (Fitjefjøra), 61.90531° N 6.71306° Ø (Stryn delta)

Notable species: Black Grouse, Goshawk, Wryneck, Grey-headed Woodpecker, Green Woodpecker, White-backed Woodpecker, Lesser Spotted Woodpecker, Three-toed Woodpecker, Redstart, Ring Ouzel, Icterine Warbler, Long-tailed Tit, Marsh Tit, Nutcracker and Parrot Crossbill.

Description: The deep fjords of the Norwegian west coast are popular tourist attractions, due to the spectacular scenery. These narrow fjords are cut deeply into the tall mountains, with several gleaming glaciers adding to the dramatic experience. The West Norwegian Fjords are on UNESCO's World Heritage list.

In a bird's eye perspective, however, the fjords are very deep and nutrient-poor with steep shorelines rising quickly to high elevations. Thus, shorebirds have a limited selection of attractive feeding grounds here. The wooded slopes do hold a

Bøyaøyri delta is beautifully located in the fjord landscape, surrounded by glaciers and tall mountains. The tower hide is adapted for wheelchairs, while also acting as a screen between birds and people in the parking area. *Photo: Bjørn Olav Tveit.*

Oslo · Interior · Skagerrak · Western · Central · Northern · Finnmark · Svalbard

Fitjefjøra wetlands by Sandane in the Gloppenfjord. *Photo: Bjørn Olav Tveit.*

few specialized species, all rather sparsely distributed.

One of the prime woodlands by the Sognefjord is Fimreiteåsen hill, a tall peninsula covered in open, mature pine forest mixed with deciduous trees. All the Norwegian woodpeckers have been found breeding here, except for Black Woodpecker. Decent wetlands can be found only in a very few locations. In the Sognefjord, the Bøyaøyri river delta in Fjærland is the most notable, to some extent Lærdalsøyri delta in Lærdal as well. Fitjefjøra by Sandane and the Stryn delta, both in Nordfjord, are also worth visiting if you happen to pass through these areas, either as a general tourist or on your way to the more bird-rich sites along the outer coast.

Other wildlife: Flocks of red deer are often seen crossing the roads or grazing on pastures from dusk to dawn.

Best season: Most of the woodpeckers are present year-round, but are easiest to find when displaying on sunny mornings in early spring. The wetlands are at their most interesting during passage and in winter.

Directions: Lærdalsøyri delta: Located along Rv5 in Lærdal village centre. Drive or walk along the river to the river mouth.

Fimreiteåsen hill: From Rv5 between Sogndal village centre and Kaupanger, exit south signposted to Fimreite. Follow this road 15 km and stop to look for woodpeckers where you pass mature forest. From the village at the end of the road a forest road leads up onto the hill itself. The best areas are near the top, at an elevation of about 600 m. You need to ask permission from the landowner in order to drive to the top. It is also possible to walk in from Sogndal airport, which is situated at a high elevation on the same hill, a bit further east.

Bøyaøyri delta: From Rv5 between Sogndal and Skei, exit onto the road signposted to *Fjærland sentrum*. After 1 km, there is a roadside resting area with a peculiar, wheelchair friendly tower hide overlooking the delta. Check the surrounding cultivated fields as well.

Fitjefjøra wetlands: The village Sandane is situated along E39. Exit onto Rv615 signposted to Hyen and Florø, a road taking you through Sandane and past the shoreline of Fitjefjøra. The surrounding farmland is worth checking too, and so are the fields around Byrkjelo, 15 km to the east along E39. White-backed Woodpecker may be found in the surrounding slopes.

Stryn delta: The village Stryn is situated at the mouth of the Stryneelva river. Waterbirds may rest on islets in the river or in the remnant areas of wetland among the populated areas and cultivated fields. A tower hide is erected by the small, vegetated pond Kjeldevatnet.

15. UTVÆR ISLAND

Vestland County

GPS: 61.03713° N 4.50991° E (Utvær lighthouse), 61.00772° N 4.67017° E (Kolgrov)

Notable species: Resting migrants, breeding and migrating seabirds; White-tailed Eagle.

Description: Utvær is a group of small and low, windswept islands and skerries covered with grass, 7 km off the coast at the mouth of the Sognefjord. The outermost islet represents the westernmost point of dry land in Norway. Several of the islands are important breeding grounds for seabirds, including Greylag Goose, Common Eider, Shag, White-tailed Eagle, gulls and Black Guillemot, probably Storm Petrel as well. The seabird populations have declined for many years, and Puffin, Razorbill and Guillemot have probably disappeared as breeding birds during the last thirty years.

The main island of Utvær is actually made up of two elongated, parallel islands interconnected with a causeway, forming an H. The largest of them being 850 m in length. This main island holds the westernmost lighthouse in Norway, automated in 2004, and a small village, now deprived of its permanent residents. A very few low bushes and scrubs are found around the houses and the lighthouse. In addition, a small garden known as *Paradise* was established more than a hundred years ago by the lighthouse warden assistant in the shelter of a crack in the rocks on the southwestern tip of the island.

Dipper was named Norway's national bird in 1963. It thrives in waterways in most of the country. The species is hardy and will survive as long as it has access to open, flowing water. *Photo: Kjetil Schjølberg.*

Oslo

Interior

Skagerrak

Western

Central

Northern

Finnmark

Svalbard

Utvær island is the westernmost point and settlement in Norway. *Photo: Bjørn Olav Tveit.*

It was later restored by his daughter. Surely, this little garden must have been a paradise to many a lost migrant flying exhausted in from the sea! The sheltered inlet on the south side has a small, shorebird-friendly tidal flat. Utvær has never really been brought to birdwatchers attention, mainly because of the lack of birdwatchers in this part of the country. It is mentioned here because of its strategic location and potential for interesting migrants. Utvær is also well worth visiting for its shear beauty.

Best season: Late spring and autumn.

Directions: The express boat to Utvær departs from either from Hardbakke or from the village of Kolgrov on Ytre Sula, the outermost large and populated island in the region. The trip takes about 30 minutes and runs only a very few times a week. You need to phone the boat crew the day before departure to let them know that you wish to attend the trip. It is possible to arrange pickup outside the regular timetables, although at a fairly high cost.

The easiest way to access Hardbakke/Kolgrov is to take the express boat from Bergen via Mjømna kai (with a corresponding express boat change) to Nåra on Ytre Sula. Here, the boat corresponds with the bus to Kolgrov in time for the Utvær express boat.

Hardbakke/Kolgrov by car: Drive route Rv57 which runs parallel with the coast west of E39. Across the Sognefjord, there is a car ferry running in a triangular connection between Rysjedalsvika at the north side of the fjord, Rutledal on the south side, and Krakhella at inner Sula island. From Krakhella, follow Rv606 across inner Sula to Hardbakke, or continue to the Daløy–Haldorsneset ferry providing a short crossing to Ytre Sula, in order to reach Kolgrov. At Ytre Sula, follow signs 15 km to Kolgrov on the southwest side of the island. You may want to arrive early, so that you have time to do some birding around the villages at Ytre Sula as well.

Tactics: The lighthouse might be rented for accommodation and is a good base from where you have the best views to the sea and can find some shelter among the buildings. From here, you can also see much of the main island, including the pastures in the southwest. Renting a private house may be possible, but if not, bring a tent and camping gear. A thorough search of the island takes about an hour.

16. ASKVIKA ESTUARY

Vestland County
GPS: 61.34335° N 5.08666° E (causeway)

Notable species: Resting wetland species; Whooper Swan (winter), Shelduck and White-tailed Eagle.

Description: Askvika is a coastal estuary surrounded by pastures, cultivated land and forest. It is perhaps the most important

staging ground for shorebirds and other waterbirds between Bergen and Ålesund. This wetland system can be divided into three sections interconnected by a river. At the mouth is Askvika inlet, a tidal zone crossed by a causeway. Further up is the brackish Lake Leira, and above it Lake Kylleren, a deeper freshwater lake. Golden Eagle, White-backed Woodpecker and Eagle Owl breeds in the vicinity.

Other wildlife: Red deer is common. Otter may be encountered.

Best season: All year, but migration periods in particular.

Directions: The site is located near Askvoll village, which is signposted to from E39 south of Førde.

Tactics: A few km before you reach Askvoll village, the fjord landscape changes character and you get large, wide farmlands on both sides of the road. You have to scan these fields for resting geese,

Askvika estuary is an important staging ground for shorebirds, ducks and more. *Photo: Bjørn Olav Tveit.*

shorebirds, pipits, thrushes and such. In the middle of these fields is a crossroads, where Askvoll is signposted to the left. You now have Lake Kylleren straight ahead. Drive back to the intersection in the field and turn towards Askvoll. After 1 km you will see Lake Leira on your right. Further towards Askvoll, the road soon crosses the riverbed. Here you can scan from the bridge. Continue for 500 m towards Askvoll and turn left onto farm road sign no. 230–238. From this road you have a view of the middle part of Askvika estuary. Continue 200 m further towards Askvoll and turn down to the left. This road leads out onto a causeway, from where you have a nice view of the outer part of the inlet and the fjord beyond. A telescope is absolutely necessary when visiting Askvika estuary.

17. BULANDET & VÆRLANDET ARCHIPELAGO

Vestland County

GPS: 61.29345° N 4.67277° E (Melværet)

Notable species: General migration including seabirds and rarities; White-tailed Eagle, Stonechat.

Description: Bulandet and Værlandet comprises two groups of islands at the outer coast, interconnected with a network of roads. A ferry crossing connects these islands to the mainland. Grazing pastures, lush gardens and a few planted coniferous woods provide feeding opportunities and shelter for resting migrants. Records of rarities include Solitary Sandpiper, Blue-cheeked Bee-eater, White's Thrush, Two-barred Greenish

Bulandet & Værlandet archipelago contains a large number of islands and skerries. This is from the outermost settlement, Sandøya island, where a number of vagrants have turned up. *Photo: Bjørn Olav Tveit.*

Red-breasted Flycatcher used to be reckoned as a speciality of Utsira island, together with other Eastern vagrants, such as Yellow-browed Warbler and Olive-backed Pipit. In recent years Norwegian birdwatchers have tried out out other islands along the coast, only to find that these "Utsira specialities" may well turn up at other locations as well, such as in the Bulandet & Værlandet archipelago. *Photo: Bjørn Fuldseth.*

Warbler, Red-eyed Vireo and White-crowned Sparrow. The Bulandet–Værlandet archipelago is an important resting area for the Svalbard population of Barnacle Geese in autumn. These birds head south along the Norwegian coast until reaching the archipelago, where they rest and feed before crossing the North Sea en route to their wintering grounds in the British Isles. Seabird passage may be appreciated from the outermost islands at Bulandet, but due to the many low islands, the birds do not come as concentrated at Bulandet as past the tall and solid headlands of Kråkenes and Stad further north. A few pairs of Stonechat breed in the heathlands.

Best season: Late spring and autumn.

Directions: Car ferry from Askvoll harbour (see directions to Askvika estuary above) to Værlandet.

Tactics: Out on the outermost islands of Bulandet you have many nice gardens, shrubbery and small plantations. It can be nice to start the day by standing on the mound at Kjempeneset at dawn, watching the movements of migrants in the air above you and out at sea. Eventually you can move around the terrain in this area, before expanding your search to other parts of Bulandet, and work your way inwards via Melværet island towards Værlandet islands. The archipelago is too large for you to search all parts thoroughly in one day, so it may be wise to have several people who split up in smaller groups. On the south side of Melværet you will find the archipelago's finest fields. They are quite remote, and geese and other shy species can therefore often roost here. It is also a nice place for pipits, plovers and other shorebirds – and for red dear. Scan surrounding islets and skerries carefully with a telescope for resting geese, shorebirds and White-tailed Eagles. Migrating divers and skuas may pass directly over the islands. It can also be a good idea to make (private) boat trips out to the outer island of Sandøya, which has large grasslands and exciting little clusters of small trees and berry bushes down in a sheltered gorge. Sit by the lighthouse or between the few houses to scan the ocean.

18. KRÅKENES HEADLAND

Vestland County

GPS: 62.03469° N 4.98576° E (Kråkenes lighthouse), 62.01603° N 5.10633° E (Vedvik inlet)

Notable species: Seabirds, resting migrants; Black Grouse, White-billed Diver, Fulmar, White-tailed Eagle, Rough-legged Buzzard, Golden Eagle, Peregrine Falcon, Water Rail (winter), Great Skua, Pomarine Skua, Glaucous Gull (winter), Razorbill, Guillemot, Puffin, Long-eared Owl, White-backed Woodpecker, Ring Ouzel, Grasshopper

Kråkenes headland with Lake Kråkenesvatnet to the right, seen from the Mehuken mountain, towering above the headland 433 m above the sea. The lighthouse hides behind the promontory on the left. The Stad headland to the north is barely visible in the distance. *Photo: Bjørn Olav Tveit.*

Warbler, Sedge Warbler, Marsh Warbler and Stonechat.

Description: Kråkenes lighthouse at the western tip of Vågsøya island is perhaps western Norway's most accessible outposts for seawatching. The site is situated in a spectacular landscape with tall mountains plunging steeply into the sea. No ferry crossings are required when driving here the 7 hrs. from Oslo, and so it is possible to respond quickly to favourable weather forecasts, predicting either a light northerly breeze in late spring or incoming low pressures and westerly strong winds in late summer and autumn. Particularly in May, substantial numbers of coastal migrants can pass by, often including several White-billed Divers and dozens of Pomarine Skuas on a daily basis. Late summer through autumn is the time for pelagic species such as Manx and Sooty Shearwaters. Kråkenesbygda village just inland from the lighthouse has nice grazing pastures, gardens and coniferous plantations, and a small lake and a very interesting reedbed, all suitable for resting migrants. Long-eared Owl probably breed annually here. White-tailed Eagle and Peregrine Falcon are year-round residents, sometimes accompanied by Gyrfalcon in autumn and winter. Just south of the village is Einevarden seabird cliff. If you have the nerve, you can stand on the edge of the cliff and look down upon thousands of seabirds including Fulmars, Shags, Kittiwakes, Puffins, Razorbills, Guillemots and a few Black Guillemots several hundred metres below you. The seabird populations have declined markedly in recent years, however. In the heathland around Kråkenesfjellet summit, Willow Ptarmigan, Black Grouse and Stonechat breed, in good rodent years Short-eared Owl and Rough-legged Buzzard as well.

Further in on Vågsøy island, Golden Eagle and White-backed Woodpecker breed. The cultivated fields and wetlands on Vågsøy can be very exciting places for nocturnal songbird trips in early summer, usually including several Sedge and Grasshopper Warblers and often with the addition of Corncrake, Quail and more. In winter, Great Northern Diver and Water Rail are often found, and the Stonechats seem to linger in the area all year. Måløy, the largest town in the area, is an active fishing community attracting large numbers of gulls, particularly in winter, usually with several Glaucous Gulls and perhaps one or two Iceland Gulls as well. The area is particularly famous for being the site of the only record of Dicksissle in Europe. Other rare species like Night Heron, Ivory Gull, Short-toed Lark, Pechora Pipit and River Warbler have been encountered as well.

Other wildlife: Red deer is common on Vågsøy island. Marine mammals are often seen from Kråkenes lighthouse.

Best season: All year, but the migration periods in particular.

Directions: From Oslo, Kråkenes is reached by following E6 towards Trondheim. At Otta, exit onto Rv15 to Måløy. From here, follow signs to Raudeberg, and then to Kråkenes.

Tactics: The wall surrounding Kråkenes lighthouse provides shelter in most conditions but even though it is located on a cliff 40 m above sea level, saltwater is sprayed well over the wall in westerly storms. On nice spring mornings it can be pleasant to sit on a camping chair in front of the wall. A telescope is a must, but you should use your binoculars actively as well, in order to discover close and high-flying birds. Check Kråkenesbygda village regularly for resting migrants. Einevarden is reached by driving up to the summit of Kråkenesfjellet. Just before the road starts going downhill towards Kvalheim village, you see the Lake Skjenevannet and a small coniferous plantation below the road. Park here and walk between the forest and the lake, and continue in the same direction 1.5 km to the edge of the 300 m tall and vertical Einevarden seabird cliff. There should be no need to warn you to watch your step! Back in the car, continue through the gardens and pastures around Kvalheim village

towards Måløy and turn left towards Refvik. Here are more interesting fields and gardens, besides the beautiful, sandy Refviksanden beach, usually supporting resting gulls and sometimes a few shorebirds. Scan the bay for auks, divers and sea ducks. The large reedbed inland from the beach is worth checking thoroughly. Continue along the main road through Refvik, leading to the next village Vedvik. Here is a sheltered lagoon and lush saltmarsh often supporting species associated with eastern Norway, such as Shelduck, Linnet and Whitethroat. Check the gulls in Måløy harbour, often roosting on the rooftops, and you may continue driving around Vågsøya island, signposted to the remarkable stone Kannesteinen. Alternatively, you can take the car ferry from Måløy to Oldeide in order to explore the coast of the next headland to the south, Bremangerlandet, which can be good at all times of year. If you follow Rv15 a few kilometres towards Oslo, you soon get to Almenningen on the south side of the road. The small pond here is a good site for ducks, and the side road past the pond leads through woodland with breeding White-backed Woodpecker and Tawny Owl.

19. STAD HEADLAND

Vestland County
GPS: 62.16467° N 5.10216° E (parking Litlehovden)

Notable species: Migrating seabirds, resting migrants; Barnacle Goose (autumn), Black Grouse, White-billed Diver, White-tailed Eagle, Golden Eagle, Peregrine Falcon, Water Rail (winter), Great Skua, Pomarine Skua, Glaucous Gull (winter), Ring Ouzel, Grasshopper Warbler, Sedge Warbler, Marsh Warbler and Stonechat.

Description: Stad or Vestkapp, the northwestern corner of southern Norway, is a landmark mountain ridge plunging into the sea. It is an exiting area for birdwatchers, particularly good for seawatching. The best point from which to enjoy the passage of seabirds is the headland of Litlehovden near Ervik village on the south side of the Stad peninsula. Here, vast numbers of seabirds have been counted, such as Brent Goose (with up to 1,895 on one day), Barnacle Goose (up to 9,875), Red-throated Diver (up to 1,005), Great Northern Diver (11), White-billed Diver (23), Fulmar (6,000), Manx Shearwater (106), Sooty Shearwater (117), Gannet (ca. 2,000), Kittiwake (15,000), Great Skua (39), Pomarine Skua (63), Arctic Skua (102) and Little Auk (3,150). Obviously, many of the birds passing Litlehovden are the same as the ones passing Kråkenes lighthouse, on the nearest headland visible to the south. At Litlehovden, however, they often pass at closer quarters than at Kråkenes. On the other hand, at Litlehovden you sit exposed in the wind after walking half an hour from the car in difficult terrain. Ducks, cormorants and gulls are often seen resting on the sea at the tip of Litlehovden and in the inlet at Ervik. The sandy beach here may hold Sanderlings and other shorebirds. Flocks of Twite often gather in the seaside vegetation. The cultivated fields of Ervik can support hundreds of Barnacle Geese, gulls, shorebirds, pipits and other birds. They are often particularly bird-rich after storms in autumn and winter. Lake Ervikvatnet is only rarely completely frozen in winter and Whooper Swan and ducks may spend the cold season here. Sedge Warbler is a common breeder in Ervik, and this can be an exciting area for nocturnal songbird trips in early summer. The heathland dominating the Stad peninsula have the densest population of Stonechat in Norway, with up to 13 pairs. They are found on south-facing slopes with heath and a few bushes. In late autumn and winter, Stonechats are found closer to the sea, such as in Ervik and in Hoddevik. Quite a few rare birds have been found at Stad, including Harlequin Duck, Balearic Shearwater, Franklin's Gull, Brown Shrike, River Warbler and Red-headed Bunting.

Stonechat is a rare but regular breeding bird in Norway, but has its most important populations in the heathers along the outer coast of western Norway. *Photo: Terje Kolaas.*

Other wildlife: Marine mammals are often seen from Litlehovden headland. Red deer is common on the Stad peninsula. Common toad can abound on the roads of Ervik on warm summer nights.

Best season: Year-round, but particularly migration periods. Few birds in winter, except after storms.

Directions: From Rv15 just east of Måløy or from Rv61 between Nore and Koparneset, exit towards Selje. Follow Rv620 to Leikanger and then follow signs to Ervik. In Ervik village, make a left before the church and park as far west in Ervika as possible. A path leads out on the Litlehovden headland from the parking area by the harbour.

Tactics: In Ervik, check the beach, fields and gardens for resting migrants. Before walking out to Litlehovden headland, check the weather forecast with information about the wind and wave height at Stad. Litlehovden is very exposed and when the wind is pushed over the headland the forces of nature can be extreme. The path takes you across a deep trench leading down to sea level. If the forecast mentions wind speeds in excess of 20 m/sec or wave heights above 5 metres, your trip should be postponed. You need good hiking shoes and should carry the telescope and all your other gear in a backpack in order to have both hands free. Bring dry clothes to put on upon arrival.

20. Runde island

Møre og Romsdal County

GPS: 62.40560° N 5.62102° E (trailhead Runde seabird cliffs), 62.32720° N 5.64302° E (Lake Myklebustvatnet)

Notable species: Black Grouse, Shag, Fulmar, Gannet, White-tailed Eagle, Peregrine Falcon, Gyrfalcon, Arctic Skua, Great Skua, Razorbill, Puffin, Guillemot, Black Guillemot, Eagle Owl, Pied Wagtail and Ring Ouzel.

Description: Runde is a tall island on the outer coast just north of Stad headland. Its precipitous cliffs support some of the largest colonies of seabirds in the country. The island is easily accessible and well-adapted for traffic, as should be expected at southern Norway's most popular bird-based tourist attraction, with ample parking and well-marked paths leading you directly to the birds. A visit to Runde during the breeding season is a truly spectacular experience. Puffin is the most numerous species with many hundred thousand pairs breeding on the grass covered slopes. Kittiwake is numerous as well, and you find several thousand pairs each of Fulmar, Guillemot and Razorbill. Brünnich's Guillemot was a regular but uncommon breeding species several years ago, but has now probably disappeared completely, as has Cormorant. Shag and Black Guillemot

Runde island. The path up towards the seabird cliffs can be seen on the mountainside above Goksøyr village. Also check bushes and gardens for passerines and the seashore for shorebirds. *Photo: Bjørn Olav Tveit.*

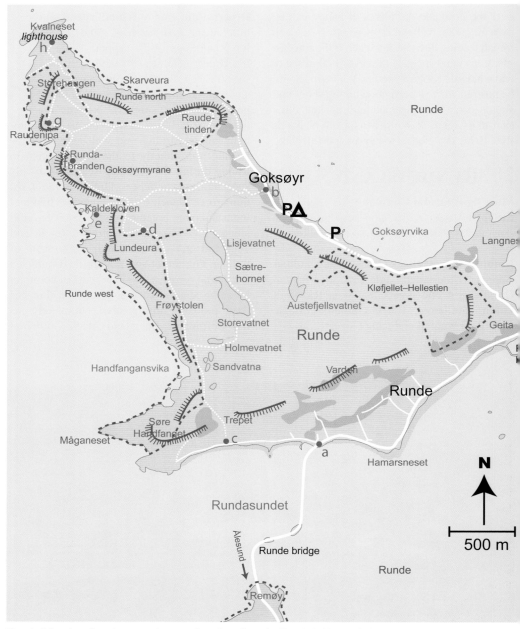

Runde island: a. Seashore, pastures and gardens from the Runde bridge to Goksøyr village (resting migrants); **b.** path from Goksøyr to seabird cliffs; **c.** path to cliffs from the south side; **d.** Lundeura (Puffins); **e.** shorline below Kaldekloven; **f.** Rundabranden (Great Skua); **g.** Raudenipa (Gannets); **h.** Kvalneset seawatch.

breed commonly along the coast, although the population of the former has decreased from about 5000 to 2000 pairs in recent years. The colony of about 5000 pairs of Gannet is the only one in southern Norway. On the grassy plateau in the middle of the island, a few dozen pairs of Great Skua nest, as one of a very few places in the country. Arctic Skua breed as well, and White-tailed Eagle, Peregrine Falcon and Gyrfalcon can regularly be seen hunting along the cliffs.

The strategic location on the coast also leads

to Runde having its share of migrants, both passing seabirds and resting passerines. More then 230 species have been recorded and with several rare birds among them, including Alpine Swift, Short-toed Lark, River Warbler, Black-headed and Red-headed Buntings. Be sure to check the shore on the south side of the island, often holding Shelduck, shorebirds and both White and Pied Wagtails in summer, and the bushes and trees in the populated areas for passerines. Seawatching can be conducted from the lighthouse.

If you visit Runde by car, you might want to stop by at the nutrient-rich Lake Myklebustvatnet, a hotspot for rails, possibly with three regularly breeding species. Shoveler and other ducks and waterbirds can also be found here.

Other wildlife: Red deer, otter, American mink, harbour seal, grey seal and harbour porpoise are often seen. The seals often rest on the skerries off Goksøyr village.

Best season: April to July for breeding seabirds and migrants. The latter can be appreciated through autumn as well.

Directions: *From Ålesund by car*, follow E39 signposted to Bergen out of town, and soon exit onto Rv61 signposted to Måløy. A ferry, usually departing every half hour, leads across to Hareid (where you soon get Grimstadvatnet lake on your left, which should be checked for resting waterbirds from the road). After the bridge to Gurskøy, 43 km from the ferry-landing at Hareid, exit Rv61 onto Rv654 signposted to Fosnavåg. You can exit just before Fosnavåg onto a road signposted to Runde, or you can continue straight ahead to Fosnavåg, where you get a view of Myklebustvatnet from the petrol station on your right. When continuing on the road to Runde, you should stop at the shallow inlet of Voldsund just before the Remøybrua bridge, and by the nice scrub along the road on Remøy. Birdwatching at Runde commences immediately after crossing the **a.** Rundebrua bridge, where you should check the shoreline and any interesting habitat on your way towards Goksøyr village.

Tactics: You really need more than one day in order to explore Runde thoroughly. The terrain is steep and requires a reasonable degree of fitness. You should bring light rain clothes (the weather changes quickly), a bottle of drinking water and an extra shirt to change into as soon as you reach the top all sweaty and wet. You should also bring your telescope, although you will see most of the breeding birds perfectly well with just binoculars. It can sometimes be particularly rewarding to visit the cliffs in the evening, when swarms of auks return from the sea. Two paths lead up to the bird cliffs: **b.** The most popular one leads up between the houses in Goksøyr village, marked with an array of signs. **c.** The other path leads up from the south side of the island. You may want to go up one path and down the other.

If you have limited time, the following route is suggested (usually taking a minimum of 4 hrs): Walk **b.** the path from Goksøyr and follow signs to **d.** Lundeura where most of

Razorbill is among the breeding birds in the cliffs on the western side of Runde island. *Photo: Tomas Aarvak.*

Great Skua is a fascinatingly brutal pirate who makes a living by stealing food from other seabirds. It only breeds regularly in a few places in Norway, including Runde. The Great Skua is known to be aggressive on the nesting ground, so be careful: It does not hesitate to attack people who approach it! *Photo: Espen Lie Dahl.*

the Puffins reside. Here you can lay on your stomach and look down upon the Puffins, the closest ones just a couple of metres away. Walk back a bit and you may want to walk down to **e.** the beach along the path from Kaldeklova passing the colony of Kittiwakes. This detour takes about two hours. From Kaldeklova continue up-hill to **f.** Rundabranden where Great Skua breed alongside the path. Continue all the way to **g.** Raudenipa where you have a nice view of the bird cliffs and down to the Gannet colony. If time allows, you can hike down to **h.** Kvalneset lighthouse, one of the best seawatching sites in this part of the country. Here, you can walk down to get close encounters with the Kittiwakes in Krykkjehola on the west side and to the Shags in Skarveura on the east side. From here, you can follow the eastern path back to Goksøyr village. You will pass Great Skuas here as well and several pairs of Fulmar breed along the ridge. A boat trip around the island is recommended; tickets can be bought locally, for instance at the camp site. At Runde there are three nature reserves (see map). In these, access is prohibited except along the marked paths from March 15 to August 31.

21. ÅLESUND AREA

Møre og Romsdal County

GPS: 62.47237° N 6.15418° E (Ålesund harbour), 62.46859° N 6.30700° E (Lake Lerstadvatnet), 62.44275° N 6.37558° E (parking Høgkubben hill)

Notable species: Waterbirds including Whooper Swan (winter), Little Grebe, Moorhen, Coot, Water Rail, Shag, Kittiwake and Sedge Warbler; White-backed Woodpecker, Grey-headed Woodpecker, Green Woodpecker, Icterine Warbler and Marsh Tit.

Description: Ålesund town has landing facilities for fishing boats, attracting large numbers of sea ducks and gulls. Check the flocks of Common Eider for King Eider and the gulls for Glaucous and Iceland Gulls in winter. Kittiwake breed on buildings in the city, particularly along Brosundet, the narrow sound leading through the city centre. Two small, nutrient-rich and overgrown lakes just outside of town, Lake Lerstadvatnet and Lake Ratvikvatnet, host several breeding waterbirds. South-facing slopes along the fjords inland, covered with mature deciduous and mixed woods, such as Høgkubben hill not far from Ålesund town, support several woodpeckers; White-backed and Grey-headed Woodpeckers are the commoner species among them.

Best season: All year.

Directions: In <u>Ålesund harbour</u>, some good viewpoints for gulls and sea ducks are from the Coastal Express (Hurtigruten) quay on the north side of town, from the dock by the main bus station on the south side, and from the Steinvågbrua bridge along E136 1.5 km west of Brosundet.

<u>Lake Lerstadvatnet</u> is situated along E136 east of town, 6 km east of the junction with the road to Vigra airport and 2.5 km west of the shopping centre Moa. It can be viewed from the petrol station by the roundabout 300 m north of the Hatlaås tunnel, or from Lerstad school at the opposite side of the lake. From this petrol station, <u>Lake Ratvikvatnet</u> is reached by crossing E136 and pulling in to Lerstadveien street leading west. You see the lake to your right after about one kilometre.

<u>Høgkubben hill</u> is reached by continuing E136 east and then heading south along E39 signposted to Bergen before heading east along Rv60

Kittiwake breeds on buildings and cliffs in Ålesund harbour. *Photo: Bjørn Olav Tveit.*

signposted to Sykkylven. Exit immediately to the west in the first roundabout, and then in the first street to the right, Blindheimsvegen. After 400 m, turn left into Skarpetegvegen and park by no. 5. On the side of this building a popular hiking path leads up the hill through perfect woodpecker habitat 1.5 km to the summit of Høgkubben hill.

22. GISKE–VIGRA ISLANDS

Møre og Romsdal County
GPS: 62.50765° N 6.03162° E (parking Makkevika inlet)

Notable species: Resting geese, ducks, divers (four species), grebes, shorebirds, gulls and other wetland species; White-tailed Eagle, Peregrine Falcon, Lesser Black-backed Gull (*intermedius*), Corncrake, Sedge Warbler, Rook and Linnet.

Description: The two islands Giske and Vigra are situated along the outer coast beyond Ålesund town. They are among the best sites for shorebirds and other wetland species in this part of the country. In autumn they represent the "last stop" before the shorebirds continue south along the rather barren coasts between here and Jæren in the south. Makkevika inlet at Giske is the most famous site in this area, with a bird observatory running since 1970. The main activity has been ringing of shorebirds and Rock Pipits. Several other good shorebird sites are found in the area, such as Rørvikvågen inlet, Synesvågen inlet and Roaldsanden beach at Vigra. Geese and ducks can be found resting in substantial numbers as well, most notably in Rørvikvågen inlet and at Roaldsanden beach, dabbling ducks also in Lake Rørvikvatnet as well as in Makkevika inlet. Both Giske and Vigra are flat islands dominated by bird friendly habitats, such as coastal meadows, marshes and cultivated fields. Grazing pastures need to be checked for geese, shorebirds, gulls and passerines, often including locally uncommon species such as Rook, Jackdaw and Linnet. Corncrake and Quail are often

Oslo

Interior

Skagerrak

Western

Central

Northern

Finnmark

Svalbard

Cormorants resting in Blindheimsvik on Vigra island, with the bridge to Giske island in the background (seen from point t. on the map on p. 202). *Photo: Bjørn Olav Tveit.*

heard calling on early summer nights. Sedge Warbler is common while Grasshopper Warbler and Marsh Warbler are annual visitors.

The two islands are situated in an area of shallow seas, rich in seaweed, making ideal resting and wintering grounds for sea ducks, divers, grebes, cormorants, gulls and auks. Red-necked Grebe is common outside the breeding season. In late winter, up to 130 individuals have been counted. Slavonian Grebe is also common but less numerous, Great Crested Grebe and Little Grebe are more sporadic. The abundance of prey in this bird-rich area attracts raptors, and thus Peregrine Falcon, Merlin and White-tailed Eagle regularly cause havoc among the flocks of resting birds. Bar-tailed Godwit winter along the coasts, usually accompanied by Grey Plover, Ringed Plover, Sanderling

and Dunlin. Winter is also a time when Glaucous Gull and Iceland Gull regularly occur. Several rare birds have been found here through the years, including Long-billed Dowitcher, Whiskered Tern, Great Spotted Cuckoo and Paddyfield Warbler.

Other wildlife: Harbour seal, American mink and otter are regularly seen.

Best season: Makkevika inlet and the other shorebird sites are particularly productive in the shorebird migration season from May to September, and particularly in the latter half of this period. The other areas are basically of interest all year.

Directions: Ålesund airport is situated on Vigra, with direct flights to Oslo and many other cities. You may rent a car here. Buses run between the airport and Ålesund, and there are local buses on the islands.

By car from Ålesund, exit from E136 just east of Ålesund onto Rv658 signposted to Vigra. The road leads through a series of subsea tunnels before surfacing at Valderøya island. In order to get to Giske island with Makkevika inlet, exit left immediately, signposted to Godøy. For Vigra, keep straight ahead along Rv658. See the map and further directions below.

Tactics: Giske island is small enough to be explored on foot, if you for instance arrive by bus. If you want to explore Vigra island as well, a car will be of great advantage. If you arrive by car from Ålesund you can explore both islands along the following suggested route:

Drive to Giske island as explained above. Stop at the end of **a.** the Giske bridge or at **b.** Staurneset headland (past the soccer court) to view Staurnessundet, a sound good for Great Northern Diver and Red-necked Grebe, with the possibility of uncommon species like Great Crested Grebe in winter. The pastures at Giske island can hold resting geese, shorebirds, gulls and passerines. Make a short detour to the right in the junction 300 m before the parking at Makkevika, down to the areas around the factory by **c.** Storevika. The inlets and beaches here can hold shorebirds and dabbling ducks. From the parking at **d.** Makkevika inlet you have a view to the general area with pastures, the muddy inlet and the shallow waters around Kvalneset headland. In winter, divers, Common Eider, Velvet Scoter and Long-tailed Duck may be found here, sometimes a few Common Scoters and Scaups as well. Raptors such as Peregrine Falcon pass by regularly and White-tailed Eagle can sit on boulders along

Makkevika inlet at Giske island with the cabin belonging to Giske Bird Observatory. Note the funnel traps by the shore for catching and ringing shorebirds and Rock Pipits. *Photo: Bjørn Olav Tveit.*

Oslo

Interior

Skagerrak

Western

Central

Northern

Finnmark

Svalbard

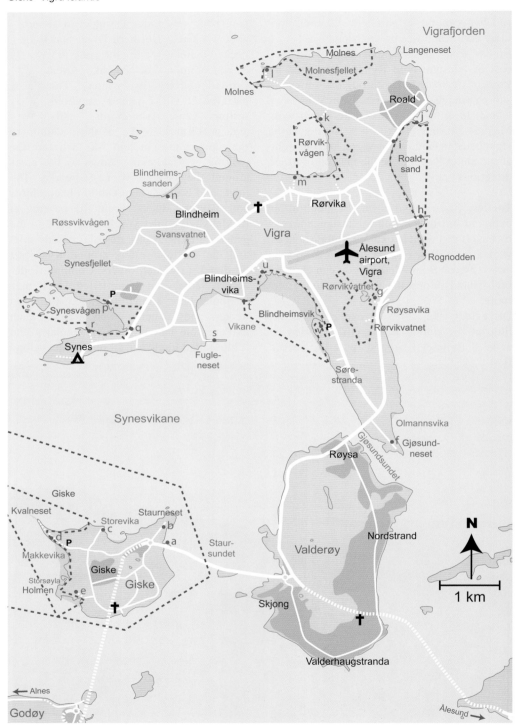

Giske and Vigra islands: a. Giske bridge; **b.** Staurneset; **c.** Storevika; **d.** Makkevika and Giske Bird Observatory; **e.** Holmen forest; **f.** Gjøsundneset, recycling facility and golf course; **g.** Rørvikvatnet; **h–j.** viewpoints Roaldsanden; **k.** Rørvikvågen north; **l.** Molnes; **m.** Rørvikvågen south; **n.** Blindheimssanden beach (*Blimsanden*); **o.** Lake Svansvatnet; **p–r.** viewpoints Synesvågen; **s.** Fugleneset; **t–v.** viewpoints Blindheimsvik.

Red-necked Grebe winters in significant numbers in the waters around Giske and Vigra islands, making this among the most important places in the country for the species. *Photo: Terje Kolaas.*

the shore, scanning for prey. The island you see out in the distance, straight out from Kvalneset headland, is Erkna island, where both Storm Petrel and Leach's Petrel breed. It is inaccessible without a private boat.

Walk the tractor tracks leading out on Kvalneset headland, or better still: walk off the path, closer to the shore, on both sides of the headland. Passerines such as uncommon pipits, Linnet and Twite are sometimes seen. The shorebirds feed on the exposed mud at low tide and rest among the pebbles and boulders on the headland at high tide. The small building here belongs to Giske bird observatory. Go back past the bird observatory and head for the marshland lining the Makkevika inlet. Check the coniferous plantation **e.** Holmen forest for passerines.

Vigra island is reached by heading back over the Giske bridge and continuing north along Rv658. After crossing the bridge, turn right towards **f.** Gjøsundneset. The golf course here can hold pipits, shorebirds and more when not in use by golfers, and particularly after rainfall. The garbage recycling facility at the end of the road can hold Twite, pipits and gulls, including Lesser Black-backed Gull in summer and Glaucous or Iceland Gull in winter.

g. Lake Rørvikvatnet can hold a few ducks and other wetland species.

h–j. Roaldsanden beach is sheltered from westerly winds, often holding shorebirds at low tide. Ducks, geese and gulls often rest here as well. Scan the sea for divers, auks and more.

k–m. Rørvikvågen inlet and Molnes headland. The grazing pastures here often support passerines. On early summer nights, Corncrake can be heard calling. The cultivated areas south of Rørvikvågen inlet are generally among the best for nocturnal songbirds. Sea ducks often gather off Molneset headland in winter, and shorebirds may rest and feed here.

n. Blindheimsanden beach and Blindheim inlet. When not crowded with bathers and swimmers, Blindheimsanden beach (also called *Blimsanden*) can be a good spot for gulls and shorebirds, often including Sanderling. The side roads here at Blindheim cross through cultivated areas around **o.** Svanvatnet wetland, a former pearl of a birdwatching site, now unfortunately almost completely overgrown. It is still an interesting location for nocturnal songbirds, though.

q–r. Synesvågen inlet is nice and shorebird-friendly and needs to be checked from three angles, preferably at low tide.

s–v. Vikane coast, including Blindheimsvik inlet. The southern shores of Vigra, known as Vikane, consists mainly of pebble beach and stranded seaweed. It can be good for ducks and other waterbirds but usually it does not hold many shorebirds. On the sea here, Slavonian Grebe and Great Northern Diver winter, sometimes accompanied by White-billed Diver as well. These species often stay until they acquire their stunning breeding plumage in April. By this time, they have been accompanied by migrating Black-throated Divers too.

Oslo

Interior

Skagerrak

Western

Central

Northern

Finnmark

Svalbard

View from Ona island towards Husøya lighthouse. Cloudy weather and light winds are the best you can wish for in order to find migrant passerines. *Photo: Bodil Gjevik.*

23. ONA ISLAND

Møre og Romsdal County

GPS: 62.86339° N 6.54361° E (Ona lighthouse), 62.80653° N 6.76727° E (Småge ferry-landing)

Notable species: Migrants, rarities and passing seabirds; White-billed Diver, Great Northern Diver, Red-necked Grebe, Cormorant, White-tailed Eagle.

Description: The beautiful and small Ona island off the coast of Møre has from the early 1990s proved to be one of the strongest rarity magnets in Norway. The jewels in the crown are the extraordinary records of Upland Sandpiper, Bimaculated Lark, White's Thrush, Grey-cheeked Thrush, Pallas's Grasshopper Warbler, White-crowned Sparrow, and many more. Ona island is also one of the best sites in the country with regards to seabird passage, with day-counts of up to 328 Sooty Shearwaters and several records each of Balearic, Cory's and Great Shearwaters and many of Sabine's Gull.

Ona island and the adjoining Husøya island constitute less than 1 km² altogether. They are mainly covered in grass but a few gardens provide shelter to bush-dwelling passerines as well. You can cover the best areas in about one hour. As with other islands of its kind, the quantities and qualities of the birds at Ona island are highly sensitive to season and weather. Late spring and autumn days with a complete cloud cover and a gentle breeze are usually considered to be the most promising conditions for turning up interesting land birds. On-shore winds are best for passing seabirds.

The other islands in the area can be interesting as well and are often visited by Ona birdwatchers in quiet periods. The larger Sandøya island 3.5 km further into the Harøyfjord is most often visited. Here are more varied habitats, including shallow inlets suitable for ducks and shorebirds. Finnøy and Harøya islands are also well worth visiting. The Harøyfjord is particularly good for resting and wintering sea ducks, grebes and divers. Several individuals each of White-billed and Great Northern Divers are often seen during the ferry crossing. Cormorants of the nominate form have their southernmost breeding colonies near Ona island. Most tourists arrive via the ferry-landing at Småge on the larger Gossa island. Exploring the airfields and pastures of Gossa island can be worthwhile when waiting for the ferry to Ona and the other islands in the Harøyfjord.

Other wildlife: Otter and harbour seal are common in the area. Harbour porpoise and other species of whale are sometimes seen while seawatching from Ona.

Best season: Late spring and autumn.

Directions: *From Molde town,* follow Rv662 west signposted to Aukra and take the ferry to Gossa island in Aukra municipality. On Gossa island, drive via the village centre Falkhytta, where you find grocery stores and other facilities, and follow signs towards Stongneset by Småge

ferry-landing on the west side of the island. The ferry runs only a few times a day and does not have room for many cars. You do not really need the car if you only intend to visit Ona and Sandøya islands. You do need a car, however, if you plan to visit Finnøy/Harøya islands.

Tactics: Search Ona and Sandøya islands thoroughly for resting migrants. Seawatching is conducted from several viewpoints: at **a.** Ona lighthouse the seabirds often pass at close range, while at **b.** Husøya lighthouse you have the advantage of overlooking the island and the strait on the inside as well. **c.** The cemetery is also good for the latter.

At Sandøya island you explore the gardens and fields. A small pond is found in the middle of the island and there are several shallow inlets around the coast. Particularly good are the two inlets on

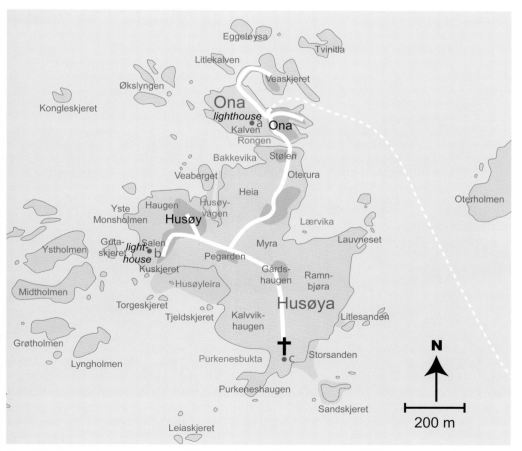

Ona island and neighbouring Husøya island: **a.** Ona lighthouse; **b.** Husøya lighthouse; **c.** cemetery.

the northeast corner of the island, the largest of which is known as <u>Nesfjøra inlet</u>.

At <u>Finnøya island</u> a good viewpoint is just before the bridge to Harøya, where you can follow a path on the right-hand side of the road leading to a small hill with a view to Lyngholman nature reserve, supporting ducks, White-tailed Eagle and shorebirds, often included Bar-tailed Godwit even in winter. The bridge itself is a good viewpoint too. At Harøya, check <u>Steinshamn</u> harbour area for gulls, ducks and grebes. Continue through the village and turn left opposite a bus stop. From here you see a tower hide down by the shore. Park roadside and follow the tracks to the hide. Afterwards, you may explore the rest of the island and check all good-looking fields and pastures from the roads. The eastern shore has important wetlands but they are not easily viewed. Breivika bay on the west side, however, can be examined from the road. Waterbirds gather here in winter, often including Great Crested Grebe.

24. Fræna peninsula

Møre og Romsdal County

GPS: 62.98015° N 7.30323° E (viewpoint Gaustadvågen), 62.96974° N 7.05456° E (Maletangen headland)

Notable species: A variety of resting and wintering waterbirds and shorebirds; passing seabirds; White-tailed Eagle, Sedge Warbler, Icterine Warbler and Great Grey Shrike (winter).

Description: The Fræna peninsula outside Molde town is one of the richest and most varied bird areas in this part of the country. It is located at the outer coastline and has a number of interesting habitats worth exploring. The Sandblåstvågen lagoon is directly connected with the brackish Lake Gaustadvågen, forming what is reckoned as the most important wetland in the region. The lagoon is lined with saltmarsh, cultivated fields and woodlands.

Sandblåstvågen lagoon on Fræna peninsula, seen from point. **m**. towards point **g**. At high tide, as in the picture, the lagoon is filled with seawater which is pushed all the way up to Gaustadvågen. The many rivers and streams that flow into the area help to create a very special brackish water system, which in combination with the surrounding landscape constitutes a very valuable bird location. *Photo: Bjørn Olav Tveit.*

At low tide, large mudflats are exposed in Sandblåstvågen lagoon, making perfect conditions for shorebirds such as Temminck's Stint, *Tringa*-waders, Ruff and other freshwater species. The access possibilities to the lagoon are not ideal however, which is an advantage to wary species like geese and Crane. Lake Gaustadvågen usually hosts a number of ducks, Whooper Swans and other waterbirds.

Hustadbukta bay is another fine site on the Fræna peninsula. Here is also Risvaet lagoon. The outer Hustadbukta bay is shallow, and large tidal flats are exposed twice a day, attracting gulls and shorebirds. The shallow waters are favoured feeding grounds for terns, sea ducks, Red-breasted Merganser and in winter, divers and grebes.

Nearby is Maletangen headland, where you find a couple of small, shallow inlets where ducks and shorebirds often gather. Several rare birds have shown up here. Maletangen headland is also good for seawatching.

In between these main sites, there are several smaller but still interesting ones. Fræna peninsula is also a fine destination for nocturnal songbird trips. The spectacular Atlantic Ocean Road (*Atlanterhavsvegen*) starts at Fræna peninsula, leading along the outer coast connecting islands and skerries with causeways and viaducts, from Molde to Kristiansund towns. White-tailed Eagles are often seen roadside.

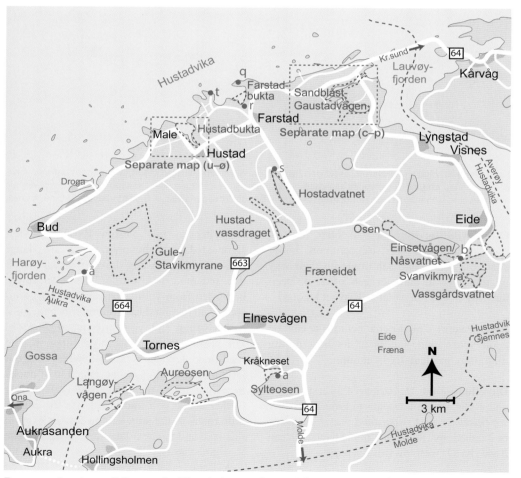

Fræna peninsula: a. Sylteosen; **b.** Nåsvatnet; **c–p.** Gaustadvågen and Sandblåstvågen; **q.** Farstadlykta; **r.** Farstadsanden; **s.** Lake Hostadvatnet; **t.** Storholmen; **u–ø.** Hustadbukta og Male; **å.** Harøysundet fish landing.

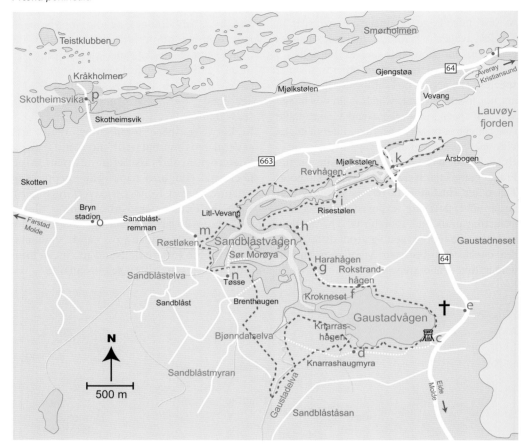

Lake Gaustadvågen and Sandblåstvågen lagoon: c. viewpoint Gaustadvågen; **d.** southern path; **e.** viewpoint Gaustad chapel; **f.** viewpoint Gaustadvågen north; **g.** Harahågen; **h–k.** viewpoints Sandblåstvågen's outer section; **l.** Atlantic Ocean Road; **m–n.** viewpoint Sandblåstvågen west; **o.** Bryn stadion; **p.** Skotheimsvika.

Other wildlife: Harbour seal, otter, red fox, roe deer and red deer are common. Large atlantic salmons and sea trouts on their way up the river in summer can often be seen jumping out of the water at Risvaet lagoon.

Best season: The Fræna peninsula can be good year-round, but is at its best during migration, particularly in autumn.

Directions: *From Molde*, leave E39 east of town onto Rv64, and follow signs to Eide. Thereafter, see below. *From Kristiansund* Fræna peninsula is reached by following the Atlantic Ocean Road (Rv64) west towards Averøya.

Tactics: Starting in Molde town you may explore Fræna as follows (see maps):

a. Sylteosen inlet. From Rv64 turn left signposted to Syltevorpa. Here are damp, lush meadows leading down to an inlet with exposed mud at low tide. The inlet should be birded from within the car in order not to spook the resting ducks, shorebirds and gulls.
b. Nåsvatnet and Vassgårdsvatnet lakes. 13 km north of the Syltevorpa junction, the reed-lined Lake Nåsvatnet appears on your left, usually holding a nice selection of ducks, in winter often including Smew. The river further ahead is a traditional site for Dipper. Side roads in the area let you further explore this lake and the next lake to your right, Lake Vassgårdsvatnet.
c –k. Lake Gaustadvågen and Sandblåstvågen lagoon (east-side). Along Rv64 some 27 km further north, exit onto the side road signposted

208

to Gaustadvågen. Here, on the east side of Lake Gaustadvågen an observation hide **c.** is situated on a hill-top overlooking the lake. Most of the birds will be in the far end of the lake, 800 m away, so a telescope is essential. The lake can be viewed form the alternative viewpoints **d.** the path along the south side, **e.** the junction at Gaustad chapel or **f.** from the side road along the north end. From the end of the latter, a path continues to **g.** Harahågen, a viewpoint to Sandblåstvågen lagoon. From Rv64 at the exit to Årsbogen you can take the roads either left or right for views **h–k.** to the outer parts of Sandblåstvågen lagoon.

l. Atlantic Ocean Road (*Atlanterhavsvegen*). A bit further ahead on Rv64 you enter a T-junction. Turning right takes you through the most scenic parts of the Atlantic Ocean Road, along which you have good chances of seeing divers, auks, White-tailed Eagle, Arctic Skua, Short-eared Owl, Great Grey Shrike and more, particularly in winter. Also make a stop after 14 km at Lake Hosetvatnet, a reliable site for Water Rail all year and where raptors may pass overhead during migration.

m–n. Sandblåstvågen lagoon (west-side). Turn left onto Rv663 leading west (signposted to Farstad), a road running parallel with the lagoon. White-tailed Eagle can often be seen sitting on the hilltops along the road. After 2.4 km, turn left onto the side road Røstløkvegen, and stop roadside at **m.** to view the lagoon. Continue and stop **n.** at any good viewpoint to overlook the lagoon and the surrounding fields. In order to walk closer to the lagoon, you need to ask permission from the landowner.

o. Bryn stadium has short grass that can host wagtails, pipits and a few shorebirds, particularly in early autumn before the surrounding fields have been harvested.

Turn right towards Skottheimsvik, and then left to Kråkholmen. This road crosses a tidal area in **p.** Skotheimsvika inlet, and from the marina at the end you have a view to the sea.

q. Farstadlykta lighthouse is the best seawatching viewpoint on Fræna peninsula. 2 km west of the Skotheimsvik junction, turn right and park down by the boat sheds.

Nedre Risvaet inlet, seen from point **x.** (overleaf). The difference between high tide and low tide in this part of the country is more than two metres. Hustadbukta and Risvaet inlets are completely flooded at high tide. Turn the page to see the same location at low tide. *Photo: Bjørn Olav Tveit.*

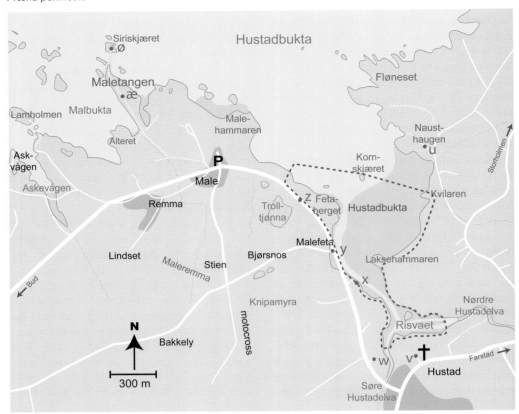

Hustadbukta and Male: u. Nausthaugen; v. Hustad church; w. old Hustad school; x. sea sheds; y. Malefeta; z. Fetaberget burial mounds; æ. Male Bird Observatory; ø. Siriskjæret.

Continue on foot 500 m to the lighthouse.

r. Farstadsanden beach. Back at Rv663, continue west 1 km and turn right signposted to Farstadsanden. This road leads to a sandy beach at the mouth of river Farstadelva. When the beach is not crowded with people, this site can support interesting shorebirds, gulls and terns.

A bit further along Rv663, you reach the main junction at Farstad. Here you keep straight ahead to get to Lake Hostadvatnet, or turn right towards Hustad/Bud to explore Farstadsanden beach from the opposite angle.

s. Lake Hostadvatnet is reached by continuing 3.5 km along Rv663. Exit to the right, discretely signposted to Tverrfjell. The lake is visible roadside further down the highway as well. Upon passing the lake, turn right onto the road signposted to Hostad. This road leads past Lake Frelsvatnet and along the west side of Lake Hostadvatnet through interesting farmlands. You end up back at

Farstad or further west, depending on which side-roads you choose to explore. This area is particularly interesting for nocturnal songbird trips in early summer, with the possibility of Long-eared Owl in the farmlands and Eagle Owl calling from the surrounding hills.

t. Storholmen headland. Back at the west side of Farstadsanden, continue west along the Hustad/Bud road. Turn right towards Storholmen, where you have a view to the outer coast. On the way out you have good views to Breivika bay, worth checking for waterbirds. On your way back from Storholmen, 300 m south of the breakwater, turn right (west) and drive through another interesting area of farmlands close to the sea, worth checking for shorebirds, geese, passerines and more.

u–z. Hustadbukta bay and Risvaet lagoon. From u. Nausthaugen you have nice views from the east-side of the tidal flats of Hustadbukta bay. From the car park by v. Hustad church

210

you can take a peak over the hill and down to the inner part of the Risvaet lagoon, in which the two rivers Søre Hustadelva and Nørdre Hustadelva empty. Good views to Risvaet lagoon can also be obtained from **w.** the old Hustad school or roadside a bit further ahead. The outer part of Risvaet and the inner part of Hustadbukta are best viewed from **x.** the boat sheds and **y.** Hustabukta west side. You can also stop at the lay-by by **z.** the Viking grave mounds at Fetaberget for a view to the outer section of Hustadbukta bay. Do not forget to check Trolltjønna pond on your left.

Maletangen headland. The small village Male is surrounded by several grassy fields, suitable for plovers, Ruff, pipits and such during migration. These fields can be explored via the roads in the area. The most famous site at Male, the Maletangen headland, is reached by parking at the community building (*samfunnshuset*) in the centre of the village (see map), and following the path across the field towards the sea. Walk carefully the last bit before the sea so you do not flush the resting waterbirds in the western inlet before

you have the chance to study them. Check the eastern inlet as well, a very interesting spot for shorebirds, good also at (normal) high tide. **æ.** The shelter belongs to the local bird club and is open to everyone, very handy during rain showers. At the tip of this rocky peninsula, a natural causeway lets you walk out to **ø.** Siriskjæret islet at low tide. This is a good viewpoint for seawatching. Just keep an eye on the causeway so that it doesn't cut off the retreat possibility behind you as the tide rises. In stormy weather, large waves may wash over the entire islet. If you would like to seawatch in this area without having to worry about such hazards, the earlier mentioned **q.** Farstadlykta lighthouse is the place to be. The road from Male to Bud is good for gulls, sea ducks, divers and more, particularly in winter. The shallow Askvågen inlet just west of Male is worth checking for shorebirds as well. In Harøysundet sound there is a **å.** fishing boat landing facility (signposted to Skjæret) often attracting substantial numbers of gulls and ducks.

Same place as on p. 209, six hours later. Check the tide table before you go on a trip to these areas. You would prefer to be here when the water is at its lowest and exposes large mudflats, or when the tide is on its way up and pushes the birds towards the shore. *Photo: Bjørn Olav Tveit.*

CENTRAL NORWAY

Central Norway has a central position in the country not only geographically, but also from historical, cultural and, of course, ornithological perspectives. Here is the Trondheimsfjord with its wetlands and surrounding agricultural fields constituting one of the most important staging areas for geese, ducks, divers, grebes and shorebirds in Scandinavia. Common Eiders and many other waterbirds originating from the Baltic Sea or even Siberia winter here. Along the outer coast, a myriad of islands and skerries in the shallow waters of Froan archipelago make equally important breeding areas for a variety of seabirds. On the islands Smøla, Hitra and Frøya off the coast of central Norway, an

Great Snipe has declined sharply in the lowlands as a result of intensified agriculture. In the mountain areas of southern and central Norway, the species probably has stable populations and is in some places quite common in marshy areas with rich willow thickets. Like Capercaillie, Black Grouse and Ruff, Great Snipe gather together for joint leks in spring, where the males put on a display in the hope of winning the females' favour. *Photo: Espen Lie Dahl.*

endemic subspecies of Willow Ptarmigan *variegata* is found. The 'Smøla Ptarmigan' differs from other Willow Ptarmigans by not turning completely white in winter. The outer islands can also be of interest during migration, often producing records of exotic migrants when the season and conditions are

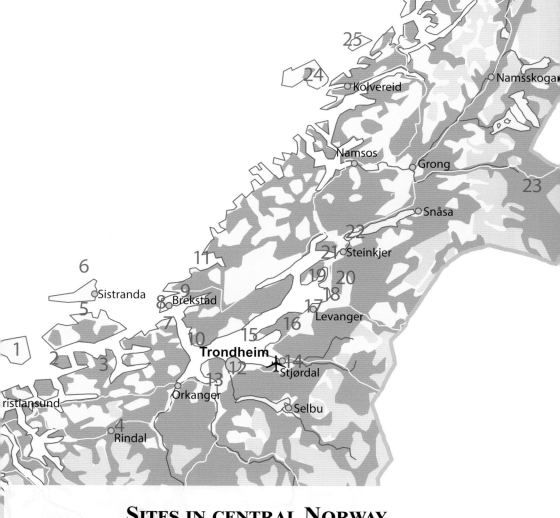

SITES IN CENTRAL NORWAY

Lake Røkvatnet on Smøla island is a nice locality, with nesting Ruff among others. The wind turbines leave their mark on the landscape on Smøla, and are particularly harmful to large birds, such as geese, Eagle Owl, White-tailed and Golden Eagles and even Willow Ptarmigans. The birds collide with both the masts and the deadly rotor blades, which at their tip can maintain a speed of up to 250 km/h. *Photo: Bjørn Olav Tveit.*

right. Central Norway also offers some of the richest forests and most spectacular mountain ranges with all the breeding species of the region. All of these can be reached within a few hours drive from Trondheim, Norway's third largest city with around 200,000 inhabitants.

THE BEST SITES

Trondheim city, similar to Oslo, is surrounded by vast forests and a significant fjord, the Trondheimsfjord. Most of the region's woodland species can be found close to the city year-round. The city itself has several good sites too, particularly productive in winter. Just 15 minutes drive south of town is one of the region's best sites, Gaulosen delta. To the north, the Stjørdalsfjord, an arm of the Trondheimsfjord, may be explored as well. It is productive all year but exceptionally so in April when the fish spawn, creating a feeding frenzy of ducks, gulls and more. During migration, tens of thousands of geese, mainly Pink-footed Goose, rest along the

inner Trondheimsfjord wetland system. There are also a few nutrient-rich lakes of great importance, perhaps with Lake Leksdalsvatnet near Verdal town as the single most important of these. Ørland wetland system at the mouth of the Trondheimsfjord is another exceptional area, just 1.5 hrs drive from Trondheim. Here you find Ørlandet with Grandefjæra, the largest continuous area of tidal flats and shallow seas in Norway. Ørland is important to a wide variety of wetland species, most notably ducks, divers, gulls and shorebirds. Further along the coast are several sites of particular interest during migration. The most interesting of these from a visiting birdwatcher's point of view is perhaps Sula island, which is easily accessible and feasible to cover sufficiently even if you travel alone or in a small party. The western part of central Norway has particularly strong populations of Capercaillie and White-backed Woodpecker. A reliable site for displaying Great Snipe at Rindal is also mentioned here, but the species can be found many places in the right habitat. The valleys inland have several important resting grounds for Crane and, in woodlands of higher elevations, Hawk Owl and Siberian Jay may be encountered. Some of the inland sites near Trondheim are covered in the Interior South chapter. One particularly interesting woodland area of central Norway is Lierne, where many of the most sought-after taiga species such as Ural Owl and Siberian Tit may be found. Rather recently, Smew has been found breeding regularly in central Norway, more specifically in Vikna archipelago, as the only location south of Finnmark.

1. SMØLA ISLAND

Møre og Romsdal County
GPS: 63.51797° N 7.95722° E (Veiholmen)

Notable species: Shelduck, Wigeon, Shoveler, Red-throated Diver, Great Northern Diver, Red-necked Grebe, Slavonian Grebe, Great Crested Grebe (winter), Willow Ptarmigan *variegata* ('Smøla Ptarmigan'), White-tailed Eagle, Golden Eagle (winter), Gyrfalcon (winter), Crane, Dunlin, Ruff, Arctic Skua, Lesser Black-backed Gull *fuscus* and Sedge Warbler.

Description: Smøla island is situated off the coast of the Nordmøre region and the large main island is surrounded by more than 5,500 smaller islands and skerries. The main island is 20 km in diameter, flat and covered in heather and marshlands, and some of these have been cultivated. Smøla is reckoned as one of the most important breeding areas for seabirds in Norway. It also has the densest population of White-tailed Eagle anywhere in the World. A large windfarm has been built here, though, with wind turbines regularly killing a number of eagles and many individuals of the endemic 'Smøla Ptarmigan', which is named after this island.

Birdwatching at Smøla is mainly concentrated to the roads crossing the fertile agricultural areas and wetlands in the middle of the main island, particularly around Frostadheia and Røkmyra. Geese and shorebirds rest here during migration, sometimes including Dotterel or even rarer shorebirds – Pacific Golden Plover has

'Smøla Ptarmigan', or Willow Ptarmigan of the subspecies *variegata*, is the only endemic bird in Norway, meaning the only bird that is not found in any other country. You will find it in small numbers on Smøla and some islands in central Norway. It differs from other ptarmigans in that it does not turn completely white in winter. *Photo: Espen Lie Dahl.*

Oslo

Interior

Skagerrak

Western

Central

Northern

Finnmark

Svalbard

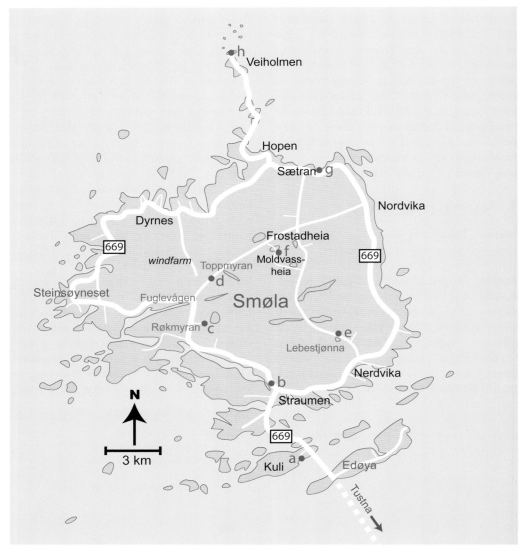

Smøla island: a. Kulisvaet sound; **b.** Lake Fløtjønnin; **c.** Røkmyran marshes; **d.** and **e.** roads through bird-rich areas; **f.** Frostadheia fields; **g.** Sætran; **h.** Veiholmen island. Protected areas are not drawn on the map.

shown up a remarkable number of times here, particularly in June. At Røkmyra, Crane and Ruff are among the regular breeders. To the north of the main island there is a road connection to the very small and northernmost Veidholmen island, an interesting site to look for passerines and more during migration.

Other wildlife: Otter, red deer, mountain hare and hedgehog are common.

Best season: Smøla island is considered to be best during the migration periods and in winter.

Directions: Express boat from Kristiansund (and Trondheim) to Edøy at Smøla. *From Kristiansund by car*, follow E39 out of town towards Molde and exit onto Rv680 to Seivika, from where you take the ferry to Tømmervåg on Tustna. Follow the highway about 14 km to the ferry landing Sandvika, signposted to Smøla.

Tactics: You can get around on Smøla by bike if you arrive by the express boat. Otherwise, a car is recommended. The most important sites are the following:

a. Kulisvaet sound. From the ferry landing at Edøy, follow Rv669 north. The Kulisvaet sound is crossed via a long causeway and a bridge. On the other side, make a left onto a side road leading down to what is one of the few accessible tidal areas at Smøla.

b. Continue on the highway to the T-junction on the south side of the main island. Here Rv669 splits up, only to meet again on the north side of the island. If you turn left towards Dyrnes, you will reach the little Lake Fløtjønnin on your left after 500 m. Wigeon and sometimes Shoveler and Pintail breed here.

c. Røkmyran is a small agricultural area with the possibility of resting shorebirds etc., as well as a wetland area with nesting Crane and Ruff. The main road runs through parts of the cultivated area. Turn onto the agricultural road to the east (on the right when coming from the south, and about 1.5 km before the entrance to Frostadheia village). Park at the end of the road and continue on foot for the last couple of hundred metres along the canal to Lake Røkvatnet.

d. and **e.** Several roads cross the island, all signposted to Frostadheia. The cultivated fields along these roads (and their side roads) are of particular interest. 'Smøla Ptarmigan' is rather common in the heathland, and may be found by shear luck, or by walking systematically through the heather.

f. The Frostadheia fields are particularly good. This is also where the Pacific Golden Plovers usually turns up. The many small lakes and pools must be checked as well. Shoveler sometimes breed here.

g. Just east of Sætran village is a shallow inlet where geese, ducks, Arctic Terns and other wetland species gather.

h. Veiholmen island. Look for resting migrants in the gardens and in weed between the houses. Also, the small islands further inland have gardens and clusters of scrub that should be examined closely during migration.

2. MELLANDSVÅGEN BAY

Møre og Romsdal County
GPS: 63.34476° N 8.51618° E

Notable species: Geese, ducks, Red-throated Diver, Great Northern Diver, White-billed Diver, Slavonian Grebe, Red-necked Grebe and shorebirds.

Description: Mellandsvågen bay is shallow and with large mudflats exposed at low tide. It is an important area for a variety of wetland species during migration and in winter, when tens of Great Northern Divers and a few White-billed Divers can be expected.

Best season: Late autumn, winter and spring.

Sedge Warbler is a characteristic species of damp areas with reeds or willow scrub in central Norway. *Photo: Espen Lie Dahl.*

Oslo | Interior | Skagerrak | Western | Central | Northern | Finnmark | Svalbard

Sengsdalen valley has fine and intact old deciduous forest on the hillsides with a large proportion of dead trees, perfect habitat for White-backed Woodpecker. *Photo: Bjørn Olav Tveit.*

Directions: *From Kristiansund*, see directions to Smøla island above, and continue along Rv680 35 km past the ferry landing to Smøla. After you cross the bridge over Torsetsundet sound, turn left signposted to Lesund.
From Trondheim, see directions to Sengsdalen forest below and keep following Rv680 some 31 km past the Aure junction before turning right onto the road signposted to Lesund.

Tactics: Mellandsvågen bay can be explored from a lay-by 500 m up the road to Lesund, or you can turn right just before you see the bay, onto a road signposted to Melland. This follows the bay past Livsneset and to Finnset farm, where you have a nice view of the cove and the fjord outside Mellandsvågen. Also check the other sounds and bays in the area, such as Dromnessundet sound 8 km further east along Rv680. A telescope is essential in this area.

3. SENGSDALEN FOREST

Trøndelag County
GPS: 63.37284° N 8.91921° E

Notable species: Capercaillie, Black Grouse, Hazel Grouse, Three-toed Woodpecker, Grey-headed Woodpecker, White-backed Woodpecker, Lesser Spotted Woodpecker, Great Spotted Woodpecker, Marsh Tit, Long-tailed Tit and Nuthatch.

Description: Sengsdalen valley is one of the very finest woodpecker sites in Norway. All the breeding species of woodpecker on the Norwegian list has been recorded here in a single day, but only the ones mentioned above breed and can be expected. Particularly good is the wooded, south-facing slope at the mouth of the valley. The dominating trees here are birch, aspen and pine, and almost 30 % of the trees are dead, creating ideal conditions for woodpeckers. Sengsdalen and the area in

218

general is also good for Capercaillie and other woodland species.

Another interesting site in the area, particularly good for White-backed and other woodpeckers, and for Capercaillie, is Seterseterdalen valley east of Kyrksæterøra village. In addition, the lakes of this general area should be checked for a variety of ducks, divers and the like.

Best season: Many of the species here are present all year. The woodpeckers are particularly active drumming and displaying during warm, sunny mornings in March and April.

Directions: Sengsdalen forest is reached by exiting from the E39 50 km west of Orkanger, onto Rv680 signposted to Kyrksæterøra. Pass through Kyrksæterøra, still on Rv680, towards Aure and continue 2.5 km past the exit to Hellandsjøen. Here, Rv680 turns north towards Tjeldbergodden while the road straight ahead is signposted to Aure (at this junction, look for White-backed Woodpecker down by the lake). Drive on towards Aure and park roadside 1.1 km from the junction, just before you reach some cultivated fields.

Tactics: After parking at Sengsdalen, the steep slope on the north side of the road is where you are headed. There is a spruce plantation at the foot of the hill, and just to the east of this is a path leading up the slope. In order to find the woodpeckers, walk slowly uphill and pay attention to any drumming or pecking noises from the woods – follow the sound! Please do not use playback to attract the woodpeckers' attention.

Several other sites are found along Rv680 between E39 and Sengsdalen forest. Just west of the exit to Rv680 is Vinjeøra inlet, which should be checked for resting ducks and other wetland birds. Vinjelia woods, just above, is good for White-backed Woodpecker. To get there, take the E39 to Vinje church and continue driving up into the housing estate. Park at the top and follow the path further up the slope. The junction between E39 and Rv680 is more or less in the middle of Stormyra marsh, but the section on the right side of Rv680 is particularly nice, and should be checked for wetland birds during migration and in early summer. A little further north you will see Lake Stavåsdammen on your right, which is also worth a short stop. A couple of kilometres further north you pass Libukta inlet, the southern end of Lake Rovatnet, on your right. This is the best place in the area for ducks and other wetland birds, and can be good all year round as long as it is not completely frozen. Libukta has several smaller bays and two deltas where the two rivers Eidselva and Lielva have their outlets. You can explore this area in more detail by taking the farm road on the right (by the farms 2.2 km after leaving the E39, just before you see Lake Rovatnet), and keep left again after you have crossed the river.

Seterseterdalen valley is reached from

White-backed Woodpecker is an endangered species that has almost completely disappeared from many parts of Norway. It is now mainly found some places in central Norway and along the western and southern coast. *Photo: Kjetil Schjølberg.*

Capercaillie has a number of traditional leks in the central Norwegian forests, where every spring the males fight ritual battles to win the favour of the females. However, many of the leks are being destroyed by logging. *Photo: Kjetil Schjølberg.*

Kyrksæterøra village by heading 5 km east towards Holla. Just after passing the exit to Holla industrial area, you turn right towards Sæter and follow the road up the valley. At Setersetra farm a road exits south to Setersetersetra, and at Asplia a road exits towards Asplisetra. Both these roads are worth exploring.

4. RINDAL VALLEY

Trøndelag County
GPS: 63.15689° N 9.26618° E (parking Austre Fossdalen)

Notable species: Whooper Swan, Tufted Duck, Goosander, Black-throated Diver, Red-throated Diver, Slavonian Grebe, Capercaillie, Black Grouse, Hazel Grouse, White-tailed Eagle, Golden Eagle, Rough-legged

Buzzard, Gyrfalcon, Crane, Green Sandpiper, Greenshank, Great Snipe, Tengmalm's Owl, Pygmy Owl, Black Woodpecker, Three-toed Woodpecker, Yellow Wagtail, Grey Wagtail, Dipper, Bluethroat, Mistle Thrush, Sedge Warbler and Siberian Jay.

Description: Rindal valley on the north side of the Trollheimen mountain range is covered in spruce-dominated mixed forest. The bird life here is much the same as in high elevation forests of the southern interior, including Great Snipe, several owls and Siberian Jay. It is also good for a variety of raptors.

To the west of Rindal valley there are a number of deep fjords, most of them with branches that form side fjords. At the head of each of these side fjords, river deltas are formed. Most of these are completely or partially filled in and developed to make room for industry and agriculture. They may still at times be rather good for resting shorebirds and ducks, and should be stopped by if you happen to pass through the area. One such place close to Rindal valley is Syltørene delta in Surnadalsøra.

Best season: Early spring for owls. Mid-May to mid-July for most other species.

Directions: From E39 in Orkanger, exit onto Rv65 signposted to Surnadal.

Tactics: Lake Lomundsjøen and Fossdalen valley: From Rv65, 8 km before Rindal village, exit onto the side road on your right, signposted to Lomunddalen and Rørdalen. In this area you can drive a loop, e.g. by turning to the right after a couple of km, onto the road towards Rørdalen and Gåsvatn. Follow this road past a number of interesting small bodies of water, the last of which, Lake Lomundsjøen, is considered to be the most bird-rich. At the T-junction at Lake Lomundsjøen, turn left towards Rindalsskogen and drive 5 km (you have then almost made a loop back to where you headed for Rørdalen and Gåsvatn). Then turn right towards Fossdalen, then immediately to the left towards Løfall, and then to the right towards Fossdalen again.

This is Fossdalsvegen, a toll road that leads up into the Austre Fossdalen valley where you might find Great Snipe lekking in the twilight hours in May and June. Follow the toll road almost to the end and park 400 m before the bridge over the Toråa stream. You can hear the birds' peculiar display sounds from the roads here, and perhaps be lucky enough to see them out on the marshes and pastures through binoculars or a telescope. Do not walk out into the terrain and disturb these vulnerable birds! There is also a Black Grouse lek in this area, which is at its most active a bit earlier in spring.

Lake Igltjønna: Back on Rv65, drive towards Rindal and drive into Rindal village centre, where you stop at the swimming area at the nice Lake Igltjønna, where wetland birds can roost and where Slavonian Grebe breed most years. A tower hide is found on the south side. Continue through Rindal village and turn left, signposted to Romundstad. This road takes you through beautiful woodland areas where you can find all the forest species mentioned above. You can explore the detours, for example the ones signposted to Haltli and to Hølstoen, after 6.5 and 11 km respectively. The latter detour is considered the best for Three-toed Woodpecker and Siberian Jay. Make trips on foot in the terrain for maximum

payoff. Check the watercourse for Grey Wagtail and Dipper, the fields for Mistle Thrush, and use your telescope towards the Trollheimen mountains to look for birds of prey.

Syltørene delta, with a potential for ducks and shorebirds, is located in Surnadalsøra village along Rv65 32 km west of Rindal. You can reach the area by following signs to Røtet industrial area and then driving down the road signposted to *Miljøstasjon*.

5. HITRA AND FRØYA ISLANDS

Trøndelag County

GPS: 63.54572° N 8.55093° Ø (Gryta), 63.66891° N 8.30445° E (Titran)

Notable species: Resting migrants and passing seabirds; vagrants; Willow Ptarmigan *variegata* ('Smøla Ptarmigan') and Eagle Owl.

Description: Hitra and Frøya are two large islands covered in heather, home to the 'Smøla Ptarmigan' and Eagle Owl. The vast Havmyran marshlands in the heart of Hitra island is a nature reserve of international significance. It

Lake Igltjønna in Rindal valley has nesting Slavonian Grebe, a variety of ducks during migration, and there is a tower hide located in the marshy area on the south side, visible in the picture. *Photo: Bjørn Olav Tveit.*

is not very rich in species diversity but it does support interesting breeding species such as Black-throated Diver and Dunlin. The cultivated areas of Hitra can be good for nocturnal songbirds, particularly around the small, nutrient-rich Lake Smågavatnet, also supporting ducks and other waterbirds. Eagle Owl can be seen or heard in the hills surrounding the lake.

Titran is a village situated at the western point of Frøya island. This is one of the best sites in central Norway for rare migrants. It is widely recognized for producing extraordinary records of species like Spotted Sandpiper, Blyth's Pipit, Eastern Olivaceous Warbler, Black-faced Bunting, Blackpoll Warbler and many more. Typical of Titran is that most birds are discovered hanging in the bird ringer's mist-nets. There is not really a tradition of ordinary birdwatching here, although it may very well be conducted, mainly along the village gardens. Usually, the best birds are found in or after periods of heavy overcast skies with a gentle breeze, preferably from the southeast. In strong onshore winds, seawatching can be conducted from headlands on Frøya.

Other wildlife: Red deer is exceptionally common at Hitra, and with them follow ticks.

Best season: May to early June, and September and October are considered best for land birds. Seawatching is most productive during late summer and autumn.

Directions:
From Trondheim, follow E6 south to Klett and then E39 to Orkanger. From here, exit onto Rv714 signposted to Hitra and Frøya. The road is free of ferries, passing instead through a series of subsea tunnels.
Havmyran marshlands: At Hitra, take the second exit to Kvenvær and Forsnes, and continue approx. 23 km to Gryta, which is in a right turn

Peregrine Falcon was on the verge of extinction in Scandinavia by the 1970s, primarily as a result of environmental toxins. Since then, the population has grown considerably. Peregrine Falcon is now a fairly regular sight in coastal areas all year round. *Photo: Terje Kolaas.*

with a rest area and a gravel road that turns left. Walk the gravel road, later dirt road/path, into the area (approx. 1 hour's walk).
Titran: Continue Rv714 to Frøya and follow Rv716 westwards, signposted to Titran.

Tactics: Along the road from Trondheim you pass several interesting sites, the most notable being Gaulosen delta just outside Trondheim. The delta at Orkanger can be viewed from the roads in the industrial area and should be checked for resting waterbirds. There is a good site for Grey-headed and White-backed Woodpeckers at Krokstadøra 38 km further on, along Rv714, reached by exiting left onto the road signposted to Krokstad/Snillfjord Camping. Pass the entrance to the campsite and park by the fjord. Walk up the road to the right just after the large farm building and follow this road through the wooded hillside.
On Hitra island, Havmyran marshlands is explored on foot. The flat, uniform terrain is easy to get lost in. A map and compass are an absolute necessity. To get to Lake Smågavatnet on the north side of the island, you drive back to

the village of Strøm and take the signposted to Melandsjø. You will see the lake on the east side of the road after about 10 km.

Titran: To reach the area with the best gardens, turn right before *Fiskarheimen cafe*, 200 m before the main access road ends at the quay. Check bushes, spruce clumps and other attractive hiding places as you would on a rarity island. The whole area must be explored, but the road that runs parallel to the main access road, and which stretches from Heia headland as far west as you can get without a boat on Titran and inland along Yttersundet sound, is particularly nice. The street is flanked on both sides by sheltered gardens, perfect for resting passerines. For seawatching, turn right just before you enter Titran village, onto a road signposted to Kjervågsund. Follow this road to the end and walk up onto the hill behind the last house.

6. FROAN ARCHIPELAGO

Trøndelag County

GPS: 63.84776° N 8.45300° E (Sula seawatching shelter), 63.79906° N 8.68163° E (Dyrøy ferry landing)

Notable species: Resting migrants; vagrants; breeding, resting and passing seabirds.

Description: Off the coast of Frøya island is Froan archipelago, a 50 km long and 10 km broad belt of small islands and skerries. These wind-beaten islands are mostly covered in grass, if anything at all. Particularly the area between Vingleia and Halten islands is considered to be of great importance to breeding seabirds, and to White-tailed Eagle and Eagle Owl. As much as 20 % of the Norwegian population of Cormorant breed here. It is also among the most important breeding grounds in the country for Shag. The strongest population of Black Guillemot anywhere in Scandinavia is found here as well, and Greylag Goose, Common Eider, Red-breasted Merganser, Turnstone and Arctic Tern all breed in good numbers. The colonies of Kittiwake and Lesser Black-backed Gull *fuscus*, however, have declined markedly in recent years. 400 km² of the area is protected as a nature reserve, which is the largest continuous area of sea protected in Norway. And this is just the core of an even larger area with a somewhat less strict degree of conservation (landscape and animal protection areas). Froan archipelago is an important area for moulting, resting and

Froan archipelago contains a myriad of small islands and islets that form important nesting areas for sea-birds, such as Shag and Cormorant. Sula island in the background. *Photo: Bjørn Olav Tveit.*

On windswept islands such as Sula in the Froan archipelago, lost migratory birds often seek churchyards, which are frequently sheltered by trees. *Photo: Bjørn Olav Tveit.*

wintering seabirds as well, particularly for Common Eider and Velvet Scoter, wintering in the thousands.

These remote islands have a great potential for rare migrants in late spring and autumn. The inhabited of these islands are the most interesting, both because of the ease of access and because they have more varied habitats to offer, including bushes and conifer plantations. The most famous islands in the Froan archipelago are Sula and Halten, although the islands Mausund, Gjæsingen, Sørburøy and perhaps Vingleia, all reachable with express boat, could be good as well. Sula is a small and easily accessible island with a population of about 50 souls, with several gardens and a variety of habitats attracting migrants, making it the perhaps the first choice for most birdwatchers. It is frequently visited by birdwatchers during the peak migration periods, and a seawatching shelter is available here. The island Halten is more remote and difficult to access except in summer, with no permanent inhabitants and very little vegetation. It is, thus, a much more extreme alternative, for good and for bad. Of several vagrants encountered at Sula and Halten, the most spectacular is perhaps the fact that the two islands have one record each of Eastern Orphean Warbler, with these two being the only records of the species in this part of Europe.

Other wildlife: Froan archipelago is one of the most important areas in Norway for grey seal. Harbour seal is common as well, and several species of whale are encountered on a regular basis, including harbour porpoise and killer whale. On some of the islands, water vole, brown rat and hedgehog are found. Otter is also common in the area.

Best season: Late spring and September–October.

Directions: *From Trondheim,* follow E6 south to Klett and then E39 to Orkanger. From here, exit onto Rv714 signposted to Hitra and Frøya. At Frøya, follow signposts to the ferry landing at Dyrøy. The ferries depart only a few times a day, so check the timetables carefully beforehand. You may bring the car on the ferry to Sula and to some of the other islands, but not to Halten.

Tactics: Visits to the seabird colonies are largely prevented by visitor restrictions, but you can see a lot from the boat trip between the islands, including *fuscus* Lesser Black-backed Gull in summer until the end of September. Black Guillemot breed in significant numbers in the breakwater at Halten island. Search thoroughly for migrants in conifer plantations, gardens, meadows and other suitable habitat. On Sula, the seawatching shelter is situated just north of the lighthouse.

7. LAKE LITLVATNET

Trøndelag County

GPS: 63.61207° N 9.65359° E (parking Lake Litlvatnet), 63.59152° N 9.65809° E (trailhead Smidalen valley)

Notable species: Whooper Swan, ducks often incl. Garganey and Shoveler, other wetland species, Sedge Warbler; forest birds incl. White-backed Woodpecker.

Description: Lake Litlvatnet is small, shallow and nutrient-rich, surrounded by deciduous forest and cultivated land close to the internationally important Ørland wetland system. Sedge Warbler is common and so are Wigeon, Teal and several other species of duck. Garganey and Shoveler may breed, and Coot has done so in the past. Substantial numbers of Whooper Swan stage here during migration in spring and late autumn.

Lake Litlvatnet is often rich in Wigeon and other ducks, and is a mandatory and convenient stop when you are in transit between Trondheim and Ørlandet. *Photo: Bjørn Olav Tveit.*

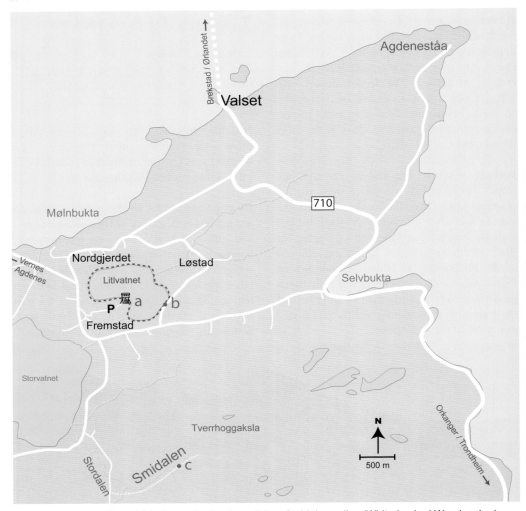

Lake Litlvatnet: **a.** Tower hide; **b.** road-side viewpoint; **c.** Smidalen valley (White-backed Woodpecker).

Smidalen valley is a nice forest area nearby where, among others, White-backed Woodpecker breed.

Best season: Lake Litlvatnet can be good all year except when frozen in winter. White-backed Woodpecker is found in Smidalen valley year-round but is easiest to locate when drumming on warm, sunny mornings in March–April.

Directions: Two roads lead to Lake Litlvatnet from Rv710 just south of the ferry landing at Valset, for the crossing to Brekstad at Ørlandet peninsula. Both roads are signposted to Vassbygda and Værnes. See the map.

Tactics: Lake Litlvatnet is best viewed from either the **a.** tower hide at Kvitneset or from the **b.** side road along the southeast corner of the lake. Both of these viewpoints are located outside the border of the nature reserve and can be used in the period of visitor restrictions within the reserve (April 20 to July 10). A telescope is of great advantage here.

In **c.** Smidalen valley, find suitable parking and walk the forest road up Stordalen valley. After 500 m you pass the bridge over Vollaelva river. Continue another 200 m to where a tractor track exits to the left, leading to the forested Smidalen valley.

8. ØRLANDET & STORFOSNA

Trøndelag County

GPS: 63.67941° N 9.57219° E (Grandefjæra tower hide),
63.65655° N 9.36684° E (Kråkvågsvaet sound viewpoint)

Notable species: Geese, ducks, shorebirds and gulls; King Eider, White-billed Diver, Great Northern Diver, Glaucous Gull (winter), Iceland Gull (winter), White-tailed Eagle and Peregrine Falcon.

Description: Ørlandet is the southwestern corner of the large Fosen peninsula, sheltering the entrance of the Trondheimsfjord from the sea. The Ørland peninsular tip consists of flat farmlands in contrast to the hilly interior, and in that respect somewhat reminicent of Lista in the extreme south of western Norway. At Ørlandet are several sites that together comprise one of the most exciting and important bird areas in the country, collectively known as Ørland wetland system. At the core is Norway's largest tidal area Grandefjæra. Its shoreline is 12 km long, and at low tide almost 6 km^2 of mud is exposed. Together with the adjacent area of shallow seas in Grandvika bay, a total of 21 km^2 is protected as Grandefjæra nature reserve. In order to get a grasp of the birds in this vast area, you need to be equipped with plenty of time – and a good telescope. The tidal flats can at times be teeming with shorebirds and other wetland species, and in Grandevika bay there are usually several thousand wintering, resting or moulting Common Eiders, Velvet Scoters and more. Among the common species you can sometimes find uncommon ones, such as King Eider, both of the large divers, and several grebes, particularly outside the summer season. At Grandefjæra, two birdwatching hides and a tower hide are available as part of the local community's efforts to make these important areas accessible and brought to the public's attention through Ørland National Wetland Centre.

Along the northern coast of Ørlandet,

Grandefjæra tidal area at Ørlandet with its surrounding fields and wetlands serve as an extremely important staging area for geese, ducks, gulls, shorebirds and passerines alike. The tower hide marked as point k. on the map of Ørlandet can be seen in the foreground. *Photo: Magne Klann.*

Oslo · Interior · Skagerrak · Western · Central · Northern · Finnmark · Svalbard

Uthaug harbour is usually host to a number of gulls, in winter often including a few of Arctic origin. Grey Phalarope are sometimes seen here, usually after northwesterly gales in late autumn. Innstrandfjæra tidal area just east of Uthaug harbour is considered to be one of the very best shorebird sites at Ørlandet. Here is also Kråktjønna brackish pond, often holding an interesting selection of ducks. The shorebirds at Innstrandfjæra tidal area are more concentrated and can often be viewed at closer range compared to Grandefjæra. At both places, pools on the beach at low tide attract some birds while most of them follow the waterline many hundred metres from the shore.

The tidal areas on the south side of Ørlandet are sheltered from the Norwegian Sea in periods of strong winds. Brekstadfjæra tidal area is suffering from industrial development but still the concentrations of shorebirds here are often higher than in the nearby and more pristine Hovsfjæra and Flatnesfjæra tidal areas. Just inland from Hovsfjæra is Austråttlunden deciduous woodland, a small area containing the original habitat of Ørlandet, from before most of it had to give way to agriculture. In these woods you can find a number of warblers and other passerines otherwise scarce at Ørlandet. Still further inland is Lake Rusasetvatnet, a nutrient-rich freshwater lake. It was dammed for several hundred years but became drained and thereby destroyed in the 1980s. The municipality and the landowners team agreed on an impressive plan for a comprehensive rehabilitation of the lake so that it has now largely been restored to its rich ornithological qualities.

Southwest of Ørlandet lies the scenic and lushly vegetated Storfosna island, which has several attractive areas for birds. Particularly important is Kråkvågsvaet

Little Stint is an Arctic breeding bird that has a small population in Finnmark, but is often seen in mixed flocks with other small shorebirds along the Norwegian coast during migration. *Photo: Espen Lie Dahl.*

sound, the large, shallow strait between Storfosna and the outlying Kråkvåg island, which is part of the Ørland Wetland System. The sound is known for being a reliable place for King Eider and Great Northern Diver, primarily in spring, autumn and winter, but the chances are also there in summer.

Best season: All year. Shorebirds from late spring through to early autumn, particularly in rainy and overcast conditions. Gulls and resting waterbirds are most numerous from late autumn through to early spring, Arctic gulls and the odd Grey Phalarope particularly in and after periods of winds from the north-west.

Directions: *By public transportation from Trondheim*, Ørlandet is reached in just over an hour with the express boat to Brekstad, or from Kristiansund in about 2.5 hrs. You can bring a bicycle or rent one at the wetland centre in Brekstad.

By car from Trondheim: You can choose between two different routes, both taking about 1.5 hrs and both having several interesting sites along the way. You may want to take one route there and the other one back. Both routes includes a short ferry crossing.

Alt. 1 via Flakk: From Trondheim, follow Rv715 west signposted to Fosen and take the ferry across the Trondheimsfjord from Flakk to Rørvik. Continue north along Rv715, and then west along Rv710 signposted to Brekstad. This road takes you close to the sites Rissa wetlands and Lake Eidsvatnet, which see.

Alt. 2 via Agdenes: From Trondheim, follow E6 signposted to Oslo. At Klett, exit onto E39 to Orkanger, from where you exit onto Rv710 signposted to Agdenes. This road leads to Valset ferry landing, from where you cross over to Brekstad. This route leads past Gaulosen delta, and Lake Litlvatnet, which see, and along the heavily industrialized Orkla delta in Orkanger.

Tactics: Ørlandet & Storfosna is a large area best covered by car (or bicycle). It is usually best to visit Grandefjæra tidal flats in the hours just before high tide, when the shorebirds are pushed towards the shore. Also check the saltmarsh and ponds along the shore and the cultivated fields inland. Inclement weather is often best, both for shorebirds during migration and for gulls in winter. In late spring and early summer, the choir of songbirds in Austråttlunden is at its most intense in early morning.

Suggested order for visiting the sites at Ørlandet & Storfosna (see map next page): Brekstad is the main village at Ørlandet and this is where you will find most of the practical facilities. The culture centre (*kultursenter*) by the ferry landing contains the **q.** Ørland National Wetland Centre, where you also may rent bicycles.

a. Brekstadfjæra tidal area can be viewed from Brekstadfjæra industrial area. **b.** Hovsfjæra tidal area is surrounded by lush saltmarsh and grazing pastures. A tower hide is found by driving down Fitjanveien road.

At Innstrandfjæra (Kråka) east of Uthaug harbour, the waders often appear more concentrated and at closer range than at Grandefjæra, which increases the possibilities of finding something exciting. *Photo: Bjørn Olav Tveit.*

Oslo

Interior

Skagerrak

Western

Central

Northern

Finnmark

Svalbard

Innstrandfjæra

e • Kråka
P gjenbruksstasjon

Kjeungan

f

Uthaug

Anddam

Ryggamyra

Neset-
fjæra
Hoøya

Juldagan

P

Øya • g

Laksklø-
holmen
h

i

Grandefjæra

Ørlandet

Grandvika

j

Brekstad

k
P

Grande

o

Flatnesfjæra

P

← Storfosna

l
Beian

Rønsholmen

Ørland
Orkland

m •
Garten

Vernes
Agdenes

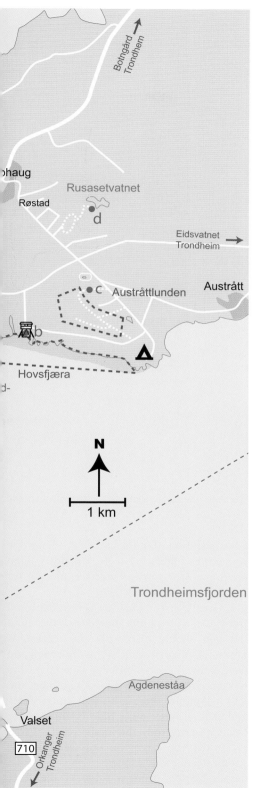

Ørlandet: a. Brekstadfjæra tidal flat; **b.** Hovsfjæra tidal flat; **c.** Austråttlunden forest; **d.** Lake Rusasetvatnet; **e.** Kråka tidal area; **f.** Uthaug harbour; **g.** Grandefjæra viewpoint north (and, from the top of the hill, to the outer sea); **h.** Lakskløholmen hide; **i.** viewpoint military base; **j.** viewpoint marina; **k.** tower hide; **l.** Rønsholmen hide and **m.** Garten viewpoint.

c. Austråttlunden forest has mature deciduous woodland adjacent to a golf course. A small pond on the golf course can be worth investigating. Park in one of two dedicated car parks for visitors to Austråtlunden. The forest is made accessible to most people with a nature trail, great for birdsong walks in spring and early summer. Also check the trotting track on the north side of the forest. Both in the middle and around the track there is a water feature that can house resting ducks and other wetland birds when there is no activity on the course. **d.** Lake Rusasetvatnet has been restored and is, together with Lake Eidsvatnet, among the most exciting bodies of fresh water at

Turnstone is a common visitor to Ørlandet during migration periods and many also spend the winter here. *Photo: Tomas Aarvak.*

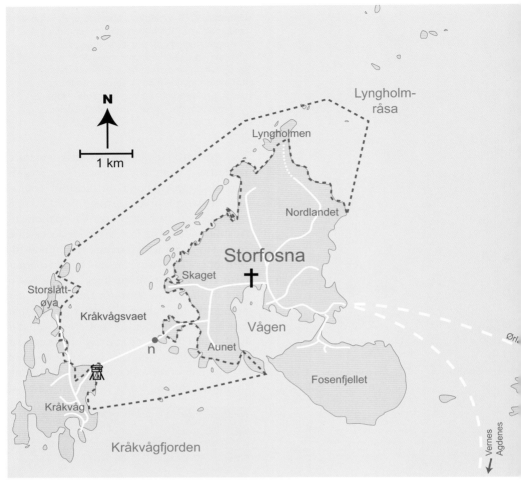

Storfosna island: n. causeway viewpoint to Kråkvågsvaet sound – the observation shelter on Kråkvåg island is not particularly useful to birdwatchers.

Ørlandet. The smaller ponds marked on the map should also be investigated, as should the surrounding farmland on the peninsula. **e.** Innstrandfjæra seashore and Kråktjønna pond often has resting waders, ducks and gulls. The road to Kråka landfill is signposted to *Gjenbruksstasjon*. Park completely out of the way by the gate and walk along the fence out to the shore and then towards the right to Kråktjønna pond. You can continue along the beach to the east and check out the many nice bays that stretch over the next couple of km. Alternatively, you can drive the main road further east and turn into farm roads on the left. **f.** Uthaug harbour can often contain a

good selection of gulls, and Grey Phalarope sometimes appear as well. Uthaug village has some nice gardens worth exploring for migrant passerines.

g–l. Grandefjæra tidal area is best explored from the viewpoints noted on the map, but you should also walk the terrain. If possible, you can walk along the beach and have someone else picking you up at the next viewpoint. At **h.** Lakskløholmen and **l.** Rønsholmen there are observation hides, and at **k.** a large tower hide (signposted to *Grandefjæra* from the main road). From **g.** Hoøya you have the best seawatching vantage point at Ørlandet.

At **m.** Garten, from a ridge at the ferry

landing, you have a view to the southern part of Grandefjæra.

Take the short ferry crossing from Garten to Storfosna island and explore gardens, conifer plantations and pastures both here and on Kråkvåg island. Scan Kråkvågsvaet sound from n. the causeway with a telescope. Back on Ørlandet, you can take the southern road back to Brekstad, and make a stop at o. Flatnesfjæra tidal area.

9. LAKE EIDSVATNET

Trøndelag County
GPS: 63.74292° N 9.81479° E (parking tower hide)

Notable species: Resting waterbirds.

Description: Lake Eidsvatnet is shallow and situated in an agricultural area on the Fosen peninsula. It is one of the most important fresh-water sites in central Norway, situated close to the important Ørland wetland system. The shallowest section with the lushest reeds and other vegetation is found in the north end of the lake. Here is also a tower hide. The lake is an important resting site for Whooper Swan, ducks and more. Sedge Warbler breeds, and formerly Coot as well.

Best season: Migration periods and early summer.

Directions: *From Trondheim*, follow Rv715 west signposted to Fosen and take the ferry across the Trondheimsfjord from Flakk to Rørvik. Continue north along Rv715, and then west along Rv710 signposted to Brekstad. In Botngård, turn south onto a road signposted to *Fuglefredningsområde* (bird protection area). Park after 500 m at a designated spot next to an information poster about the area. Follow the path along the field to the tower hide.

Lake Eidsvatnet, just like Lake Litlvatnet, is an essential stopping place when visiting Ørlandet and Storfosna from Trondheim, not least when choosing the northern route via Flakk. *Photo: Bjørn Olav Tveit.*

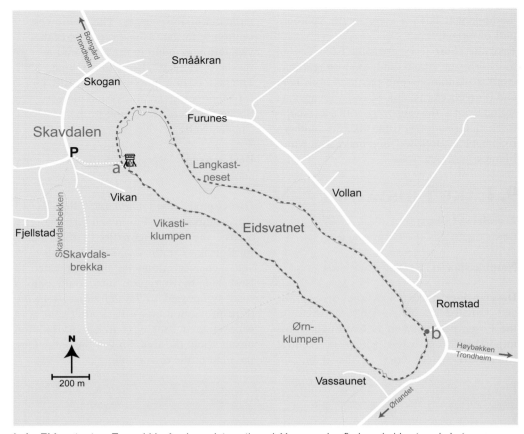

Lake Eidsvatnet: a. Tower hide; **b.** viewpoint south end. You can also find road-side stops in between.

From Brekstad (Ørlandet) you can reach Lake Eidsvatnet by following Rv710 east and turning right after 1.5 km and then follow signs towards Ottersbo and later to Høybakken. You will see the south end of Eidsvatnet on your left just before the last exit to Høybakken.

Tactics: See the map. **a.** The tower hide is the best starting point for exploring the north end of the lake. From the road along the east side, you get a view of the middle section of the lake. In the **b.** south end of the lake is another sign to the bird protection area (*Fuglefredningsområde*), pointing you to a viewpoint of this section of the lake. You can drive further on to Ørlandet by following the signs for Brekstad from here.

10. RISSA WETLANDS
Trøndelag County
GPS: 63.49175° N 9.99226° E (Bingberget, Grønningsbukta)

Notable species: Resting waterbirds.

Description: Straumen river, in the area known as Rissa, is a 2 km long river connecting the brackish Lake Botn to the Trondheimsfjord. At high tide, saltwater flows in through Straumen and brackish water flushes out again at low tide. Straumen meanders through the flat cultivated landscape. In the lower part close to the fjord, Straumen river splits in two runs before reconnecting 500 m further downstream. One of them is only flooded at very high tide. The area between the two runs is covered in lush saltmarsh and meadows. At low tide, mud

234

is exposed along the edges, adding to the area's value as feeding ground to a variety of ducks, shorebirds and other wetland species. The tidal current prevents Straumen river from freezing in winter, a time when Little Grebe and other uncommon species may be found amongst the more numerous common species.

Grønningsbukta inlet by Stadsbygd village just south of Straumen is one of a very few areas of shallow water and mudflats along the northern side of the Trondheimsfjord, and thus worth visiting to look for shorebirds, ducks and other wetland species. The inlet is equipped with a tower hide.

Best season: Migration periods and winter.

Directions: *From Trondheim*, follow Rv715 west towards Fosen and take the ferry across the Trondheimsfjord from Flakk to Rørvik ferry landings. Continue west along Rv717, signposted to Stadsbygd. At the roundabout with the exit to the *Kystens arv*, continue 1.2 km along Rv717 towards Rissa, and turn down a farm road to the left signposted to *Grønningsbukta parking*, where you will pass through the farmyard and reach the tower hide by the shore.

After another 9 km on Rv717 towards Rissa, you see Lake Botnen on your right. The road signposted to Sørbotn encircles the lake. Rv717 crosses Straumen sound a bit further ahead.

11. BINGSHOLMSRÅSA SOUND

Trøndelag County
GPS: 63.93394° N 9.94793° E (Lauvøya island)

Notable species: Resting waterbirds; White-tailed Eagle, Peregrine Falcon, Gyrfalcon (winter); migrating and wintering passerines.

Description: Bingsholmsråsa is a coastal sound between Lauvøya island and Tårnes headland with a 5 km² area of tidal zones and shallow water. The area is of great importance to geese, ducks, divers, grebes and auks, particularly in winter. The shoreline

Bingsholmsråsa sound is a shallow strait, attractive to resting and wintering ducks, grebes and divers. *Photo: Bjørn Olav Tveit.*

Gaulosen delta at low tide, as seen from Byneset headland to the northwest. *Photo: Bjørn Olav Tveit.*

consistency is alternating between mud, sand and pebbles, some places with large build-ups of decomposing seaweed where insects and other small animals thrive, providing an important source of food for a variety of bird species. There are several small islands and skerries covered in heather and grass in the sound. The largest area of shallow water is offshore from the agricultural fields at Tårnes headland. Lauvøya has a varied cultivated landscape with several habitats suitable for resting migrants.

Best season: Winter, although migration periods can be interesting as well.

Directions: *From Trondheim*, follow Rv715 west signposted to Fosen and take the ferry across the Trondheimsfjord from Flakk to Rørvik. Continue north along Rv715 to Åfjord.

Lauvøya: From Åfjord village centre, follow the road signposted to Lauvøy. When you reach Lauvøya island, the road splits up to loop in a circle around the island. Keep right here and turn right again after 1.5 km by a large mailbox rack. This road leads down to

the sea. Walk past the small inlet and onto the headland on your left to overlook the sound.

Tårnes headland: From Åfjord village centre, follow Rv723 north signposted to Stokkøya. After 13 km turn left towards Ratvik, pass the last farm and continue to Tårnes chapel at the end of the road. Investigate the sound, seashore and fields of the headland, but do not walk in the fields during the growing season.

Tactics: Scan the sound with a telescope and look for resting migrants in the surrounding terrain.

12. GAULOSEN DELTA

Trøndelag County

GPS: 63.34548° N 10.21673° E (tower hide), 63.32711° N 10.21388° E (parking Øysand)

Notable species: Whooper Swan, Pink-footed Goose, ducks, Little Ringed Plover, Temminck's Stint, other shorebirds, White-tailed Eagle, Peregrine Falcon, Wood Warbler, Marsh Tit, Hawfinch; resting migrants in general.

Description: Gaulosen is a delta at the mouth of the river Gaula at the head of one of the Trondheimsfjord's southern arms. This is the last intact – or nearly so – estuary of central Norway and one of the very finest birdwatching sites in the region. It is a part of the larger Trondheimsfjord wetland system together with the sites further up the fjord. At low tide, large mudflats are exposed, creating favourable conditions for a variety of wetland species. In May and September, large numbers of Greylag and Pink-footed Geese often rest in Gaulosen and the cultivated fields along river Gaula south to Melhus. Sometimes, other species of goose may sneak in amongst them. The banks of river Gaula and the shores of Gaulosen delta are lined with lush vegetation, including the berry-carrying sea-buckthorn bush. On the fjord outside the delta, particularly at Buvika bay, large numbers of ducks, divers, grebes and gulls may spend the winter.

The south-facing slopes along the northern shore of the fjord are covered in deciduous forest supporting species like Wood Warbler, Nuthatch, Marsh Tit and Hawfinch. Several rare birds have been found in the Gaulosen area over the years, including Barrow's Goldeneye, Sharp-tailed Sandpiper, Bonaparte's Gull and Great Spotted Cuckoo.

Other wildlife: River Gaula is one of the richest in Atlantic salmon and sea-trout in Norway.

Best season: All year, particularly during the migration periods.

Directions: *By public transportation*: From Trondheim city centre, take a bus to Øysand bus stop at the southern side of Gaulosen. The same bus also stops at Buvika a bit further on. A bus runs along Rv707 along the northern side of Gaulosen delta, and most of the sites mentioned here are between Leinstranda and

Leinøra in Gaulosen delta has several lagoons and coves with exposed mud at low tide, good for ducks, *Tringa*-waders and more. The orange berries on the sea-buckthorn bushes are snacks for thrushes, *Sylvia*-warblers and the like. *Photo: Bjørn Olav Tveit.*

Oslo · Interior · Skagerrak · Western · Central · Northern · Finnmark · Svalbard

Spongdal bus stops (you might ask the driver to kindly let you off in between these stops as well). Some of the buses to Gaulosen leave from the city centre while others leave from Heimdal on the south side of Trondheim.

By car from Trondheim, follow E6 south towards Oslo to Klett. From there, see the map.

Tactics: The area is large and must be explored from several angles, preferably with a telescope. Two tower hides gives views of different parts of Gaulosen delta, but other viewpoints can be good as well. From **a.** Øysand tower hide you get a view of the delta from the south. It is approximately 1.5 km to walk from the car park by the campsite reception. A small parking fee is to be expected. Drive as far as possible into the camping area if you want to shorten the walking distance somewhat. You can walk along the beach towards the river outlet to the tower hide, or choose to follow the roads and a path. Look for resting migratory birds on land, shorebirds and ducks out in the wetland and in the small brackish ponds, as well as divers, grebes and ducks out on the fjord. When the tourist season is at its worst, it may be more convenient to scan the delta from **b.** Bynes tower hide on the north side of the delta. You can drive all the way to the tower hide, and parking is free. The view from this tower is best in the morning, as you often get a difficult backlight here during mid-day. In any case, there is less heat haze in the air and fewer people on the beaches in the morning hours. Otherwise, it

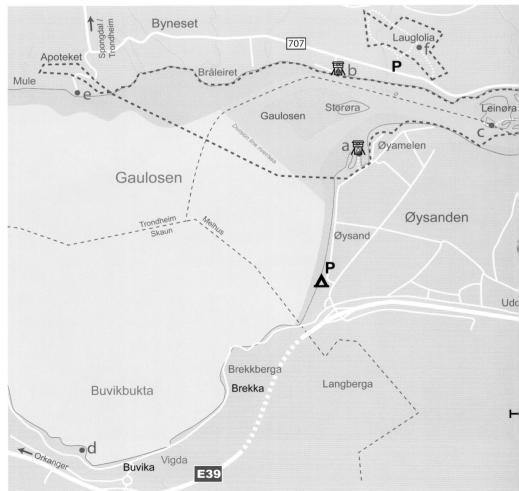

is best to check the shorebirds in Gaulosen delta when the tide is rising, the large mudflats are shrinking and the birds are concentrated closer to land. It is permitted to walk along the beaches, but you should, of course, avoid flush resting and nesting birds. There is a traffic ban on the Storøra gravel bank in the period 1 April–15 July. From **c.** Leinøra you have a view of the inner parts of Gaulosen delta and the lower part of the river Gaula. **d.** Buvika bay is best explored with a telescope from land, and the same applies at **e.** Apoteket nature reserve on the north side, where you have a view of the outer part of Gaulosen and the fjord. You can look for woodland birds in the deciduous forest both here and in **f.** Lauglolia hill.

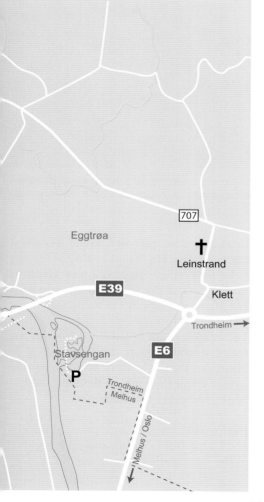

13. Trondheim Area

Trøndelag County

GPS: 63.43157° N 10.38491° E (BirdLife Norway HQ), 63.38395° N 10.3958° E (Nedre Leirfoss river site), 63.41767° N 10.26096° E (Skistua lodge)

Notable species: Ducks, gulls, White-tailed Eagle, Lesser Spotted Woodpecker, Grey Wagtail, Long-tailed Tit, Rook and Hawfinch. *In surrounding woodlands and lakes:* Black-throated Diver, Capercaillie, Hazel Grouse, Willow Ptarmigan, White-tailed Eagle, Rough-legged Buzzard, Green Sandpiper, Hawk Owl (uncommon), Tengmalm's Owl, Pygmy Owl, Three-toed Woodpecker, Black Woodpecker, Ring Ouzel, Dipper, Siberian Jay and Nutcracker.

Description: Birdwatching can sometimes be rewarding even in the middle of Trondheim city, particularly in winter and spring. The lower part of Nidelva river, up to Stavne bridge, often supports a number of ducks and gulls. Glaucous and Iceland Gulls are among the species you can hope for, and even greater rarities have been found, such as Ring-necked Duck, Glaucous-winged, Caspian and Ring-billed Gulls. Wintering passerines may be found in the vegetation along the banks. Nedre Leirfoss power plant further up the same river can also host a nice selection of wintering waterbirds. Grey Wagtail breeds here in summer, and the nearby deciduous forest is home to Lesser Spotted Woodpecker, Long-tailed Tit and Hawfinch. Trondheim harbour and the Trondheimsfjord is sometimes teeming with ducks and gulls. White-tailed Eagle is often seen resting at Munkholmen island, just off the harbour. The Lade peninsula east of the harbour has a range of habitats suitable for resting migrants. Here you

Gaulosen delta: a. Øysand tower hide; **b.** Byneset tower hide; **c.**Leinøra, lagoons and bushes; **d.** Buvika bay (one of several viewpoints here); **e.** Apoteket woodlands and fjord view; **f.** Lauglolia woodland hill.

Oslo
Interior
Skagerrak
Western
Central
Northern
Finnmark
Svalbard

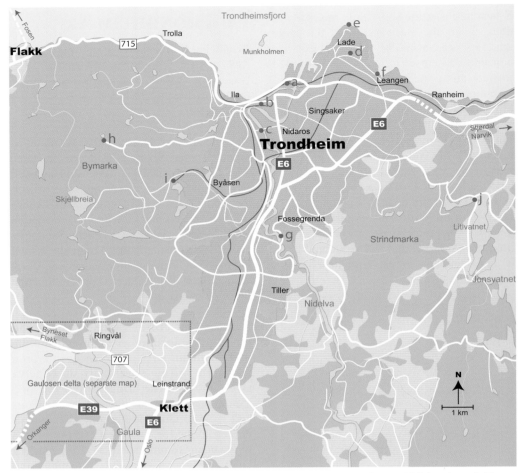

Trondheim area: a. Trondheim railway station; **b.** BirdLife Norway's main office; **c.** Nidelva river; **d.** Ringve botanical garden; **e.** Østmarkneset viewpoint; **f.** Leangenbukta inlet; **g.** Nedre Leirfoss; **h.** Skistua (bus); **i.** Lian (tram); **j.** Lake Jonsvatnet.

can often see Rooks from the breeding colonies in the area. The coast further east towards Stjørdal town has several areas of tidal flats and shallow waters, attracting a diversity of waterbirds.

The surrounding coniferous woodland, particularly Bymarka forest to the west of the city, is extremely popular with the outdoor-minded citizens of Trondheim. Even though the traffic of hikers at times can be a bit too heavy for a birdwatcher's liking, you can easily find tranquil spots and many of the forest species of the region. There are even several Capercaillie and Black Grouse leks here, although you may need to team up with a

local birder in order to find the less-noisy Capercaillies. On the other side of the city, Lake Jonsvatnet can host a nice selection of ducks, and the surrounding woodland can support many of the same species as in Bymarka forest.

The headquarters of BirdLife Norway are found downtown, at Sandgata 30 B in Midtbyen borough. Here you can ask for advice about the birds in the area and buy books and other merchandise.

Best season: Winter and spring in the city and by the fjord. Spring and early summer in the surrounding forests and lakes.

Directions: Trondheim is situated on the E6 about 7 hrs drive north of Oslo. It can also be reached by train or air. Trondheim airport is in Stjørdal town, a 30 min drive east of Trondheim city centre.

In Trondheim city, a bicycle or a combination of public transportation and walking are the best means of transportation.

Tactics: You may walk up along **c.** Nidelva river, or from Fagerheim bus stop at Lade peninsula along the path via **d.** the arboretum at Ringve botanical garden to **e.** Leangenbukta inlet. With a bike (or a car) you can explore the coast along the Stjørdalsfjord east to Midtsandan or even all the way to Stjørdal.

f. Nedre Leirfoss along the Nidelva river is best reached by car: Follow the new E6 south towards Oslo to Exit 37 signposted to Fossegrenda. Keep following signs to Fossegrenda until 1.7 km from E6, where you turn right onto the road signposted to N. Leirfoss. Park in the picnic area on your right 1 km past this junction.

Bymarka forest: From the central terminal, take a bus to **g.** Skistua or the tram to **h.** Lian and follow paths into the forest. A detailed map or a GPS will come in handy. Lake Jonsvatnet: Take a bus to *Solbakken bru* bus stop, and walk from there. By car, leave E6 at Exit 39 and follow signs to Jonsvatnet. Overlook the lake from several viewpoints, such as from **i.** the bridge.

Black-throated Diver is typically found on large lakes in the woodlands and mountains of central Norway including in the forests around Trondheim city. *Photo: Ingar Jostein Øien.*

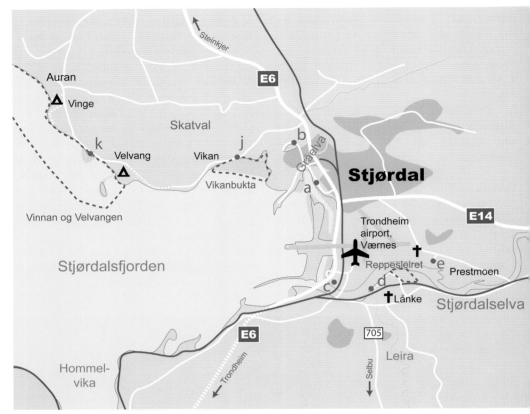

Stjørdal area: a. Halsøen estuary; **b.** Leca pond; **c.** Sandfærhus estuary; **d.** Leksosen; **e.** Notenghølen; **f.** Bergshølen; **g.** Hofstadøra; **h.** Trøiteshølen; **i.** Bjertemsøra; **j.** Vikanbukta inlet; **k.** Vinnan and Velvangen.

14. STJØRDAL AREA

Trøndelag County

GPS: 63.47930° N 10.79452° E (Vinnan/Velvangen), 63.46193° N 11.11927° E (Hegramo river view)

Notable species: Ducks, divers, grebes, gulls and shorebirds incl. Little Ringed Plover.

Description: The Stjørdalsfjord is an arm of the Trondheimsfjord reaching east to the mouth of river Stjørdalselva at the head of the fjord. The Stjørdal delta is for the most part developed into Stjørdal town, industrial areas and an airport. Fragments of the original delta habitats are more or less intact, however, still attracting a number of wetland species during migration both at the estuary and up along the river. Gravel banks in the river support breeding Little Ringed Plover. The fjord itself is interesting too, particularly in winter and early spring. In March–April spawning fish create a feeding frenzy of ducks, divers and gulls in the fjord. Several rare birds have been recorded in the Stjørdal area, including Black Duck, Greater Yellowlegs, Marsh and Stilt Sandpipers.

Best season: Autumn, winter and early spring.

Directions: Stjørdal is situated along the E6 east of Trondheim. It can be reached by train, and holds the main airport for Trondheim city.

Tactics: a. Halsøen estuary is the remains of the original Stjørdalselva river mouth and is now more of a lagoon with tidal flats and areas of shallow water teeming with algae, good for ducks and shorebirds. Now, only the Gråelva river empties here. The **b.** Leca pond behind the industrial area of Sutterøleiret has remnants of the original delta meadows.

a. <u>Sandfærhus estuary</u>. Because the natural mouth of the Stjørdalelva river is cut off by the airport runway, a new outlet has been established on the runway's south side. Sediment built up here has created an area of shallow waters attractive to swans and ducks. The salt marsh, meadows and tidal zone can be good for shorebirds and a variety of other migrants. Up along the Stjørdal river are several accumulations of sediment, shallow backwaters and lush vegetation, of which the most interesting are marked on the map as points **d–i**. Little Ringed Plover breeds on some of the gravel islands in the river.

j. <u>Vikanbukta inlet</u> by the Stjørdalsfjord has large tidal flats surrounded by lush saltmarsh and cultivated fields, a good spot for *Tringa*-waders, among others. **k.** <u>Vinnan and Velvangen bird sanctuary</u> often hold ducks, gulls and more. When fish are spawning in spring it can be just as good or better then the more widely known Tautersvaet sound by Tautra island. The flocks often include uncommon species such as King Eider and Iceland Gull. Park seaside by Velvang campsite.

The Stjørdalsfjord buzzes with life during the herring spawning period in early spring, with gulls and other sea-birds feasting on fish and roe, such as here at Vinnan and Velvangen bird sanctuary. *Photo: Bjørn Olav Tveit.*

15. TAUTRA ISLAND

Trøndelag County

GPS: 63.57156° N 10.61161° E (tower hide)

Notable species: Gadwall, Garganey, Shoveler, Common Eider, Common Scoter, Velvet Scoter, Long-tailed Duck, geese, Red-throated Diver, Black-throated Diver, Great Northern Diver, White-billed Diver, Slavonian Grebe, Red-necked Grebe, Coot, gulls, auks, shorebirds, White-tailed Eagle, Peregrine Falcon and Goldfinch (winter).

Description: Vast numbers of ducks, divers and grebes rest and winter near the small Tautra island in the Trondheimsfjord. The area is considered to be one of the most important bird sites in the country, just a bit more than an hour's drive from Trondheim. Tautra island is connected to the mainland via a 2.5 km causeway across Tautersvaet sound. The causeway was built in the 1970s and was originally constructed as a continuous breakwater, blocking the vital natural flow of water through the sound. This had a dramatically negative effect on the birdlife in

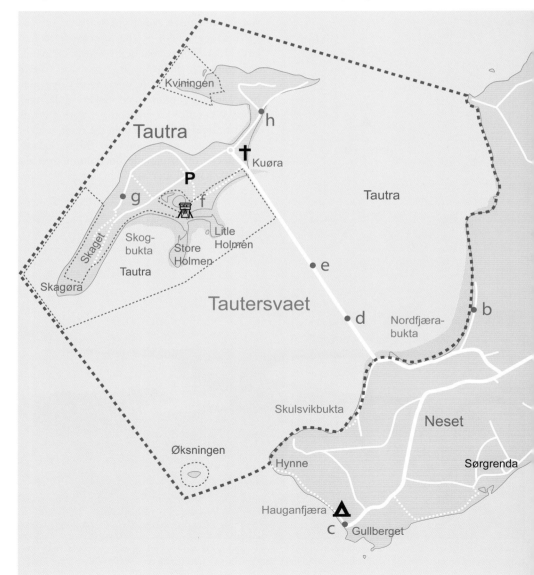

the area. The causeway also made it possible for terrestrial predators such as red fox and badger to walk freely out to the seabird colonies on the island. So, in 2003 a 350 m wide opening in the causeway was made and replaced by a bridge, letting the water flow through again. The bridge is equipped with a gate, reducing the risk of letting unwanted animals come across. The effect of this operation was immediate and positive, primarily in terms of increased numbers of wintering ducks. The winter flocks here are dominated by Common Eider but contain significant numbers of Velvet Scoter and Long-tailed Duck as well. Also numerous are Common Scoter, Mallard, Teal, Wigeon, Shelduck, Red-breasted Merganser and Goldeneye. In between the common species, uncommon visitors such as King Eider, Steller's Eider and Surf Scoter are sometimes found. Red-throated Diver is the most common diver closely followed by Black-throated Diver. Both are often encountered in the dozens while Red-throated sometimes are counted in the hundreds. The two large divers, Great Northern and White-billed, are usually also present with a few individuals. Great Crested Grebe, Red-necked Grebe and Slavonian Grebe are usually present in the dozens. Large numbers of Cormorant, gulls, terns and auks can be seen here as well, and sometimes flocks of geese and shorebirds. The flocks of birds can move around the fjord quite a bit and, when not feeding around Tautra, they are often found at the head of the Stjørdalsfjord. Tautra island itself is also worth exploring for resting and wintering birds. The areas on the island holding the largest numbers of breeding (and resting) waterbirds are subject to visitor restrictions during spring and summer. Måsdammen pond supports a colony of Black-headed Gull. Shoveler, Coot and in some years Gadwall breed as well. A tower hide overlooking the wetlands on the east side of the island can be entered during the period of visitor restrictions (April 1 to July 15) as long as you keep to the marked path. On some days, a noticeable visible migration of passerines can be appreciated. A few rare birds have been encountered in the area, including Marsh Sandpiper, Bittern and Sabine's Gull. King Eider has, surprisingly, bred on Tautra island a couple of times several decades ago.

Other wildlife: Otter and harbour seal are common.

Tautra island: a. Storleiret inlet; **b.** Nordfjærabukta tidal area; **c.** Hauganfjæra viewpoint; **d.** and **e.** lay-bys on the causeway, good for views over the Tautersvaet sound; **f.** tower hide and Måsdammen pond; **g.** fields, bushes and clusters of trees; **h.** bushes and a public restroom.

Tautra island

Tautra is one of central Norway's finest birdwatching sites. The picture shows the area of shallow sea towards Litle Holmen islet, as seen from the tower hide. The causeway is seen in the back. *Photo: Bjørn Olav Tveit.*

Best season: Tautra can be good all year but is at its most spectacular during April and early May. Mid-summer may be relatively quiet save for a few flocks of moulting ducks, but the numbers start to increase already from late summer.

Directions: Tautra is reached by car from Trondheim in just over an hour by following E6 past Stjørdal, exiting in Åsen onto Rv753 signposted to Frosta. Then, see the map.

Tactics: On your way out, stop at **a.** Storleiret inlet 1 km after passing Frosta village centre and **b.** Nordfjærabukta tidal area further out. **c.** Hauganfjæra is a good viewpoint to the outer section of Tautersvaet sound, but the best view is obtained from the **d.** and **e.** lay-bys on the causeway. **f.** The tower hide at Tautra gives views to the saltmarshes on the east side of the island. Outside the period of visitor restrictions, you may explore the meadows and areas of shallow ponds on foot. Måsdammen pond can be explored from behind the tower hide or from the road along the opposite side of the pond. Also check **g.** all woods and cultivated fields during migration. There is an area of nice scrub by the **h.** soccer field on the north side of Tautra.

16. Lake Hammervatnet

Trøndelag County
GPS: 63.62352° N 11.06621° E (tower hide)

Notable species: Whooper Swan, Shoveler, Tufted Duck, Slavonian Grebe, Coot, Sedge Warbler; resting waterbirds in general.

Description: Lake Hammervatnet is one of the most bird-rich fresh-water lakes of central Norway. It is 6 km long and surrounded by damp woods and cultivated fields, supporting good numbers of Whooper Swans, ducks and other species of waterbird, particularly during migration and as long as it is not completely frozen in winter. In late summer, thousands of Swallows roost by the lake. Uncommon species like Pochard and Little Grebe or even rarer birds are regularly found. Shoveler, Slavonian Grebe and Coot regularly breed with several pairs each, and Marsh Harrier and Grey Wagtail have bred. Water Rail is sometimes seen, particularly just before the lakes freezes over in late autumn.

Lake Hammervatnet is the last of a series of lakes along the Hopla watercourse. Some of the other lakes may be worth exploring as well, particularly during migration or on nocturnal songbird trips in early summer. Among these, the reed lined Nesvatnet, Lynvatnet, Hoklingen and the eastern end of Lake Movatnet are considered to be the most rewarding.

Other wildlife: Otter, otherwise mainly found in saltwater environments, are regular at Lake Hammervatnet.

Best season: All year is good, except perhaps when completely frozen in winter. Lake Hammervatnet is at its best just after the ice thaws in spring and in early summer.

Directions: If arriving from the south, exit from E6 in Åsen and continue north towards Hammer. If arriving from the north, exit from E6 at the *Hammerkrysset junction* north of Åsen and continue just a few hundred metres back towards Hammer. Follow signposts

leading to the tower hide on the east-side of Lake Hammervatnet. The other lakes in the Hopla watercourse are reached by exiting from E6 in the centre of Åsen and following signs to Markabygd. Lake Nesvatnet can be seen from the E6 a few km north of Lake Hammervatnet.

17. LEVANGER WETLANDS
Trøndelag County
GPS: 63.70514° N 11.14037° Ø (Hotterbukta inlet), 63.76749° N 11.34245° Ø (Tynesfjæra tidal area)

Notable species: Pink-footed Goose, Shelduck, ducks, divers, grebes, gulls and shorebirds.

Description: In the area around Levanger town, you will find large riparian areas that make up several more or less interconnected bird sites. The areas are dominated by tidal flats with soft mud and occasional pebbles, surrounded by salt marshes and cultivated land, partly also industrial areas. These are nationally important

Lake Hammervatnet and its nice cove in the southeast have been restored to prevent overgrowth. The tower hide can be seen on the far right of the picture. *Photo: Magne Klann.*

Shoveler breeds scattered and few in number in Norway, typically in vegetated lakes in the lowlands over most of the country, such as in Lake Hammervatnet. *Photo: Bjørn Fuldseth.*

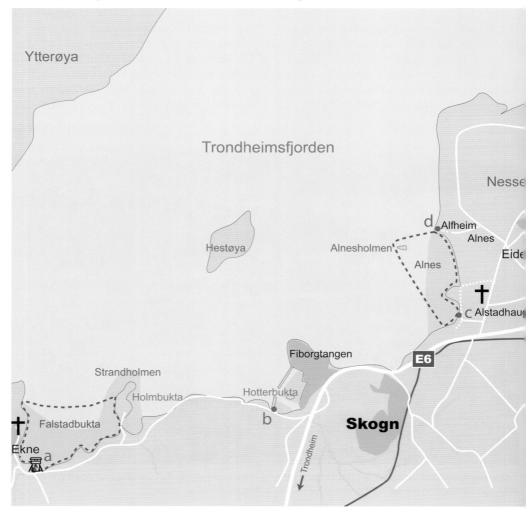

staging grounds for geese, ducks and shorebirds.

Best season: The areas can show interesting species and a good number of birds all year, but are considered to be at their peak during migration, especially in autumn.

Directions: The areas are located by Levanger town and can be reached via the E6. See the map.

Tactics: a. Falstadbukta inlet is the southern-most of these large tidal areas. An observation shelter has been set up at the outlet of river Byaelva in the south-west. Further north is **c.** Alnesfjæra tidal area, which is particularly well used by resting Pink-footed Geese that commute between the shore and the pastures on

Nesset north of Alstadhaug church. Just off the shore from Alnesfjæra is the small grassy Alnesholmen islet, which is an important nesting place for several species of gulls and ducks, formerly also Arctic Skua. The islet is best seen from **d.** Alfheim. Between Falstadbukta inlet and Alnesfjæra tidal area, at the outlet of the Hotterelva river, on the heavily industrialized Fiborgtangen industrial area, lies **b.** Hotterbukta inlet. This is a small area with great potential, where the species diversity and density of resting wetland birds can be on a par with the larger surrounding areas. **e.** Eidsbotn lagoon is connected to the Trondheimsfjord through the 1.5 km long **f.** Levangersundet sound. This is an important resting, moulting and wintering area for Common Eiders and other ducks, as well as a staging ground for geese, shorebirds and other wetland species. Eidsbotn lagoon has limited protection, and the area has gradually deteriorated, including with a trotting track out in the finest riparian area in the south. However, it still has good and important qualities as a bird site. The horse shelter on the beach at Eidesøra is equipped with windows and functions as an observation shelter for bird watching. **g.** Tynesfjæra and Sørleiret wetlands just east of Levangersundet sound's mouth, are less often visited by birdwatchers but still interesting sites with many of the same qualities as Falstadbukta and Alnesfjæra wetlands, albeit on a smaller scale and with a lower proportion of ducks.

18. VERDAL WETLANDS

Trøndelag County
GPS: 63.79605° N 11.44630° Ø (parking Ørin), 63.81297° N 11.43924° Ø (Kåra tower hide)

Notable species: Geese, ducks, divers, shorebirds; Oystercatcher (winter); Lesser Spotted Woodpecker, Sedge Warbler and Marsh Tit.

Levanger wetlands: a. Falstadbukta hide; **b.** Hotterbukta inlet; **c.** Alnesfjæra tidal area; **d.** Alfheim viewpoint; **e.** Eidsbotn lagoon, with observation hatches in horse stable; **f.** Levangersundet sound; **g.** Tynesfjæra and Sørleiret wetlands.

Pink-footed Geese have just a few main resting places along the route between the breeding area on Svalbard and the wintering areas on the European Continent, and one is precisely around the Trondheimsfjord. This picture is from Alnesfjæra tidal area (point c. on the map on the previous page). *Photo: Terje Kolaas.*

Description: Around the town of Verdalsøra you will also find wonderful coastal wetland areas, such as Rinnleiret, Ørin and Bjørga. Rinnleiret is located where the Rinnelva flows into the Trondheimsfjord. The large salt meadow area, which is one of the largest in the country, is of international importance for its botanical qualities. Unfortunately, large parts of the wetlands and salt meadows here have been filled in and turned into industrial areas. It is still a nationally important area as a breeding ground for a variety of wetland birds and a resting place for shorebirds and other species all year round. A number of common and unusual bird species have been recorded here, dominated by wetland birds, but also many passerines. Among old rarities enthusiasts, Rinnleiret is particularly remembered for its record of Pallas's Sandgrouse in 1990, an event which mobilized bird enthusiasts from all parts of the country. The quality of the area later gradually deteriorated, partly because of the expansive industrial activities, but also because the area was no longer used as grazing pasture. Large parts of the coastal meadow were thereby regrown with buckthorn, grey alder and other trees, which made it less attractive as a nesting site for wetland birds. From 2008, however, grazing cattle has been introduced to recreate the former favourable conditions. Although Ørin also has nice salt marsh and floodplain forest, it is primarily the river delta's riparian and shallow water areas that are important for birds here. Thousands of geese and ducks use the area during migration, and significant numbers of shorebirds can be found as well. In the vegetation lining the wetland you will often also find a number of passerines, not least when the orange berries on the sea-buckthorn bush ripen in autumn. Compared to Rinnleiret, Ørin is more important as a resting place and breeding ground for scoters, eiders and other ducks, and less important as

a nesting place for shorebirds. A three-digit number of Oystercatcher winter in the Rinnleiret–Ørin area, which are significant numbers being this far north. Three tower hides have been erected in the area, two large and well-placed on either side of the Verdalselva estuary Ørin, and a slightly smaller one at the outlet of the Rinnelva creek at Rinnleiret. By the Ørin tower hide on the south side of the Verdalselva estuary is also the Ørin field station, connected to the Trondheimsfjord Wetland Centre, where bird ringing is conducted regularly. A short distance up the Verdal river is the Langnes nature reserve, a floodplain forest dominated by grey alder with a high density of passerines and Lesser Spotted Woodpecker.

The next bay north from Ørin is Tronesbukta bay, protected as Bjørga bird sanctuary. The tidal zone in the inner part of the bay is particularly notable for having good numbers of *Tringa*-waders during migration periods. Among others, Sedge Warblers nest in the reedbed that encircles the cove. In winter, the area is good for wintering ducks and more,

Ørin is a combined wetland and industrial area at the outlet of the Verdal river. Despite the nasty aesthetics, this is an area with a lot of birds. The tower hide is practically designed and well placed. *Photo: Bjørn Olav Tveit.*

Oslo · Interior · Skagerrak · Western · Central · Northern · Finnmark · Svalbard

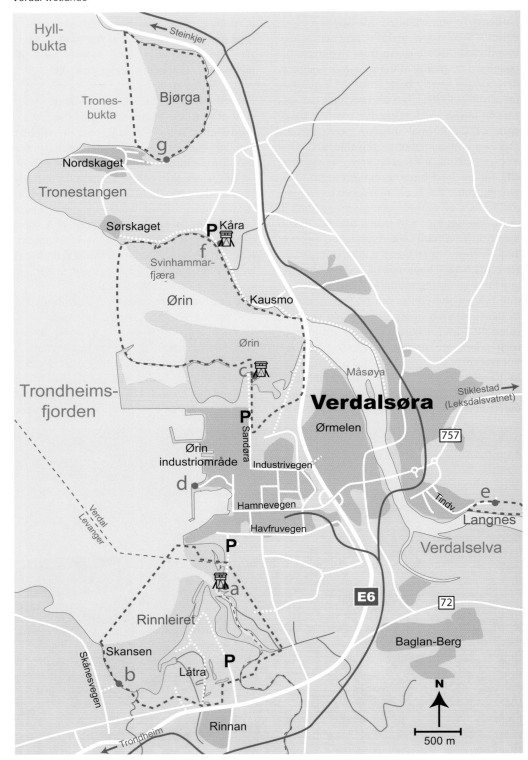

Verdal wetlands: a. Rinnleiret tower hide; **b.** Låtrabekk creek outlet; **c.** Ørin tower hide; **d.** Verdal marina viewpoint; **e.** Langnes riparian area; **f.** Kåra tower hide; **g.** Bjørga viewpoint.

especially the northern part of the area, which is called Hyllbukta inlet.

Best season: All year. May has the greatest diversity of species, and the period from May to September is particularly good for shorebirds. In winter, the area holds substantial numbers of wintering ducks and shorebirds, including Oystercatchers.

Directions: You reach the central parts of this area starting from the E6, where you take the exit to Verdal and follow signs for the Ørin industrial area. Then, see the map.

Tactics: This area can be explored just as effectively by bicycle as by car. A telescope is essential in most places. You can get to **a.** Rinnleiret tower hide at on the south side of the industrial area, by parking at the Havfrua swimming area at the end of Havfruvegen road and following the path 500 m south. **b.** Låtrabekk creek outlet on the opposite side of Rinnleiret may also be worth exploring. You must then drive south on the E6 and take the exit and follow the signs for Skånes. **c.** Ørin tower hide, on the south side of Verdalselva river mouth, can be found by driving in to the industrial area and taking Neptunveien road and then drive to the end of the Sandøra road. You can park at the Ørin field station and walk 250 m further to the tower hide. You get a view of the fjord from Verdal harbour. **e.** Langnes nature reserve is located inside the densely populated area of Verdalsøra, in the innermost part of Tindvegen road. **f.** Kåra tower hide on the northern side of the Verdalselva river mouth can be reached by driving north on E6 and taking the exit to Trones. About 700 m after turning off the E6, turn down towards the sea onto a dead-end road. Park at the end and follow the footpath to the tower hide. This tower hide is well placed in relation to the resting birds here, but on sunny days the birds have a strong backlight for large parts of the day. By continuing the footpath past the tower and on towards the centre of Verdalsøra, you get closer to the river outlet. Small shorebirds

in particular can be found quite far up towards the E6 bridge on both sides of the river. If you follow the walkway in the opposite direction, you get a better view of resting ducks and other birds in the outer parts of the bay.

g. Tronesbukta bay is best appreciated from the south, from the buildings on the north side of Tronestangen headland. Hyllbukta bay (see also the map on next page), can be seen by continuing on the E6 for a further 2 km to the north and taking the exit to Koa campsite. You can also exit to Hylle a little further north to get around to the north and west side of the same bay.

19. INDERØYA PENINSULA

Trøndelag County
GPS: 63.8573° N 11.3138 ° Ø (Vikaleiret tidal area)

Notable species: Geese, ducks, divers, grebes, shorebirds and gulls.

Description: Lake Børgin is a 10 km long brackish lake connected to the Trondheimsfjord via the narrow **a.** Straumen sound which has a strong tidal current. Straumen does not freeze over in the cold season and is an attractive wintering ground for ducks and gulls. Lake Børgin with Straumen sound is next to Tautersvaet sound the most important wintering location for Common Eider in the Trondheimsfjord. The whole of Lake Børgin is of ornithological interest, as it is rich in nutrients and shallow, surrounded by agricultural land. This makes it particularly attractive for the many thousands of Pink-footed Geese that regularly stage in the inner part of the Trondheimsfjord in spring and autumn. The geese commute between the tidal areas in Lake Børgin and the surrounding fields, especially the ones around the farm Mære on the north-east side of the fjord. The best areas for geese, ducks and shorebirds in Lake Børgin are **c.** Lorvikleiret tidal area on the east side and **d.** Gjørv at the north end.

Just outside Straumen sound, lies **b.** Vikaleiret tidal area in the Trondheimsfjord, which has many of the same characteristics as Lorvikleiret in Lake Børgin. Vikaleiret is

Oslo

Interior

Skagerrak

Western

Central

Northern

Finnmark

Svalbard

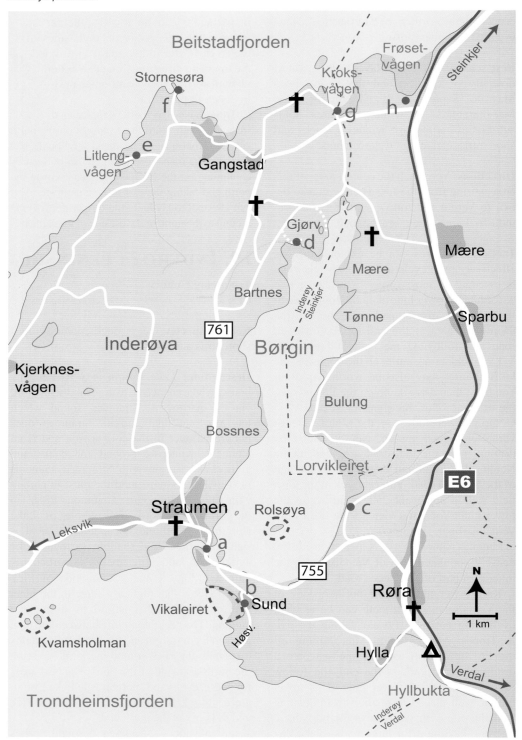

Inderøy peninsula: a. Straumen sound; **b.** Vikaleiret tidal area; **c.** Lorvikleiret viewpoint; **d.** Gjørv viewpoint; **e.** Litlengvågen inlet; **f.** Stornesøra. The map overlaps with the map of Steinkjer wetlands, p. 258. For discussion of Hyllbukta bay in the southeast of this area, see the previous site, Verdal wetlands.

particularly known for housing good numbers of Lapwing during spring migration.

Worth mentioning on the Inderøy peninsula is also **e.** <u>Litlengvågen inlet</u> northwest of Lake Børgin, a small brackish water pool surrounded by lush salt marsh, as well as **f.** <u>Stornesøra</u> at the northern tip, which can house resting shorebirds and more, preferably at times when the area is not congested by human activity linked to the sports facility nearby.

Best season: April–June and July–September for geese and shorebirds, and in winter for ducks and gulls. Spring is generally considered to be somewhat better than autumn here.

Directions: From the E6 at Røra between the towns of Verdal and Steinkjer, take the westbound Rv755 signposted to Inderøy.

20. LAKE LEKSDALSVATNET

Trøndelag County
GPS: 63.81415° N 11.61938° E (Zone 1)

Notable species: Whooper Swan, Pink-footed Goose, other geese and ducks, Black-throated Diver, Red-throated Diver, Slavonian Grebe, Great Crested Grebe, Coot, Little Gull, Black-headed Gull, Whitethroat, Sedge Warbler and Marsh Tit.

Description: Lake Leksdalsvatnet is almost 12 km long, and is considered one of the most important freshwater sites in central Norway. The lakes has several inlets rich in vegetation, the most important of which are incorporated into three conservation areas, sometimes referred to as three zones: The restored Lyngsås–Lysgård bird sanctuary at Ausa in the southeast (Zone 1), Lundselvoset nature reserve in the middle section (Zone 2), and Figgaoset bird sanctuary at Leksdalsvatnet's outlet through river Figgja at the northern end (Zone 3). The lake is important as a breeding ground for ducks and as a resting place for migratory wetland birds in general, with up to 1,000 resting Whooper Swans in spring. Pink-footed and Canada Geese are the most numerous among the geese, but other species, such as Taiga Bean Goose, are also regularly found here. Slavonian Grebe nest with 40–50 pairs in the water, and a few pairs of Great Crested Grebe nested until the end of the 1990s. Shoveler, Garganey, Coot, Crane and Little Gull have also been found nesting (Zone 1). Smew is sometimes seen here outside the breeding season, especially in Zone 2. Water Rail is also occasionally seen, especially just before the ice settles in October–November.

Slavonian Grebe breeds in small numbers in nutrient-rich ponds in the lowlands, especially in central and northern Norway. Lake Leksdalsvatnet is one of the most important breeding sites for this species in Norway. *Photo: Terje Kolaas.*

Oslo · Interior · Skagerrak · Western · Central · Northern · Finnmark · Svalbard

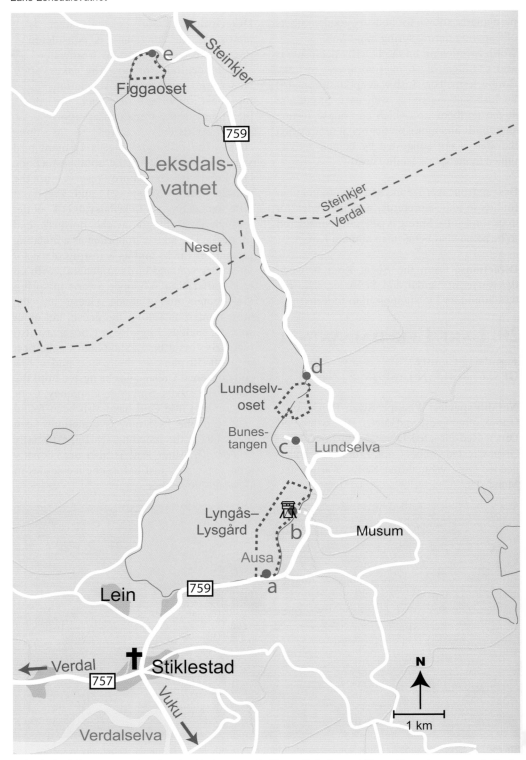

Lake Leksdalsvatnet: a. Lay-by view to Ausa south (Zone 1); **b.** tower hide; **c.** Bunestangen, periodically flooded fields; **d.** lay-by view to Lundselvoset delta (Zone 2); **d.** Figgaoset delta (Zone 3).

Best season: Spring, summer and autumn as long as the lake is not completely frozen over. Greatest activity and diversity during April– June and September.

Directions: From E6 in Verdal, exit onto Rv757 east to Stiklestad, and from there continuing along Rv759 north towards Steinkjer and Henning. Then, see the map.

Tactics: Lake Leksdalsvatnet can be explored from a number of observation points along the road network that surrounds the lake on all sides. You get the best view of **a.** Zone 1/Ausa from a lay-by in the south. A **b.** tower hide has been set up at the outlet of the Tømmerås stream. You can find it by driving further north along Rv759 on the east side of the lake. Immediately after the exit to Musum, there is a farm on the left. Drive in between the barn and the farm building, and park below by an information board about the conservation area. From here, a nice path leads 200 m

to the tower hide. At **c.** Bunestangen headland, there are occasionally flooded agricultural areas, which a number of species appreciate. There are also nice wetlands that must be checked at **d.** Lundselvoset delta and at **e.** Figgaoset delta at the north end of the lake.

21. STEINKJER WETLANDS

Trøndelag County

GPS: 63.9690° N 11.3726° Ø (Kroksvågen inlet), 64.0193° N 11.4460° Ø (Lundleiret inlet), 64.1070° N 11.3869° Ø (Vellamelen inlet)

Notable species: Ducks, shorebirds and gulls.

Description: The innermost part of the Trondheimsfjord is called the Beitstadfjord, and here are several good resting places and nesting sites for sea and wetland birds, especially in the eastern part, near the town of Steinkjer.

Lake Leksdalsvatnet seen from the south-eastern corner, **a.** Ausa. *Photo: Bjørn Olav Tveit.*

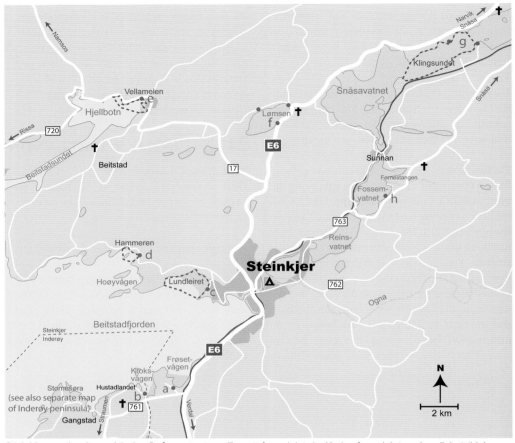

Steinkjer wetlands and Lake Snåsavatnet: a. Frøsetvågen inlet; **b.** Kroksvågen inlet; **c.** Lundleiret tidal area;
d. Hammeren; **e.** Vellamelen; **f.** Lake Lømsen; **g.** Klingsundet sound and **h.** Lake Fossemvatnet, the latter three sites
described under the Lake Snåsavatnet section (next page). See previous pages for sites on Inderøya peninsula.

Best season: Migration periods and winter.

Directions: These areas are located near
Steinkjer town and can be reached from the
E6. See the map.

Tactics: In the south-eastern corner of the fjord
lies **a.** Frøsetvågen inlet and **b.** Kroksvågen inlet,
both having small but often very bird-rich tidal
flats during migration. Outside Kroksvågen are
the two islets Hustadøya and Kalven, which
together with the two inlets form important
nesting areas for Common Eider and gulls. The
exit from E6 to Frøsetvågen inlet is signposted
to Inderøy, and the exit to Kroksvågen inlet
from Rv761 is signposted to Hustadlandet.
Rv761 takes you on to Lake Børgin (which
see). Just north-west of Steinkjer town lies the
large **c.** Lundleiret tidal area, one of the most
important resting places for migratory wetland
birds in the Trondheimsfjord. Thousands
of Pink-footed Geese, ducks and shorebirds
can gather here, and several rare and unusual
species have been encountered over the years.
The exit to Lundleiret is from the E6 in Steinkjer
signposted to Sør-Beitstad. After a couple of
km, turn left towards Egge and then the first
road on the right into the residential area, and
follow Lundsengvegen road to the end. You
then get a view of Lundleiret from the southeast.
Drive back where you left to Egge and continue
towards Sør-Beitstad. After half a km you get a
view of the north side of Lundleiret tidal area.
Continue 3 km to the west to reach
d. Hammeren nature reserve, which has
potential for ducks and shorebirds.

258

In Hjellbotn, the innermost corner of the Trondheimsfjord, you will find **e.** Vellamelen. This is a shallow water area in line with many of the other tidal areas of the Trondheimsfjord, and is a correspondingly good resting place for wetland birds during migration. It is one of the most important staging areas for Pink-footed Geese in central Norway, not infrequently with up to 15,000 individuals at the same time in the latter half of April and the beginning of May. You reach Vellamelen by continuing via Beitstad, or from Steinkjer by following the E6 4 km north and turning left on Rv17 towards Namsos. In the village of Vellamelen, drive down the slipway towards the fjord, signposted to Stadion, for the best view of the tidal area. You can also continue through the settlement and get a new view from Rv17 on the north side of the Moldelva river outlet. A tower hide is planned here.

22. LAKE SNÅSAVATNET

Trøndelag County

GPS: 64.13959° N 11.74153° E (Kvam, Klingsundet sound)

Notable species: Whooper Swan, geese, ducks, Slavonian Grebe, Crane and Sedge Warbler.

Description: Lake Snåsavatnet is the largest fresh-water lake in central Norway. It has several areas of shallow water, attractive to waterbirds during migration and in winter. Slavonian Grebe, Sedge Warbler and several others breed. Klingsundet sound is the narrowest, shallowest and most bird-rich part of Lake Snåsavatnet. This is a particularly important place for Cranes during migration. Lake Fossemvatnet is an expansion of Lake Snåsavatnet's outlet river, often supporting swans, geese, ducks and Cranes. Just

Crane breeds on marshes, mainly in central and south-eastern Norway. During migration in spring and autumn, the birds move in flocks and gather at fixed resting places, often in the cultural landscape in the lowlands, such as in the wetlands and fields by Lake Snåsavatnet. *Photo: Kjartan Trana.*

southwest of Lake Snåsavatnet is Lake Lømsen. It is densely lined with reeds, which is an uncommon feature of lakes in this region, making it important to a number of species. Several pairs of Crane and good numbers of Slavonian Grebe breed. Red-necked Grebe has bred in the past, and may well breed here again.

Best season: Birds are usually found in April–May and in September but all year can be good, except when it is frozen during winter.

Mistle Thrush is the least common of the breeding thrush species in Norway. Here it is associated with open pine forest, mainly in eastern Norway. In Lierne you will find certain birds native further south and east than elsewhere in central Norway, including species such as Mistle Thrush, Hobby and Honey Buzzard. Elsewhere on the Continent, the Mistle Thrush has begun to adapt to a life in the cultural landscape, which may also be about to happen in Norway. *Photo: Terje Kolaas.*

Directions: See map of the Steinkjer wetlands on the opposite page. Lake Snåsavatnet (and Lake Lømsen) are located just north of Steinkjer town centre and can be reached from here by taking a loop around Lake Snåsavatnet, either by following the E6 north towards Namsos, or the opposite way, by following the Rv763 signposted to Snåsa.

Tactics: Lake Snåsavatnet west and Lake Lømsen: From Steinkjer, follow E6 north towards Namsos for 9 km, and you will see Lake Lømsen on the left. The lake can be partially viewed from the resting area along the E6. Drive a little further and turn left onto the road signposted Ulven, and you will immediately have a view again from the pump house by the outlet stream in the northern end of the lake. Continue towards Ulven and turn left onto the road signposted Røsegg, which takes you to a quarry with a view from the west side.

Back on the E6, continue north and you will soon see Lake Snåsavatnet on your right. If you just want to stop for a quick overview of **g.** Klingsundet sound, you can turn off in Kvam to Kvam motel and drive down to the water's edge to the left of the church. If you have more time and want a better view from this side of the sound, you can turn off at Haugan farm and walk out onto Veines headland (private road) 1.5 km southwest of Kvam.

Lake Snåsavatnet east: From Steinkjer, take Rv763 towards Snåsa and you will first see Lake Reinsvatnet on the right-hand side and – after crossing the bridge 6 km from Steinkjer – **h.** Lake Fossemvatnet on the left. You have some kind of view from the road, but it is better to walk along the edge of the field out onto Fornestangen headland. Further north on Rv763, turn left towards Sunnan for a view of Bergsbukta inlet, still in Lake Fossemvatnet. Continue on Rv763, after 5 km take the sign Friskgården to the left. Follow this road (called Klingvegen) over the railway, and immediately turn right into Solnesvegen, which you follow all the way to Solnestangen headland, where you get a nice view of Klingsundet sound. You can possibly continue Rv763 to the east end of the long Lake Snåsavatnet and make stops here and there, and take detours, e.g. towards Grønøra airport and Grønøra swimming area, where you get a view of this part of Lake Snåsavatnet. At the very eastern end of the lake, the road runs along the shallow water area of Semsøra. Continue on to E6, where you can turn south-west back towards Steinkjer to make the full circuit around Lake Snåsavatnet.

23. LIERNE WOODLANDS

Trøndelag County
GPS: 64.15501° N 13.80683° E (Ulen delta)

Notable species: Whooper Swan, Long-tailed Duck, Common Scoter, Velvet Scoter, Hazel Grouse, Willow Ptarmigan, Rock Ptarmigan, Black Grouse, Capercaillie, Red-throated Diver, Black-throated Diver, Slavonian Grebe, Honey Buzzard (uncommon), Hen Harrier, Goshawk, Golden Eagle, Osprey, Kestrel, Hobby (scarce), Gyrfalcon, Crane, Dotterel, Great Snipe, Whimbrel, Greenshank, Green Sandpiper, Wood Sandpiper, Broad-billed Sandpiper, Temminck's Stint, Purple Sandpiper, Ruff, Red-necked Phalarope, Long-tailed Skua, Arctic Tern, Great Grey Owl (uncommon), Hawk Owl, Long-eared Owl, Short-eared Owl, Tengmalm's Owl, Wryneck, Three-toed Woodpecker, Shore Lark, Yellow Wagtail, Grey Wagtail, Dipper, Bluethroat, Redstart, Ring Ouzel, Mistle Thrush, Icterine Warbler, Wood Warbler (uncommon), Siberian Tit (unreliable), Great Grey Shrike, Siberian Jay, Pine Grosbeak, Lapland Bunting, Snow Bunting and Rustic Bunting (uncommon).

Description: In central Norway the national border makes an eastward bend into Sweden, incorporating an area known as Lierne.

Lierne has varied landscapes, from dense forests to high mountains, and with a correspondingly high diversity of birdlife. In addition, you really get the feeling of being in untouched nature. *Photo: Kjartan Trana.*

Elk is a rather common sight when birdwatching many places in Norway, particularly if you are out and about at dawn. Roadside in Lierne woodlands, you might also get to see brown bear or lynx. *Photo: Kjartan Trana.*

This is a vast and remote wilderness with wooded valleys and high mountains host to two national parks, Lierne and Blåfjella–Skjækerfjella, both primarily covering the montane areas. However, it is the woodlands in particular that makes Lierne of special interest to birdwatchers. Here you have good populations of grouses, owls and raptors, besides Siberian Jay and Pine Grosbeak. It is also a place where Great Grey Owl breed in some years. The rivers in the area run east and empties in the Bothnian Bay in Sweden. This gives an eastern touch to the avifauna and a chance to find birds which have a more northerly distribution in Sweden than they have in Norway such as Osprey, Hobby, Honey Buzzard, Buzzard, Mistle Thrush, Wood Warbler and Rustic Bunting. The latter is generally scarce and seemed to have disappeared completely for a few years. In recent years, however, the population in Scandinavia has shown signs of recovery, and so there is still hope of finding Rustic Bunting in Lierne as well.

Other wildlife: In Lierne you may encounter a variety of exciting mammals, including muskrat, lynx, wolverine, Arctic fox and brown bear. The most common mammals seen by birdwatchers here are usually elk, roe dear, red squirrel and mountain hare besides various small rodents.

Best season: March and April for calling owls and displaying Capercaillie. May to July has the greatest species diversity. The resident species can be found year-round.

Directions: From the E6 20 km south of Grong, turn east onto Rv74 signposted to Nordli.

Tactics: You can drive around the network of forest roads in the area, and stop at suitable places and then explore the terrain on foot. Spend as much time as possible outside the car. You can find Capercaillie and other grouses by driving the forest roads early in the morning, while they sometimes eat gravel on the side of the road. Lierne is also a perfect area for nocturnal trips searching for owls and mammals.

In <u>Nordli</u>, either continue east along Rv74 signposted to Gäddede (in Sverige), or exit onto Rv765 signposted to Sørli. Both roads take you through highly productive forests. If you go east from Nordli, turn right before Lake Murusjøen, just before the exit signposted to Kvelia. This road leads to the <u>Murubekk cabin</u> (*Murubekkhytta*), which is a good vantage point for hiking into <u>Lierne national park</u>. The creeks here on the south side of <u>Lake Murusjøen</u> should be searched for Rustic Bunting.

If you exit south from Nordli you reach the village of <u>Sørli</u> after 35 km. The 20 km stretch of road onwards from Sørli to the national border, including the side roads, is among the most productive in Lierne. Siberian Jay, Pine Grosbeak and many more may be encountered here. A popular vantage point for hiking into <u>Blåfjella-Skjækerfjella national park</u> is <u>Berglia</u>, signposted to from the road. Here you also pass the <u>Ulen delta</u> in the north end of <u>Lake Ulen</u>, a terrific area for wetland species, also supporting breeding Great Grey Shrike, although the delta is not easy to explore from the roads.

24. VIKNA ARCHIPELAGO

Trøndelag County
GPS: 64.86417° N 11.56272° E (Kanalen)

Notable species: Resting shorebirds and other migrants; breeding seabirds and Smew.

Description: Vikna archipelago includes a great number of islands, comprising an area of immense importance to breeding seabirds. On the larger islands, several lakes and cultivated fields provide suitable resting habitats for a variety of migrants. Rather surprisingly, a few pairs of Smew started to breed here in 2003, which is the only place in Norway south of Finnmark. Just inland from Vikna archipelago, the area of Kolvereid also has several sites worth exploring, particularly for migrants. Several rarities have been encountered at Vikna and Kolvereid over the years, including Long-billed Dowitcher and Red-headed Bunting.

Best season: May to September.

Oslo

Interior

Skagerrak

Western

Central

Northern

Finnmark

Svalbard

Capercaillie, such as this female, can sometimes be seen roadside in remote forests, filling their crop with gravel in the morning. *Photo: Kjetil Schjølberg.*

Vikna archipelago contains a multitude of hilly islands full of small lakes, a handful of which have breeding Smew. *Photo: Bjørn Olav Tveit.*

Directions: *From Trondheim by car:* In order to reach these areas without a ferry crossing, follow the E6 north (towards Narvik) to a few kilometres north of Grong. Here, exit onto Rv775 signposted to Høylandet and later onto Rv17 towards Rørvik, until finally heading west along Rv770 signposted to Rørvik/Kolvereid.

Tactics: Along Rv770 1.5 km past Kolvereid village, the road crosses a reed-lined creek called <u>Kanalen,</u> a nature reserve often supporting wetland birds including Little Grebe. Kanalen drains the roadside <u>Lake Mulstadvatnet,</u> which should be checked as well. Side roads to the left before (signposted Abelvær) and after Lake Mulstadvatnet (signposted Horvereid) leads to <u>Abelvær</u> fishing village in the Foldafjord at the very tip of the Kolvereid peninsula, with a potential for autumn migration of land and seabirds, besides resting gulls. On the way to Abelvær, turn left signposted to Salsbruket in order to view <u>Skagabukta inlet</u>. Both here and, further towards Abelvær, along the road to <u>Øksninga island</u> (signposted to Øksninga), there are tidal areas good for shorebirds and ducks. On the

way back from Abelvær fishing village, you can follow signs to Rørvik and explore tidal areas and farmland at <u>Arnøya island</u> and the nutrient-rich lakes north of <u>Lundringen church</u> (signposted to Flosand), before continuing on Rv770 past Rørvik to Austafjord in outer Vikna. At <u>Vikna</u>, about 2 km after passing the exit to Lysøya, look for Smew in the roadside <u>Lake Svarthammarvatnet</u> and other lakes in this area. At outer Vikna, look for migrants for instance at <u>Valøya</u>, the outermost settlement.

25. LEKA ISLAND

Trøndelag County
GPS: 65.0996° N 11.7575° Ø (parking Skeisneset)

Notable species: Resting ducks and shorebirds; White-tailed Eagle.

Description: Leka island is fairly large and a really scenic island on the Namdal coast, with a geological peculiarity that gives the mountains an exotic and beautiful reddish hue. The island has a rich bird life. Skeisneset wetland area in the north-east of the island stands out as a particularly great place for resting ducks, shorebirds and others. It is protected as a bird sanctuary and covers an area of approx. 3.7 km². A trail through the area leads to an observation shelter on a hill overlooking the wetland from the east. White-tailed Eagles also nest on the island. An old story out here is the one about the so-called "Eagle Abduction on Leka", where a little girl is said to have been attacked by a White-tailed Eagle and carried up to just below the eagle's nest in the mountainside, but the child was unharmed. Scientists have denied that this story could be true, it is simply not physically possible for a White-tailed Eagle to fly up a mountainside with something that heavy. Experiments have also shown that it is possible for a child of the girl's age to climb on her own up to the point where she was found. It is sad that this legend is kept alive from the official side at Leka, e.g. on information signs in the terrain that show where the episode is supposed to have taken place. This partly contributes to creating unnecessary fear among hikers with small children, and to fueling hatred of raptors and carnivores – which is a big problem in Norway. Leka has so many positive truths that it should be unnecessary to use such nonsense in the marketing of the island.

Best season: Late spring and early summer.

Directions: From Rv17 on the border between Trøndelag and Nordland counties, exit and follow signs for Leka. The road leads to Gutvik ferry landing, where you take the car ferry 20 minutes across to Leka island.

Tactics: You reach Skeisneset wetland area by turning right at the first crossing after the ferry landing. Park at the end of the road and walk the nature trail out into the area to the observation shelter called *Ivarshallaren panorama*. A telescope is required. You can enter the area to get a closer look at the birds, but be careful not to flush resting and nesting birds in this open terrain. You should also drive around the island and investigate fields and wetlands that look promising. There are particularly fine fields in the area by the ferry landing. Also drive out to the small neighbouring Madsøya island. The beautiful red mountains are best appreciated on Leka island's west side.

Leka island with its beautiful reddish rock formations contrasting with grazing pastures and lakes, and with a large and bird-rich wetland on the northern tip. *Photo: Bjørn Olav Tveit.*

NORTHERN NORWAY

Northern Norway encompasses the three northernmost counties, Nordland, Troms and Finnmark. In this book, however, the latter has been crowned with its own separate chapter. The two remaining counties of northern Norway have a lot to offer birdwatchers though, and to top it off, Nordland and Troms present their birds in some of the most spectacular scenery on earth. Tourists from all parts of the world come to appreciate the beautiful landscape. Many of them choose to do so by driving along *The Coastal Highway*, route Fv17

White-tailed Eagle has northern Norway as its most important breeding area world wide. *Photo: Kjetil Schjølberg.*

from Steinkjer in central Norway to Bodø in Nordland. Other popular tourist destinations are Lofoten and neighbouring archipelago Vesterålen and Andøya. The northern lights – *aurora borealis* – has also become a very popular tourist attraction in recent years, primarily in winter. Tromsø city with its 79,000 inhabitants and Bodø with 54,000 are the largest towns in the region.

SITES IN NORTHERN NORWAY

27

26

Tromsø

29 **30**

28

Andenes

23

25

Bardufoss

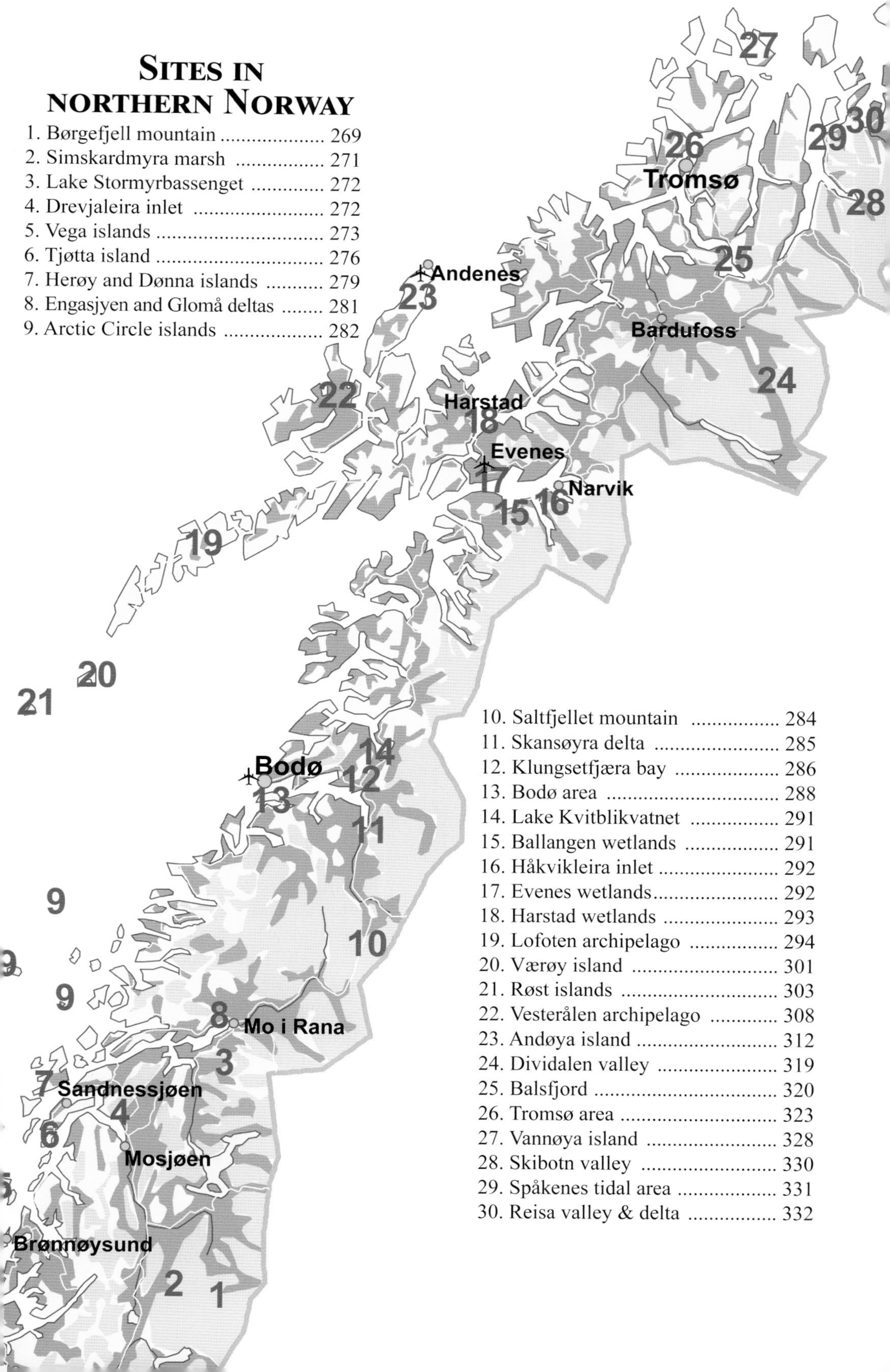

22

Harstad

18

Evenes

17

15 **16**

Narvik

24

19

20

21

Bodø

13

14

12

11

10

9

9

9

8 **Mo i Rana**

3

7

Sandnessjøen

4

6

Mosjøen

Brønnøysund

2 **1**

Wolverine is a carnivore mainly associated with mountain ranges where it follows reindeer herds in search of injured animals. You have the best chance of seeing one by putting out bait, or if you happen to catch sight of this predator as it crosses snowdrifts in the distance, e.g. at Børgefjell mountain. *Photo: Terje Kolaas.*

Norway is host to about 50 % of the world population of White-tailed Eagle, and more than half of these are found in Nordland and Troms counties. A small number of Black-tailed Godwits of the northern subspecies islandica breed in northern Norway, but the number is steadily decreasing, and the total population is now probably between 5 and 10 pairs. Important resting grounds for geese are found several places along the coast, as are some of the few remaining colonies of Lesser Black-backed Gull of the northern subspecies *fuscus*. There are also areas supporting Eagle Owl, with some of the densest populations anywhere in the world. Northern Norway is also of year-round importance for a variety of seabirds and wetland species. Important breeding areas include the islands of Vega, Herøy, Træna, Røst, Lofoten, Vesterålen and Andøya. The Svalbard populations of Pink-footed and Barnacle Geese are dependent on the resting areas along the Helgeland coast and Vesterålen archipelago to reach their breeding grounds

in good condition. The interior of northern Norway has many spectacular mountain ranges of great importance to breeding birds, particularly Saltfjellet and Børgefjell and the interior mountains of Troms County. Some of these used to be home to breeding Lesser White-fronted Goose, a species now sadly close to extinction in Scandinavia, and which in Norway now breeds in Finnmark.

The vast Saltfjellet mountain range represents a natural barrier for many plants, animals and birds. South of Saltfjellet is dominated by tall, mature spruce forest with much the same birds as in central Norway. To the north of Saltfjellet is a more Arctic climate where pine is the predominate species of coniferous tree. The Arctic Circle crosses Saltfjellet, dividing the country in a northern section where the sun shines day and night in summer and never rises above the horizon in winter, while people south of the Arctic Circle can enjoy birdwatching in sunlight at least for some hours every day of the year.

THE BEST SITES

The undoubtedly best single site of Nordland and Troms is Røst, the outermost group of islands in the Lofoten archipelago. With its multitude of exquisite birding features, Røst ranks among the very best sites in the country. Here are large seabird colonies, attractive resting grounds for migrant geese, ducks, shorebirds, gulls and passerines, and a strategic geographical position both for general migration and for producing vagrants. The inner parts of Lofoten and Vesterålen archipelagos have interesting birds all year and are important wintering areas for waterbirds, including White-billed Diver, Glaucous Gull and King Eider. A very good all-round site is Andøya island with its vast peat marsh supporting breeding wetland birds, and a strategic position with many outstanding resting habitats for migrants. Along the coast are many other splendid migration sites, such as Tjøtta, Herøy and Dønna, and Tisnes near Tromsø. The most shorebird-friendly tidal flats of the region, besides Røst, are perhaps Klungsetfjæra bay east of Bodø and Balsfjord east of Tromsø. Many of the islands of the outer coast have a potential for turning up vagrants. Again besides Røst, the most notable being Træna, Myken and Værøy. Northern Norway also offers vast areas of wilderness combined with supreme birdwatching. Reisadalen valley and Børgefjell mountain are perhaps the most rewarding of these, both in terms of good birds and giving you the sense of encountering serene, wild and beautiful nature.

1. BØRGEFJELL MOUNTAIN

Trøndelag and Nordland counties

GPS: 65.31343° N 14.35785° Ø (Storvollen, Susendalen)

Notable species: Taiga Bean Goose, Scaup, Long-tailed Duck, Common Scoter, Velvet Scoter, Hazel Grouse, Willow Ptarmigan, Rock Ptarmigan, Black Grouse, Capercaillie, Red-throated Diver, Black-throated Diver, Golden Eagle, Rough-legged Buzzard, Kestrel, Gyrfalcon, Dotterel, Temminck's Stint, Broad-billed Sandpiper, Great Snipe, Red-necked Phalarope, Long-tailed Skua, Eagle Owl, Snowy Owl, Hawk Owl, Short-eared Owl, Tengmalm's Owl, Black Woodpecker, Three-toed Woodpecker, Yellow Wagtail, Bluethroat, Redstart, Great Grey Shrike, Siberian Jay, Lapland Bunting and Snow Bunting.

Description: Børgefjell is a vast mountainous wilderness, one of the most pristine in Scandinavia. It is protected as a national park and is considered to be among the most important areas in the country for owls, raptors and carnivorous mammals. Few people visit this area, especially in early summer, which gives you the feeling of having it all by yourself. You do run the risk, however, of encountering Sami reindeer shepherds on off-road motorcycles popping up in the middle of nowhere. This, perhaps, only adds to the exotic experience.

Børgefjell is one of the most reliable areas to find Snowy Owl, with a double figure number of breeding pairs in good rodent years although absent in recent years. Lesser White-fronted Goose, formerly a common breeder here, bred until 1993. A few pairs of Taiga Bean Goose may still breed in the mountain valleys, as one of a very few places outside Finnmark. Many Taiga Bean Geese come to the marshes of Ovrejohken–Jallah in the east end of Lake Namsvatnet to moult in summer, although most of these may breed on the Swedish side of the border. Of great importance to the birdlife of Børgefjell are also the marshlands between the two lakes Austre Tiplingvatn and Vestre Tiplingvatn in Hattfjelldal in Nordland, a three hour's walk from the nearest drivable road. Here is a beautiful terrain sporting charming species like Red-necked Phalarope and Temminck's Stint in the wetlands, and Dotterel, Long-tailed Skua and, in some years, Snowy Owl on the dry moraine ridges. Jack Snipe possibly breeds in the dampest marshes. Another interesting area, only an hour's hike from the nearest road, is the Storelvdalmyrene marshes just east of Lake Tomasvatn. In addition to supporting several of the above-mentioned species, this is a regular site for Great Snipe.

Other wildlife: Børgefjell mountain is an important area for carnivores and one of Scandinavia's few remaining breeding areas for Arctic fox. In several places, old dens can be seen as large, well-fertilized and green mounds in the barren moraine landscape. Particularly important for this species are the areas south of river Orrvasselva, east of Lake Namsvatnet, in and around the biotope protection area for Taiga Bean Goose. In Børgefjell you may also encounter wolverine, domestic reindeer and lemming. In recent years, a few records of the invasive species raccoon dog have been made in the area. An ornithological expedition to Børgefjell mountain can be combined with good trout fishing, although the bird life is often at its busiest in early summer, while the trouts are more cooperative somewhat later.

Best season: June and July.

Directions: The four most important vantage points for trips into Børgefjell mountain are the following – and, of course, you may combine these if you prefer hiking across Børgefjell instead of walking back to your starting point.
1) Lake Namsvatnet is reached from the E6 58 km north of Grong and by exiting east onto Rv773 signposted to Røyrvik. In Røyrvik, follow signs to Børgefjell. In Namsvassgrenda at the end of the road you can get a boat ride across Lake Namsvatnet.
2) Tomasli. Exit from E6 signposted to Tomasvatn, 4 km south of the exit to Simskardmyra (which see). Park at Tomasli and walk a few hundred metres past the cabin village before making a left, onto a path called *Jengelvegen*, signposted to Jengelen. From here it is a 4 km hike to Storelvdalmyrene marshes along a good-quality path with walkways over the wettest parts and a bridge across Storelva river south of Storelvdalmyrene marshes.
3) Simskardet is reached by following directions to Simskardmyra below. Follow the

road up the pass as far as possible and walk on along the marked path 3 km to an open mountain cabin. Beyond the cabin, the path gets less clear but keep the direction straight east past Lake Simskardvatnet to reach the Båttjørnhytta mountain cabin (open for all) and Lake Vestre Tiplingvatn.
4) Susendalen valley is reached by driving to Hattfjelldal and then following Rv804 southwards, signposted to Susendal. You can then either exit to Sørmo and park near Lille Susna river and walk the side-valley south from here, or your can continue straight ahead to Storvollen at the end of the road – contact the landowner at the farm for permission and parking payment. From Storvollen, follow the path along the south side of Tiplingelva river.

Public transportation: Tomasvatn can be reached by train to Majavatn railway station. Susendalen can be reached by train to Trofors followed by bus via Hattfjelldal to Susendalskroken. After leaving the bus, continue 3 km along the gravel road past Kroken, and make a left onto the path leading to a bridge across Susna river. Next, the path leads through mature conifer forest with good populations of Capercaillie, and across Susenfjellet to the area between the two lakes Østre and Vestre Tiplingen. It is important to find the bridge over the river that runs between these lakes, as the river can be very difficult to cross otherwise.

Tactics: If you are headed further into Børgefjell mountain than to the Storelvdal marshes, the trip must be planned as a wilderness expedition with associated equipment and safety arrangements. There are far fewer cabins and marked trails here than in the southern Norwegian mountain areas. Hike in and set up a base camp with all the equipment in a suitable area and explore the surrounding area with lighter packing. You should bring a telescope with a light stand. The mosquitoes can be intense up here in summer, so come prepared. The Taiga Bean Goose moulting area of Ovrejohken–Jallah is closed to traffic in the period June 20 to July 25.

2. SIMSKARDMYRA MARSH

Nordland County

GPS: 65.29595° N 13.50439° E (viewpoint)

Notable species: Scaup, Long-tailed Duck, Common Scoter, Velvet Scoter, Hazel Grouse, Black Grouse, Capercaillie, Goshawk, Golden Eagle, Gyrfalcon, Temminck's Stint, Ruff, Red-necked Phalarope, Eagle Owl, Hawk Owl, Tengmalm's Owl, Three-toed Woodpecker, Yellow Wagtail, Dipper, Bluethroat, Redstart, Great Grey Shrike, Siberian Jay, Lapland Bunting and Snow Bunting.

Description: In the outskirts of Børgefjell are several interesting sites, including Simskardmyra marsh between the lakes Øvre Fiplingvatn and Nedre Fiplingvatn. The marshes and lakes of this area thaw relatively early in spring, providing important resting grounds for geese, ducks, Cranes and shorebirds waiting for the breeding areas at the higher elevations of Børgefjell

mountain to become free of ice and snow. Simskardmyra marsh and the surrounding woodlands have interesting breeding birds as well.

Best season: May and June, to some extent throughout summer until August.

Directions: Simskardmyra marsh is reached by leaving E6 on the road signposted to Fiplingdal. This exit is 13 km north of the county border between Trøndelag and Nordland. Follow this road about 12 km, on the way passing Bjortjønnlimyrene marshes close to E6, which are also worth a stop. Just after you pass a short stretch with barriers on both sides of the road, there is a roadside information poster about Simskardmyra nature reserve. From here you have a narrow view to the marsh in between the trees. Continue 500 m, turn right and turn right again after another 2 km before parking by a sand quarry. Walk 150 m further along the road and you will see

Simskardmyra marsh is an important staging ground in spring for migrant waterbirds waiting for winter to ease its harsh grip up in Børgefjell mountain, which here looms in the background. *Photo: Bjørn Olav Tveit.*

Oslo

Interior

Skagerrak

Western

Central

Northern

Finnmark

Svalbard

a bridge leading across river Simskardelva. Cross the bridge and follow the path a couple of hundred metres to overlook the marsh from the opposite side. If you drive further up the road along the Simskardelva river, you reach one of the vantage points to Børgefjell mountain (which see).

Tactics: Scan the marsh from the road with a telescope. Access within Simskardmyra nature reserve is prohibited from May 15 to July 20. Outside this period, you may walk along the path on the other side. Walk very carefully not to spook any resting birds on the marsh.

3. LAKE STORMYRBASSENGET

Nordland County
GPS: 65.97079° N 13.77772° E (Flatmoen), 66.16325° N 13.79707° E (Røssåauren headland)

Notable species: Whooper Swan (winter), Pintail, Scaup, Common Scoter, Velvet Scoter, Hazel Grouse, Willow Ptarmigan, Black Grouse, Capercaillie, Red-throated Diver, Black-throated Diver, Slavonian Grebe, Goshawk, Golden Eagle, Rough-legged Buzzard, Crane, Ruff, Red-necked Phalarope, Arctic Tern, Pygmy Owl, Long-eared Owl, Tengmalm's Owl, Black Woodpecker, Yellow Wagtail, Dipper (winter), Redstart and Siberian Jay.

Description: The river Røssåga runs north from Røssvatnet lake, through beautiful conifer forests to its mouth in Sørfjorden, an arm of Ranfjorden. Several places along this watercourse are of interest to birdwatchers. Particularly well known is the dammed Lake Stormyrbassenget, which is of great importance to the region's wetland species, both for resting and for breeding. Among the regular breeders are Slavonian Grebe, Crane, Ruff, Red-necked Phalarope, Black-headed Gull and Arctic Tern (probably Common Tern as well), in some years also Shoveler, Temminck's Stint and Green Sandpiper. Black-throated and Red-throated Divers breed in the vicinity and are often seen in Lake Stormyrbassenget. Another

area worth exploring along the watercourse is Bleikvassli, which has nice woodlands and where Pintail has been found breeding. The Røssåga delta down by Sørfjorden is also worth exploring during migration.

Best season: Late spring and early summer, and to some extent autumn as well. A few Whooper Swans, ducks and Dipper winter in these areas.

Directions: From the E6, Lake Stormyrbassenget is reached by exiting at Korgen onto Rv806 signposted to Bleikvassli. Almost 14 km from E6 a gravel road exits to the right, discretely signposted *Almannsvegen*. This gravel road leads around the lake and closer to the usually most bird-rich southwestern corner. You can also continue along Rv806 a few hundred metres to overlook the lake from the highway or head even further and exit onto the gravel road signposted to *Kvern*, leading to the waters edge and to a well-marked network of hiking paths.
The Røssåga delta is reached from Korgen by continuing 7 km north along the E6 (to 1 km south of Bjerka, which is also worth a roadside scan), and exiting west signposted to Vallabotn. Do not turn right across the railway but continue to the T-junction. Here you turn right, signposted to Mula. Go another 2.5 km and turn right once again, signposted to *Badeplass* and continue 800 m to the end of the road. Walk out onto the sandy headland Røssåauren.

Tactics: Scan the lake from the above-mentioned viewpoints with a telescope. A canoe can be a nice means of exploring Lake Stormyrbassenget.

4. DREVJALEIRA INLET

Nordland County
GPS: 65.93774° N 13.13877° E (gravel road entrance)

Notable species: Resting wetland birds; breeding Hazel Grouse and Slavonian Grebe.

Description: Drevjaleira is a shallow tidal inlet along the Vefsnfjord at the mouth of river Drevja, worth checking if you happen to pass by. The inlet is lined with lush saltmarsh and sheltered pools, surrounded by deciduous forest and cultivated fields, making a nice stop-over site for migrant wetland species. Lake Motjønna is a small, nutrient-rich lake further up the Drevja valley, with breeding Slavonian Grebe and several ducks, and with Hazel Grouse in the forest behind the lake.

Directions: Drevjaleira inlet is located along Fv78 between the towns of Mosjøen and Sandnessjøen, just south of Holandsvika. From the E6 north of Mosjøen, exit onto Fv78 signposted to Sandnessjøen. You see the inlet after nearly 10 km. Turn left onto the road signposted to Utnes, and then turn left again onto a dead-end road. It can be scanned roadside from the hilltop along this road, or from walking along the gravel road crossing the railway at the end of the dead-end road.

Lake Motjønna is reached by driving out the dead-end road and going left 1 km and then making a right onto the road signposted to Elsfjord. After about 20 km, just after passing the larger Lake Drevvatnet and Drevvassbygda village, you see Lake Motjønna on your left.

5. Vega Islands

Nordland County
GPS: 65.70077° N 11.85088° E (Vallasjøen)

Notable species: Whooper Swan, Barnacle Goose (April–May), Shelduck, Shoveler, King Eider (winter), Common Scoter, Velvet Scoter, Willow Ptarmigan, Rock Ptarmigan, Black Grouse, Red-throated Diver, White-billed Diver (winter), Great Northern Diver (winter), Cormorant, Shag, White-tailed Eagle, Coot, Spotted Redshank (late summer), Arctic Skua, Lesser Black-backed Gull (*fuscus*), Eagle Owl, Short-eared Owl, Grey-headed Woodpecker, Bluethroat, Redstart, Ring Ouzel, Sedge Warbler and Icterine Warbler.

Description: Vega is an island surrounded by 6500 smaller islands and skerries forming an archipelago covering 127 km² off the coast of Helgeland. Vega is reckoned as a core area in Norway for Greylag Goose and

Vega islands have diverse nature. The main island, which is often referred to as Fast-Vega, has a central marsh area, here seen with the mountains on the west side of the island in the background. You find nutrient-rich wetlands in the north and lush woodlands in the south of Fast-Vega. *Photo: Bjørn Olav Tveit.*

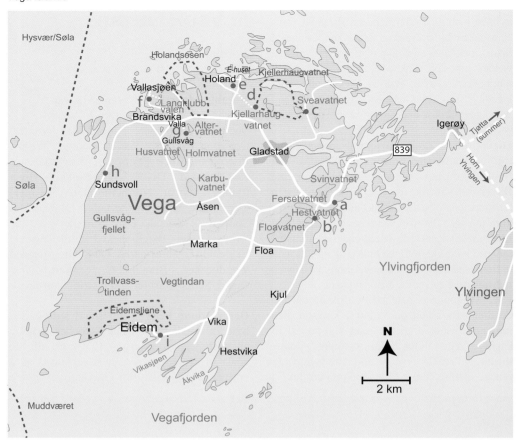

Vega – main island (Fast-Vega): a. Rørøya tidal area; **b.** Lake Hestvatnet; **c.** Lake Sveavatnet; **d.** Lake Kjellarhaugvatnet; **e.** Holandsosen delta; **f.** Vallsjøen wetlands; **g.** Lake Altervatnet; **h.** Sundsvoll seaview; **i.** Eidem.

it also supports the largest colony in Europe of Cormorant of the nominate race *carbo*. Eagle Owl breed as well. The archipelago is a very important area for moulting, resting and wintering sea ducks, and for auks. Vega was in 2004 put on the UNESCO World Heritage List, mainly based on the thousand-year-old, now unique, practice of Common Eider down harvesting. People, mainly the women, gave the Eiders breeding boxes and protection in exchange for some of the down collected from their nests. You still get to buy expensive eiderdown products such as pillows from the few down harvesters still active. The population of Common Eiders actually declined when the industrial-scale eiderdown harvesting stopped. Now, only a few hundred pairs of Common Eider breed, although as many as 12,000 moulting Eiders

have been counted during summer in recent years, and up to 31,000 during winter. King Eider has bred a couple of times, and up to 2,000 individuals winter in Vega, usually in deeper waters than Common Eider. Lesser Black-backed Gull of the northern race *fuscus* breeds here as well, although its population has declined from about 500 pairs in 1977 to just a few pairs today.

There are several interesting birdwatching sites on the main island of Vega, particularly for wetland species. A few pairs of Coot, Red-throated Diver and Shoveler breed regularly, and Pintail, Gadwall and possibly Garganey has bred at least once. Vega is also an important stop-over site for migrants. Large numbers of Barnacle Goose arrive around April 20 and stay in the area until late May. Black-tailed Godwit is seen almost every year in the end of May. Red-necked Phalarope and Spotted Redshank are regular visitors in late summer. The Eidemsliene slopes are covered in deciduous woodland and coastal pine forest where Grey-headed Woodpecker breed.

Best season: Vega is important to birds year-round, but birdwatching is best during migration in spring and autumn, and early summer for many of the breeding birds.

Directions: Vega is reached via the car ferry from Horn, which is along Fv17 north of Brønnøysund. In summer the car ferry leaves from Tjøtta. Vega can also be reached by express boat from Brønnøysund and Sandnessjøen.

Tactics: The main island of Vega can be explored by bicycle or car, using the following suggested route:
From the ferry landing at Igerøy, drive west towards Gladstad, which is the administrative centre in Vega municipality. After just over 7

km, turn left to **a.** Rørøya to investigate the tidal area by the harbour. Continue 200 m along the main road and make a left again to check the shore and the small **b.** Lake Hestvatnet. You may walk upstream along the creek until you get a view of the larger Lake Fersetvatnet. Continue along the main road to Gladstad village and make a right signposted to *Museum*. After 2 km on this road, turn left and you will after another 1 km come to **c.** Lake Sveavatnet on your left. Back in Gladstad, continue on the main road north towards Sundsvoll and exit right onto the road signposted to the eider museum *E-huset* (E is the local name for Common Eider!). Make a stop in the junction 1 km ahead, just before the *Nes* sign, where there is a car park on your right. From here you can walk 1 km east along the track to Kjellarhaugen, from where you can view the nutrient-rich **d.** Lake Kjellarhaugvatnet. If you turn left at this junction, signposted to Holand, and then just before Holand village make a right

Common Eider plays a key role in Vega's nature and culture. Since the dawn of time, eiderdown from wild birds has been collected from the nests. In return, the birds have been protected and well looked after by the eider farmers, e.g. by offering them attractive nest boxes. This tradition is globally unique. *Photo: Kjetil Schjølberg.*

Lake Stovatnet on Tjøtta island has rich edge vegetation and is surrounded by pastures and several smaller lakes. The mountains of Fast-Vega can be seen in the background. *Photo: Bjørn Olav Tveit.*

onto the gravel road, you will reach the upper side of **e.** Holandsosen delta where gulls, ducks and shorebirds gather. Another view of Holandsosen tidal area is obtained by following the paved road around the village, out onto the causeway on the north side. You may also want to visit the *E-huset* eider museum and the Vega World Heritage Centre while in this area.

Back at the main road, head north towards Sundsvoll, and make a right after another 2 km, signposted to **f.** Vallsjøen. Here are several small pools of water, making this site one of the best in Vega for resting and breeding shorebirds and ducks, often including Shoveler. Here is also a nice view to the sea. Back on the main road, continue 200 m and make a left onto the road signposted to Guldsvåg, where you soon get the little Lake Barnvatnet on your right. There are several more lakes in the area, for instance the

overgrown **g.** Altervatnet, besides Husvatnet and Holmvatnet with large sections of open water. The main road takes you all the way west to **h.** Sundsvoll, where you find cultivated fields and pastures. Here is again a nice view out to the sea, and to the shoreline and the tall island Søla.

You can next cross the main island and drive south to **i.** Eidem which has several interesting fields, wetlands and Eidemsliene woodland slopes.

6. TJØTTA ISLAND

Nordland County
GPS: 65.82885° N 12.40801° E (parking Ostjønna)

Notable species: Shelduck, Long-tailed Duck, Common Scoter, Velvet Scoter, Red-throated Diver, Black-throated Diver, Slavonian Grebe, Lesser Black-backed Gull (*fuscus*); migrants.

Description: Tjøtta is an island by the mouth of the Vefsnfjord. It appears as a southern extension of the prominent and mythological mountain range known as *The seven sisters*. By contrast, Tjøtta is rather flat and dominated by cultivated fields and nutrient-rich lakes and inlets, making up perfect habitat for a variety of ducks and shorebirds. The commoner species are often accompanied by uncommon species like Garganey and Shoveler. The outer section of the Vefsnfjord support good numbers of moulting sea ducks in summer, including several hundred Velvet Scoters. Grebes and divers are also found in good numbers in winter and spring. A few small colonies of Lesser Black-backed Gull *fuscus* are found here.

During migration, a prominent rush of passerines can be noted passing through the area, particularly in mornings with heavy overcast and a light breeze from the southeast. On such days, southern species like Icterine Warbler and Whitethroat are often encountered. Sometimes even rarer birds as well.

Best season: All year in the Vefsnfjord. The wetlands at Tjøtta are at their best from April to October.

Directions: Tjøtta is situated along Fv17 25 km south of Sandnessjøen.

Tactics: Explore the fjord, inlets, lakes and fields in the area. If you arrive along Fv17 from Sandnessjøen to the north, the road turns over to Offersøy 6 km after Alstahaug church. Here, **a.** Hamnes on the north-east side of Offersøy island provides a good viewpoint to the outer part of the Vefsnfjord in the east. In the past, there were often good numbers of shorebirds on the alluring mud banks and salt flats on **b.** Hamnesvalen but this is now primarily a nice place for dabbling ducks. If you follow the Fv17 further 6 km further south, 500 m past a rest area on the left, you will get the nice bays at **c.** Storvollhalsen on the right, where there are often shorebirds to be found. Park well on the side of the road and walk out into the area. If you drive further 1 km south, you will come to the **d.** Russian memorial cemetery, which is located opposite the small Lake Kråkvikvatnet, which has nesting Wigeon among others. Park by the cemetery and walk up past the bauta for a view of the outer strait, where divers, auks and ducks can migrate and stage, and where there are often quite a few gulls, not least in winter. During migration times, you should check the plant field north of the cemetery for resting passerines and more. Continue a couple of km further to the small **e.** Lake Ostjønna, visible on the left. Park in a suitable place and follow the path along the edge of the field for a view from the north. Lake Ostjønnna is good for dabbling ducks, and shorebirds can often roost on the floating algae mats. You get **f.** a different

Arctic Tern is the world's longest-travelling migrant. It breeds along most of the Norwegian coast and on Svalbard. In northern Norway, it also breeds inland. From the breeding grounds, the migration journey goes back and forth every year to the wintering areas in the Antarctic seas. *Photo: Kjetil Schjølberg.*

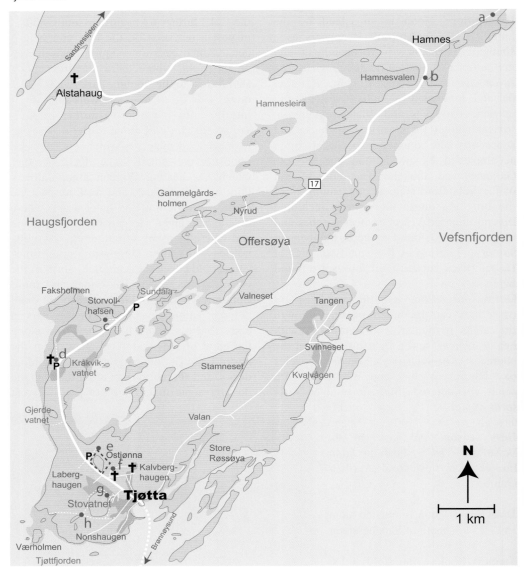

Tjøtta island: a. Hamnes, viewpoint Vefsnfjorden; **b.** Hamnesvalen; **c.** Storvollhalsen; **d.** Russian memorial cemetery; **e.** Lake Ostjønna; **f.** Ostjønna east, migration viewpoint; **g.** Lake Stovatnet; **h.** pools and pastures.

view of Lake Ostjønna by continuing Fv17 further south and taking the road signposted to Tjøtta church to the left. Immediately bear left again and drive to the bend. Here, you have a good view of Lake Ostjønna, while at the same time it is a good vantage point from which to monitor the overhead migration of e.g. passerines.

You reach **g.** Lake Stovatnet by continuing Fv17 350 m past the Tjøtta church exit and turning down to the right. Keep to the right,

park, and follow the footpath down to the north-east side of the lake. Follow the path along the north and west sides of the lake. Walk over to the west side to get a better view of the edge areas, for nesting ducks and the possibility of Coot and more. Feel free to continue the path over the fence splitters to **h.** ponds and fields south of Lake Stovatnet, and you can follow a tractor track up to the guest house on the east side of Lake Stovatnet from here.

7. Herøy and Dønna Islands

Nordland County

GPS: 66.18357° N 12.55595° E (parking Lake Altervatnet)

Notable species: Wetland birds, resting Barnacle Goose (April–May) and other migrants; Shoveler, Water Rail, Black Grouse, White-tailed Eagle and Grey-headed Woodpecker.

Description: Herøy and Dønna are two island communities along the outer coast near Sandnessjøen, just south of the Arctic Circle. The two main islands and several smaller ones are interconnected with roads, but you need a ferry to get there from the mainland. Herøy and Dønna are strategically positioned along the migration routes both over land and at sea, and migrants in need of a rest have many different habitats to choose from here. Herøy island is a most important staging ground for the Svalbard population of Barnacle Goose, particularly in spring. Several rare birds have been encountered through the years here, including Pied-billed Grebe, Upland Sandpiper and Bearded Tit. Dønna island is rockier and more barren than Herøy, and at a distance, the contours of the island resembles the silhouette of a man's face,

known as *The Dønna man*. However, on the north side of the island are gentler areas with lush meadows, forests and fields, and several lakes and wetlands. One of the most interesting of these is Lake Altervatnet, which is good for a variety of waterbirds, regularly holding unusual species for this part of the country. Shoveler is regular and has bred. Grey-headed Woodpecker is rather common in the woods of Dønna. Smew has proven to be quite regular in late spring in Lake Liss-Gleinsvatnet. In winter, both islands can host Water Rail and Jack Snipe. Look for them in wet ditches and along creeks. A variety of shorebirds usually winter, particularly along the coast. Arctic Redpoll is also often found in the cold season.

Other wildlife: Both Herøy and Dønna islands have good populations of roe deer and recently also elk. Hedgehog is common on Herøy island.

Best season: All year, but particularly during migration and in winter.

Directions: Car ferry from Sandnessjøen to Bjørn village at Dønna island, or from Søvik along Fv17 in Alstahaug to Flostad at Herøy island.

Oslo

Interior

Skagerrak

Western

Central

Northern

Finnmark

Svalbard

Lake Altervatnet (right) and Lake Bladvatnet on Dønna island, as seen from NE. *Photo: Bjørn Olav Tveit.*

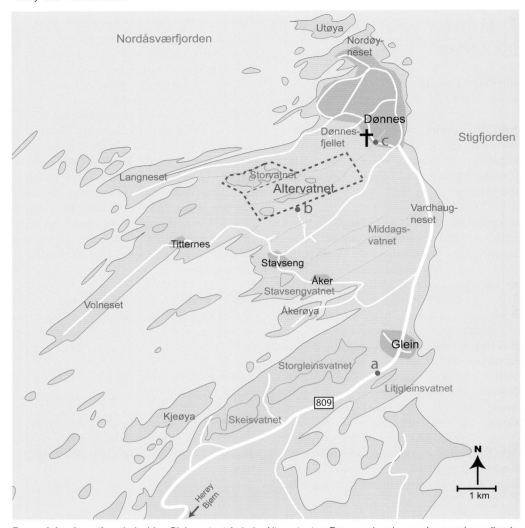

Dønna island, north: a. Lake Liss-Gleinsvatnet; **b.** Lake Altervatnet; **c.** Dønnes church, meadows and woodlands.

Tactics: From the island <u>Tenna</u> at the southern tip of Herøy you have a view to the <u>Husværfjord</u> where sea ducks, gulls and terns gather. This area is also good for shorebirds. Drive north along Fv828 towards Bjørn and check the fields along the way, preferably including detours along the side roads, for geese and other species associated with farmland. At northern Herøy, exit from Fv828 and continue 100 m towards Engan before turning right onto a road signposted Hokleppan, leading close to the western section of <u>Lake Salsvatnet</u>. You will need to find parking after a few hundred metres on this road and walk past the trees blocking the view to the lake on your left. Back on the road towards Engan, continue another 500 m. To the right is a track leading through deciduous woodland where Garden Warbler, Blackcap, Icterine Warbler and Wood Warbler can be found singing in summer. Walk onto the little hill between the two lakes Vikvatnet and Salsvatnet.

See map of Dønna island, north. From Herøy, follow Fv828 to the northern part of Dønna and stop wherever it looks nice, such as by the nutrient-rich **a.** <u>Lake Liss-Gleinsvatnet</u> which can be viewed roadside just before you reach Glein village. To get to **b.** <u>Lake Altervatnet,</u> continue 5 km past Glein and exit on the second road signposted to Titternes. Keep straight ahead (past the exit to Dønnes church), and after 3 km turn right onto a gravel road just after a large circle of antennas on your right. Park just before the field, 500 metres up this road (where Bluethroat and Black Grouse are often seen). Bring your telescope and walk along a poorly marked path towards the northwest, between two hills and across a marsh, to an information poster concerning Altervatnet nature reserve. From the hill above the information poster, you have a view to Lake Altervatnet. Scan the area thoroughly from here – closer access is prohibited from May 1 to July 15. Near **c. Dønnes church,** and several other places at Dønna, are lush and damp pastures surrounded by dense deciduous forest.

8. ENGASJYEN AND GLOMÅ DELTAS

Nordland County
GPS: 66.34006° N 14.14042° E (Engasjyen delta)

Notable species: Pintail, Scaup, Common Scoter, Velvet Scoter, Willow Ptarmigan, Black Grouse, Capercaillie, Black-throated Diver, Red-throated Diver, Slavonian Grebe, White-tailed Eagle, Goshawk, Crane, Temminck's Stint, Ruff, Arctic Tern, Pygmy Owl, Hawk Owl, Tengmalm's Owl, Wryneck, Three-toed Woodpecker, Yellow Wagtail, Bluethroat, Redstart, Sedge Warbler, Great Grey Shrike and Siberian Jay.

Description: By the town of Mo i Rana is Engasjyen delta in the Ranfjord, at the mouth of river Ranelva. Here are resting opportunities for ducks, gulls, shorebirds and more. Uncommon species such as Gadwall, Shoveler and Garganey are encountered annually, and this is one of the most reliable staging sites for Temminck's Stint in the region, particularly in May. Just north of Engasjyen delta is the larger Glomå delta in Lake Langvatnet, which is an important breeding area as well as a resting area for a variety of waterbirds, particularly in spring. Birds breeding in the Saltfjellet mountain range can rest here while awaiting the snow and ice to thaw at higher altitudes. The delta surroundings are densely forested, mainly with spruce, and several patches of cultivated land. Greenshank and Wood Sandpiper are common breeders, Whooper

Engasjyen delta in Mo i Rana has fine salt meadows and shallow water lagoons. *Photo: Bjørn Olav Tveit.*

281

Lovund island with its characteristic profile, here seen directly towards the Puffin colony at Lundeura on the northern side of the island. *Photo: Bjørn Olav Tveit.*

Swan and Crane breed as well, possibly still also Ruff. Capercaillie, Hawk Owl, Wryneck, Three-toed Woodpecker and Great Grey Shrike can all be found in the surroundings.

Best season: Engasjyen delta is at its best from mid-April to early June. Temminck's Stint is usually found at the end of this period. The Glomå delta is at its prime in spring and summer.

Directions: Engasjyen delta is reached from along the E6 in the town of Mo i Rana. Exit onto E12 (later Fv810) signposted towards Nesna. In the roundabout at the exit to Ytteren church, turn left and park by the breakwater, from where you have a view to the eastern section of the delta. Along Fv810 for the next kilometre or so there are several bus stops on both sides of the road where you can make short stops.

Glomå delta is reached by heading back to E6 and continuing 10 km north before exiting to Røssvoll airport, and then turn right towards Røvassdalen. Continue 4 km and make a left, signposted to Langvassgrenda. Check the river mouth by this junction, and there is also a couple of small oxbow lakes, reached by walking along the tractor track to your left. Then continue along Lake Langvatnet 15 km to the Glomå delta. Some parts of the delta can be scanned from the road.

9. ARCTIC CIRCLE ISLANDS
Nordland County
GPS: 66.33922° N 13.00272° E (ferry landing Stokkvågen), 66.51097° N 13.00580° E (ferry landing Tonnes)

Notable species: Shelduck, White-tailed Eagle, Golden Eagle, Peregrine Falcon, Razorbill, Puffin and Eagle Owl; migrants; vagrants.

Description: Along the northern part of the Helgeland coast are several archipelagos containing a myriad of small islands. The Arctic Circle runs right across them. Many of the islands here can provide interesting birds, the most famous being Træna and Myken islands, both proven rarity magnets. Lovund island is well-known for its large Puffin colony and Sleneset/Solvær with its dense population of Eagle Owls.

Solvær archipelago, with its main village Sleneset, comprises more than 300 small islands and skerries. In a total land area of 13.5 km², one pair of Eagle Owl is found for every km², resulting in a total of 60-70 individuals by the end of a successful breeding season. This makes the Eagle Owl population in Solvær the densest in Norway, possibly in the world. Lovund a bit further out to sea is a peculiar hat shaped island rising 625 m above the ocean.

The island supports one of the largest colonies of Puffin in Norway, counting about 150,000 pairs. Tradition has it that the Puffins arrive at the colony on April 14th every year. The Puffin colony is situated in an area of scree and boulder on the island's northwestern slopes, easily accessible on foot. Storm Petrel probably breeds here as well. Lovund also supports a few pairs of Eagle Owl, Shag and Kittiwake, the latter breeding on building walls in the harbour. The 500 inhabitants on Lovund are concentrated in the village on the east side, hosting many lush gardens suitable for resting migrant passerines.

Træna, the archipelago to the north of Lovund, consists of about 1000 small islands, most of which are flat and grass covered. The exceptions are mainly the island of Sanna, with the 334 m tall Trænstaven and several other prominent rock formations together creating a characteristic outline and a trollish atmosphere. Træna is positioned 38 km from the mainland and 16 km from the nearest large island, Lovund. The archipelago has a good population of White-tailed Eagles and several seabirds, although no classical cliff-breeding species, save for a few pairs of Kittiwake in the harbour. Træna is most of all noted for being a highly potent rarities islands, with records of exotic species like Dusky Thrush, Sykes's Warbler, Eastern Bonelli's Warbler and Black-headed Bunting and several each of Olive-backed and Pechora's Pipit, not to mention the cryptic record of a Red-necked Grebe of the North American subspecies *holboellii*. Birdwatching is for the most part concentrated to the main island Husøya, where most people live, providing gardens where migrants can rest and feed, and Selvær further north, on the opposite side of the Arctic Circle. Selvær too has several gardens and more pastures than Husøya island. It is an important resting place for Barnacle Goose. Sanna island has a few gardens and provides the best view to the outer sea. A walk-through tunnel leads up

Husøya island in the Træna archipelago has lovely gardens and pastures that create high expectations for the diversity of bird species during migration. Trænstaven rock and the other peaks on the neighboring Sanna island can be seen in the background. *Photo: Bjørn Olav Tveit.*

Oslo
Interior
Skagerrak
Western
Central
Northern
Finnmark
Svalbard

Myken island near the Arctic Circle is an incredibly exciting place to be during migration times. The small size and limited vegetation result in an island that is easily traversed by individuals or a small group of birdwatchers. *Photo: Terje Kolaas.*

the mountain Trænstaven to a breathtaking view of the archipelago and the inland mountains and glaciers.

Myken island is a very small fishing community with about 15 inhabitants more than 30 km off the mainland, 20 km north of Træna and the Arctic Circle. It is situated on a small group of islands, of which the largest is only 1800 by 300 metres. Vagrants like Black-browed Albatross and Hume's Warbler have been recorded. It is a perfect place for a lone or a small party of birdwatchers.

Best season: Late spring and autumn has the greatest diversity of birds. White-tailed Eagle and Eagle Owl can be found all year.

Tactics: You can do very well without a car on these small islands. On Lovund you may walk to the Puffin colony by following the road to the cemetery and follow the path uphill from here, but access to the colony itself is prohibited from April 15 to July 31. At Træna you can rely on ferries to travel between the islands, or you may negotiate transportation with local fishermen.

Directions: Lovund, Sleneset and Træna are reached by car ferry from Stokkvågen along Fv17 or by express boat from Sandnessjøen. Myken is reached by express boat from either Tonnes or Vågaholmen further north along Fv17. In summer, you can take the express boat between Sandnessjøen and Bodø, stopping by all these islands on the way.

10. SALTFJELLET MOUNTAIN

Nordland County

GPS: 66.55147° N 15.32084° E (Arctic Circle Centre)

Notable species: Scaup, Long-tailed Duck, Common Scoter, Velvet Scoter, Willow Ptarmigan, Rock Ptarmigan, Black Grouse, Capercaillie, Goshawk, Golden Eagle, Rough-legged Buzzard, Gyrfalcon, Dotterel, Temminck's Stint, Purple Sandpiper, Ruff, Great Snipe, Red-necked Phalarope, Long-tailed Skua, Eagle Owl, Snowy Owl, Hawk Owl, Pygmy Owl, Long-eared Owl, Short-eared Owl, Tengmalm's Owl, Lesser Spotted Woodpecker, Yellow Wagtail, Bluethroat,

Redstart, Ring Ouzel, Sedge Warbler, Icterine Warbler, Great Grey Shrike, Parrot Crossbill, Lapland Bunting and Snow Bunting.

Description: Saltfjellet is an impressive mountain range on the Arctic Circle stretching from the coast across to the Swedish border, constituting an ecological barrier. Svartisen, the largest glacier in Northern Scandinavia, covers 370 km^2 of the western part of the mountain massif. Saltfjellet has a wide variety of montane habitats, suitable for most high-altitude species of the region. The E6 and the railway crosses the mountain, both of which have contributed to reducing its qualities for birds, particularly along the Lønselva river where Lesser White-fronted Goose and Bean Goose used to breed. Saltfjellet is a popular recreational area with an elaborate network of marked paths and mountain cabins open to the public.

The forested valleys on the outskirts of Saltfjellet, particularly Junkerdalen valley to the north, can offer several forest species, including Eagle Owl, Hawk Owl and Pygmy Owl. On south-facing slopes of the upper valley, several pairs of Icterine Warbler breed, and the deciduous forest along the river in the bottom of the valley holds Lesser Spotted Woodpecker. Junkerdalsura scree slope is of great interest to botanists and can also boast a single record of Greenish Warbler. Virvassdalen valley southeast of Saltfjellet has several interesting wetlands.

Other wildlife: Wolverine and lynx have relatively strong populations in the Saltfjellet area, whereas Arctic fox, having become nearly extinct, has been reintroduced to the area.

Best season: Late spring and early summer.

Directions: By train from Bodø or Oslo/Trondheim to Lønsdal station makes a nice vantage point for hiking trips in Saltfjellet mountain. *By car:* E6 crosses Saltfjellet north

of Mo i Rana and south of Rognan towns. Popular trailheads are the Arctic Circle Centre, Stødi or the Polar Camping. For trips to the adjacent valleys, Virvassdalen is signposted to from E6 70 km north of Mo i Rana, and to Junkerdalen from E6 on the north side of Saltfjellet, 35 km south of Rognan.

Tactics: You get a taste of Saltfjellet when crossing it by car, stopping every now and then to explore the surrounding terrain. Even better is to park the car and do a real hiking trip, preferably over the course of several days.

11. SKANSØYRA DELTA

Nordland County
GPS: 67.10219° N 15.41914° E (Skansøyra viewpoint b.)

Notable species: Staging ducks, shorebirds and gulls; Red-throated Diver, Slavonian Grebe, Temminck's Stint, Green Sandpiper, Bluethroat, Redstart, Sedge Warbler, Icterine Warbler and Long-tailed Tit.

Description: Skansøyra delta at the mouth of Saltelva river in the Skjerstadfjord is readily accessible close to E6 and well worth

Eagle Owl has the densest population in Europe at Sleneset on the Solvær archipelago, near Træna and Lovund islands. *Photo: Kjartan Trana.*

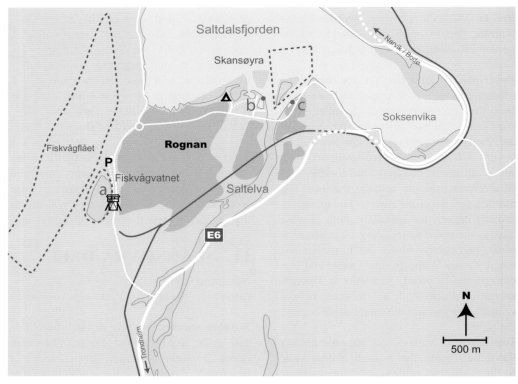

Skansøyra delta: a. Lake Fiskvågvatnet tower hide; **b.** and **c.** Skansøyra delta viewpoints. Also, see map p. 289.

a stop if passing by. Resting waterbirds can be found by the river mouth, and divers and grebes can be seen on the fjord. The nearby Lake Fiskvågvatnet is shallow and nutrient-rich, supporting breeding Slavonian Grebe and often a variety of ducks. The deciduous woodland mixed with pine surrounding the lake, particularly along Fiskvågflåget nature reserve on the slope to the west, support forest species otherwise uncommon in this part of the country, such as Great Spotted Woodpecker and Icterine Warbler. Green Sandpiper breeds in wetlands upstream along river Saltelva.

Best season: Spring, summer and autumn.

Directions: From along E6, exit to Rognan. Then, see the map.

Tactics: Lake Fiskvågvatnet is best viewed from the tower hide, reached via a path leading from the parking area in the northern end of the lake. A telescope comes in handy both here and at Skansøyra delta.

12. KLUNGSETFJÆRA BAY

Nordland County

GPS: 67.26211° N 15.35147° E (parking tower hide), 67.39724° N 15.38616° E (parking Østerkløft valley)

Notable species: Whooper Swan, Shelduck, Long-tailed Duck, Velvet Scoter, Red-throated Diver, Black-throated Diver, White-billed Diver, Red-necked Grebe, Slavonian Grebe, White-tailed Eagle, Bar-tailed Godwit, Knot and Great Snipe.

Description: Klungsetfjæra is a 2 km wide tidal area in the Skjerstadfjord near Fauske village, a one hour's drive from Bodø. At low tide the mudflat stretches 600 m from the shore, creating an el dorado for shorebirds during migration, often with several hundred each of Bar-tailed Godwit and Knot. Broad-billed Sandpiper is sometimes encountered. A tower hide is erected by the mouth of Leireelva river. The shallow waters of the fjords attract a variety of ducks, divers and grebes. Red-throated Diver and Black-throated Diver can in May sometimes being seen by the hundreds. White-billed Diver is regularly encountered from December to April, sometimes Great Northern Diver as well. There are a few other tidal areas along the Skjærstadfjord, particularly Røvika west of Klungsetfjæra, and at Stokkland and Stemland by Valnesfjord village.

Østerkløft is a mountain valley inland from Valnesfjord. Here Whooper Swan, Red-throated Diver, Black-throated Diver, Slavonian Grebe and Great Snipe can be found breeding in summer.

Best season: Klungsetfjæra bay is at its best in spring and early summer, with shorebirds predominantly in May and August. The fjord is good all year. Østerkløft valley is best visited in June and early July.

Directions: The railway stops at Fauske, and there are express buses leaving from Bodø. See map of the Bodø region. For directions to **j**. Klungsetfjæra bay by car from along route E6, exit onto Rv80 in Fauske, signposted

to Bodø. *From Bodø*, follow Rv80 east signposted to Fauske.

Tactics: j. Klungsetfjæra bay: Just west of Fauske village, find parking in the Rv80/Tareveien road junction. Walk carefully 150 m along Rv80 to just before the Leirelva river crossing, and follow the path down to the tower hide. A telescope is mandatory by the fjord.

i. Røvika inlet is 3 km further west along Rv80. In the right-bend on the headland where the tidal area comes into view, a small road exits to the left, leading to a nice viewpoint. Another 9 km to the west, at **h**. Stokkland just before you reach Valnesfjord, you see a roadside inlet surrounded by pastures. The inlet can be explored from the road and side-roads, and from a path leading down to the shore from behind the church. (More habitats like these are found if you exit from Rv80 a bit earlier, to Stemland on Alveneset, signposted to Nes).

h. Østerkløft mountain valley is reached from Valnesfjord by exiting onto the road signposted to Kosmo. Continue 15 km

Klungsetfjæra bay has shallows and extensive mudflats, often holding good numbers of sea ducks, divers, grebes and shorebirds. *Photo: Bjørn Olav Tveit.*

Seinesodden peninsula is perhaps the most exciting migration site near Bodø town. *Photo: Bjørn Olav Tveit.*

to the parking area just past *Valnesfjord helsesportsenter* (recreational centre). This is a good vantage point for hiking trips to the wetlands stretching from here and up to Lake Hømmervatnet (about 2 km) and beyond.

13. Bodø area

Nordland County

GPS: 67.27805° N 14.43418° E (parking Rønvik fields), 67.20424° N 14.36945° E (Seinesodden peninsula)

Notable species: Whooper Swan, Shelduck, Gadwall, Shoveler, King Eider, Willow Ptarmigan, Black Grouse, Red-throated Diver, Black-throated Diver, White-billed Diver, Slavonian Grebe, White-tailed Eagle, Gyrfalcon, Whimbrel, Ruff, Great Snipe, Red-necked Phalarope, Glaucous Gull, Lesser Black-backed Gull *fuscus*, Little Auk (winter), Short-eared Owl, Bluethroat, Sedge Warbler, Icterine Warbler, Arctic Redpoll, Lapland Bunting and Snow Bunting.

Description: In Bodø, birdwatchers do not

need to travel far to enjoy their favourite past time. The town is surrounded by several good sites, and besides, one of the country's best birding islands, Røst (which see), is only a ferry crossing away. The Bodø area has the world's densest population of White-tailed Eagle. The birds are seen in good numbers around the town at all times of year, except perhaps for the dead of mid-summer. Bodø harbour is the easiest place to see White-tailed Eagle, and this is also a very good site for sea ducks and gulls. In summer you may find Lesser Black-backed Gull of the northern race *fuscus* here (sometimes other subspecies as well). In late winter, King Eider, Glaucous Gull and perhaps Iceland Gull may be encountered.

The Rønvik fields is a popular recreational area just outside of town with cultivated farmland and pastures lined with lush vegetation along the Bodøgårdselva creek. These fields thaw rather early in spring and are where many birdwatchers go to see the first returning migrants. A few pairs of Whitethroat breed, and geese may stop by here in autumn.

Saltstraumen sound connect the Skjerstadfjord to the sea. It has the world's strongest tidal current, attracting many of both tourists and birds, the latter particularly in winter when several thousand Common Eiders and other ducks, gulls and auks gather here. The flocks often include a few uncommon species like King Eider, Glaucous and Iceland Gulls. Gyrfalcons regularly sweep by to pick up a Little Auk or some other easy meal.

Seinesodden peninsula on Straumøya island on the south side of the Saltfjord has several nice marshes, lakes and brackish ponds surrounded by grazing pastures, making it a prime location for resting migrants in spring and autumn. It is also an interesting summer site, as Gadwall, Shoveler, Red-necked Phalarope and Short-eared Owl breed here. White-tailed Eagle breeds nearby. Lake Seinesvatnet support breeding Whooper Swan, Slavonian Grebe, Whimbrel and Ruff. Lake Loddvatnet is a small, nutrient-rich lake

in Brekke, south of Misvær, considered to be an important resting and breeding site for wetland species in the southern part of the Bodø area, including Slavonian Grebe.

Mjønesodden peninsula in the Skjerstadfjord is another traditional local migration site, particularly good in spring. Red-throated and Black-throated Divers often rest on the fjord, and from December to April, regularly White-billed Diver as well. This peninsula is also worth a visit in early summer when several local breeding species can be found, including Black Grouse, Grey Heron, Woodcock, Whinchat, Sedge Warbler, Icterine Warbler and Whitethroat. Together with the four common species of thrush, i.e. Blackbird, Song Thrush, Redwing and Fieldfare, the choir of birdsong can be quite impressive here.

Best season: All year.

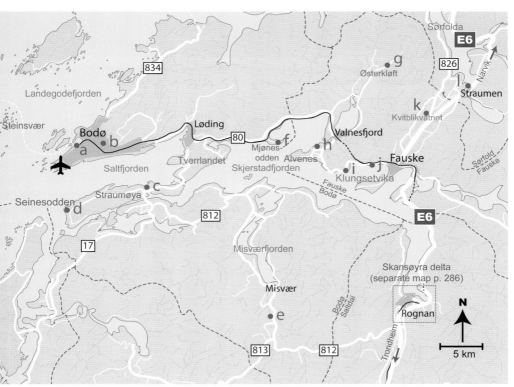

Bodø area, with Klungsetfjæra bay (p. 286) and Skansøyra delta (p. 285): **a.** Bodø harbour; **b.** Rønvik fields; **c.** Saltstraumen sound; **d.** Seinesodden peninsula; **e.** Lake Loddvatnet; **f.** Mjønesodden peninsula; **g.** Østerkløft mountain valley; **h.** Stokkland inlet; **i.** Røvika inlet; **j.** Klungsetfjæra bay; also **k.** Lake Kvitblikvatnet and **l.** Straumbukta inlet (see p. 291).

Bluethroat is a quite common and characteristic bird in northern Norway, particularly associated with willow thickets. In southern Norway, you will find the species in similar terrain in the mountains. *Photo: Kjartan Trana.*

Directions: Bodø is reached by plane, train, boat or car. Local public transportation is well developed, in terms of buses.

Tactics: By car you can reach the mentioned sites as follows:
a. The harbour is best viewed from along the docks in the town centre and from Burøya in the northern section of the harbour.
The **b.** Rønvik fields are reached by heading east along Rv80, signposted to Fauske. Just before the Bodø tunnel, exit towards Hunstadmoen and follow signs to Bodømarka parking area. A path leads from here up along the Bodøgårdselva creek.
Continue east along Rv80 and exit south onto Fv17 signposted to Saltstraumen. The road passes by several interesting sites, such as the Tverrlandet nature reserve inlet by Bodø golf course. **c.** Saltstraumen sound can be viewed from the side roads just before

and after the bridge crossing the sound. Saltstraumen is at its best two times a day, when the current leads out of the fjord after high tide.
Back on Fv17 after crossing the bridge over Saltstraumen sound, exit right onto the road signposted to Seines and continue to **d.** Seinesodden peninsula 11 km ahead. Park by the small community building at the end of the paved road. Walk 300 m further along the gravel road and turn onto the second road to the right, leading past a farm and down to the sea. Walk along the shore south to the wetland on the headland (just before the tall military antennas). If you drive further along the gravel road past the community building, you will soon reach a T-junction. If you park here and walk on along the road to your right, this is an alternative access point to Seinesodden peninsula (keep right just before the gate to the military area). If you

turn left at the T-junction and continue 700 m, you will reach a rock on your left from where you have a view to Lake Seinesvatnet.

e. Lake Loddvatnet is reached by driving back to Saltstraumen sound and following Fv17 1 km further south, and then turn left onto Fv812 signposted to Rognan. Follow this for 44 km via Misvær and take the exit signposted to Vestvatn and *Skianlegg* (ski resort). You will see Lake Loddvatnet on your right after 200 m. (You may now complete a lap around the Skjerstadfjord via Rognan.)

f. Mjønesodden peninsula is reached by heading back to Rv80 and continuing towards Fauske. A few hundred metres past the Mjønes sign, exit right onto a gravel road by a bus stop, leading out onto the Mjønesodden headland. Park in the designated area and continue on foot, either along the road or along the shore.

14. LAKE KVITBLIKVATNET

Nordland County
GPS: 67.31900° N 15.47589° E (parking tower hide)

Notable species: Whooper Swan, Pintail, Black-throated Diver, Slavonian Grebe, Greenshank, Wood Sandpiper, Yellow Wagtail and Bluethroat.

Description: Lake Kvitblikvatnet is a shallow lake lined with lush vegetation in a forested area with several large marshes, protected as Fauskeidet nature reserve. Whooper Swan, Black-throated Diver and Slavonian Grebe breed. In Straume village nearby, gulls and ducks often gather, particularly in winter. This also used to be a reliable spot for wintering Purple Sandpipers, until the seashore was partially destroyed.

Best season: Late spring and summer.

Directions: See map on previous page. *To* **k.** *Lake Kvitblikvatnet*: From along E6 6 km north of Fauske, exit onto Fv826 signposted to Røsvik. Immediately after, turn left onto a dirt road signposted to *Fugletårn* (tower

hide). Follow the red-marked path through the woods, a 10-minute walk. To view a different section of the lake, continue along Fv826 and park roadside after 1.4 km by an information poster concerning Fauskeidet nature reserve. **l.** Straumbukta inlet in Straume is reached by continuing E6 north 6 km. Here, exit to Røsvik and continue 400 m.

Tactics: A telescope is useful at the tower hide.

15. BALLANGEN WETLANDS

Nordland County
GPS: 68.28537° N 16.71472° E (Lake Grunnvatnet, east)

Notable species: Whooper Swan, Garganey, Red-throated Diver, Black-throated Diver, Slavonian Grebe, Little Gull, Yellow Wagtail and Bluethroat.

Description: Near Ballangen village is a watercourse, with Lake Grunnvatnet as the centre piece. Here you find good populations of many of the wetland species of the region, including Teal, Slavonian Grebe, Black-headed Gull and Wood Sandpiper. A few pairs of Whooper Swan, Black-throated Diver and Red-throated Diver breed annually, possibly Garganey and Little Gull as well.

Directions: From along E6 3 km southwest of Ballangen village centre, exit south signposted to Melkedalen. After about 2 km you come to the Børselva river on your left. 4 km from the Melkedalen junction, the road makes a sharp bend to the left and a gravel road continues straight ahead. Here, you can either follow the main road leading along the east-side of Lake Grunnvatnet, or you can enter the gravel road leading past the north end of Lake Grunnvatnet on your left, and – by turning right at the next exit – get a view of Lake Djupvatnet. The latter is also visible from E6 8 km southwest of the Melkedal junction.

16. HÅKVIKLEIRA INLET

Nordland County
GPS: 68.40199° N 17.29361° E (viewpoint)

Notable species: Shelduck, Scaup, Velvet Scoter, Common Scoter, Knot, Long-eared Owl, Three-toed Woodpecker, Bluethroat, Redstart, Great Grey Shrike, Lapland Bunting and Snow Bunting.

Description: Håkvikleira is a tidal inlet just south of Narvik town, with perfect conditions for ducks and shorebirds. Large numbers of the latter can occur from May to August, dominated by Knot in May. In summer, notable numbers of moulting Velvet and Common Scoters are found.
The narrow Skjomen fjord to the south, and particularly at the Skjoma river mouth should be checked for resting wetland species. If you continue up the Skjomdal valley east of the fjord you may find several forest species.

Best season: Spring and summer.

Directions: Håkvikleira is along E6 9 km south of Narvik, and can be viewed from several places along the way. Exit from E6 a couple of km further south, onto a road signposted to Skjomdal.

17. EVENES WETLANDS

Nordland and Troms counties
GPS: 68.46382° N 16.66490° E (Stunesosen)

Notable species: Whooper Swan, ducks, Slavonian Grebe, shorebirds, Bluethroat and Redstart.

Description: The main airport for the towns of Narvik and Harstad is located midway between

Lake Store Trøsevatnet with Sætertinden mountain, on Hinnøya island in the Lofoten archipelago, in the background. *Photo: Bjørn Olav Tveit.*

them, at Evenes. The airport area is noted for having several interesting lakes and marshes surrounded by lush and damp deciduous woodland. These sites are of importance to breeding, moulting and resting wetland birds and has a particularly strong population of Slavonian Grebe. There have been several occasions with suspected breeding of Red-necked Grebe, and Shoveler is seen quite often. The surrounding woodland is rich in passerines and other birds. The last lake downstream along this watercourse is Lake Kjerkvatnet, which is regularly flooded with salty sea water resulting in a brackish environment. The watercourse enters the Ofotfjord via the tidal rivers Stunesosen and Tårstadosen, both having sand and gravel banks where Temminck's Stint has bred several times. Another interesting tidal area is Trøselvosen across the county border to Troms. It is situated in the narrow Tjeldsundet sound which bites Lofoten archipelago off from the mainland, so to speak, providing a shortcut for waterbirds migrating up and down along the coast.

Best season: Summer.

Directions: The area is in the vicinity of Harstad/Narvik airport, Evenes, with direct flight connections from Oslo. *By car from Narvik*, follow the E6 north to Bjerkvik and continue along the E10 west for about 40 km.

Tactics: Park just before the exit to Evenes church. Here is an information poster concerning Nautå nature reserve. Explore the wetlands here and the woodland surrounding Lake Nautåvatnet to the north of the E10. Also, look for ducks, breeding Whooper Swan and more in Lake Svanvatnet on the south side of the E10. Leave the E10 and continue past Evenes church. In the upcoming T-junction, turn right towards Tårstad. Continue 1.7 km and stop in the lay-by at Stunesosen. Drive 500 m further on and stop at Tårstadosen. Drive back to the E10, and make a right, going back a few hundred metres to the exit to Kvitfors, leading along the Nautå watercourse to Lake Kjerkhaugvatnet and Lake Sommervatnet, besides the two lakes Store Trøsevatnet and

Lille Trøsevatnet. If you continue past these lakes and turn left towards Elvemo, you will soon reach Trøselvosen tidal inlet. The latter can also be reached by continuing along E10 about 10 km past Evenes before exiting onto Fv824 signposted to Ramsund/Trøsemarka. Follow signs to Ramsund and you will reach Trøselvosen after about 1 km.

18. HARSTAD WETLANDS

Troms County
GPS: 68.67702° N 16.56627° E (Lake Kvannesvatnet)

Notable species: Resting migrants; Black Grouse, Slavonian Grebe and Long-eared Owl.

Description: Near Harstad town are several sites of interest if you happen to be in the area. Just north of Tjeldsundet sound just north of Trøselvosen (see previous site) are several grassy headlands that jut out into the Vågsfjord, creating nice resting grounds for a variety of migrants. The larger of these, Grasholmen headland, is particularly interesting, and so is Lake Brokvikvatnet and adjoining farmlands just inland from the headland. Long-eared Owl has been found breeding here. Several lakes can be explored in this area, with Lake Kvannesvatnet the most promising. Here, however, there is a traffic ban in the best birdwatching season. Just north of Harstad is Trondeneset headland protruding into the Vågsfjord with the small Lake Laugen. Several pairs of Slavonian Grebe breed here and you may find ducks and other wetland species, particularly during migration. Lake Møkkelandsvatnet northwest of Harstad can be good as well. Its northern and western shores are lined with farmland. In the west-end, several channels cut into the fields, creating suitable habitats for shorebirds, dabbling ducks and more. At the head of Kasfjord a bit further west is Mølnelvosen tidal area, at the mouth of the river running from Lake Kasfjordvatnet.

Best season: Late spring and early summer.

Oslo Interior Skagerrak Western Central **Northern** Finnmark Svalbard

Slavonian Grebe has its original core areas in Troms and the northern part of Nordland counties. In recent decades, the species has increased in number and spread to several areas, from eastern Norway to Finnmark. It thrives best in nutrient-rich lowland waters, where it nests in shallow coves with a mosaic of reed clusters and open water. *Photo: Kjetil Schjølberg.*

Directions: Harstad town is reached by car, plane or the Coastal Express (Hurtigruten). *From Harstad by car,* Lake Kvannesvatnet and Grasholmen headland are reached by following Rv83 south 14 km. Check the roadside Sørvika inlets. In Sørvika village, exit from Rv83 and follow signs to Brokvik. After 300 m, a road turns down to the right, signposted to Grasholmen. Do not drive down here yet but continue straight ahead for 800 m and find suitable parking. Here you can follow the tractor road and walk down to Lake Kvannesvatnet. There is a traffic ban here in the period May 1 to July 15. Now drive back and turn down towards Grasholmen. You will immediately see the small Lake Brokvikvatnet on your right, which can be viewed from the side of the road. The road further leads out to Grasholmen headland, but you must park the car and walk the last km onto the headland.

Lake Laugen is reached by driving Rv83 north in Harstad and turning right towards Trondenes. You see the lake on your left just before Trondenes church. Lake Møkkelandsvatnet is reached by following Rv83 1 km further and then turning right onto Rv867 and after another 2 km keeping straight ahead towards Kasfjord. After about 1 km you will see the lake on your left. Continue along this road another 3 km and turn right towards Kasfjord. This road passes Mølnelvosen tidal area.

19. LOFOTEN ARCHIPELAGO

Nordland County

GPS: 68.13752° N 13.57218° E (parking Lake Storeidvatnet), 68.30819° N 13.65290° E (Borgen, Eggum)

Notable species: Whooper Swan, Pink-footed Goose, Barnacle Goose, Shelduck, Scaup, King Eider, Willow Ptarmigan, Rock Ptarmigan, Red-throated Diver, Black-throated Diver, Great Northern Diver, White-billed Diver, Slavonian Grebe, Cormorant, Shag, Gannet, White-tailed Eagle, Golden Eagle, Kestrel, Gyrfalcon, Peregrine Falcon, Coot, Dotterel, Black-tailed Godwit *islandica*, Spotted Redshank, Temminck's Stint, Red-necked Phalarope, Lesser Black-backed Gull, Puffin, Short-eared Owl, Bluethroat, Sedge Warbler, Lapland Bunting and Snow Bunting.

Description: Lofoten, the row of beautiful mountainous islands with visible traces of Viking heritage and ongoing traditional fishing communities, is one of Norway's top tourist destinations. Lofoten is also of great importance to birds, with a long list of good birdwatching sites of which the outer two, Værøy and Røst islands, are treated separately below. Several interesting species breed in the inner Lofoten islands as well, including Scaup, Slavonian Grebe and about 100 pairs of White-tailed Eagle. Black-tailed Godwit of the northern race *islandica* used to breed here as well, but it seems to have disappeared in recent years. It might reappear, though, so be on the lookout.

In winter, substantial numbers of sea ducks, divers and gulls gather along the coasts of Lofoten, particularly on the northern coasts facing the Norwegian Sea and in the narrow sounds (*straumene*) in between the islands. About 14,000 King Eiders, 100 White-billed Divers and several Great Northern Divers winter in these waters, in addition to those wintering near Værøy and Røst (see below). The southern coast facing the deeper Vestfjorden strait, may in contrast appear rather devoid of birds in winter, except at active fishery harbours such as Ballstad, Mortsund and Henningsvær. Along the northern coast are also several good sites for enjoying seabird passage, particularly Eggum, Kvalnes and Laukvik. The many brackish lagoons (*pollene*) in Lofoten have wintering swans and ducks, and can be teeming with birds in spring, before the ice thaws on the freshwater lakes. Several vagrants have been found in the inner sections of Lofoten, although such rare migrants are more often encountered on the outer two islands, Værøy and Røst.

The contrasts in Lofoten are spectacular, from the shallow fjords to the towering mountains. If you take the E10 and drive the back road around Leknes, you will be taken through areas with pastures and sheltered wetlands, such as here at Saltisen. *Photo: Bjørn Olav Tveit.*

295

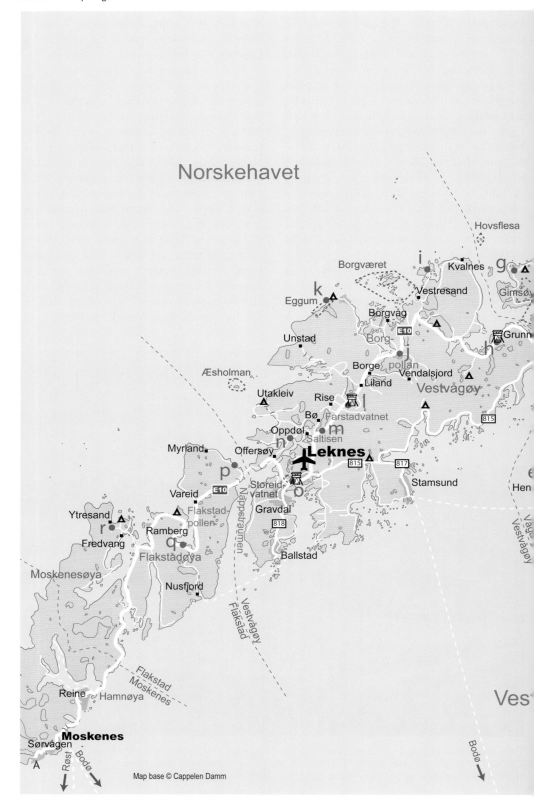

Norskehavet

Hovsflesa

Borgværet

i

Kvalnes

g

k

Eggum

Vestresand

Gimsø

Borgvåg

Borg-

Grunn

Unstad

h

Æsholman

Borge

pollan

Vendalsjord

Utakleiv

Rise

Liland

Vestvågøy

Bø

Farstadvatnet

815

Myrland

Oppdøl

m

Offersøy

Saltisen

n

p

Leknes

Vareid

Storeid-

vatnet

Gravdal

815

817

Stamsund

Hen

Ytresand

r

Flakstad-

pollen

818

Ramberg

Fredvang

q

Ballstad

Flakstadøya

Moskenesøya

Nusfjord

Vestvågøy

Flakstad

Flakstad

Moskenes

Reine

Hamnøya

Ves

Moskenes

Sørvagen

A

Røst

Bodø

Bodø

Hadseløya
Sortland
Melbu

Hadselfjorden

Seløya

Hadsel
Lågan

Grunnfør
/Lågen

Hadselsand

Fiskebøl

Evenes
Narvik

Straumnes

Morfjorden

aukvik

ikøyene

E10

Sildpollen

Austvågøy

Svolvær

Kabelvåg

Lille Molla

Skrova

Fuglbergøya
/Nautøya

Skutvik

N

5 km

Best season: All year can be good, except for the darkness of the polar night in December. Spring and early summer is probably the best period for general birdwatching. Most King Eiders arrive in mid-winter and leave again in spring.

Directions: *By air:* A convenient way to visit Lofoten as a birdwatcher, is by plane to Harstad/Narvik airport, Evenes, and from there drive a rental car westbound along the E10. There are also flight connections from Bodø to Leknes and Svolvær airports in Lofoten. The network of buses within Lofoten is fairly well developed.
By boat: Hurtigruten – The Norwegian Coastal Express – lands in the towns of Stamsund and Svolvær.
By car: From the E6 in Bjerkvik north of Narvik, exit onto the E10 signposted to *Å i Lofoten.* If you arrive via Vesterålen, take the ferry from Melbu to Fiskebøl.

Tactics: Rather great distances make a car invaluable. The network of cycle paths is also good throughout Lofoten, and many tourists prefer to explore the archipelago on bicycle. If you arrive by car from the east along the E10, the sites will appear in the following order (see the map):
Austvågøy island. Along the E10, after the Sløverfjord tunnel, exit right onto Fv82 to Fiskebøl (the ferry landing to and from Vesterålen, which see). Just before the ferry landing, exit towards Laukvik and continue 19 km to **a.** Hadselsand. Pull down the side road signposted *Kirke* (church), leading to the seashore where shorebirds and gulls gather at low tide, often including Sanderling in June. Gannets can often be seen over the sea in summer – they breed at Ulvøyholmen islet across the fjord. Hadselsand

Lofoten archipelago: a. Hadselsand. **b.** Grunnfør and Lågen; **c.** Laukvik; **d.** Lake Sandslettvatnet; **e.** Henningsvær; **f.** Gimsøystraumen sound; **g.** Gimsøy island; **h.** Lake Gårdsvatnet; **i.** Sandøya island; **j.** Borgpollan lakes; **k.** Eggum headland; **l.** Lake Skjerpvatnet; **m.** Lake Farstadvatnet; **n.** Holandsveien road; **o.** Lake Storeidvatnet, tower hide; **p.** Nappstraumen sound; **q.** Lake Flakstadpollen; **r.** Fredvang.

Oslo

Interior

Skagerrak

Western

Central

Northern

Finnmark

Svalbard

is also a nice place for seabirds in winter. Another 3 km further west, stop at the roadside pond **b.** Lågen, supporting breeding Red-throated Diver, Tufted Duck, Teal and other wetland species. They can be appreciated at close range if you stay in the vehicle. The pastures at Grunnfør on the opposite side of the road are regularly exploited by Pink-footed Geese in May.

Keep going west to **c.** Laukvik and drive down to the harbour. Make a left to view the lagoon Osen, often supporting *Calidris* shorebirds and Black-tailed Godwit in summer, or keep right onto the breakwater. The little lighthouse here makes a good viewpoint for seabirds. Continue on the main road along Nordpollen, the lagoon on your right after 3 km. The roadside **d.** Lake Sandslettvatnet is reached 7 km from Laukvik. This is one of the best lakes in Lofoten, supporting Slavonian Grebe, Red-throated Diver, Tufted Duck, Sedge Warbler and Bluethroat.

e. Henningsvær harbour, one of the most reliable sites for King Eider and Iceland Gull i January–March, is signposted to from E10.

Gimsøya island. From the E10, exit north onto the road signposted to Barstrand/Vinje. Stop to check **f.** Gimsøystraumen sound for ducks, divers, gulls and more. Puffin is often seen here in summer, and sometimes King Eider, Great Northern Diver, White-billed Diver or Brünnich's Guillemot. White-tailed Eagle and Peregrine Falcon are regular. Continue north and check the wetlands, lakes and grazing pastures of **g.** Gimsøya island, most of them can be scanned from the road. Gimsøya is an important resting area for Pink-footed Goose in May and the Gimsøya marshes support breeding Willow Ptarmigan, Black-throated Diver, Red-throated Diver, Arctic Skua, Whimbrel, Red-necked Phalarope and Bluethroat. Whooper Swan and Temminck's Stint have also been found breeding here. Lapland Bunting, uncommon other places in Lofoten, is regular in the drier areas.

Vestvågøy island. Further west along the E10, in the village of Grunnstad, is **h.** Lake Gårdsvatnet, a nutrient-rich lake with a variety of ducks, Black-throated Diver, Bluethroat, Sedge Warbler and more. Ruff can be found displaying here in May. Less charming, but perhaps still interesting, is the fact that gulls from the nearby Haugen landfill come here to bathe. A tower hide is put up in the

Sandøya island is an important staging ground for geese. Temminck's Stint nests here in some years. Just in front of the mountains, the large Høynesvøda tidal flats can be seen. *Photo: Bjørn Olav Tveit.*

Scaup nest in Indre Borgpollen. The two brackish lakes Borgpollen are also important resting and wintering grounds for Whooper Swan, ducks, divers and Slavonian Grebe. *Photo: Bjørn Fuldseth.*

southwestern corner of Lake Grunnstadvatnet but it is equally nice to scan the lake with a telescope from the car park.

The next sites coming up to the west are **i.** Sandøya and Høynesvøda, a grassy island and a vast tidal area, respectively. Exit the E10 north signposted to Vestresand and at the upcoming T-junction, keep heading towards Vestresand. When in Vestresand, follow the road signposted *Kirkegård 400 m*. This road leads past the cemetery and onto a causeway leading across Høynesvøda tidal area. Stop to look for shorebirds, geese, ducks (often including Shelduck) and such. By the end of the causeway is Sandøya island, attractive to geese during migration. Explore the island from within the car when geese are present, in order not to spook them. Temminck's Stint has bred several times here. Drive back to the mentioned T-junction and head towards Kvalnes, a road leading to the east side of Høynesvøda tidal area and out to Kvalnes headland, where nice grazing pastures line the shore. Here, you can also seawatch from the road and you have a view to the small colonies of Gannet and Cormorant on Kvalnesflesa islet, 1.5 km to the northeast.

E10 continues west between the two Borgpollan **j.** Indre and Ytre Borgpollan. These brackish lagoons often support divers, Slavonian Grebe (sometimes in the dozens during September) and ducks, including Scaup, which breeds here with up to 28 pairs. The two Borgpollan lagoons are particularly good in early spring. Several Whooper Swans winter in Ytre Borgpollen,

arriving in October. Exit onto the side road signposted to Vendalsjord for closer inspection of Indre Borgpoll lagoon, and exit to Borgvåg for exploring the eastern side of Ytre Borgpoll. Further along E10, exit onto the first of two exits signposted to **k.** Eggum headland. Note the little pond Sevtjørna between the two exits, a reliable site for Red-throated Diver. The road to Eggum leads along the western side of Ytre Borgpoll. Eggum headland has beautiful white sandy beaches often attracting gulls, terns and a few shorebirds. Red-throated Pipit has been suspected to breed here. Eggum is perhaps the best seawatching site in Lofoten, often with tens of White-billed Divers and Pomarine Skuas on good days in May and in September–October. In October, tens of Glaucous Gulls may pass as well. Eggum is pretty reliable for Gannet and Great Skua in summer. The most commonly used viewpoint at Eggum is at Borgen, the tourist service building at the end of the road, an architectural gem containing public restrooms and an activity room. Back on the E10 heading west, stop roadside by Lake Ostadvatnet on your right, just past the exit to Unstad/Tangstad. The little Skjerpholmen islet in the lake host a colony of gulls, including a few pairs of Lesser Black-backed Gulls of unknown heritage, possibly *intermedius* or hybrid *graellsii x fuscus*. If you do not see the gulls here, look for them in the surrounding fields.

The small lake **l.** Skjerpvatnet on the opposite side of E10, a bit further west, is equipped with a tower hide. Enter a gravel road across from the exit to Ostad. Park in the designated spot and walk along the path to the tower hide 200 m through the forest. Here is often a nice selection of wetland species, and Sedge Warbler is often found in the ditch along the gravel road.

m. Lake Farstadvatnet. Continue along the E10 and turn right onto the second of two gravel roads signposted to Rise. Turn left just after the bridge and drive down to Lake Farstadvatnet. This used to be one of the most reliable sites in Lofoten for Black-tailed Godwit, and they can be hard to spot in the dense grass lining the lake. There are

Oslo Interior Skagerrak Western Central **Northern** Finnmark Svalbard

Værøy island, the next outermost island community in Lofoten, seen from the helicopter that effectively connects the island with the mainland. You can also arrive from Bodø by ferry. *Photo: Bjørn Olav Tveit.*

many pairs of Slavonian Grebe here. To view the southern section of the lake, continue along the E10 a bit further and stop in a bus stop lay-by just before the exit to Bø.

Leknes town. Further on, you must choose between two roads past Leknes (or drive them both in turn), either continuing along E10 or along the back road **n.** *Holandsveien*. The latter is reached by exiting to Bø and then turning left signposted to Uttakleiv before turning left again towards Offersøy. This route leads through a beautiful landscape with meadows, small ponds and lagoons and even past a conifer plantation with a Grey Heron rookery, before you reach E10 again just west of Leknes.

When driving E10 through Leknes, you will soon pass the 1.5 km long Leknesfjæra tidal inlet and the Haldsvågen lagoon, particularly good at hight tide, often supporting resting Spotted Redshank in late summer. It is difficult to find safe places to stop along the heavily trafficked E10, so you will need to park at one end and walk to explore these tidal areas thoroughly. Take a short detour towards Stamsund town to reach Fyglefjæra inlet, having much the same qualities as Leknesfjæra/Haldsvågen. From along E10 past Leknesfjæra, turn right onto the road signposted to Storeidet. Keep left at the upcoming T-junction, park in the designated area on your right, and walk over to

the tower hide overlooking the small but very bird-rich **o.** Lake Storeidvatnet. This shallow lake is lined with lush water vegetation, another traditional Black-tailed Godwit site, although now probably abandoned by this species. A variety of ducks are found here as well, some years including Shoveler. Tens of Slavonian Grebes can be crammed into this little lake, still leaving room for several pairs of Coot. On your way further west, make a quick stop to check the pond in the middle of the racetrack on the south side of E10.

Flakstadøya island. Just after the Nappstraumen tunnel leading to Flakstadøya island, exit to Myrland. From this road you can view **p.** Nappstraumen sound with many of the same qualities as Gimsøystraumen sound (above), besides being a stronghold for King Eider in winter and early spring. Further west, E10 passes goose-friendly pastures by Vareid and the shallow fjord **q.** Flakstadpollen, often supporting ducks, shorebirds and both giant divers. You should also make a detour to the farmlands at **r.** Fredvang, clearly signposted from E10, followed by signs to Ytresand. Here are very interesting, cultivated fields and meadows. Some of them are dry and steppe-like, others are moist, particularly those lining the meanders of the Sandelva river that cuts through the area. Lesser Black-backed Gull,

Arctic Tern, Short-eared Owl and Whimbrel are often found here, besides resting geese.

Before you reach the ferry landing at Moskenes along E10 (with connections to Bodø, Værøy and Røst), note the roadside Kittiwake colony on the little <u>Hamnøya</u> island, just before Reine village.

20. VÆRØY ISLAND

Nordland County

GPS: 67.65486° N 12.72538° E (helipad Sørland)

Notable species: King Eider (winter), Great Northern and White-billed Diver (winter), Cormorant, Shag, Fulmar, White-tailed Eagle, Golden Eagle, Peregrine Falcon, Arctic Skua, Great Skua, Glaucous and Iceland Gull (winter), Arctic Tern, Guillemot, Razorbill, Black Guillemot and Puffin; migrants; vagrants, including regularly occurring Olive-backed Pipit, Pechora Pipit, Red-flanked Bluetail and Yellow-browed Warbler.

Description: Værøy is a high and isolated island in Lofoten, 15 km from Moskenes and 20 km from Røst. It is actually an archipelago, the main island being almost 10 km long at the most and dominated by a massive mountain rising 450 metres above the sea. Several small islands and skerries are found to the northeast. Among them is also a larger, uninhabited island, Mosken, 2 km across. Large colonies of seabirds nest in the scree slopes and cliff walls at Måstad on the southwestern headland. The cliffs here are protected as the Måstadfjellet nature reserve, with a traffic ban in the period April 15 to July 31.

For bird enthusiasts, Værøy was for a long time in the shadow of neighbouring Røst, partly because the seabird colonies at Værøy are smaller, but also partly because few rare migrants had been encountered here. From 2010, however, birdwatchers' investment in Værøy has shown that it is a rarity magnet that certainly does not

stand behind Røst or even Utsira in that respect. Particularly striking are the eyebrow raising occurrences of Yellow-browed Warbler in the autumn, with day-counts of up to 160 individuals, the highest in Europe. This Asian species tends to occur a couple of weeks earlier in autumn in northern Norway compared to southern Norway. Olive-backed Pipit and Red-flanked Bluetail have also shown striking regularity at Værøy. Mugimaki Flycatcher, Baltimore Oriole and Common Yellowthroat are also on the island's constantly growing list of rare birds. Spectacularly, a flock of seven Blue-cheeked Bee-eaters has been documented at Værøy.

On the south and east side of the island you will find lush *Locustella*-friendly pastures and fields, and in the town of Sørland, where most of the

White-tailed Eagles were hunted on Værøy in the old days. The picture shows a restored trapping site. A sheep's jaw is laid out to show where a carcass was placed to attract the eagles. The trapper sat in a room inside the pile and waited. When the eagle landed, he grabbed hold of the eagle's feet. White-tailed Eagle was protected in Norway from 1968. *Photo: Bjørn Olav Tveit.*

Oslo Interior Skagerrak Western Central **Northern** Finnmark Svalbard

Puffin breeds in large colonies in burrows in rocky and grassy slopes facing the outer sea. Each pair lays only one egg per year. Ringing on Røst has shown that they can be more than 40 years old. *Photo: Tomas Aarvak.*

680 inhabitants live, you will find attractive gardens for resting sparrows and other things. Habitats similar to those in Sørland are found on the north side of the island, albeit in a much smaller scale. Here, around the old airstrip, are also a couple of ponds worth checking for ducks and shorebirds. Breeding birds include quality species such as Pintail, Garganey and Gadwall. The people of Værøy are grand producers of stockfish and have a very active harbour, attracting a large number of gulls. Glaucous and Iceland Gull are regular among the commoner species, and several of each can be found in winter and early spring. About 2000 King Eiders winter in the waters surrounding the island together with quite a few Great Northern and White-billed Divers.

The community of Værøy has been based upon reaping whatever nature had to offer, including seabirds. A now abandoned village, Måstad, with 150 inhabitants was localized near the cliffs. Large numbers of Puffins were hunted and conserved in salt. The Værøy islanders even had their own breed of dog, locally called the Måstad dog. It is a Spitz type dog with several peculiar physical features, including extra toes and extremely flexible joints, specifically developed for digging Puffins out from their nesting burrows. The people of Værøy also trapped White-tailed Eagles by hiding in small cavities made from rock in the scree slopes in the mountain. When the eagles landed to feed on the carcass laid out in front of the cavity, the hunter would grab the bird with his bare hands. Of course, hunting birds of prey has long been banned in Norway, but some of these old trapping sites have been restored and can be seen on Værøy today.

Best season: Værøy can offer great birdwatching all year, except perhaps for the darkest month, December. The most notable periods for migrants and vagrants are late May–June, and mid-August–October.

Directions: Værøy is reached via the helicopter route from Bodø to Sørland, or by the car ferry from Bodø, Røst or Moskenes which offers a chance to watch seabirds including Storm Petrel.

Tactics: The seabird colonies can be appreciated either from guided boat trips from Sørland or by hiking to the cliffs from the north side of the island via the ghost town of Måstad. For general birdwatching, you should have access to a car or bicycle if you have ambitions to cover all the most important areas on the island every day in the migration periods. The lushest fields and gardens are found in Sørland. In the heart of the valley here, there is also a very exciting forest plantation. However, many of the rare migrants have been seen in gardens and fields of Nordland, where the terrain is easier to monitor.

21. RØST ISLANDS

Nordland County

GPS: 67.50535° N 12.07200° E (ferry landing)

Notable species: Barnacle Goose (Apr–May, Sep–Oct), Pintail, Shoveler, King Eider (winter), Great Northern Diver and White-billed Diver (winter), Cormorant, Shag, Sooty Shearwater and Manx Shearwater (Aug–Sep), Fulmar, Storm Petrel (Aug–Oct), Leach's Petrel (few), White-tailed Eagle, Peregrine Falcon, Gyrfalcon, Spotted Redshank, Knot, Ruff, Red-necked Phalarope, Arctic Skua, Great Skua, Glaucous Gull and Iceland Gull (winter), Arctic Tern, Guillemot, Razorbill, Black Guillemot, Puffin; Olive-backed Pipit (Sep–Oct), Yellow-browed Warbler (Sep) and Lapland Bunting; migrants, rarities.

Description: The outermost group of islands in the Lofoten archipelago is Røst, 100 km from Bodø and more than 20 km from the closest major island, Værøy. Røst consists of about 1000 rather low islands and skerries besides five prominent cliff islands, lying like a row of pearls on a string towards the Norwegian Sea. The vegetation on the islands is sparse, mainly consisting of heather and grass. The flat, 4 km² main island Røstlandet is where most of the 460 inhabitants live, and here are also gardens, cultivated fields and large wetlands.

Røst islands have a location and a selection of habitats that make them among Norway's best birdwatching sites. Røstlandet island in the foreground, with the tall outermost islands in the distance. *Photo: Terje Kolaas.*

Oslo
Interior
Skagerrak
Western
Central
Northern
Finnmark
Svalbard

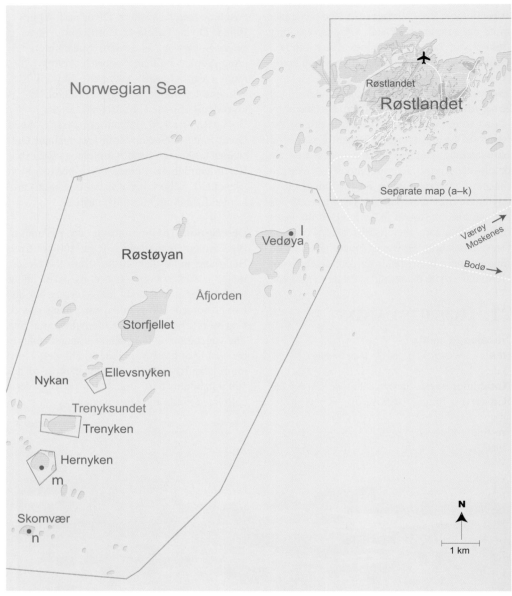

Røst islands: a–k. Røstlandet island (separate map p. 306); **l.** Vedøya island, seabird cliff without traffic ban during breeding season; **m.** Hernyken island, ornithology field station; **n.** Skomvær lighthouse.

Røst has long been internationally noted for its large colonies of seabirds, primarily Puffin but also Common Eider, Fulmar, Cormorant, Shag, Guillemot, Razorbill, Black Guillemot, Kittiwake, Great Skua, Arctic Tern and Grey Heron. The latter normally breeds in trees, but on Røst it nests on barren skerries just like a proper seabird. Great Skua is represented with 2–3 breeding pairs, and Røst is the single most important breeding site in Norway for Storm Petrel. These birds can often be seen at dusk close to their breeding islands in late summer and autumn, a unique experience attracting birdwatchers from far afield. Exactly how many pairs of these nocturnal and highly pelagic species breed on Røst is not known, but probably several hundred pairs. An estimated population of about 100 pairs of Leach's Petrel used to breed as well, but in recent years this species appears to have almost completely vanished. The large colonies of cliff-nesting seabirds can be appreciated through regularly conducted boat trips from the main island. Seabird research is conducted from a base camp on Hernyken island.

Unfortunately, much of the attention Røst has received lately has concerned the catastrophic decline in many of its seabird populations over the last decades. This is partially because of excessive fishing by humans, but also a result of climatic changes, causing the food chains in the ocean to collapse. In 1979, the population of Puffin on Røst counted 1.5 million pairs, whereas it has now plummeted down to about 300,000 pairs. The populations of Fulmar and Kittiwake have also shown substantial declines. The 30,000 pairs of Kittiwake that nested on Vedøya island are now gone, and the original 12,000 pairs of Guillemots were suddenly reduced to only 2 % of this. The few pairs of terns that nested here in the past have long since disappeared. The few pairs of Brünnich's Guillemot formerly breeding here are long gone. On the brighter side, the population of Puffin is slowly recovering, Cormorant has re-established itself as a breeding bird, and the number of summering White-tailed Eagles has increased noticeably. Ironically, the latter is probably one reason why some cliff-nesting seabird populations have a hard time recovering. The wetlands of Røstlandet have several breeding ducks, gulls and shorebirds. Among the regular breeders are Arctic Skua, Whimbrel, Dunlin, Red-necked Phalarope and, in some years, Pintail, Shoveler and Gadwall. A Kittiwake colony containing a few hundred pairs breed in the Kårøya sound by the harbour, partly on a specially built Kittiwake hotel. Curiously, in the 1930s, three species of penguin was released at Røst in the hope that they would establish sustainable populations there. Fortunately, they did not succeed.

In addition to being a prominent site for breeding birds, Røst has also proved to be one of the most exciting migration sites in the country. As opposed to most other offshore islands in Norway, it offers vast areas suitable for resting shorebirds. For instance, day counts of up to 3750 Knots, 100 Spotted Redshanks, 250 Curlew Sandpipers and 40 Red-necked Phalaropes have been encountered in August. Ruff, Dunlin and Little Stint can be numerous on migration as well. Several thousand Barnacle Geese may rest here in spring and autumn. The numbers of resting birds vary from one year to the next, being highest in years of much rain through the summer, creating pools of water in the terrain. Passerines and other land birds can occur in good numbers and with great diversity. They may do so particularly after a change in the weather, preferably to heavy cloud-cover accompanied by a calm breeze from a wide sector between southwest and northeast. These birds are mostly attracted to the fields and gardens of Røstlandet, but the other islands as well, particularly the outermost one, Skomvær island, have great potential as well. Migrating

Storm Petrels spend their lives on the open sea and only come ashore during the breeding season, and then under the cover of darkness. Because of the midnight sun in the North, it postpones nesting until the nights are dark and it can crawl into the nest holes without risking attack from gulls and other diurnal predators. *Photo: Tomas Aarvak.*

seabirds can pass by in good numbers, especially in strong winds from the northwest. A seawatching hut is available on the north side of Røstlandet.

Røst has proven to be spectacularly attractive to rare migrants, now challenging Utsira in being the hottest rarity magnet in Norway. The list of species recorded at Røst includes mega rarities such as Swinhoe's Storm Petrel, Northern Harrier, Oriental Plover, Hudsonian Godwit, Stilt Sandpiper, Baird's Sandpiper, Franklin's Gull, Pied Wheatear, Pallas's Grashopper Warbler, Eastern Bonelli's Warbler, Two-barred Greenish Warbler, Black-headed Bunting and White-throated Sparrow, to name but a few. There are also many records each of Pacific and American Golden Plovers, Pechora Pipit and Citrine Wagtail. Sabine's Gull and

Røstlandet island: a. Klubben headland, tidal area; **b.** Langneset headland, grasslands and pebble beach; **c.** Øyran islets and tidal area; **d.** Røstlandet nature reserve; **e.** water works viewpoint; **f.** grassy fields; **g.** Sandbergan Seawatch Hut; **h.** Nordneset headland, gardens and pastures; **i.** Kalvøya stockfish racks; **j.** Ystneset stockfish racks; **k.** Grimsøya gardens. For an overview of all the Røst islands, see the previous page.

Grey Phalarope are seen quite often, the latter with an all-time high autumn season total of up to 14 individuals, with nine in one day. Pectoral Sandpipers are also quite regular, and Røst is the Norwegian record holder with up to five individuals in one flock.

Compared to Utsira, Røst is significantly more isolated out in the sea and thus has a greater 'island effect', which should indicate a greater potential for vagrants. Røst's myriad of islands and reefs reduce the concentration of birds somewhat, but here, in return, there is less vegetation for the birds to hide. Røst also has more attractive habitats for shorebirds and other wetland birds than Utsira, which contributes to Røst almost becoming a year-round birdwatching location. Røstlandet also has more practical facilities than Utsira, including an airport.

In winter, both Great Northern and White-billed Divers are regular, King Eider is numerous, and up to 30 Glaucous Gulls and 10 Iceland Gulls can be found in the harbour at Røstlandet.

Other wildlife: Harbour and grey seals are common year-round. Other regularly occurring marine mammals include several species of whale. Brown rat is the only wild-living terrestrial mammal on the archipelago.

Best season: Most breeding seabirds arrive in March/April, except for Storm Petrel (and perhaps still Leach's Petrel), which come ashore to breed in late summer and autumn. Migration can be particularly good from March to the end of October. The best period for rarities is June, September and the first half of October. All winter can be good, although the polar night causes almost complete darkness from early December to early January. Even in this period, though, it is possible to watch birds for a couple of hours at midday on clear days.

Directions: Røst has an airport with daily flights to Bodø and Leknes. Røst also has a car ferry connection to Bodø, and in some days (every day through the summer) to the rest of Lofoten. The Bodø–Røst ferry takes 4 hrs one way, with a couple more hours if making stops at other destinations along the way.

Tactics: A trip to the bird cliffs is a highly memorable experience. At dusk in late summer and early autumn, or even on foggy days, a boat trip between the islands gives the chance of seeing Storm Petrel. On organized boat trips, you can choose to go ashore on l. Vedøya, where there are no visitor restrictions in the breeding period. Here you can experience Puffins and the cliff-breeding seabirds, or you can stay on-board to continue all the way to n. Skomvær lighthouse. The latter will be the most varied alternative, in terms of more species seen, with chances even in mid-summer of encountering King Eider, Glaucous Gull, Great Skua and dozens of White-tailed Eagles. In calm seas the boat travels close to the seabird cliffs and past some impressive caves at Vedøya and Trenyken islands. Skomvær island might offer surprises in terms of resting migrants. In summer, the boat often continues further out to sea to do some fishing, after the people who wishes to do so have gone ashore on Skomvær. During these fishing trips, the chances of encountering Storm Petrels are particularly good, besides the chances of getting to see Gannet, skuas, shearwaters, and perhaps Leach's Petrel, up close. For maximizing the effect, throw some finely chopped fish liver on the sea and wait 30 minutes or more. There are several commercial options for boat trips to the outer islands.

Røstlandet island is not bigger than that it may be explored on foot. A bicycle or a car does come in handy, however. Bikes can be rented from some of the accommodation providers. Røstlandet has the best habitats for resting shorebirds. They can turn up anywhere, but particularly good are the tidal areas stretching west from the airport. From the terminal entrance, walk north along the shore outside the airfield fence, around a. Klubben and b. Langneset headlands, and continue as far as you like along c. Øyrene. A telescope (and a water bottle) is well worth bringing along. The marshlands in d. Røstlandet nature reserve make up a mosaic of brackish ponds, perfect for resting geese,

Barnacle Geese depend on the staging grounds in Vesterålen archipealago to gather strength for the long journey to Svalbard in spring as do Pink-footed Geese. *Photo: Kjetil Schjølberg.*

ducks and more. Be particularly considerate when visiting the area in early summer, because it is an important breeding area. An easily accessible viewpoint to obtain a general overview of these wetlands, is **e.** the water works in Klakkveien road. **f.** Kvalvågsletta fields between Kvalvåg and Sankthanshågen are particularly good for plovers and geese. Seawatching is best conducted from **g.** Sandbergan Seawatch Hut.

Most gardens and thereby bush-dwelling passerines are found along the southern road, between **h.** Nordneset headland and **i.** Kalvøya island, including the many side roads, and on **k.** Grimsøya island. There are also several interesting pastures, shallow inlets and water pools in these areas. The residents of Røst have been accustomed to having their gardens scrutinized by birdwatchers, but please keep to the roads and beaten paths, and respect private property. The total lack of forest on Røst makes the many stockfish racks a surrogate for woodland species gone astray – particularly interesting are those at **i.** Kalvøya island and **j.** Ystneset headland.

22. VESTERÅLEN ARCHIPELAGO

Nordland County

GPS: 68.68804° N 14.47160° E (parking Straume), 68.92021° N 15.21863° E (Sørvågen, Bussingvika)

Notable species: Whooper Swan, Pink-footed Goose, Barnacle Goose (May), Scaup, King Eider (winter), Red-throated Diver, Black-throated Diver, White-billed Diver, Slavonian Grebe, Cormorant, Shag, Gannet, White-tailed Eagle, Gyrfalcon, Peregrine Falcon, Black-tailed Godwit *islandica*, Purple Sandpiper, Ruff, Red-necked Phalarope, Razorbill, Guillemot, Puffin, Bluethroat, Sedge Warbler and Snow Bunting.

Description: Vesterålen does not stand behind Lofoten when it comes to spectacular nature and bird life. However, Vesterålen is less frequently visited by birdwatchers, and is therefore less well-known and with fewer rarity records. Vesterålen also includes the marvellous Andøya island, which is presented here separately (next site).

Vesterålen archipelago is mountainous and picturesque and has a number of wetland and farmland areas strategically facing the Norwegian Sea. Wetlands such as Straume, Lake Selnesvatnet and Grunnfjorden inlet are important to breeding and resting birds alike, the former representing one of the traditional breeding grounds for Black-tailed Godwit in northern Norway. Slavonian Grebe, Whooper Swan, Tufted Duck, Scaup, Ruff, Black-headed Gull and Sedge Warbler breed there as well. Grunnfjorden nature reserve is an important resting ground for wetland birds during migration and is the northernmost regular wintering site for Whooper Swan. Red-throated Diver, Black-throated Diver, Greylag Goose and several species of ducks and shorebirds breed, probably including Black-tailed Godwit. Vesterålen in general is a core breeding area for White-tailed Eagle. Along the coast are also seabird cliffs such as Nykan and Anda, and colonies of Gannet. The Gannet colonies of this region have moved around quite a bit over the years, and they may well move again in the future. If you want to see seabirds out at sea, the whale safaris conducted from Stø can be very rewarding both for birds and marine mammals. The pastures of Vesterålen thaw early in spring, becoming attractive to the Svalbard populations of both Pink-footed and Barnacle Geese in late April and May. Unfortunately, local farmers have started organized goose chasing to prevent the geese eating their crops. This is, of course, illegal and may cause the Svalbard population of Pink-footed Geese to change their migration habits. King Eider, White-billed Diver and Glaucous Gull winter along the coast in good numbers.

Other wildlife: The area is rich in marine mammals all year, many of which can be appreciated on the whale safaris.

Best season: All year. Black-tailed Godwits arrive in late April and leave again in August–September. Large flocks of Pink-footed and Barnacle Geese arrive around early May and leave by the end of the month. Whale safaris are conducted from June to August.

Directions: The most convenient way to visit Vesterålen is by plane to Harstad/Narvik airport, Evenes, and from there drive a rental car westbound along E10 well past Lødingen before exiting onto Rv85 signposted to Sortland. Stokmarknes airport Skagen has connections to Bodø and Tromsø, and the Coastal Express (Hurtigruten) lands in Stokmarknes and Sortland. From Lofoten, Vesterålen is reached by car ferry from Fiskebøl to Melbu, or by driving E10 east to Rv85.

Tactics: Vesterålen is most effectively explored by car, although Stormyra and Grunnfjorden should be investigated more closely on foot. If you are arriving from Lofoten via Melbu, follow Fv82 north past Stokmarknes to Sortland. Stop at **a.** Hakaneset just before Stokmarknes Airport, Skagen. There are good parking options at the exit to Sandnes on the north side of the Hadsel bridge, signposted to Sandnes, and from here you get a nice overview of the wetland area on Hakaneset. Several pairs of Lapwing, Ruff and other wetland birds breed here. On the fields and marshes on the inside, Whimbrel, Arctic Skua, Dunlin, Golden Plover, Red-throated Diver and more nest. The road runs along the Hadselfjord and Sortland sound, a stretch with a number of fine salt meadows and ponds, as well as some tidal areas. It is particularly nice at **b.** Rognan, southwest of Sortland town centre. For several years, hundreds of pairs of Mew Gulls have nested here, and Gadwall has nested as well. Other dabbling ducks such as Shoveler are regularly seen, and several other species of duck breed, in addition to Turnstone on the grass embankments towards the sound.

In Sortland, follow the Fv820 signposted to Bø, later Straume. From the roadside you should check out all the good-looking places you pass, perhaps especially **c.** Vikosen inlet 9 km after Sortland and **d.** Lake Selnesvatnet 2.5 km west of the junction with Fv821, and **e.** Lake Kringelvatnet in Bø, which has a large Black-headed Gull colony, several pairs of Slavonian Grebe and nesting Whooper Swan.

Oslo

Interior

Skagerrak

Western

Central

Northern

Finnmark

Svalbard

Vesterålen archipelago: **a.** Hakaneset; **b.** Rognan; **c.** Vikosen inlet; **d.** Lake Selnesvatnet; **e.** Lake Kringelvatnet; **f.** Straume; **g.** Nykvåg, viewpoint to the sea and to the seabird colony at Nykan; **h.** Hovden seawatch; **i.** Alsvåg, Bussingvika inlet south; **j.** Gisløy, Bussingvika north; **k.** Stø whale watching harbour.

In **f.** Straume, drive in behind the gas station on the left side of the road, and walk over to Lake Saltvatnet with the peculiar looking tower hide. The hide is normally locked, but a key can be borrowed on request at the petrol station. This used to be northern Norway's most important nesting area for Black-tailed Godwit, but these have probably not nested

here the last several years. Slavonian Grebe, Whooper Swan, Tufted Duck, Scaup, Ruff and Sedge Warbler currently nest here, and during migration a number of other species appear, especially ducks, divers and shorebirds, still often including Black-tailed Godwit.

If you drive back along Fv820, after 7 km you can turn left towards Eide church and continue on to **g.** Nykvåg village. From here you have a view of a seabird cliff, and both here and at **h.** Hovden a little further north you have a view of the ocean for observing passing seabirds. Along this road you also drive along the bird-friendly beach and coastal areas in the Nyke–Tussen nature reserve.

Further back on Fv820, 10 km from Sortland, turn left on Fv821 signposted to Myre. Turn right towards Asvåg after 6.5 km. This road leads along the small Lifjorden inlet and later out into the Gavlefjorden, which is an extension of the Sortland sound, and has many grazing areas and small coves that should be explored. In Alsvåg, make a right signposted to Meløy. The road to Meløya island soon crosses a fine and shallow lagoon. Continue all the way to the cemetery at the far end of Meløya, as there is also a nice tidal inlet here. Back in Alsvåg, continue to the right for a couple of km in the direction of Myre. Be sure to check Lake Alsvågvatnet via the slip road on the left and then Lake

Sørvågvatnet on the south (left) side of the road, before turning onto the gravel road to the right at the sign for *Båtbyggeri* (boat constructor). This road leads down to **i.** Sørvågen village with a view of the beautiful inner shore of Bussingvika bay at the entrance to the Grunnfjorden nature reserve. This is a very important resting place for ducks and geese during migration, and is among the country's northernmost permanent wintering sites for Whooper Swan. It is also an important moulting area for the swans, with up to 100 individuals in midsummer. Breeding here is Arctic Skua, both Black- and Red-throated Divers, Greylag Goose, several species of ducks and shorebirds, including Whimbrel, Red-necked Phalarope, Ruff and, at least in the recent past, Black-tailed Godwit.

Continue to Myre village and turn right onto Fv821 leading through the village, and then keep right, signposted to Stø. On the right, you catch a glimpse between the birch

Grunnfjorden nature reserve in Øksnes has some inviting saltmarsh areas, like here in Bussingvika inlet near Alsvåg. *Photo: Bjørn Olav Tveit.*

Whale watching is an experience you absolutely must have when you are in Vesterålen. On these boat trips, which sail from both Stø and Andenes, you get the opportunity to see orca (depicted) and other marine mammals, in addition to the chance of exciting seabirds. *Photo: John Stenersen.*

trees of the nutrient-poor Stormyra peat bog, which resembles a small copy of the huge Skogvollmyran marshes on Andøya island (which see). At the crossing at Strengelvåg, the road to the left takes you to **l.** Stø, where whale safaris are arranged, while the road signposted to Gisløy takes you over a patch of Stormyra peat bog, from where it is possible to walk out onto the marsh. Wetland restorations have been carried out here, between **j.** Lake Hattavatnet and Lake Grunnvatnet, which has immediately led to an increase in nesting shorebirds and ducks. A little further along the road you get a new view of Bussingvika bay and a bit of Grunnfjorden nature reserve, on the right side of the road. On the opposite side lies Strengelvågen inlet with a fantastically beautiful tidal area in the innermost part. After the road has crossed the narrow strait at Valen, after 1 km you will see Husvågen bay on your left. This is a shallow bay surrounded by northern Norway's largest saline marsh. Here lies **k.** Husjordmyra marsh, where several of the most sought-after shorebird species nest in a small area, and all of them can be enjoyed from the road.

23. ANDØYA ISLAND

Nordland County

GPS: 69.16783° N 15.97063° E (Skogvollmyran marshes), 69.30948° N 16.09749° E (Lake Kleivvatnet, Andenes)

Notable species: Whooper Swan, Barnacle Goose (April–May), King Eider (winter), Common Scoter, Willow Ptarmigan, Black Grouse, Black-throated Diver, Red-throated Diver, Fulmar, Storm Petrel, Cormorant, Shag, Gannet, White-tailed Eagle, Golden Eagle, Puffin, Razorbill, Guillemot, Black Guillemot, Black-tailed Godwit, Bar-tailed Godwit, Knot, Red-necked Phalarope, Glaucous Gull, Iceland Gull, Short-eared Owl, Ring Ouzel and Bluethroat.

Description: Andøya is a 490 km² island protruding north into the Norwegian Sea almost like a peninsula. It is considered to be a part of Vesterålen, but it appears as a separate entity. Here are large tidal flats, small inlets and brackish pools, seaside pastures, and in the middle of the island, a vast peat marsh with several small lakes. All these habitats are suitable for wetland birds, and when combined with its strategic geographical position, makes Andøya one of the most important bird areas in Norway. As a bonus, Andøya and its surroundings has a breathtaking scenery. Here are plains and sandy beaches facing the ocean, in stark contrast to snow capped mountains with jagged peaks.

Natural tall vegetation is scarce. However, Andenes town at the northern tip and Bleik village along the west coast provides gardens suitable for resting passerines. Andenes has a very active harbour often supporting a great number of gulls and sea ducks, including many of exotic Arctic heritage such as King Eider, Iceland Gull and Glaucous Gull, the latter sometimes in the hundreds in late winter. Whale safaris are regularly conducted from Andenes harbour in summer, and these trips can be very rewarding when it comes to pelagic birds. At the rock Bleiksøya just off the coast of Andøya are large colonies of seabirds, with Puffin, Kittiwake, Herring Gull and Shag as the dominating species. Razorbill, Guillemot,

Fulmar and Storm Petrel breed as well. The enormous peat marsh of Skogvoll nature reserve makes an important breeding ground for Whooper Swan, Common Scoter, Black-throated Diver, Red-throated Diver, Slavonian Grebe, Golden Plover, Whimbrel, Dunlin, Red-necked Phalarope, Arctic Skua and Short-eared Owl. Little Gull has been suspected to breed as well. The marsh and surrounding cultivated fields, particularly along the east coast, are important resting grounds for Pink-footed and Barnacle Geese in spring and as moulting grounds for Greylag Geese in late summer. The east coast is also good for shorebirds, sometimes with thousands of Knot and other calidrids from May to August. Black-tailed Godwit breed in some years. Several rare birds have been encountered on and off Andøya island over the years, including Black-browed Albatross, Wilson's Phalarope, several Ivory Gulls and Corn Bunting.

Sooty Shearwater is a true pelagic species that breeds in the Southern Hemisphere and migrates north via the Norwegian Sea during its winter, mainly from July to October, when it is summer up here. They are then seen from land exclusively during or after periods of strong onshore winds. If you travel out to sea, however, such as on a whale watching vessel, you have the chance to see it up close even in calm conditions. *Photo: Kjetil Schjølberg.*

The vast Skogvollmyran marshes on Andøya island can be highly interesting both during the breeding season and during migration, supporting a number of wetland species, Short-eared Owl and more. Photo: Bjørn Olav Tveit.

Other wildlife: Marine mammals abound in the waters surrounding Andøya, including sperm and killer whales. Whale safaris are conducted in summer, and so are seal safaris along the coast (from Stave village south of Bleik). The seal safaris usually visit the seabird colonies of Bleiksøya as well. Elk is common on Andaøya island. These animals are said to be rather aggressive in the open terrain here, and so you'd better be careful not to approach them too closely.

Best season: Late winter is best for Arctic ducks and gulls, May for geese, and May to August/September for shorebirds. There are a few weeks of very little daylight during early December to early January.

Directions: Andøya island can be reached by air from Tromsø and Bodø to Andøya airport at Andenes. The Coastal Express (Hurtigruten) lands in Risøyhamn twice a day.
By car along Fv82 from Sortland in Vesterålen. In summer a car ferry traffics the route Andenes to Gryllefjord on Senja island.

Tactics: A telescope is a necessity on Andøya island. The areas are large and are best explored by car. But as always, it pays off to investigate several of the sites on foot. In the following, the sites are described as you reach them if you arrive along Fv82 from Sortland to Andenes towns and then continue south again along the west side of Andøya island:
Just before you reach Forfjord village, a road turns off to the right, signposted to Langvassdalen. In this valley you will find the Forfjorddalen nature reserve with some of Scandinavia's oldest pine trees, which are attractive to a number of bird species, including Hawk Owl, Tengmalm's Owl and Redstart. Back north along Fv82 you soon pass Buksnesfjorden inlet, where Little Grebe sometimes is found, especially in May. Great tidal flats and strong fjord currents in the strait over to Andøya make **a.** Risøysund sound

well worth a stop. **b.** Lake Risøyvatnet is on the right-hand side 2.5 km north of Risøysund bridge. Mute Swan (being far north) and dabbling ducks are often found here. 5 km further up is Åse at the outlet of the Åseelva creek. Park beyond the bridge and walk back along the road for a view of the lake with the possibility of gulls and shorebirds. At low tide, it can also be worthwhile to drive out to Lilandsholmen peninsula. 3.8 km further on, turn right onto a parking space for visitors to **c.** Åholmen marsh. This wetland has been restored and, as a result, most breeding ducks and shorebirds have increased in numbers. There is a traffic ban within the nature reserve during May. At Å village 2 km further north, stop at the rest area just beyond the bridge, where you have a view of **d.** Åholmbukta estuary where you often will find geese and shorebirds.

Perhaps the single most rewarding wetland at Andøya, in terms of numbers and species diversity seen from along the road, lies 2.5 km further on, just before you reach Sellevoll village. Here, the road crosses **d.** Sellevollvalan lagoons, an exciting tidal area with brackish lagoons, marshes and puddles – an el dorado for shorebirds and ducks. However, there is only one traffic safe lay-by to park your car in along this stretch of road. You might want to walk along the road a bit (be careful!) to get a satisfying view of all the nice roadside lagoons and puddles. You might walk out into the terrain, but then again, you risk scaring the birds away. This site could really benefit from some kind of arrangements to facilitate birdwatching.

A couple of km further on, just before reaching Dverberg village, the **f.** Rognan wetland appear to the right. Besides the fact that this too is a great place for shorebirds, dabbling ducks and nesting Whooper Swan, there are often White-tailed Eagles perched out on the headland. In the village **g.** Myre, you can turn left and scan the pastures for geese, plovers and more. In Dverberg harbour and at **h.** Hestneset peninsula north of Dverberg church, shorebirds and gulls can be found. Take a look with the telescope further out in the Andfjord as well, as there are often large flocks of gulls and terns foraging in the nutrient-rich updrafts out here, besides divers, grebes and sea ducks. During winter

Lake Kleivvatnet on Andøya island is easily accessible and can offer high-quality birdwatching in a concentrated area in the outskirts of Andenes town. *Photo: Bjørn Olav Tveit.*

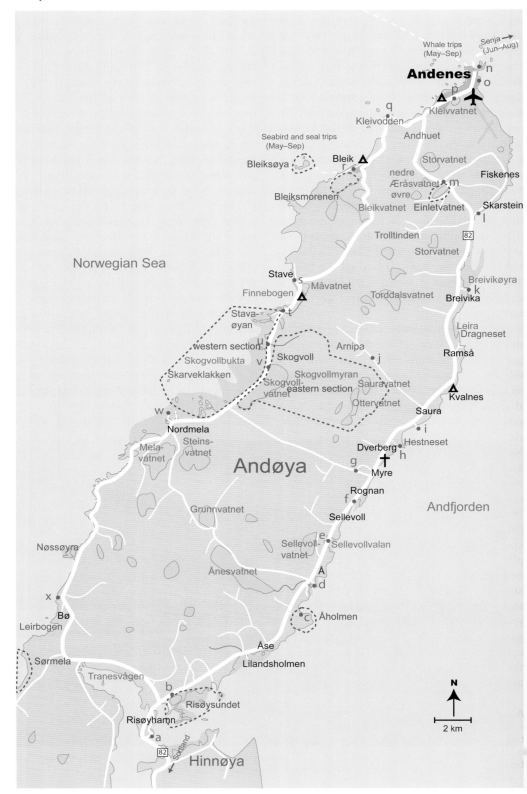

and sometimes well into spring, there can be dozens of King Eiders in this area. In Saura village, you can drive down to the breakwater that juts out into the **i.** <u>Saurabogen bay</u>. The beach here is more weather-beaten than further south, catering primarily to Purple Sandpiper and Turnstone. Out here, however, you can also find gulls, ducks and divers, and the fields lining the shore are good for geese and plovers. There may also be some bare black soil fields in this area, which must be checked extra thoroughly for plovers, wagtails, larks and pipits.

In the middle of Saura village, 1.7 km north of the descent to the jetty, a gravel road exits to the left by a large farm building and with bus stop signs on both sides of the main road. This gravel road takes you into the seemingly endless **j.** <u>Skogvollmyran marshes</u>. You reach the marshland after 2.5 km, but it is better to continue a couple of km further in, almost to the foot of the mountain Arnipa. Here you can get an overview of the peat landscape with a telescope, but for maximum outcome you should also take a stroll in the terrain.

Back on Fv82 northbound, you pass the large peat plain at Ramså, where there is a small river oasis and sand dunes with nesting Sand Martins. A little over 4 km further north (1.4 km after arriving in Breivika village), a dirt road goes down to the right which takes you out to **k.** <u>Breivikøyra bay</u>. This is a large and beautiful tidal area with the opportunity to get shorebirds up close. The area stretches a couple of km to the north, with views several places along Fv82. The coast is quite rocky here, but you will find coves with potential for shorebirds and gulls in Skarstein village and at the exit to **l.** <u>Fiskenes</u>. In this small settlement, you also have a number of nice gardens, pastures and fields, which may be worth exploring for passerines etc. If you continue the road to Fiskenes, across the heather fields and past the airfield, you will come back out

on Fv82 again a bit further north. Here you will see some larger bodies of water, including **m.** <u>Lake Einletvatnet</u> and <u>Lake Nedre Æråsvatnet</u> on the left side of the road and then <u>Lake Storvatnet</u> on the right side, all with the potential for ducks, Black-throated Diver, Grey Heron and more. When Fv82 passes under the steep <u>Andhue mountain massif</u>, you can stop to look and listen for Ring Ouzel and raptors, before continuing into Andenes town. Follow the main road out to **n.** <u>Andenes harbour</u> to look for gulls, terns, ducks and more. From the harbour jetties you get a view of the sea, with Puffins and other seabirds passing the northern tip of Andenes peninsula. For joining <u>whale watching boat trips</u>, follow signs to *Turistinfo* (Tourist Information) to purchase tickets. It can be a good idea to order in advance. The providers offer a 'whale guarantee', which means that if you do not get to see whales on the trip, you get to join a new trip for free. If the weather allows the ship to go out to sea, however, this is rarely a problem – you'll usually see whales of several species on these trips, including sperm and killer whales (orca). And more importantly, it gives you extraordinarily good opportunities to encounter a variety of seabirds, sometimes including Storm Petrel, Sooty Shearwater and other highly pelagic species.

In <u>Andenes town</u>, look for resting migrants in the gardens and on the soccer pitch. The beach by **o.** <u>Skjeggholmen island</u>, just south of Andenes school, often has Sanderling. Large mixed flocks of shorebirds can gather here if a carpet of kelp has been washed ashore.

One of the best all-round birdwatching sites in Andøya island is the small **p.** <u>Lake Kleivvatnet</u> just west of Andenes airport. The lake often holds good numbers of breeding Slavonian Grebe, a variety of ducks, and often Ruff, Spotted Redshank and other shorebirds having a stop-over on migration. Black-tailed Godwit

Andøya island: a. Risøysundet sound; **b.** Lake Risøyvatnet; **c.** Åholmen marsh; **d.** Åholmbukta estuary; **e.** Sellevollvalan lagoons; **f.** Rognan wetland; **g.** Myre pastures; **h.** Hestneset peninsula; **i.** Saurabogen bay; **j.** Skogvollmyran marshes; **k.** Breivikøyra bay; **l.** Fiskenes; **m.** Lake Einletvatnet; **n.** Andenes harbour; **o.** Skjeggholmen beach; **p.** Lake Kleivvatnet; **q.** Kleivodden seawatch; **r.** Bleik gardens and harbour; **s.** Stave beach; **t.** Staveelva river mouth; **u.** Kvanndalsbekken river mouth; **v.** Lake Skogvollvatnet; **w.** Nordmela; **x.** Marmelkroken and Bø salt marshes.

*Side margin text (top to bottom): Oslo · Interior · Skagerrak · Western · Central · **Northern** · Finnmark · Svalbard*

may breed here in some years. The scrub lining the lake often holds Bluethroat and other passerines.

To continue the round trip south along the west side of Andøya island, drive out of Andenes town and take the road to the right, signposted to Bleik. After almost 3 km, pull over to the **p.** Kleivodden rest area. This is probably the best land-based place on Andøya for observing passing seabirds. Here you also have a nice view of Forøya island, where for several years there has been a fairly large harbour seal colony. There is also a fairly large Cormorant colony on the island, and a few pairs of Kittiwake.

r. Bleik village lies at the foot of a threatening ridge of a mountain massif, and must be the most contrasting and idyllic settlement north of the Arctic Circle. The gardens here must appear as welcome resting places for migrating passerines. From Bleik, you can see the pyramid shaped Bleiksøya island with its colonies of seabirds. You can hire a canoe or take part in guided trips out there by contacting the campsites/lodges in Bleik or **s.** Stave. Puffin and seal safaris are also arranged along the coast and to Bleiksøya island, with the opportunity to see harbour seals and otter at close range, as well as Puffin and a range of other seabirds. In the Stave area you will find pastures and several small lakes. The coast here is characterized by small islets and rocky cliffs with white sandy beaches in between. One km south of the entrance to Stave Camping is a particularly fine riparian area that stretches several hundred meters down to the outlet of the Stave river. The road continues for almost 10 km along a tidal area several hundred meters wide with stones and seaweed, all included in the Skogvoll nature reserve. The shorebirds are scattered in this large area at low tide, and it can also be difficult to spot them between the rocks. It is therefore best to be here about two hours before the tide is at its highest, when the birds are

Iceland Gull regularly spend the winter in Lofoten and Vesterålen archipelagos, including at Andøya island. Andenes harbour in particular can produce some of the most interesting collections of gulls, particularly in the cold season but often well into spring and summer. *Photo: Bjørn Fuldseth.*

pushed closer to the shore. At the outlet of **u.** Kvanndalsbekken creek, just before Skogvoll village, some shorebirds and gulls may gather also at low tide. Immediately after arrival (and just before you drive out of) Skogvoll village, you can park and walk up the 100 m narrow embankment that separates the 3 km long **v.** Lake Skogvollvatnet from the sea. There may be interesting gulls, ducks and divers in this lake. The fields in this area can be good for plovers, Skylark and more. Also look out for Short-eared Owl. The large tidal range ends approximately at **w.** Nordmela town, and the small harbour here can hold a nice selection of gulls.

You can then take one of the roads that cut across to the east side of the island, or you can take the scenic circuit around the south side of Andøya island. Be sure to make a stop at **x.** Marmelkroken in Bø village and the surrounding salt meadows. Arrangements are made for birdwatchers here, including hides. A variety of wetland species breed in the area, stretching south to Leirbogen inlet and Bømyra marshes, in some years including Black-tailed Godwit. Even more species stop by during migration and can be appreciated at close range.

24. DIVIDALEN VALLEY

Troms County

GPS: 68.79873° N 19.67430° E (bird observatory)

Notable species: Scaup, Willow Ptarmigan, Black Grouse, Capercaillie, Golden Eagle, Rough-legged Buzzard, Goshawk, Gyrfalcon, Temminck's Stint, Purple Sandpiper, Red-necked Phalarope, Long-tailed Skua, Hawk Owl, Pygmy Owl, Long-eared Owl, Short-eared Owl, Tengmalm's Owl, Lesser Spotted Woodpecker, Three-toed Woodpecker, Yellow Wagtail, Red-throated Pipit, Bluethroat, Redstart, Sedge Warbler, Long-tailed Tit, Siberian Tit (scarce) and Parrot Crossbill.

Description: Dividalen valley in the scenic interior of Troms County leads to Øvre Dividalen National Park covering the barren mountains along the border with Sweden. In this pristine wilderness are several bird-rich marshes and watercourses, in some areas with good populations of Red-throated Pipit and – more variably – Long-tailed Skua, always with the possibility of encountering Snowy Owl. Dividalen valley is dominated by open pine forest with a sparse heather undergrowth. The river on the valley floor is lined with dense woods dominated by birch and willow. Dividalen Bird Observatory is situated here. Quite a few sought-after species such as Capercaillie, Hawk Owl, Three-toed Woodpecker and Siberian Tit can be found in the valley. Great Grey Owl may be seen in years with good rodent populations, and a few records of Ural Owl have been made over the years. You can look and listen for these species also in the neighbouring valleys to the east and west, Rostadalen and Kirkesdalen, respectively. Målselvdalen valley to the north is also worth exploring, particularly with passerines in mind. Icterine Warbler and Wood Warbler are regular here in early summer, and the odd Arctic Warbler and Little Bunting have been found singing here, once even Yellow-breasted Bunting.

Other wildlife: The Dividalen area has one of Europe's densest populations of wolverine, and there are populations of lynx and brown bear here as well.

Best season: Spring for calling owls and displaying Capercaillie. Late spring and summer for other birds. Siberian Tit and Yellow-browed Warbler have been caught almost annually during ringing at the bird observatory in autumn.

Directions: *Fastest from Tromsø by car (about 2 hrs)*: follow the E8 south to Nordkjosbotn and turn left along the E8/

Hawk Owl nests in open conifer or birch forests, particularly in the northern parts of the country. As with many other owls and birds of prey, Hawk Owls may travel long distances to reach areas with good food supply. In years when rodents abound, this is usually the most common owl in northern Norway, often seen conspicuously perched on roadside poles or treetops. *Photo: Espen Lie Dahl.*

Oslo · Interior · Skagerrak · Western · Central · **Northern** · Finnmark · Svalbard

Sørkjosleira tidal area in Balsfjord. This wetland area is of great importance for many species and is one of the most important staging grounds in Europe for Knot. *Photo: Bjørn Olav Tveit.*

E6 signposted to Kirkenes and continue for about 10 km. Here, exit to the right along Fv87 signposted to Øverbygd, a road leading through Tamokdalen valley. After another 30 km, in Holt village, turn left signposted to Dividalen.

Alternatively, from Tromsø, you can travel via the sites at Balsfjord (next section) and thereafter continuing along the E6 south (signposted to Narvik) to Buktmoen, from where you can follow the lush and bird-rich Målselvdalen valley up along Fv87 to Holt, where you exit right onto the road signposted to Dividalen.

Tactics: Cover as much terrain as possible on foot. A hiking trip over a few days in the mountains of the national park is highly recommended. Drive as far up the valley as possible, 5–6 km past the toll booth at Frihetsli. At Øvre Divifoss is a bridge enabling you to explore the mature conifer forest on both sides of the river. Park at Gambekken by the end of the road and walk up the valley and up above

the tree line, for instance to the marshes at Luovnevadda/Storflåan and east towards the Skaktardalsjuvet canyon. Owls can be heard calling at night in early spring along the road up Dividalen and its side-valleys, Rostadalen to the east and Kirkesdalen to the west, both signposted from along Fv87. Dividalen Bird Observatory's cabin is along the road just north of Frihetsli, 28.4 km from Fv87 (and 6.6 after the road crosses Divielva river). It is the small cabin just visible from the road, 50 m past a small bridge after a hilltop.

25. BALSFJORD

Troms County

GPS: 69.23106° N 19.31370° E (Strandnes, Sørkjosleira) 69.48577° N 18.87584° E (Kobbevågen inlet)

Notable species: Ducks, Red-throated Diver Black-throated Diver, White-billed Diver Slavonian Grebe, Red-necked Grebe, White tailed Eagle, Knot and other shorebirds and wetland species.

Description: Balsfjord is a magnificently beautiful, 50 km long fjord cutting inland from Kvaløya island and Tromsø city. The surrounding terrain is dominated by gentle slopes with forest and cultivated fields, but the backdrop is that of tall, pointed and snow-capped mountains. Where rivers meet the fjord, large amounts of sediment have built up shallow waters and vast tidal flats over the years. Wetland birds gather here at all times of year. The most important of these is Sørkjosleira, with its 5 km long stretch of muddy beach, and with tidal flats stretching 1 km out from the shore. Nearly equally important is Kobbevågen inlet further out the fjord. The tidal flats of Balsfjord are particularly noted for being one of the two single most important spring stop-over sites in Northern Europe for the North American population of Knot, the other site being the Porsanger fjord (which see). Flocks of 25,000 birds have been reported from Balsfjord, although in recent years, numbers have rarely exceeded 4000 individuals. The Knots arrive in early May and numbers keep building until all of them leave in a matter of a couple of days at the end of the month. Several other species exploit these tidal flats, with Dunlin and Ringed Plover as the next-most numerous shorebirds. In autumn, Little Stint, Temminck's Stint and Curlew Sandpiper are among the commoner species. Balsfjord is also a very important resting area for ducks, divers and grebes, particularly from March to June, while they wait for the ice and snow to melt up in the mountains. A few thousand Long-tailed Ducks, Velvet Scoters and Common Eiders are commonly encountered in May–June, besides a few hundred Slavonian Grebes and a few dozen Red-necked Grebes, Black-throated Divers and Red-throated Divers. In summer, Wigeons in the hundreds moult in these waters. The outer part of the tidal zone of Balsfjord never freezes completely

White-billed Diver winters scattered along the Norwegian coast and is seen on migration to an increasing extent the further north you go. Balsfjord is probably the most reliable place in the country to experience this majestic species in winter and early spring. *Photo: Terje Kolaas.*

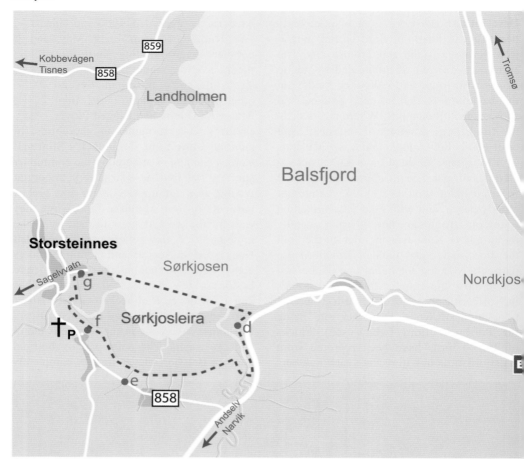

over in winter, but instead provides supreme conditions for species like Whooper Swan, Shelduck, King Eider, Red-necked Grebe and Purple Sandpiper. Balsfjord is among the most reliable winter and spring sites for White-billed Diver in Norway, particularly from February to May.

Best season: All year, except for the darkest days in December to early January.

Directions: E6 runs along Sørkjosleira near the exit to E8, 60 km south of Tromsø. See the map.

Tactics: Sørkjosleira can be viewed from the roads between **d.** Strandnes and **g.** Storsteinnes but you should walk along the shore to gain better control of the area. The shorebirds are pushed towards the shore at high tide, making the hours leading up to the peak level the best time to visit. You can park at Sagelv church and walk down to **f.** Sagelva river mouth. The tidal area continues north of Storsteinnes, viewable from bus lay-bys along Fv858. Lake Sagelvvatn is reached by heading 5 km west from Storsteinnes along the road signposted to Andselv. This road leads alongside the lake, which can hold a nice selection of wetland birds. Kobbevågen inlet is reached by continuing Fv858 north from Storsteinnes. The road strays past the neighbouring fjord Malangen but comes back along Balsfjord again later. Kobbevågen inlet is best explored from along the highway. This road leads north to Kvaløya, providing an alternative route between Balsfjord and Tromsø via Tisnes (featured in the following section).

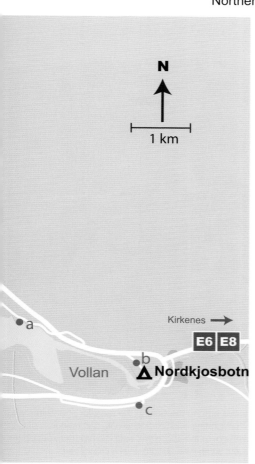

N

|—————————————|
1 km

●a

Kirkenes →

E6 **E8**

Vollan ●b
▲ **Nordkjosbotn**

●c

Oslo · Interior · Skagerrak · Western · Central · Northern · Finnmark · Svalbard

26. TROMSØ AREA

Troms County

GPS: 69.60207° N 18.83187° E (Tisnes), 69.73318° N 19.09470° E (Tønsneset)

Notable species: Shelduck, King Eider, Willow Ptarmigan, Rock Ptarmigan, Black Grouse, Slavonian Grebe, White-tailed Eagle, Golden Eagle, Gyrfalcon, Knot, Temminck's Stint, Purple Sandpiper, Ruff, Red-necked Phalarope, Lesser Black-backed Gull, Glaucous Gull, Iceland Gull, Puffin, Short-eared Owl, Red-throated Pipit, Bluethroat, Sedge Warbler, Arctic Redpoll and Snow Bunting.

Description: In Tromsø city, you hardly need to leave town to achieve high-quality bird experiences, no matter the time of year. The city is situated on Tromsøya island in the

Balsfjord: a–c. viewpoint to Vollan at the Nordkjoselva river mouth; **d.** Strandnes, viewpoint to Sørkjosleira at the Tømmerelva river mouth; **e.** pastures near Sørkjosleira; **f.** Sagelva river mouth; **g.** viewpoint from Storsteinnes industrial area.

sound between the mainland and the large and mountainous Kvaløya island. During migration, skuas (often including Pomarine), gulls, ducks and divers can pass through the sound in good numbers. At the harbour, interesting gulls and ducks may gather, with King Eiders in the hundreds from December through to April. They often come close to the docks to feed on fish scraps. Keep an eye out for Steller's Eider in these flocks as well. In the middle of town is the small Lake Prestvannet, surrounded by deciduous woodland and a park. This is a very popular recreational area with the residents, but if the lake is not covered in ice during winter, it is of great interest to birdwatchers. Several ducks, gulls and other wetland species can be found, including Arctic Tern and 5–6 breeding pairs of Red-throated Diver. Lake Prestvannet is probably the most accessible site in the country providing close encounters with the latter species. During migration, displaying Ruff and *Tringa*-shorebirds can be found by the lake, particularly in the marsh in the southeastern corner. Willow Ptarmigans are found commonly in this and other undeveloped areas of Tromsøya, and they are often very approachable. On the mainland just northeast of Tromsø is Tønsneset headland and Tønsvika inlet, both strategically positioned along the Tromsøya sound. They represent easily accessible migration sites close to town.

Tromsø is sheltered from the rough Norwegian Sea by the massive and beautiful Kvaløya island, in itself containing many interesting sites for both breeding birds and resting migrants. All year, you can find Willow Ptarmigan, White-tailed Eagle, Golden Eagle, Gyrfalcon and Arctic Redpoll here. Red-necked Phalarope breed in marsh ponds and lakes, such as at Rakkfjordmyra and Hansmyra marshes, where you also

Langneset headland by Tromsø Airport. The tidal area here is among the best in the area for shorebirds and gulls. Out in Sandnessundet sound, you can experience both migrating and feeding skuas, divers, ducks etc., in late winter often including King Eider. The sculpture *Else* towers over the shore and sets the nature experiences in a cultural context. An untrained eye might mistake it for a stockfish rack. *Photo: Bjørn Olav Tveit.*

may find Whooper Swan, Purple Sandpiper, Red-throated Pipit and Bluethroat. Hawk Owl breeds in some years in the forests and Snowy Owl is seen every now and then. Rock Ptarmigans may be found on barren ground, mainly at higher elevations. Resting migrants can be looked for along the coast, particularly the east coast where suitable habitats for a variety of wetland species are found. The foremost site at Kvaløya is Tisnes, a headland protruding into the sound outside of the mouth of Balsfjord. On the opposite side of Kvaløya you can obtain good views to the sea from the picturesque villages of Sommarøy and Hillesøy. Hillesøy is the easiest accessible seawatching site from Tromsø, and its harbour can showcase large amounts of gulls and sea ducks including King Eiders when incoming fishing vessels deliver their catch. The smaller Vengsøya

just off the west coast of Kvaløya is a reliable Puffin location.

North of Kvaløya are Ringvassøya, Reinøya and Vannøya, all large islands with qualities of interest to birdwatchers. The sounds between them hold sea ducks, grebes and gulls. Dåfjorden at Ringvassøya island has a good tidal zone and a productive fish landing facility. Lake Skogsfjordvatnet on the same island has several colonies of Lesser Black-backed Gull, apparently of mixed subspecific heritage. At Hansnes is another good tidal zone and you can take the ferry to Reinøya island, which has a particularly high density of Golden Eagles. Or, you can take the ferry to Vannøya island, presented separately below. Several rare birds have been encountered in the Tromsø area over the years, including Squacco Heron, Lesser Yellowlegs, many Ivory Gulls and Pechora Pipit.

Other wildlife: Musk ox have been released on Ryøya islet in Rysundet sound as part of a scientific program.

Best season: All year, but with greatest variation during migration periods.

Directions: Tromsø airport Langnes has direct connections to Oslo and is also the main hub for the local flight network in northern Norway. Tromsø city has well-developed public transportation.

By car: Tromsø is located at the western end of E8. The network of roads on Tromsøya, where the city centre and some of the birdwatching sites covered here are located, is very complex and difficult to navigate for first time visitors. The situation is normalized, however, as soon as you leave town, and it is no problem, for instance, to follow the main road through to Kvaløya. See the map and the suggestions below.

Tactics: Most of these sites are easily accessible by car and by bus. Whale safaris, also good for seabirds, are conducted from Tromsø harbour and other nearby places.

By car, the sites in the Tromsø area can be visited as follows:

a. Lake Prestvannet is on top of Tromsøya island, in the middle of Tromsø town and can be approached from many directions. It is convenient to park by Tromsø Geophysical Observatory (*Nordlysobservatoriet*) on the northern side of Lake Prestvannet. Follow foot paths into the area and around the lake.

b. and **c.** Tromsø harbour can be explored from several viewpoints.

d. Tønsneset headland and Tønsvika inlet. From the E8 on the east side of Tromsø exit north on Fv53, signposted to Kroken/Oldervik. 8 km after the last roundabout, after a long stretch of straight road where the road makes a bend to the right, an unmarked

track exits to your left just before a private house. This track leads the 200 m out onto Tønsneset headland. Continue on Fv53 1.5 km further and make a left on a paved road signposted *Ørretholmen* leading to views of **e.** Kjosen, a muddy brackish area at the Kjoselva river mouth in Tønsvika inlet. Continue another km along the highway and you will see the Tønsvikelva river mouth with its accompanying tidal banks on your left.

Back at the E8 just east of Tromsø, follow E8 signposted towards the airport. At the roundabout just before the road passes underneath the airport runway, make a left onto Fv862 signposted to *Sentrum*. At the next roundabout, go right towards *Folkeparken* and at the upcoming roundabout, make a right onto a simple, unmarked road leading to a viewpoint of the small Giæverbukta inlet where you may find shorebirds. Go back to E8 and continue underneath the runway (signposted to Kvaløya) and make a left in the upcoming roundabout. Park and walk out to

Willow Ptarmigan is quite common in the green areas of Tromsø town, such as around Lake Prestvannet, and they are often quite approachable. *Photo: Terje Kolaas.*

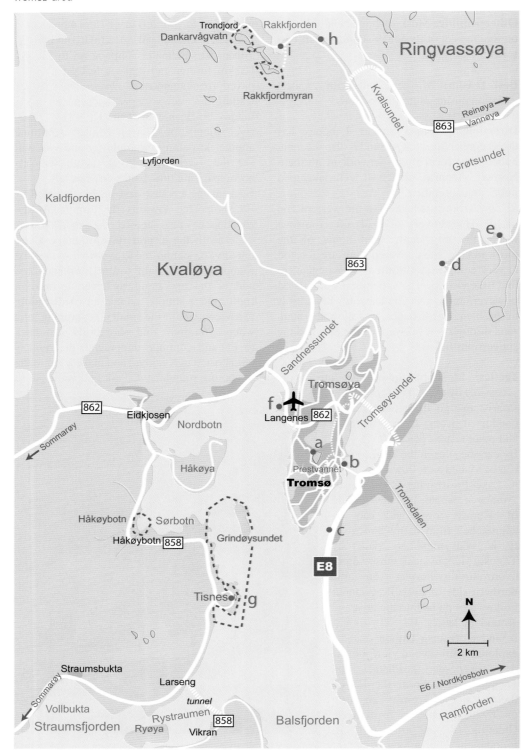

Tromsø area: a. Lake Prestvannet; **b.** Tromsø harbour; **c.** fish landing facility; **d.** Tønsneset; **e.** Kjosen river mouth; **f.** Langneset headland; **g.** Tisnes headland; **h.** Kvalsundet sound; **i.** Rakkneskjosen tidal area.

obtain views of the tidal areas on both sides of the small **f.** Langneset headland. Keep driving towards Kvaløya and cross over to Kvaløya via the Sandnessund bridge.

Over at Kvaløya island you are met by a roundabout where you must choose one of two options: **1)** Go straight ahead along Fv862 towards Sommarøy and exit left in Eidkjosen along Fv858 signposted to Larseng if you want to visit Tisnes headland and the other sites on southern Kvaløya or to Kobbevågen in Balsfjord (which see), or **2)** you can make a right onto Fv863 signposted to Hansnes to get to Vengsøya island, Rakkfjorden and Rakkfjordmyra marsh, and the islands north of Kvaløya island.

1) When choosing the first option, the road leads along the coast with several fields and tidal areas well worth checking. The exit to **g.** Tisnes headland is clearly signposted Tisnes from Fv858. This headland has a few gardens but is mostly dominated by cultivated fields and marshes. The road leads out on the headland and can be explored from the car or, preferably, on foot. Please respect private property. The little pond to your right, visible from the road 1.4 km from Fv858, can often hold ducks and displaying Ruff in late spring. Black-tailed Godwit are recorded almost annually. Keep an eye out for passing and resting seabirds and check the tidal zone thoroughly.

Further south along Fv858 you pass the strong tidal current in Rystraumen sound via an underseas tunnel to reach Kobbevågen inlet in Balsfjord. For musk ox, scrutinize the small Ryøya island through a telescope. Further south is Straumsbukta, where you may find shorebirds along the shore and by the river mouth. In June and July, you can park by the school and walk up along the eastern side of the river about a kilometre or so to Hansmyra marsh, one of the most bird-rich wetlands of Kvaløya island. Continue the main road 30 km to Sommarøy and Hillesøy to obtain views to the Norwegian Sea and to look for shorebirds along the shore and

Tisnes headland near Tromsø juts out into the sound like an oasis of lush saltmarsh. *Photo: Bjørn Olav Tveit.*

Skipsfjorddalen valley on Vannøya island is a pearl of a bird locality. Here you have a sheltered sea bay, beautiful shorebird beaches and huge marshy areas that stretch into the valley. To get here, you either have to hike over the mountain ridge or be transported in by boat. *Photo: Tor Olsen.*

passerines in the gardens and meadows. From here, you can take a quicker route back to Tromsø across Kvaløya island.

2) Back at the roundabout at Sandnessundbroa bridge, follow Fv863 north signposted to Hansnes. Stop and check the sound wherever suitable. The exit to Vengsøya island (involving a ferry) is clearly signposted. Further along Fv863, just before the tunnel leading underneath Kvalsundet sound, exit onto the road signposted to Kvaløyvågen, a road leading along the often rewarding **h.** Kvalsundet sound. After 5 km this side road crosses the inner section of Rakkfjorden with **i.** Rakkneskjosen tidal area on your left. Just before the fjord crossing, a gravel road exits to the left, leading about 1 km to Rakkfjordmyra marsh, where you may find Red-necked Phalarope and Red-throated Pipit in early summer. You can continue along Fv863 to Ringvassøya island. From Hansnes you can go by car ferry to Reinøya and Vannøya islands (next section).

27. VANNØYA ISLAND

Troms County
GPS: 70.2476° N 19.5031° Ø (Torsvåg village)

Notable species: Staging wetland species and more during migration; Pintail, Common Scooter, King Eider, White-billed Diver, Black-throated Diver, Red-throated Diver, Golden Eagle, Whimbrel, Black-tailed Godwit *islandica*, Dunlin, Red-necked Phalarope, Ruff, Red-throated Pipit, Bluethroat, Twite and Lapland Bunting.

Description: Of the many interesting birdwatching sites around Tromsø, the strategically located Vannøya island is particularly exciting. The island stretches to the north and out towards the Norwegian Sea. A distinctive landscape element here is the Skipsfjorden bay which extends on land as Skipsfjorddalen valley, a combined structure which penetrates deep into the island from the north. High mountains surround the fjord and valley in a horseshoe shape on three sides. Skipsfjorddalen valley, and the flat Slettnes headland just to the east of the fjord mouth, both have very good habitats for wetland birds. Here you will find a mosaic of small lakes and ponds in a vast marshland. This is now considered to be the core area for Black-tailed Godwit of the northern subspecies *islandica* in Norway, but not even here more than a very few pairs are found. Whimbrel is a characteristic species in these areas, and you will also find nesting Ruff, Red-necked Phalarope, Dunlin and Common Scooter. Red-throated Pipit is common along the shore, and Lapland Bunting is found on the drier peat bogs. The wetland areas are also good for resting shorebirds and more during migration. By Lake Haugvatnet at Slettnes headland, there is a small grove of planted pine trees that provide shelter for resting passerines. The Vannareid marshes, more centrally located on Vannøya island, also have potential for some of the mentioned wetland species, including Black-tailed Godwit. The beach meadows at Vannareid are also worth checking for resting shorebirds. Torsvåg village in the far north of the island is a good place for observing migrating seabirds, and it is also a favourable area to look for resting passerines and such in late spring and autumn. The same applies to the gardens in Burøysund. Willow Ptarmigan is numerous on Vannøya, and you can find Rock Ptarmigan by scanning the mountain sides with a telescope in early morning. There are particularly good numbers of King Eider and White-billed Diver around Vannøya in winter. Vannøya island has not often been visited by birdwatchers in the past, but rarities such as Glossy Ibis and Franklin's Gull have nevertheless been encountered here.

Other wildlife: Northern hare is a common sight on Vannøya island.

Best season: Late spring and early summer generally have the most to offer. Winter for gulls, divers and sea ducks. Both spring and autumn for general migration.

Directions: From Tromø by car, follow Fv862 towards Kvaløya and cross over the

Sandnessund bridge. From here, follow Fv863 northwards signposted to Hansnes. The ferry to Vannøya island leaves from Hansnes.

Tactics: Vannøya island is large and should be explored by car – preferably also by boat. To reach the most exciting wetlands, Skipsfjorddalen and Slettnes, you can team up with local fishermen etc., who can pick you up and bring you, e.g. starting from Burøysund or Vannvåg villages. The alternative is to walk for a couple of hours each way. For the fit and adventurous, it is a fantastic option to bring some camping equipment and walk the island from the west side via Skipsfjorddalen valley to Slettnes and then on to Vannvåg. There are taxis on the island for driving back again. Seabird watching and searching for resting passerines is best conducted in Torsvåg village.

Black-tailed Godwit is a very scarce breeding bird in Norway. It is found as two subspecies, of which the southern *limosa* breeds on Jæren and the northern *islandica* (depicted) in northern Norway. Vannøya island is considered to be the most important site for the species up north. *Photo: Tomas Aarvak.*

Lake Nedrevatnet in Skibotn valley is one of several fine wetland locations in this area. *Photo: Bjørn Olav Tveit.*

28. SKIBOTN VALLEY

Troms County

GPS: 69.36810° N 20.29721° E (Lake Nedrevatnet)

Notable species: Ducks incl. Scaup and Velvet Scoter; Willow Ptarmigan, Capercaillie, Slavonian Grebe, Golden Eagle, Rough-legged Buzzard, Gyrfalcon, shorebirds incl. Green Sandpiper, Temminck's Stint and Red-necked Phalarope, Long-tailed Skua, Hawk Owl, Pygmy Owl, Short-eared Owl, Tengmalm's Owl, Yellow Wagtail, Red-throated Pipit, Waxwing, Bluethroat, Redstart and Sedge Warbler.

Description: Skibotn valley has good populations of raptors, owls and other mountain and forest dwelling species, much the same as Dividalen and Reisa valleys (which see). The relative closeness to Tromsø city makes Skibotn valley more frequently visited by birdwatchers compared to Reisa

valley. The two small lakes Nedrevatnet and Øvrevatnet can hold Slavonian Grebe, Little Gull, ducks and other wetland species. Black-headed Gull and Slavonian Grebe breed in Lake Nedrevatnet. By the fjord, Skibotnelva river has deposited a large amount of gravel and sand, producing a delta with qualities appreciated by shorebirds during migration. Further in along the fjord, by the Signaldalelva and Kitdalselva river mouths is the 1.5 km² Storfjordmelen delta which a number of migrant shorebirds, ducks and other wetland species also find suitable for resting. Rarities encountered in this area over the years include Purple Heron and Bonaparte's Gull.

Other wildlife: Raccoon dog, introduced to Russia and not welcome to spread to Norway, has been seen in Skibotndalen valley, as one of a very few sites in the country.

Best season: Late spring, summer and early autumn.

Directions: *From Tromsø,* follow the E8 south. Storfjordmelen delta is reached after 90 km and can be viewed from a lay-by 700 m north of the exit signposted to Signaldalen. In order to view the Skibotn river mouth, exit onto the E6, follow it through the settlement and turn left just before Skibotn Camping. Skibotn valley is reached by continuing along the E8 east from Skibotn, signposted to Åbo/Turku in Finland. Lake Nedrevatnet is visible from the E8 on your right about one km from the E8/E6 junction, and a lake-side gravel road exits to the right just before the lake comes into sight. 2 km further along the E8, Lake Øvrevatnet appears on your right, readily viewable roadside. Continue on the E8 to explore the valley further, preferably with many stops to scan the mountains for raptors, and on foot through a variety of habitats. Helligskogen lodge (30 km from Skibotn village) makes a nice trailhead for hiking trips in the area.

29. SPÅKENES TIDAL AREA

Troms County

GPS: 69.76263° N 20.50735° E (Parking, ridge viewpoint)

Notable species: Ducks, shorebirds and gulls; White-tailed Eagle and Bluethroat.

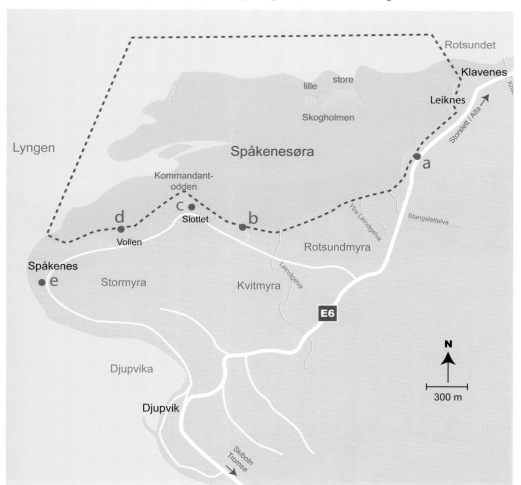

Spåkenesøra tidal area: a. Leiknes, viewpoint from the east; **b.** ridge viewpoint; **c.** Slottet viewpoint; **d.** Stranden; **e.** Spåkenes, viewpoint to the Lyngen fjord.

Description: Spåkenesøra is a 3 km² tidal area on the northern side of Spåkeneset headland, protruding into the Lyngenfjord. Large numbers of shorebirds can gather here during migration. The area is remote, however, and rather seldom visited by birdwatchers. There used to be a hide on the ridge, overlooking the shore, but a landslide brought the structure down recently. You may still scan the shore from the ridge, though, just watch your step. The habitats on the headland itself are mainly those of heather, willow scrub and birch woods with a strong population of Bluethroat. Spåkenesøra is situated with a backdrop of the impressive Lyngsalpene mountain peaks on the opposite side of the fjord.

Best season: May to September.

Directions: From the E6 about 20 km west of Storslett and 90 km north of Skibotn,

exit onto the northernmost of the two exits signposted to Spåkenes. Continue 1 km and park in the small lay-by on your right. A path leads through marshy terrain 150 m to **b.** the ridge.

Tactics: View the area with a telescope from the viewpoints **a–e.** marked on the map. At low tide you may walk along the shore and even out on the tidal flats.

30. Reisa Valley & Delta

Troms County

GPS: 69.78192° N 20.96240° E (Solbakken, Reisa delta)

Notable species: *Reisa delta:* ducks, shorebirds and gulls. *Reisadalen valley and Reisa National Park*: Long-tailed Duck, Willow Ptarmigan, Rock Ptarmigan, Black Grouse, Capercaillie, Rough-legged Buzzard, White-tailed Eagle, Golden Eagle,

The Reisa delta has large areas of grassy tidal banks, great for shorebirds, ducks, etc. However, it can be challenging to find good vantage points at this site. A telscope is mandatory here. *Photo: Bjørn Olav Tveit.*

Osprey, Kestrel, Gyrfalcon, Goshawk, Temminck's Stint, Hawk Owl, Pygmy Owl, Short-eared Owl, Tengmalm's Owl, Snowy Owl (scarce), Lesser Spotted Woodpecker, Three-toed Woodpecker, Dipper, Bluethroat, Ring Ouzel, Sedge Warbler, Siberian Tit and Parrot Crossbill.

Description: At the mouth of Reisaelva river in the Reisafjord is a delta with up to 8 km² of tidal flats lined with meadows, fields and woodland. The Reisaelva river drains the western part of the Finnmarksvidda plateau (which see) and forms a natural migration flyway for the wetland birds breeding there. Many of them stop by to rest at the Reisa delta. Also, substantial numbers of Goosander and Red-breasted Merganser spend the summer moulting in the delta. Still, because of the remote location, few birdwatchers visit the sites here. Up the Reisadalen valley are good populations of raptors and owls, Hawk Owl in particular. Several other woodland species can be found, including Three-toed Woodpecker and Capercaillie. Further up the valley, in Reisa National Park stretching all the way to the border with Finland, you can hike up the barren mountain terrain to one of the most remote areas of wilderness in Scandinavia, with a long list of montane species including Long-tailed Skua and – although scarce – Snowy Owl. Spotted Redshank has been found breeding here, as the only Norwegian location outside Finnmark County.

Other wildlife: Reisadalen valley, with side valleys and surrounding mountains, has some of the strongest populations of wolverine and lynx in Scandinavia. Both may be encountered along the drivable road through the valley. Brown bear is also possible, particularly south of Bilto. Arctic fox has a small population in Reisa National Park, while grey wolf may stray by. Reisaelva river has sea trout and Atlantic salmon. Trout fishing in the mountain lakes can be very good.

Best season: Late spring, summer and autumn is best in the Reisa delta. Owls are easiest found in early spring (March) while calling at night. Hiking for birds in Reisa National Park is best in June and the first half of July.

Directions: Sørkjosa village has a small airport with flight connections to Tromsø and Hammerfest. There is a bus route along Reisa valley.
By car: The Reisa delta is situated along E6 by Sørkjosa and Storslett villages. The road up Reisa valley exits from E6 in Storslett, signposted to Bilto and to *Reisa Nasjonalpark*. Reisa National Park can alternatively be approached from Kautokeino on the Finnmarksvidda plateau, by following the road signposted to Biedjovaggi 33 km.

Tactics: A telescope is necessary in the Reisa delta and is a great advantage when looking for raptors in the valley. A hiking trip in Reisa National Park demands planning as a proper expedition into the wilderness, with clothes and equipment accordingly, including (analogue) maps and a compass.
The Reisa delta: From the E6 at Sørkjosa, exit to Sørkjosa airport and turn right on the road Solbakkmelen just before the terminal building. Drive through the residential area 400 m to the viewpoint at the end of the road at Solbakken. Continue E6 past Storslett and explore the side roads leading into the delta, such as Kipernesvegen and the road signposted to Nordkjosneset.
Reisadalen valley: The first 3–4 km of the road up the valley from Storslett passes the meander bends of the river, with suitable habitats for ducks. Further up, look for raptors over the ridges and forest species in the woodlands. When hiking into the national park, head for Geatkkutjavri lake west of Buntadalen valley west of Bilto or Geatkevuopmi. You may even go as far in as to Bavdnjaleamsi marshlands east of Njallaávzi valley or the marshes north and south of Ráisjávri lake.

Oslo

Interior

Skagerrak

Western

Central

Northern

Finnmark

Svalbard

FINNMARK

Finnmark is by far the most famous birdwatching destination in Norway. It includes the vast Varangerfjord, the deep taiga forest of Pasvik valley and spectacular Arctic seabird migration at Slettnes headland. It is the northernmost county in Norway and is technically a part of northern Norway. In this book, however, it has been granted its own separate chapter because of its iconic status as a world-leading birdwatching destination. With its 48,618 km² it is larger than the entire country of Denmark. The population, however,

Ruff is a good candidate for the title of Norway's most eccentric-looking bird. In spring, the males fight for the females' favor in traditional leks, which take place on grassy marshes and moist pastures. In southern Norway, Ruff is in sharp decline as a breeding bird, but in Finnmark there are still good populations of the species. *Photo: Tomas Aarvak.*

is as low as 75,000 and so it goes without saying that the infrastructure and practical facilities are marked by the fact that few people live here. This is really a remote wilderness. This area is dominated by Sami and Kven

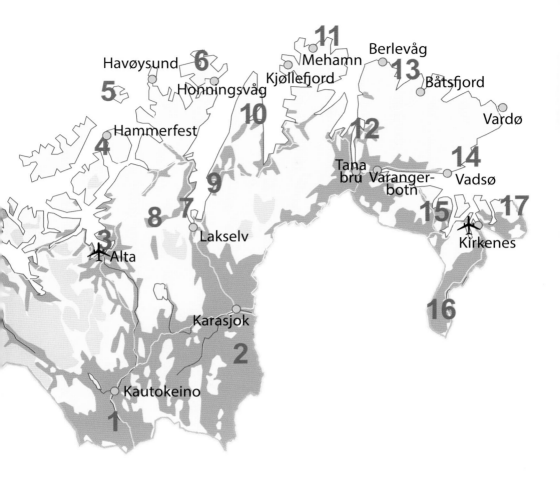

SITES IN FINNMARK

Hawk Owl is active during the daylight hours and hunts small rodents and shrews primarily with the help of its sight. It often sits and scouts well exposed, although it may blend in with the birch trees. *Photo: Terje Kolaas.*

culture, which is also reflected in place names and associated distinctive letters.

Finnmark has been coined "an accessible part of Siberia" because it is sporting an Arctic climate with tundra and taiga habitats similar to those you find further east. Hence, it is home to a long list of breeding birds often associated with the northern parts of Russia, such as Tundra and Taiga Bean Geese, Lesser White-fronted Goose, Smew, Jack Snipe, Bar-tailed Godwit, Spotted Redshank, Little Stint, Broad-billed Sandpiper, Brünnich's Guillemot, Snowy Owl, Great Grey Owl, Red-throated Pipit, Waxwing, Arctic Warbler, Pine Grosbeak, Rustic and Little Buntings. Additionally, Finnmark is home to an equally long list of species associated with the Scandinavian mountain ranges, such as Gyrfalcon, Dotterel, Red-necked Phalarope, Ruff, Long-tailed Skua and Hawk Owl. It also offers enormous seabird breeding colonies and important resting areas for migrant geese, ducks, divers and shorebirds.

Along the coastline, massive movements of Arctic seabirds – including vast numbers of White-billed Diver and Pomarine Skua – take place in spring and autumn, a spectacle that can be appreciated from headlands along the coast. To top it off, Finnmark is host to Europe's largest numbers of wintering Steller's Eider and King Eider, alongside substantial numbers of Glaucous and Iceland Gulls. It has even proven to be a real hotspot for vagrants, e.g., with several records of the enigmatic Spectacled Eider.

When to go

Mid-January to March is best for Arctic gulls and spectacular rafts of Steller's and King Eiders in smart breeding plumages along the Varangerfjord. The numbers remain high until mid-April in most years, but these birds gradually migrate back to their breeding grounds further north and east from the end of March. Substantial numbers may still be present well into May, though, particularly in years with predominantly strong northerly winds. The seabird colonies are busy from mid-March to early August.

Mid-May sees the peak of the White-billed Diver and Pomarine Skua migration. This is also the best time for staging Lesser White-fronted Geese and thousands of Knots in delightful, brick-red summer-plumages.

From mid-May and onwards, Finnmark gets large influxes of Arctic shorebirds on migration, often including vagrants.

Mid-June to mid-July is best for Arctic Warbler, Little Bunting and shorebirds, both breeding and resting. Major rarities have turned up in this period as well.

Autumn until mid-October can be good for migration and may turn up rarities as well, but there are traditionally a lot fewer birdwatchers here at this time of year and hence fewer records. However, this is the time when you share these magnificent bird sites with few other birdwatchers, which increases the chances of you finding something extraordinary by yourself.

Birdwatchers visiting Finnmark for a short period of time will have difficulties finding all the specialities in one single trip. Some of the species will fail to turn up simply because they are sparsely distributed in homogeneous taiga forest, making them hard to track down when time is limited. Other species will be missed out on because they only appear in Finnmark at certain times of year. For example, it can be challenging to find both Lesser White-fronted Goose and Arctic Warbler on the same one-week trip, because the warbler often arrives at the breeding grounds after the middle of June, when the geese have left their regular staging areas for the few remaining and secretly kept breeding grounds in the mountain marshes.

THE BEST SITES

The Varangerfjord is one of the best general areas for birdwatching in Norway and among the most spectacular sites in Europe. It is a rather large area with a wide range of habitats and birds at all times of year, although difficult to appreciate in the darkest days of mid-December. This is the prime location in the

World for both King and Steller's Eiders. Pasvik is a wooded valley sandwiched between Finland and Russia where you can find all the taiga species occurring in Norway. In combination, the Varangerfjord and Pasvik adds up to become a complete Arctic tundra and taiga experience. The Porsangerfjord is one of the most important resting areas in Scandinavia for shorebirds, divers, ducks and geese, and is the only regular staging site in Scandinavia north of the Gulf of Bothnia for Lesser White-fronted Goose. The Tana and Neiden deltas also have qualities that place them firmly on the list of the most important bird localities in Norway, particularly in terms of staging areas for wildfowl and shorebirds. The Finnmarksvidda plateau is a prime locality for Arctic breeders in early summer, while the headlands of Slettnes, Kjølnes and Hamningberg have proven to be among the best in Europe for coastal migration of Arctic seabirds.

1. FINNMARKSVIDDA PLATEAU

Finnmark and Troms counties
GPS: 69.09817° N 23.20851° Ø (Niittojávrrit), 68.73972° N 23.30218° Ø (Áidejávri rest area), 69.83073° N 25.18779° Ø (Luostejohka)

Notable species: *Residents:* Hazel Grouse, Willow Ptarmigan, Rock Ptarmigan, Black Grouse, Capercaillie, Gyrfalcon, Hawk Owl, Snowy Owl, Great Grey Owl, Lesser Spotted Woodpecker, Siberian Tit, Arctic Redpoll and Pine Grosbeak. *Summer:* Whooper Swan, Tundra and Taiga Bean Geese, Lesser White-fronted Goose (scarce), Pintail, Scaup, Long-tailed Duck, Common Scoter, Velvet Scoter, Smew, Red-throated Diver, Black-throated Diver, Hen Harrier, Rough-legged Buzzard, Golden Eagle, Osprey (locally), Crane, Dotterel, Broad-billed Sandpiper, Little Stint, Temminck's Stint, Spotted Redshank, Ruff, Bar-tailed Godwit, Jack Snipe, Red-necked Phalarope, Long-tailed Skua, Short-eared Owl, Shore Lark, Red-throated Pipit, Yellow

Long-tailed Skua nests in tundra habitat on mountain plateaus. The population fluctuates with the rodent populations. *Photo: Kjetil Schjølberg.*

Wagtail, Waxwing, Dipper, Bluethroat, Sedge Warbler, Arctic Warbler (scarce), Great Grey Shrike and Lapland Bunting.

Description: Finnmarksvidda is a vast highland plateau in the interior of Finnmark. Large portions are tree-bare, but the slopes and valleys are covered in birch and pine forest. It is mainly the birch-wooded parts you experience when you cross Finnmarksvidda along the main roads. The many lakes and marshes are interconnected through creeks and rivers and lined with willow scrub, forming suitable habitats for a variety of montane birds. Early in the season the birds gather in ice-free parts of the water systems and on snow-free spots. During the short and hectic breeding season, Finnmarksvidda is teeming with life. Particularly popular with wetland species are the wettest marshes surrounded by natural dykes with permanently frozen cores. These dykes can be up to 7 m tall and 100 m long, providing dry nesting ground protected from floods, besides a good overview of the flat marshy landscape. This is typically where species like Bean Goose and Broad-billed Sandpiper place their nests.

You may get a good taste of Finnmarksvidda from the roads. However, you will get a complete wilderness experience by hiking Finnmarksvidda with a backpack and a tent, for instance to Stuorrajavri, Luostejohka, Náhpolsaiva, Goahteluoppal, Iesjavri or Anárjohka. In this manner and at these sites you increase your chances of encountering the less densely distribuated species of the mountains and forests. Canoe is another excellent means of exploring Finnmarksvidda, and many of the rivers are well-suited for this kind of activity. Remember to be adequately equipped when entering Finnmarksvidda by foot or canoe.

It is possible to be transported into the wilderness by helicopter or seaplane from towns such as Alta.

Other wildlife: Several mammals can be seen, but the most commonly encountered are mountain hare, red fox, elk, red squirrel and several smaller rodents, including Norway lemming, endemic to northern Scandinavia. The Sami tradition of keeping domesticated reindeer has led to a near-extinction of the four large Scandinavian carnivores on Finnmarksvidda, although bears still roam the forests of Anárjohka and the population of wolverine has increased somewhat in recent years.

Best season: From early June through July. The first migrants, including Bar-tailed Godwit and Spotted Redshank, arrive as soon as the first snow-free patches emerge in May. Broad-billed Sandpiper, Jack Snipe, Red-necked Phalarope and Long-tailed Skua usually arrive in the last few days of May, while Arctic Warbler and the odd Little Bunting is not to be hoped for until the last week of June. The marshy areas are easier to walk when the ground is still frozen until early June. Most migrants leave in early August, but the taiga forest holds many of its species year-round.

Directions: *By car from Finland*, choose route E8 north from Tornio, and continue towards Kautokeino along Rv93. The best birding is between the border crossing into Norway and Kautokeino. *From Varanger or Lakselv* take route E6 to Karasjok and continue Rv92 (later Rv93) towards Kautokeino.

Tactics: You can get a complete wilderness experience by hiking with a tent, binoculars and a telescope, e.g., into Stuorrajavri, Luostejohka, Náhpolsaiva, Goahteluoppal or Iesjavri, places which are described in more detail below. The chance of close contact with species such as Tundra Bean Goose, Lesser White-fronted Goose, Long-tailed Skua, Broad-billed Sandpiper and Snowy Owl increases in these more remote areas. And the nature experiences and wilderness feel put birdwatching in a very special context. It is possible to be transported into the wilderness by seaplane or helicopter, e.g., from Alta, but be aware of safety regulations

that prevent landing in some protected areas. Particularly destructive to birdlife is driving on bare ground with an ATV, which is offered as transport in several places. The wheel ruts drain the wetlands and pose the biggest threat against the Norwegian population of Broad-billed Sandpiper. Canoeing is a great way to travel in wetlands, and several of the waterways on the Finnmarksvidda plateau are very well suited to just this. The Karasjoka river in particular is widely known among paddling enthusiasts. Do the preparatory work thoroughly before paddling off, as several stretches of river can be demanding. Do not go hiking on Finnmarksvidda plateau at the end of June and in July without a mosquito net and plenty of first-class mosquito oil. The uniform landscape is easy to get lost in, so remember a map and compass in addition to a GPS. These are real wilderness and sparsely populated areas, in some places with deep sinkholes, so report where you are going and how long you will be gone, and preferably go with several people.

If you have limited time at your disposal, you can get a taste of the Finnmarksvidda plateau

Finnmarksvidda plateau has many interesting sites of which some require lengthy hiking trips in the terrain. The wetlands around Lake Leamšejávrrit (depicted) are situated close to the road, offering many of the most attractive species of the area, e.g. Long-tailed Skua and Broad-billed Sandpiper. *Photo: Bjørn Olav Tveit.*

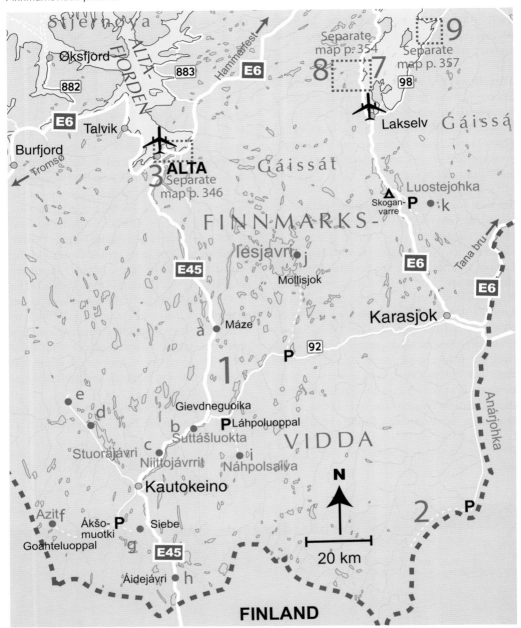

Finnmarksvidda plateau and nearby sites: 1a. Alta river canyon and watercourse; **b.** Suttášluokta cove; **c.** Niittojávrrit pond; **d.** The Biedjovaggi road; **e.** Leamšejávrrit marshland; **f.** Goahteluoppal wetland; **g.** Opmoáhpi wetland; **h.** Lake Áidejávri; **i.** Náhpolsaiva wetland; **j.** Lake Iešjávri; **k.** Luostejohka watercourse; **2.** Anárjohka valley; **3.** Altaosen delta; **7.** The Porsangerfjord; **8.** Stabbursdalen valley; **9.** Børselvosen delta.

by crossing it by car and taking in an overview from the side of the road and perhaps some detours out into the terrain on foot. The most productive stretch of road is along the E45 between Kautokeino and the border with Finland, a stretch that is easy to reach if you are on your way to or from southern Norway by car. Here you pass e.g., right past Opmoahpi and through Áidejávri, both very bird-rich wetlands where you can find most of

Oslo

Interior

Skagerrak

Western

Central

Northern

Finnmark

Svalbard

the vidda's specialties barely without having to leave the car. Alta river canyon is also conveniently located close to the road.

The best – and some of the most easily accessible – sites on the Finnmarksvidda plateau can be summarized as follows, in the order you reach them if you drive south from Alta town:

a. Alta river canyon and watercourse is located on the north side of Finnmarksvidda plateau and is the narrowest and most spectacular part of the Alta river run. The river is regulated, but the steep cliffs along each side makes excellent conditions for raptors, including Golden Eagle and Gyrfalcon, with a density of breeding pairs which is among the highest in Europe. The most popular vantage points are southwards from Máze, signposted from Rv93 68 km south of Alta, or from Alta by driving as high up as possible along Stillaveien road in Tverrelvdalen valley.

b. Suttášluokta cove in the long Lake Heammojávri east of Kautokeino town. Here there are often nice, moist pastures down to the water's edge, with resting dabbling ducks, *Tringa* waders and others. This site is located along the E45 3.3 km south of the Gievdneguoika junction between the E45 and Rv92, and is well suited for a short roadside stop.

c. Niittojávrrit pond is a small, half-overgrown, roadside lake that can hold a nice selection of resting wetland birds. In June, it may hold Red-necked Phalaropes in the hundreds. The site is located along the E45 11 km before Kautokeino town. Stop at the rest area on the right-hand (west) side of the road.

d. The Biedjovaggi road leads past wetlands of great importance to geese, ducks and shorebirds. From Kautokeino town, follow the road northwest towards the old Biedjovaggi gold mine. After 10 km, take a 1 km detour on the side road signposted to Čunovuohppi which leads to a mountain lodge. Here, in the marshland on the south side of Lake Stuorajávri, displaying Spotted Redshank, Jack Snipe and more can be found. Continue further in on the Biedjovaggi road. Along the way, several interesting lakes and wetlands can be explored, and you might want to hike 5 km east to the bird rich north end of the Lake Stuorajávri. On the left side of the Biedjovaggi road, 24 km from the Čunovuohppi junction, is the **e.** Leamšejávrrit marshland with its small lakes and mosaic of pools holding Broad-billed Sandpiper and Jack Snipe. Their display may be appreciated roadside.

f. Goahteluoppal wetland is a core area for Broad-billed Sandpiper and an important nesting and moulting area for Tundra Bean Goose, so special attention must be paid here. You reach the area by driving from Kautokeino 12 km on the road signposted to Ákšomuotki. Smew, Little Bunting and Arctic Warbler may be found in suitable habitat along the Ákšomuotki road. 400 m before the road crosses the Kautokeino river, take the unmarked gravel road to the right (towards Gálaniito). From here it is about 6 hrs of efficient walking (about 20 km) along a clear path to the old Goahteluoppal mountain lodge. The lodge are located in the south-eastern part of the wetland area, which stretches 11 km to the north-west and includes the marshland around lakes Azit and Muvrisjavri.

g. Opmoáhpi wetland can also be reached via the Ákšomuotki road. From the end of the paved section of the road, just after Kautokeino river, you can walk for about 2 hrs to the southeast. Alternatively, drive from Kautokeino 15 km south on the E45 to Siebe village, where you can cross the small Lake Siebejávri, that is if you brought a canoe or inflatable boat.

h. Lake Áidejávri is located along the E45 26 km south of Kautokeino, about 10 km north of the border with Finland. The areas around Áidejávri mountain lodge as well as the entire 28 km long stretch of the E45 from Siebe village to the border have a rich and regionally representative bird life. The fields on the north side of the mountain lodge is particularly favourable for resting geese, shorebirds, Skylark, Yellow Wagtail and other farmland species of the area. The rest area at the war memorial on the west side of the road, a little over a km south of the mountain lodge, has a nice view of Lake Áidejávri, in addition

Brown bear is a possible bonus in Anárjohka and in Pasvik valleys. The chance of finding them is greatest by searching systematically with a telescope over large areas of bog and along the banks of rivers and lakes. *Photo: Terje Kolaas.*

to a public restroom. Broad-billed Sandpiper and Jack Snipe are among the species you can hear displaying from the road in this area, e.g., 1–2 km north of Áidejávri mountain lodge.

i. Náhpolsaiva wetland is a magnificent site and among the few places on the Finnmarksvidda plateau where Crane and Little Stint have been found nesting. You can reach the location on foot starting from Láhpoluoppal mountain lodge, which is located along Rv92 9 km east of the Gievdneguoika junction. Park by the school on the south side of the road and follow the path under the power line along the west side of the Náhpoljohka river approx. 13 km straight south. Náhpolsaiva wetland spreads southwards from Náhpolsaiva hut and includes the marshes Buollánáhpi/ Sarvvagasjeaggi, Goddejeaggi, Gállojeaggi and the wetlands eastwards from Lake Lávželuoppal.

j. Lake Iešjávri is Finnmark's largest lake and is considered a core area in Norway for

several wetland birds. The area is most easily reached by walking north from Šuoššjávri mountain lodge, which is located along Rv92 33 km east of Lappoluobbal mountain lodge, 55 km west of Karasjok town, signposted to *Fjellstue* (mountain lodge). The path is marked with the Norwegian Trekking Association's (DNT) red Ts, and first takes you to Mollisjok mountain lodge at the southern end of Lake Iešjávri. This first stage is estimated at 5 hrs of effective walking. From here it is another 2-3 hrs' walk to the core areas along the east side of the lake.

k. Luostejohka watercourse is known for its rich abundance of geese, ducks and waders. The lush swamp forest along the river also has Lesser Spotted Woodpecker and a rich passerine fauna. The area is reached from the E6 between Karasjok and Lakselv towns, 3 km south of Skoganvarre campsite. Take the gravel road eastwards, signposted to Gaggavatn. Continue inwards to the end of the road – keep right at the dam and right at the large car park. Follow the watercourse upwards on foot from here. It may be worth contacting the staff at the campsite for further access details. In addition to being one of Finnmarksvidda plateau's most bird-rich wetland areas, the Luostejohka watercourse is popular with trout fishermen.

2. ANÁRJOHKA VALLEY

Finnmark County
GPS: 68.91044° N 25.64945° Ø (Basevuovdi)

Notable species: Taiga Bean Goose, Smew, Hazel Grouse, Willow Ptarmigan, Black Grouse, Capercaillie, Red-throated Diver, Black-throated Diver, Hen Harrier, Rough-legged Buzzard, Golden Eagle, Osprey, Gyrfalcon, Spotted Redshank, Jack Snipe, Red-necked Phalarope, Hawk Owl, Great Grey Owl, Short-eared Owl, Great Spotted Woodpecker, Yellow Wagtail, Waxwing, Dipper, Bluethroat, Redstart, Whinchat, Arctic Warbler, Siberian Tit, Great Grey Shrike, Siberian Jay, Arctic Redpoll, Parrot Crossbill and Pine Grosbeak.

Description: Anárjohka National Park covers 1,409 km² and is connected to the 2,850 km² Lemmenjoki National Park on the Finnish side of the border. Anárjohka National Park includes the south-eastern part of the Finnmarksvidda plateau (previous site), where all the typical birds of the plateau can be found. In addition, the national park includes large areas of pine forest in the east. It is usually this pine-clad valley that is visited by birdwatchers. This area has much in common with Pasvik valley (which see) and houses several sought-after forest species. Anárjohka valley is less accessible than Pasvik, being very marshy, having almost no roads or marked paths, and thereby much more difficult to manoeuvre in. However, Anárjohka appears more as true, untouched wilderness, if that is what you are seeking.

Other wildlife: A small Finnish-Norwegian population of brown bears lives in the area.

Best season: May to July.

Directions: *From Karasjok:* Drive Rv92 eastwards for 17 km in the direction of Finland, and take the gravel road signposted to Iskurasjok just before the border crossing.
Via Finland: Take the E75 towards Karasjok on Rv92, a few kilometres north of Inari. Take off from Rv92 on the Norwegian side of the border and follow the increasingly poor road signposted to Iskurasjok south as far as the car can handle – a good four-wheel drive will be able to take you the 65 km all the way to Basevuovdi (Helligskog hut). Alternatively, it is possible to follow the better road signposted to Angeli on the Finnish side of the river. If the water level is not too high, the river may be crossed on foot at Angeli. It may be a good idea to stop and ask for permission and advice from the guards at the border crossing.

Anárjohka valley on the border between Norway and Finland represents pure wilderness. This site has some of the same qualities as Pasvik valley but with significantly less human traffic. *Photo: Bjørn Olav Tveit.*

343

Tactics: You should set aside several days to explore Anárjohka and expect to hike some distance. There are few cabins and no tourist-marked trails, so the trip must be considered and prepared as a wilderness expedition. A map and compass as well as copious amounts of mosquito repellent are mandatory. Be prepared for the snow to cover much of the terrain until well into June. Basevuovdi is an open *gamme* (Sami-style hut) set up and used by gold diggers, and you can apparently find small amounts of gold in the Anárjohka river gravel without much effort. If you avoid becoming obsessed with gold fever, you move south on foot through mature, crooked pine forests with the opportunity to encounter most of the above mentioned birds. Three kilometres up along the river you reach *Andreas Nilsen hut*, which is also an open gamme. It marks the border of the national park, and beyond this you are completely left to yourself in the wilderness.

3. ALTAOSEN DELTA

Finnmark County

GPS: 69.96733° N 23.42373° Ø (trailhead to Rørholmen)

Notable species: Staging ducks, shorebirds and other wetland species; Black Guillemot, Tengmalm's Owl, Lesser Spotted Woodpecker, Yellow Wagtail, Redstart and Ring Ouzel.

Description: The Altaosen delta is formed at the outlets of the grand Altaelva and the smaller Tverrelva rivers. The delta spreads out into the Altafjord in a 4 km wide fan shape. At low tide, the mudflats extend continuously as far as Rafsbotn, 11 km to the east. This makes Altaosen delta the best place for staging shorebirds in western Finnmark, and you can also find geese, ducks and sometimes Crane here, the latter especially at the turn of April/May. In recent years, a few Lesser White-fronted Geese have also been found

Altaosen delta in the midnight sun in the last half of May. The large open areas here on Rørholmen islet and elsewhere in Finnmark often provide long distances that make a telescope a prerequisite if you want to see as much of the bustling birdlife as possible. Alta airport can be seen in the background. *Photo: Bjørn Olav Tveit.*

Pintail is one of the most common dabbling ducks in Finnmark. *Photo: Bjørn Fuldseth.*

forest with the same birds in the Tverrelvdalen valley just east of Aronnes, where you in addition may find Tengmalm's Owl and Redstart. If you travel further up either of these two valleys, the landscape soon becomes steeper and wilder. Here you will find among the country's densest populations of raptors, in addition to several mountain birds, including Ring Ouzel (see the section on Alta river canyon and watercourse in the chapter on the Finnmarksvidda plateau). A few rarities have been encountered in Altaosen delta, including Ring-necked Duck, Avocet and Red-rumped Swallow.

staging here in May. The area is particularly important as a moulting ground for Goosander in summer, often with significant numbers in winter as well.

In the east, the Altaosen borders the Lathari nature reserve, an untouched area of heather pine forest on a dry sandy moraine by the fjord. There are few bird species here, but this is the only place in Finnmark where a large pine forest borders the sea. The outlet of the Transfarelva river further east has good qualities as a shorebird site, and it is also well worth checking the several smaller river outlets further east along the fjord. To the west from Alta Airport you, can get a view of the outer part of the Altafjord from Amtmannsnes. Here there can be several terns and gulls in summer, especially in connection with strong onshore winds. At the seashore around Bossekop further west, you often find gatherings of gulls and ducks.

If you follow the Altaelva river up to Aronnes, you'll find a number of nice backwaters and oxbow lakes, often with nice gatherings of ducks and shorebirds. The floodplain forest in this area contains Lesser Spotted Woodpecker and many of the common deciduous forest species of northern Norway, such as Sparrowhawk, Woodpigeon, Blue Tit and Yellowhammer, species that have their northern limit about here. You have similar

Best season: April to September.

Directions: Alta town is located along the E6 and with direct flights to Oslo and several other towns.

Tactics: The Altaosen delta can be seen from the airport area if you arrive by air. But you will need a car and a telescope here. On large mudflats like this, it is best to look for birds in the hours just before the tide is at its highest. The birds often concentrate at the outlet of rivers and streams. At high tide, **d.** Rørholmen and the other islets in the delta is recommended, and you can then also find shorebirds in the surrounding agricultural areas. At low tide, you can search the backwaters, like the one at **c.** Teglplana.

The sites around Alta can be reached as follows: Alta Airport is located on the western bank of the river delta, and you can find migrants, such as geese, plovers and pipits along the runway. This can be viewed from several points adjacent to **a.** Elvebakken residential area just outside the airfield fence and from **b.** Kronstad on the opposite side of the river. Further a little over a kilometre east on the E6, turn left just after the E6 crosses the Tverrelva river, signposted Teglbakken. This leads down to **c.** Teglplana backwater, a nice place for *Tringa* species and other shorebirds, dabbling ducks and more. Here

Oslo · Interior · Skagerrak · Western · Central · Northern · **Finnmark** · Svalbard

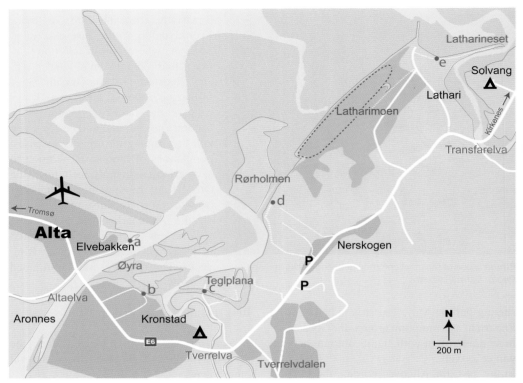

Altaosen delta: a. viewpoint from Elvebakken; **b.** viewpoint from Kronstad; **c.** Teglplana backwater; **d.** viewpoint to Rørholmen islet; **e.** Lathari beach at the Transfarelva river mouth.

you can get them at comfortable photo range from the farm road that is built over the backwater, or from the tractor road that crosses a little further east. Out on the E6 again, continue a km east, until 50 m before a pedestrian crossing. Here, a tractor road goes down to the left between the residential buildings. Park here, put on rubber boots and walk 400 m down to the shore overlooking **d.** Rørholmen islet. Further 1.7 km east along the E6, turn left onto Latharimoen road. This leads to Lathari beach and the **e.** Transfarelva river mouth, with its shallow lagoons and sand banks. This is also a good starting point for a stroll in the Lathari pine forest. Feel free to follow the E6 further east to Rafsbotn, making stops especially where small rivers and creeks meet the fjord.

Westwards on E6 from Alta Airport, follow signposts to Amtmannsnes to find the viewpoint here. You find interesting seashore by continuing on the E6 through Bossekop town, turning right into Bossekopveien road (keep to the right again along the same road). When you reach the fjord, turn left onto Strandveien street.

Further west along the E6 there are several bays and inlets along the fjord worth investigating. In Kvenvik (7 km west of Bossekop) Slavonian Grebe often is found in May, while Melsvik inlet by Talvik (a further 30 km further west) has a regular occurrence of Shelduck and potential for shorebirds.

If you want to investigate the Altaelva river valley and Tverrelvdalen valley, there are roads signposted to Aronnes and Tverrelvdalen from the E6 in Alta. To reach the upper part of the Alta river watercourse, as well as the locations on the Finnmarksvidda plateau, head for the E45 towards Karasjok.

4. Hammerfest area

Finnmark County

GPS: 70.66720° N 23.69472° Ø (Mollafjæra), 70.44802° N 24.31314° Ø (Repparfjordbotn)

Notable species: King Eider, Golden Eagle, Black Guillemot, seabird migration; migrants.

Description: Hammerfest town is sheltered from the Norwegian Sea by the massive Sørøya island. The town's surroundings are barren and hilly, with few typically bird-friendly habitats. During migration times, however, Hammerfest can be favoured with a direct migration of skuas, divers and other Arctic seabirds that take the shortcut through Sørøysundet, the sound between Sørøya island and Hammerfest. The harbour is ice-free throughout the winter and can house exciting gulls all year round, as well as King Eider in winter. Ivory Gull has been found here a couple of times. Some small and nutrient-poor freshwater lakes in and around the city can accommodate gulls, ducks and more. The lack of natural forest and scrub vegetation makes the city's sparse gardens potentially attractive to resting passerines.

From Hammerfest it is only about half an hour's drive to Repparfjorden. In this fjord you can find resting ducks, divers and other seabirds. Repparfjordbotn, the head of the fjord, with the Skaidejokka river mouth, also has decent conditions for resting dabbling ducks, shorebirds, gulls and such.

Further out the fjord, at Kvalsund sound, there can be a nice gathering of gulls. Three-digit numbers of Red-necked Phalarope on autumn migration have been seen here on particularly good days.

Hammerfest otherwise has an express ferry connection to the fascinating sites Ingøy and Rolvsøy islands (which see). Seiland National Park is located south of Hammerfest, can be reached by boat and may also be worth a visit, primarily because of its magnificent nature and scenery, but also because it may have potential for raptors and several mountain species. Sørøya island off Hammerfest also has magnificent nature and exciting bird habitats to offer, especially regarding resting migrants around the settlements.

Directions: Hammerfest is located at the end of Rv94. The town can also be reached by plane or boat.

Lake Storvatnet in the centre of Hammerfest town. *Photo: Bjørn Olav Tveit.*

Ingøy island is exposed to the harsh climate along the Barents Sea coast. There is little vegetation other than occasional patches of waste-high herbiage. Here on the west side of Inga village, you will encounter such vegetation, as well as a small beach with seaweed and a few fishing huts, which together form one of the best areas for resting passerines on the island. *Photo: Bjørn Olav Tveit.*

Tactics: Rv94 passes by Hammerfest harbour. At the junction signposted to (Lake) Storvatnet, the Storvatnelva river flows into the sea at Mollafjæra seashore. Both here and in Lake Storvatnet, gulls and ducks often gather. If you drive west towards the airport, turn right signposted to Forsøl. This road takes you past three small lakes in a row: first Lake Mellomvatnet followed by Lake Storvatnet (same name as the lake in the town centre), both on the left, and then Lake Rundvatnet on the right. These can also host gulls, ducks and perhaps some shorebirds.

If you drive south out of the town, you pass Lake Jansvatnet after just over a km, just before Rypefjord village. There seems to be a certain potential for shorebirds in the small marina in Rypefjord and a little further south along Rv94. Repparfjordbotn is explored from suitable stopping points along the roads around the river mouth at the head of the Repparfjord. There are roads along both sides of the Repparfjord, for exploring the other sections of the fjord.

5. INGØY AND ROLVSØY ISLANDS

Finnmark County

GPS: 71.08447° N 24.06205° Ø (Ingøy harbour)

Notable species: Migrating seabirds, resting shorebirds and passerines; King Eider, White-billed Diver, Ruff, Red-necked Phalarope, Glaucous Gull, Black Guillemot and Red-throated Pipit.

Description: Ingøy is an archipelago of 18 km², which includes Fruholmen lighthouse, referred to as the world's northernmost lighthouse. The island has some gardens and shorebird pools but is otherwise very weather-beaten and has very sparse vegetation. Vikermyran marshes in the south is a rather unique, coastal peat bog with a good population of e.g., Red-throated Diver and Arctic Skua. Lake Storvannet and Lake Steinvannet have nesting Ruff and Red-necked Phalarope, often with roosting gulls including

Glaucous Gull in Lake Storvannet. The areas around the island's centre (Inga) can be good for resting passerines. The same applies to the salt meadows eastwards from Kuhelleran on the southern tip of the island. Also check the stockfish racks on the island, which are the only thing here that can resemble a forest. There are otherwise very few willow thickets on the island, but instead in several places lush, contiguous meadows which make the search for passerines rather demanding. Even the modest ornithological activity carried out here indicates that Ingøy and Fruholmen have great potential for seabird migration. It is probably the best accessible migration site between Slettnes headland and Andøya island – with the possibility of rare passerines as well. Neighbouring Rolvsøy is larger and has nice marsh areas for nesting wetland birds, including Pintail, Red-throated and Black-throated Divers and Lapwing. Shoveler and Black-tailed Godwit have been found breeding here, as the northernmost place in the world for the latter species.

Direction: The two islands have express ferry connections from the towns of Hammerfest (1 hour 45 minutes) and Havøysund (45 minutes).

The ferry can take just a few cars, and so there may be a risk of not being able to bring it in busy times. There is only 12 km of road on Ingøy, so a bicycle can be a good alternative.

Tactics: Private boat ride is needed to reach Fruholmen lighthouse, which is best for seabird watching. Alternatively, you can sit on the hill Bakken just west of Inga settlement, but then with a 1.5 km further distance to the open sea.

6. The North Cape

Finnmark County

GPS: 71.16954° N 25.78190° Ø (North Cape), 71.09699° N 25.37197° Ø (Gjesvær harbour)

Notable species: Fulmar, Gannet, Shag, Storm Petrel, White-tailed Eagle, Willow Ptarmigan, Rock Ptarmigan, Long-tailed Skua, Glaucous Gull, Kittiwake, Puffin, Razorbill, Guillemot, Black Guillemot, Snowy Owl, Red-throated Pipit, Bluethroat and Snow Bunting.

Description: The North Cape is visited every summer by thousands of tourists from home and

Gannet colony at Gjesværstappan. *Photo: Tomas Aarvak.*

abroad, mainly because it claims to be Europe's northernmost point. However, Knivskjellodden peninsula, which is 4 km further west, sticks out a little further north, but getting there requires you to walk this distance. In any case, the nature and scenery here is spectacular and certainly worth the trip. If you look at the map, it seems as if North Cape is strategically located in relation to the huge seabird migration that passes along the coast both in spring and autumn. However, the North Cape plateau is actually a full 300 m above sea level, so that the distance is long even to the birds passing close to shore. Slettnes headland further east is therefore better suited, and it also represents Europe's most northerly mainland – the North Cape is, after all, situated on the Magerøya island.

But don't despair if your trip is limited to only visiting North Cape, because the rest of Magerøya island has good populations of attractive bird species such as Rock Ptarmigan, Long-tailed Skua and Snow Bunting. This is also one of the best places to look for Snowy Owl. There are several important seabird colonies in the Gjesværstappan nature reserve on the west side of Magerøya island. Gjesværstappan consists of several peculiar, tall and slim islands, called -*stappen*. Storstappen and Kirkestappen are high and steep classic seabird cliffs, where Fulmar, Kittiwake, Razorbill and Guillemot nest in the steep parts. A large colony of Puffin can be found on the gentler grassy slopes. Storm Petrel, Gannet, Shag and Arctic Skua breed as well. Bukkstappen is flatter, with nesting Greylag Goose, Eider, gulls and Black Guillemot. Gjesværstappan nature reserve is available on seabird safaris by boat from Gjesvær village. The small Valan wetland just east of Honningsvåg airport may be worth a check for nesting shorebirds, Red-throated Pipit and Bluethroat.

Best season: May to August.

Directions: Magerøya island can be reached by car from the E6 between Alta and Lakselv by taking the E69 at Olderfjord. From here it is 130 km to the North Cape (Nordkapp) and Gjesvær, both of which are clearly signposted from the main road.

7. THE PORSANGERFJORD

Finnmark County

GPS: 70.16629° N 24.91283° Ø (hide parking)

Notable species: Tundra Bean Goose, Lesser White-fronted Goose, Shelduck, Velvet Scoter, Common Scoter, Smew, Red-throated Diver, Black-throated Diver, White-billed Diver, Slavonian Grebe, White-tailed Eagle, Peregrine Falcon, Lapwing, Knot, Broad-billed Sandpiper, Jack Snipe, Bar-tailed Godwit, Little Gull, Black Guillemot, Bluethroat and Arctic Redpoll.

Description: The inner part of the Porsangerfjord represents one of Norway's largest contiguous shallow water areas and ranks as one of the country's most important bird localities. The area is particularly known for being the only remaining permanent staging ground in Norway for the critically endangered Lesser White-fronted Goose, and it is considered – together with Balsfjord in Troms County – as the most important stop-over area for Knot of the subspecies *islandica* on their spring migration. The Porsangerfjord is surrounded by Arctic salt marshes and large mudflats are uncovered at low tide, which together form very favourable resting places for wetland birds. The core of the area is Stabbursneset headland with Valdakmyra marsh, which is the primary haunt of the Lesser White-fronted Geese. From being an abundant species in Fennoscandia a hundred years ago, with perhaps 10.000 individuals in Norway, the population has declined dramatically, and it was at its lowest level at 30 individuals in 2007. Measures to combat red foxes and other predators in the breeding area, and against illegal hunting in the migration and wintering areas, have reversed the trend. The population was up to 137 individuals in 2023. The Lesser White-fronted Geese on Valdakmyra marsh are thought to represent almost the entire Norwegian population. Most of these birds nest in a secretly kept area on the Finnmarksvidda plateau and spends a week or so at Valdakmyra marsh before nesting. After breeding, they stay at

Valdakmyra marsh for about a month. From time to time, they can use other rest areas in the vicinity, such as Oldereidneset and Seines in the Porsangerfjord, in the Altaosen delta or along the coast of the Varangerfjord. In the autumn, most of the Lesser White-fronted Geese migrate from Valdakmyra marsh and directly to the wintering area in Greece.

Several other geese and dabbling ducks are regular visitors at Valdakmyra during migration. Tundra Bean Goose arrive on average a week earlier than the Lesser White-fronted Geese and can appear here with up to a couple of hundred individuals, from the beginning to the end of May. These birds breed on the Finnmarksvidda plateau and the Varanger Peninsula, in contrast to the forest marsh-breeding Taiga Bean Goose which breeds in Anárjohka and Pasvik valleys and is only rarely seen at Valdakmyra marsh.

An elaborate observation shelter has been set up on a ridge with a nice overview of Valdakmyra marsh. A telescope is neccessary here, as some of the birds will be more than 1 km away. Closer views may be obtained at Valdakbekken creek.

Flocks of many thousands of Knot in their brick-red summer plumage can roost on the extensive mudflats in the Porsangerfjord mid-May, before heading across the sea towards Greenland. A number of other shorebirds are also active users of the wetland. The most numerous are Greater Ringed Plover, Dunlin and Bar-tailed Godwit, the latter sometimes in four-digit numbers. Other attractive shorebirds such as Sanderling and Curlew Sandpiper are also seen in good numbers during migration. Broad-billed and Pectoral Sandpiper occur infrequently but regularly. The inner Porsangerfjord often has substantial numbers of White-tailed Eagle. Various diving ducks actively use the area both as a resting place during the migration and as a nesting area in summer. In particular, Common Eider, Goosander, Common and Velvet Scoters can be numerous at times. Divers are often seen here, including White-billed Diver.

The Porsangerfjord at the Stabburselva river mouth, with Stabbursnes headland seen from the north. *Photo: Ingar Jostein Øien.*

Valdakmyra marsh in August, seen from the c. observation hide (see map, p. 354). The a. viewpoint at Valdakbekken can be seen by the masts to the right in the picture. *Photo: Ingar Jostein Øien.*

North of Stabbursneset headland lies Lake Aigirvannet and a couple of other interesting freshwater lakes where Slavonian Grebe nests, and which quite often harbour Smew and Little Gull during migration.

Several vagrants have shown up in and around the Porsangerfjord, including Sandhill Crane, Cackling Goose and the first records in Norway of both Cattle Egret and White-tailed Lapwing.

Best season: The Lesser White-fronted Geese are usually in place at Valdakmyra marsh from the second week of May to the beginning of June – in some years until a week or two later. They are back again by the end of August and normally disappear before mid-September. May to September is the season for Shorebirds. However, it is only for a short period around mid-May that the huge flocks of Knot roost in the Porsangerfjord area. In autumn and winter, the area holds significant numbers of ducks.

Directions: The site is located along the E6 just north of Lakselv town, which has an airport.

Tactics: The most rewarding tactic for exploring the Porsangerfjord is to drive along the fjord and stop and examine the surroundings with a telescope at suitable places. Starting from Lakselv town, you can start by exploring the inner parts of the Porsangerfjord and the lower part of the Lakselva river. In Lakselv you can also follow the lush valleys and waterways leading inland towards the south, in search of Little Bunting and Arctic Warbler in suitable terrain in early summer. Here you also have some hope of finding certain species that are unusual this far north, such as Dunnock, Blackbird and Wood, Garden and Icterine Warblers. Lakselvdalen valley also forms a nice gateway to the Finnmarksvidda plateau, with the Luostejohka watercourse as the nearest recommended destination, 26 km along the E6 in the direction of Karasjok.

For further exploration of the Porsangerfjord,

follow the E6 out of Lakselv in the direction of Alta along the western bank of the fjord. The entire fjord here is very shallow, and large mudflats are exposed at low tide. You can make several stops at suitable places along the 15 km stretch from the Lakselv river mouth to Valdakmyra marsh. Once you see Stabbursneset headland, make a stop at the lay-by with a small power transformer, just before the road turns away from the fjord and starts to climb uphill. Here, at the **a.** Valdakbekken creek outlet you have one of the best places in the area for close encounters with resting ducks, shorebirds and some of the other species you can find on Valdakmyra marsh. Check out the local tide table and try to arrive a couple of hours before the tide is at its highest. The shorebirds are then pushed closer to the shore. You may also be lucky enough to see the Lesser White-fronted Geese from here, often at close range, so be careful with your movements when standing on the ridge by the power transformer. There is a traffic ban on Valdakmyra in May and June, and from August 10 to September 20. Drive further along the E6 and up to the **b.** hilltop viewpoint, marked with an information poster

about the bird life in the area. Here you get a good overview of the southern part of Valdakmyra marsh. Note that here, the gravel road to Stabbursdalen valley (which see) starts on the opposite side of the road. For a view of the northern part of Valdakmyra marsh, continue for another 2 km on the E6. Turn right onto a gravel road just after the blue sign marking the arrival at Stabbursnes. Continue to the parking area 400 m from the main road and follow the path marked *Fuglekikking* (Birdwatching) on the marked path out to the **c.** observation hide on the ridge. This is the most reliable viewpoint to Lesser White-fronted Geese. A telescope is mandatory here. As there is a traffic ban down on the marsh during the most important weeks for the geese, you can instead continue further north along the escarpment (which here is a geological ice edge terrace) on foot or by car to the end of the gravel road at the far end of **c.** Stabbursnes headland. Here you get a view over large parts of Valdakmyra and to the delta area where the Stabburselva river flows into the Porsangerfjord. Be careful not to show yourself unnecessarily over the ridge in order not to scare away the birds.

Lesser White-fronted Goose has its only remaining regular staging ground in Norway at Valdakmyra marsh in the Porsangerfjord. *Photo: Ingar Jostein Øien.*

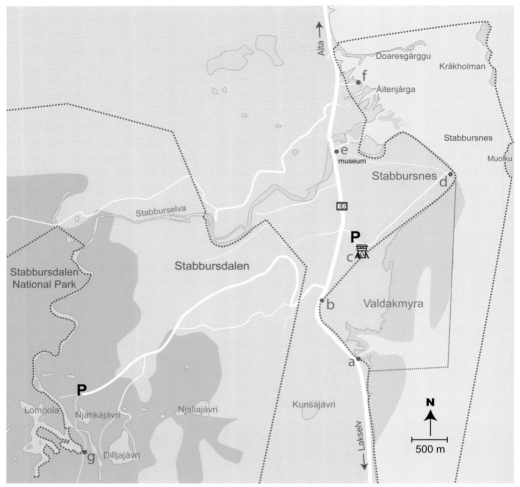

The inner Porsangerfjord with Valdakmyra marsh and Stabbursdalen valley: a. Viewpoint at Valdakbekken; **b.** hilltop viewpoint; **c.** observation hide; **d.** viewpoint Stabbursnes; **e.** Stabbursnes museum; **f.** viewpoint Áitenjárga; **g.** Lompola wetland in Stabbursdalen valley.

e. Stabbursnes museum - which is also the Stabbursdalen National Park Centre - is located along the E6, two km north of the exit to Valdakmyra's observation hide. But you can also walk to the museum from the tip of Stabbursnes by following the ridge for 2.8 km on foot along a marked nature trail that takes you along the south side of the Stabburselva river outlet.

Further along E6 in the direction of Alta, a few km after the road crosses Stabburselva river, a gravel road turns off to the right. By following it 400 m outwards to an old gravel soccer pitch, you get a view from **f.** Áitenjárga to the saltmarshes in the northern Stabbursnes delta. The area can house some of the same species

as at Valdakmyra marsh, but they can often be viewed here at closer range. There is no traffic ban here during migration but it is, of course, important to be careful not to disturb resting birds here either. After yet another 4 km along the E6, the small Igeldas inlet appears on the right side of the road. This is one of the best places to see Knot and Bar-tailed Godwit up close in May, especially good for photography and colour ring reading. The place is reliable for Turnstone, and you can also find other wetland species here.

Lake Aigirvannet. Almost 10 km further along the E6 from Igeldas, you will see the small, nutrient-rich lake on the right. Drive into the side road just before the lake and scan it from the roadside. Several pairs of Slavonian Grebe nest here and you can often find Smew – especially in late summer – and Little Gull, in addition to a selection of the more common wetland species. Lake Gåradakvannet. Continue a few hundred metres north on the E6 and take the first road on the left, signposted to Gåradakvatn. This fairly large body of water, without significant edge vegetation, comes into view after just over a km. You can find resting ducks and divers here. The lush birch forest that flanks Lake Gåradakvannet in the north-west is worth investigating for species that are otherwise not so common this far north. Wren, for example, has been found nesting here a couple of times.

Lake Kolvikvannet, which is divided by the E6 one km north of the exit to Lake Gåradakvannet, is perhaps the easiest place to see Slavonian Grebe up close, and several species of duck nest here, including Velvet Scoter.

Indre Billefjord, yet a couple of km further north, is a side fjord to the Porsangerfjord where you find salt meadows, occasionally with resting geese, shorebirds and more. The marshy areas below the soccer pitch housed Norway's first White-tailed Lapwing. The fjord itself can be good for resting ducks and divers here. The same applies to the areas a little further north in the Porsangerfjord, such as the salt meadows and fields at Kistrand. In this part of the fjord, you can find considerable numbers of Knot in May. It is also worth looking for the large divers where the fjord bends into the Olderfjord.

8. STABBURSDALEN VALLEY

Finnmark County
GPS: 70.14979° N 24.78481° Ø (trailhead parking)

Notable species: Scaup, Capercaillie, Golden Eagle, Rough-legged Buzzard, Osprey, Spotted Redshank, Red-necked Phalarope, Hawk Owl, Three-toed Woodpecker, Lesser Spotted Woodpecker, Dipper, Waxwing, Redstart, Bluethroat, Siberian Jay, Siberian Tit, Arctic Redpoll, Pine Grosbeak and Little Bunting (scarce).

Description: In Stabbursdalen valley you will find what is considered the world's northernmost pine forest. The harsh climate means that the tree trunks in most places are short and robust, but in some places the trees are taller, especially in the lower part. The Stabbursdalen valley also forms the northern border for several conifer species, such as Capercaillie and Siberian Jay. The many old trees make good nesting places for raptors and hole-nesting ducks and owls. The pine forest borders lush birch forest down towards the Porsangerfjord. Stabbursdalen valley has magnificent scenery and gives a distinct wilderness feel. The Stabburselva river with its many branches forms picturesque waterfalls and ravines. Along the banks in the quiet parts of the river, Goosander and Goldeneye breed in cavities in the trees. In the marshy areas you will find several shorebird species. Little Bunting can also nest in Stabbursdalen valley some years.

The area is protected as Stabbursdalen National Park, which includes both the pine forested valley floor and the mountain range to the south. The mountains here are characterized by being rather low and rounded, with barren and blocky land with gravel deposits from the Ice Age. In this part of the national park, you can find some of the species associated with the Finnmarksvidda plateau, but it is mainly the forested part of Stabbursdalen that appeals to birdwatchers.

The most varied and bird-rich – and at the same time most easily accessible – part is the area known by the Kven name Lompola, which means river expansion. Here you will find floodplain and deciduous forests and marshland in addition to the pine forest.

Other wildlife: Red squirrel has its northern limit in Stabbursdalen, and you can also come across elk, domestic reindeer, wolverine, lynx and red fox. A bird trip to Stabbursdalen valley can be combined with salmon and trout fishing.

Best season: Some of the species are present all year round, but June and July have the richest bird life and the easiest access.

Oslo
Interior
Skagerrak
Western
Central
Northern
Finnmark
Svalbard

Børselvosen delta (seen from **e.**) has huge mudflats to offer shorebirds at low tide. When the tide rises, the birds are pushed closer to shore. *Photo: Bjørn Olav Tveit.*

Directions: See the map of the Porsangerfjord. From Lakselv, follow E6 north in the direction of Alta. About 11 km after crossing Lakselv river, the road goes up a hilltop with a gravel lay-by on the right. Here, Lompolavegen road goes off to the left (it is not signposted). Follow this winding gravel road for 6 km to a car park in the pine forest at the end of the road, with an information sign about the national park.

Tactics: From the car park, two marked footpaths continue into the forest. The red- and blue-marked path leading towards the southwest, signposted to Stabbursfossen (waterfall), takes you after 500 m over the river Diljohka (if the bridge still is intact after the spring flood) and further on for just over a km to g. Lompola, the wetland area by the quiet part of Stabburselva river. The green-marked path towards the west, signposted to Rørkulpen, takes you 1 km from the car park through a large pine forest to an open hut by Rørkulpen. Take your time and move quietly through the pine forest and wetlands.

9. BØRSELVOSEN DELTA

Finnmark County

GPS: 70.31256° N 25.50161° Ø (viewpoint Hestnesfjellet), 70.28697° N 25.50492° Ø (Lake Surbuktvannet)

Notable species: Ducks and shorebirds; Whooper Swan, Rough-legged Buzzard, White-tailed Eagle, Peregrine Falcon, Lesser Black-backed Gull, Black Guillemot, Bluethroat, Whinchat and Sedge Warbler.

Description: Børselvosen delta on the east side of the Porsangerfjord consists of huge mudflats, which at low tide cover more than 3 km² and are partly surrounded by cultivated land. Adjacent to this area is also the small brackish Lake Surbuktvannet, where Whooper Swan nest, and which can attract quite a few ducks, especially during migration. Little Stint and Whinchat have been found nesting here, and a few pairs of Lesser Black-backed Gull breed on the islets out in the fjord, which is the easternmost permanent breeding area in Norway. The first record in the country of White-winged Scoter was made in the Børselvosen delta.

Best season: May to August/September.

Directions: Børselvosen delta is located near Børselv village along Fv98 between Lakselv and Tana.

Tactics: You get an overview of the area from several lookout points – see the map. A telescope is necessary here. If you are coming from Lakselv, turn left 5 km before Børselv village on the road signposted to Viekker, to get to <u>Lake Surbuktvannet</u>. Drive down to the lake, where shorebirds can roost at low tide. A low pavilion has been set up just beyond, but it offers only a limited view of the lake. Rather, **a.** <u>climb up the hill</u> west of the lake. If you continue this side road 1 km further, you will come to **c.** <u>the boat sheds</u> at the end of the road. Here you get a view of the outer part of Børselvosen delta, from the south side. This far out in the delta, it is best for shorebirds at low tide. From the **b.** <u>Kokkosokka height</u>, you have a good overview of the fjord, where there are often significant quantities of Velvet and Common Scoters, grebes and divers.

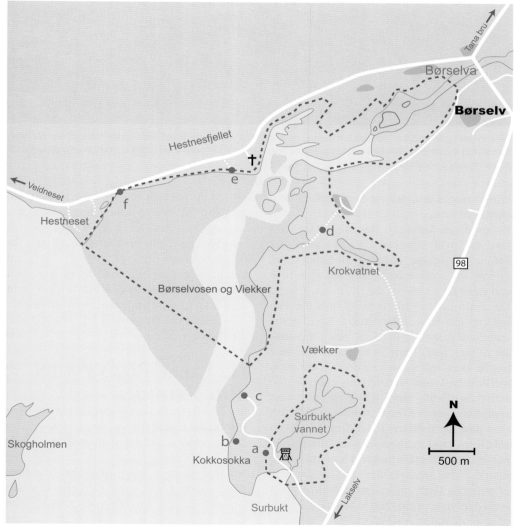

Børselvosen delta: a. hill viewpoint at Lake Surbuktvannet; **b.** viewpoint to the fjord; **c.** boat shed viewpoint to outer delta; **d.** inner delta (be careful not to spook the birds here); **e.** northern ridge viewpoint; **f.** pools and flashes.

Lille Porsangen and the Veidnes peninsula are wonderful sites with varied habitats that almost always have something exciting to offer birdwatchers. These areas are also important as nesting and staging grounds for shorebirds and wildfowl. *Photo: Bjørn Olav Tveit.*

White-winged and several Surf Scoters have been seen here.

Follow Fv98 further to Børselv village, and take the road signposted to Svartnes, just before the bridge crossing Børselva river. Follow the road downstream and stop here and there for a view of the riverbank. After 2 km, find suitable parking and follow the path out into the delta at the estuary. Here, in the innermost part of the delta, the best conditions are at high tide. There is no traffic ban here but be extra careful so you don't scare away the shorebirds and ducks resting here.

Back on Fv98, turn left onto the road signposted to Veidnes. Continue 2.7 km (just past the cemetery) and turn left onto a small dirt road marked with an information poster about Børselvosen and Viekker nature reserve. This leads to the suitably elevated **e.** northern ridge viewpoint, which gives a good overview of the huge mudflats and areas of shallow water. Also check the **f.** shoreline just below, with its small coves, puddles and saltmarsh. The lighting conditions are best

here in the evening and early morning. If you continue the tarmac road 1 km further, you can take these puddles closer in sight from the car. From here, you can also walk out onto the mudflats on low tide.

You have now come a short distance on your way out to Lille Porsangen and Veidnes (next section). The road there runs along the eastern coastline of the Porsangerfjord, which is also well worth exploring.

10. Lille Porsangen and Veidnes

Finnmark County
GPS: 70.66800° N 26.58932° Ø (Veidnes)

Notable species: Geese, ducks incl. Shelduck, White-tailed Eagle, Rough-legged Buzzard, Golden Eagle, breeding shorebirds incl. Lapwing, Woodcock, Bar-tailed Godwit, Greenshank, Ruff, Knot, Dunlin, Temminck's Stint, Little Stint, Turnstone and Red-necked

Phalarope; Black Guillemot, Red-throated Pipit, Bluethroat and Arctic Redpoll.

Description: Lille Porsangen is an 11 km deep side fjord to Laksefjorden. At the head of Lille Porsangen you will find mudflats, salt meadows and large marshes surrounded by willows and birch forest along the Lille Porsangerelva river. The area is particularly rich in nesting shorebirds. The salt meadows are home to Oystercatcher, Great Ringed Plover, Curlew and Lapwing. Common Sandpiper breeds along the streams, while Dunlin, Wood Sandpiper, Common Redshank, Ruff, Common Snipe, Bar-tailed Godwit, Whimbrel and Red-necked Phalarope are all found in the marshlands. Temminck's Stint, Little Stint and Turnstone breed in the dryer sections, particularly along the border between marshland and saltmarsh. At the edges of the forest, you will find Woodcock and Greenshank. During migration you will find many of these and several more species out on the mudflats. Knot can appear in numbers of several thousand individuals, especially in the second half of May. This general area is also rich in birds of prey. Out on Veidnes headland, near the end of the road, and in the strait along the road there, you will find nice shallow water areas and pastures, favourable for geese, ducks and shorebirds. Pink-footed Geese are sometimes seen here in spring, and the species has attempted to breed here a couple of times. A few vagrants have turned up here, including Isabelline Shrike.

Best season: In principle, interesting all year but mainly in May–September.

Directions: The area is reached by a 75 km detour from Fv98. The road is asphalted all the way, but poorly maintained in some

places. In Børselv, 167 km west of Tana and 41 km east of Lakselv, take the road signposted to Veidnes. Here you pass Børselvosen delta, which see. The road runs the first 40 km along the east coast of the Porsangerfjord, before it turns and cuts across the Sværholt peninsula towards Lille Porsangen, which you reach about 65 km from Børselv. The last kilometres down to the head of the fjord run along the Lille Porsangerelva river.

Tactics: You get a good overview of the area in several places along the road, which runs along the marsh area and then along the entire southern coast of the Lille Porsangen fjord and out onto Veidnes peninsula. In order to come sufficiently close to the resting shorebirds at the head of the fjord, it is necessary to walk a little along the beach and perhaps a bit out on the mud if necessary. At high tide, the Knots feeding in Lille Porsangen often rest in the outer Porsangerfjord. Here, they roost on the moraine ridge islands in the sea between Sløkevika bay and Kjæs settlement, which you can see along the road just before you cross over from the Porsangerfjord to

Knot has one of its most important staging grounds here. Many of the birds are colour ringed for research purposes. Please report findings of such birds on the cr-birding.org website. *Photo: Tomas Aarvak.*

Arctic Skua has one of its largest and densest colonies in Norway at Slettnes headland. The skuas can be seen robbing food from Kittiwakes, Arctic Terns and other seabirds. *Photo: Espen Lie Dahl.*

Lille Porsangen. White-billed Diver is often seen out here in the deeper parts of the fjord. Even though there is a birdwatching shelter at Veidnes, that site is best explored from the car, scanning the fields around the village and the tidal area from several angles.

11. SLETTNES HEADLAND

Finnmark County

GPS: 71.08901° N 28.21802° Ø (Slettnes lighthouse)

Notable species: Pintail, Velvet Scoter, Common Scoter, Long-tailed Duck, King Eider, Steller's Eider, Red-throated Diver, Black-throated Diver, White-billed Diver, Ruff, Little Stint, Turnstone, Red-necked Phalarope, Great Skua (scarce), Pomarine Skua, Arctic Skua, Long-tailed Skua, Glaucous Gull, Brünnich's Guillemot (migration), Black Guillemot, Snowy Owl (scarce), Red-throated Pipit, Bluethroat, Arctic Redpoll, Lapland Bunting and Snow Bunting.

Description: Slettnes headland is probably the best migration site for Arctic seabirds and waterfowl, and is also an important nesting area for a number of wetland species. Located at 71 degrees north, at the far end of the Nordkinn peninsula that juts out into the Barents Sea, Slettnes is strategically placed in relation to the huge migration of ducks, divers, terns and gulls that pass the Finnmark coast in spring and autumn. Certain species appear here in numbers that are sensational in an international context, such as daily numbers in the hundreds of King Eider, White-billed Diver, Glaucous Gull and Long-tailed Skua, and numbers in the thousands of Pomarine Skua. You can also have Brünnich's Guillemot passing by in considerable numbers. This naturally comes on top of significant numbers of more common species, of which Razorbill, Common Guillemot, Puffin and Fulmar appear on several days in the tens of thousands. Many of the migrating seabirds also pass Slettnes at close range and in good light conditions most of the day, something photographers appreciate.

Slettnes itself is a flat coastal plain in contrast to the surrounding mountain landscape. The habitats are dominated by barren tundra with several bogs and small lakes and can resemble the mountainous landscape in the interior of southern Norway, only that it is situated by the seashore. There is hardly any vegetation above ankle height, only some low willow thickets in the depressions. The majority of the area is protected as a nature reserve with Ramsar status due to its great importance as a breeding ground for gulls, ducks and shorebirds. Arctic Skua is one of the dominant species, and the population here is considered one of Norway's largest and densest. The skuas mainly live by parasitizing the local Arctic Terns and passing Kittiwakes. A couple of Great Skuas have also nested in recent years. There are good breeding populations of Red-necked Phalarope and Ruff, as well as Pintail, Common Scoter, Long-tailed Duck, Red-throated and Black-throated Divers and Turnstone, some years also Little Stint and Lapwing. Among the commonly occurring

Transportation by ship

Finnmark is a large county with many deep fjords, and it can be very time-consuming to travel by car between the sites. On some stretches, you can also risk convoy driving over the mountain passes because of drifting snow, even in early summer. An alternative may be to take the car on board the Norwegian Coastal Express (*Hurtigruten*). There can often be exciting birds to see on the journey, especially on the stretches crossing the open sea. Particularly good is the stretch Slettnes–Varanger, which by car involves an arduous trip of around 5 hrs on a winding road. You can cover the same stretch (Mehamn–Berlevåg) with the ship in 2.5 hrs of relaxation. Other relevant sections are Vardø–Båtsfjord, Kjøllefjord–Honningsvåg and Honningsvåg–Havøysund. On some departures, it may be a good idea to book in advance.

Slettnes forms the outer tip, has proven to be one of the most reliable areas for Snowy Owl in recent years.

Slettnes has limited favourable habitats for resting migrants. There are only a couple of small shorebird-friendly inlets here and you must go into Gamvik village to find any real bushes. Even in the most cultivated gardens, willow thickets dominate in this harsh climate. However, such a lack of habitats can have the positive effect that the birds are concentrated in these limited areas. However, they are unlikely to settle there for long. The potential for rarities is great in this area, proven by records of Black Scoter, Ross' Gull, Black-browed Albatross, Greater Sand Plover, Calandra Lark, Semipalmated and Stilt Sandpipers.

Traffic in the nature reserve has been arranged with prepared, marked paths as well as simple practical facilities. It is possible to rent a room at the lighthouse. There is high traffic here from tourists who want to visit this northernmost, accessible mainland point in Europe, marketed as the *Little North Cape*.

species are also Red-throated Pipit, Wheatear, Arctic Redpoll, Lapland and Snow Buntings, with a few pairs of Bluethroat in the willow thickets. The surrounding mountain areas are home to good populations of birds of prey, including Golden Eagle and Peregrine Falcon. The large Nordkinn peninsula, on which

Slettnes lighthouse as seen from the nature trail in the nature reserve. *Photo: Bjørn Olav Tveit.*

Oslo
Interior
Skagerrak
Western
Central
Northern
Finnmark
Svalbard

Only Cape Kinnarodden nearby is situated slightly further north, but a visit here requires a few hours of hiking in rather demanding terrain. Slettnes lighthouse is considered the world's northernmost mainland lighthouse.

Other wildlife: Marine mammals occur in considerable numbers and species diversity. Killer whale (orca) is quite common. Other whale species are also seen regularly.

Best season: May–September for nesting and resting ducks and shorebirds. Migrating birds at sea can be seen most of the year, but the intensity and species selection is greatest from the end of April to the beginning of June and in September and October. The best time

for Pomarine Skua and White-billed Diver is mid-May, although large numbers of Pomarine Skuas have been seen as early as late April. The most intense migration often occur during low pressure passages with winds turning towards the western sector. However, onshore winds in general (northern sector) are considered to be good. The autumn migration is less concentrated than the spring migration, but good day-counts can occur then as well. King Eider appears regularly in Gamvik harbour, and White-billed Diver resting on the sea is not unusual in winter and early spring.

Directions: *By plane or ship:* Mehamn is a port for the coastal express ship. It also has a small airport with daily flights to and from Tromsø

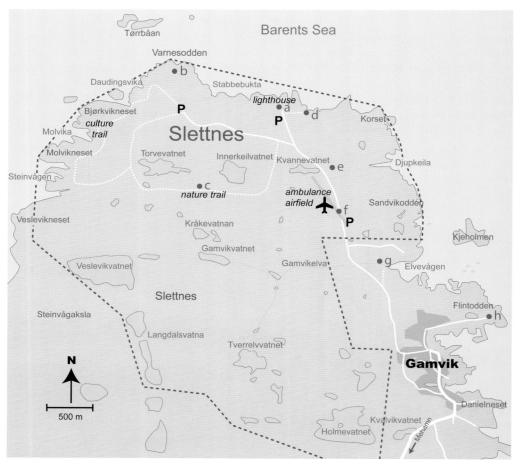

Slettnes headland: a. Slettnes lighthouse; **b.** Varnesodden, close seabird encounters; **c.** trails through the nature reserve; **d.** *Lighthouse Bay*, shorebirds; **e.** bird observatory; **f.** airfield; **g.** Elvevågen inlet; **h.** Flintodden headland, alternative seawatch.

Snowy Owl has become rare in Norway but you still have the opportunity to come across one, especially here in the north. The tundra areas on the large peninsulas of Finnmark are the places to look for them. The most effective method is to stop regularly at ridges along the road and look over the terrain with a telescope. If you are lucky and find a snowy owl, you should wait until the breeding season is over to announce it to the general public. *Photo: Bjørn Fuldseth.*

and Kirkenes. Take a bus, a taxi or hire a car to get via Gamvik village to Slettnes headland. *By car:* From Fv98 in Ifjord (85 km west of Tana, and 123 km east of Lakselv) exit onto Fv888 signposted to Mehamn. Follow the road 100 km to Mehamn town, and further 20 km to Gamvik village. In the centre of Gamvik, follow the signs to Slettnes.

In winter and spring, access to Slettnes by car from Eastern Finnmark can be hampered by the fact that the road across the Ifjord highlands is sometimes affected by convoy driving or short-term closures. However, a detour will usually be possible on the E6 via Karasjok and Lakselv. This detour is 380 km long and takes just over 5 hours, while the normal route Tana–Ifjord along Fv98 is 87 km and takes 1.5 hrs. Check with the local road authorities for an update on the road's status, which can change at short notice.

Tactics: In calm weather conditions, the seabird migration is most easily observed from **a.** Slettnes lighthouse. You can park at

the lighthouse gate and find shelter behind the buildings. The wind is usually bone-chilling in Slettnes even in the middle of summer, so show up in your complete winter plumage! Out on **b.** Varnesodden headland, you can get very close to some of the migrating birds, perfect for photography and close-up studies. In winter and early spring, if the road to the lighthouse is closed by snow, you can seawatch from **h.** Flintodden headland in the northern part of Gamvik village.

Nesting and resting birds are best observed along the roads and by walking the marked nature and cultural trails. The trails are prepared in places with stone slabs and boulders, with several information posts along the route. There is no traffic ban in the reserve, but out of consideration for the birdlife and the vulnerable vegetation, visitors are asked to stick to the marked and prepared paths. The culture trail takes you around the outer areas, while the nature trail goes through the wetlands further in on the coastal plain. The nature trail takes you around the lakes Torvevatnet and Innerkeilvatnet and is usually the most rewarding for birdwatchers. The two paths merge on parts of the route. The combined nature and cultural trails are approximately 5 km long, and you should expect to spend a couple of hours at a leisurely pace. Remember warm clothes and waterproof footwear. Red-throated Pipit can be found in the wettest parts, Snow and Lapland Buntings in the barren parts, and Bluethroat in the low willow thickets. It can also be nice to walk out along the old, gravel-covered **f.** ambulance airfield, but be considerate, as the north-west end of the airstrip is a nesting area for several species. Resting gulls and shorebirds can be found in the **d.** Lighthouse Bay, 200 m east of the lighthouse, as well as in **g.** Elvevågen inlet just outside the border of the nature reserve. This inlet has willow thickets in the innermost part, which should be checked for Bluethroat and migrant passerines. Birds of the latter category can also be searched for in the gardens of Gamvik village. Check Gamvik harbour for gulls and shorebirds – there is exposed mud in the coves here at low tide. There are often small flocks of Steller's and King Eiders both in Gamvik and Mehamn in winter. The river outlet near the airport in Mehamn is also worth checking if you're driving past anyway.

When driving to and from Slettnes along the Fv888, stop and scan regularly for Snowy Owl on the tundra of the Nordkinn peninsula, particularly between the Hopseidet isthmus and the Kjøllefjord junction. Although the landscape here at first glance seems barren and lifeless, a closer look will reveal that there are several lush marshlands where species such as Long-tailed Skua and Little Stint can be found.

12. TANA DELTA

Finnmark County

GPS: 70.51184° N 28.45164° Ø (Høyholmen), 70.49840° N 28.36250° Ø (Kaldbakknes), 70.13821° N 28.16464° Ø (Mannsholmen, Tana river)

Notable species: Ducks incl. large numbers of Goosander and Red-breasted Merganser, shorebirds, often incl. Knot, Little Stint and Temminck's Stint; White-tailed Eagle, Rough-legged Buzzard, Gyrfalcon . Peregrine Falcon and Black Guillemot.

Description: At the Tana river mouth, deep in the Tanafjord, a large delta has formed. Sediment drift in the river and form extensive sand and mud banks at the mouth, changing position and shape from one year to the next. Along the shores there are large salt meadows with rich Arctic vegetation. The Tana delta is an important staging ground for geese and ducks, and of particular significance for the northern Scandinavian population of Goosander. Large flocks, consisting mostly of males, stay here during the moulting season from July to late autumn. More than 21,000 Goosanders have

White-tailed Eagle has become common in eastern Finnmark. When they are not out flying, you often see them sitting out on the shore along more remote stretches of coast. In winter, dozens can be seen in the Tana delta. *Photo: Ingar Jostein Øien.*

been counted. Red-breasted Merganser also appears in significant quantities. In autumn, this site is considered the best in Finnmark for shorebirds but can also be productive for this species group in late spring, a time when Vadsøya and Nesseby by the Varangerfjord are considered somewhat better. The flocks of shorebirds usually consist of several species, and are often dominated by Greater Ringed Plover, Dunlin and Little Stint. Høyholmen often provides close contact with displaying Temminck's Stint. In May, significant numbers of Knot can also roost in the delta. Several rarities have been encountered in the Tana delta, including Greater Sand Plover and Mediterranean Gull. Divers and large numbers of ducks winter here, a resource that a significant number of White-tailed Eagles know how to exploit. The species has increased in number in recent years and is now seen all year round with up to about 30 individuals at the same time in October. Golden Eagles are also seen frequently, and they are often chased by the White-tailed Eagles.

The Tana river is formed by the confluence of the rivers Kárášjohka and Anárjohka east of Karasjok village. It is the second longest river in Norway. The Tana watercourse and valley represent an important migration flyway both in spring and autumn. In several places along the river, you can find nice resting places for wetland birds, although the qualities of these varies from one year to the next with the water flow, the season and the agricultural activities along the banks. A few vagrants have been encountered along the Tana watercourse, including Spotted Sandpiper and Griffon Vulture.

Other wildlife: The area has a population of resident harbour seals, with breeding sites on the sandbanks at the far end of the delta. Grey seal also resides in the delta all year round and otters are common. The river Tana is the largest and historically most important salmon river in Norway.

Best season: All year, but greatest diversity during migration times of late spring, summer and early autumn. However, the number of birds is greatest in late autumn, when tens of thousands of ducks gather here.

Directions: See the Varanger overview map, p. 372. Exit from E6 at Tana bru onto Fv890 northwards, signposted to Berlevåg. This road will also bring you to Varanger north, which see.

Tactics: After 30 km on the Fv890 there is a a. lay-by on the left side of the road. Stop here to scan the mud bank out in the river as well as the steep cliff face above, with the possibility of birds of prey including large falcons.
Further 7 km along Fv890 take a road off to the left, signposted to **b.** Høyholmen. It takes you the 3 km out to the small islet at the far end of the river delta, scanning for birds along the way.
You reach **c.** Leirpollnes by continuing a short km further on Fv890 to an unmarked gravel area on the left side of the road, with a nice view over the inner part of the bay on the east side of Høyholmen. A further 4.5 km further away you get a view of lagoons as well as

The Tana delta, seen from **b.** Høyholmen in July, with the characteristic Giemasvárri rock formation in the background. *Photo: Ingar Jostein Øien.*

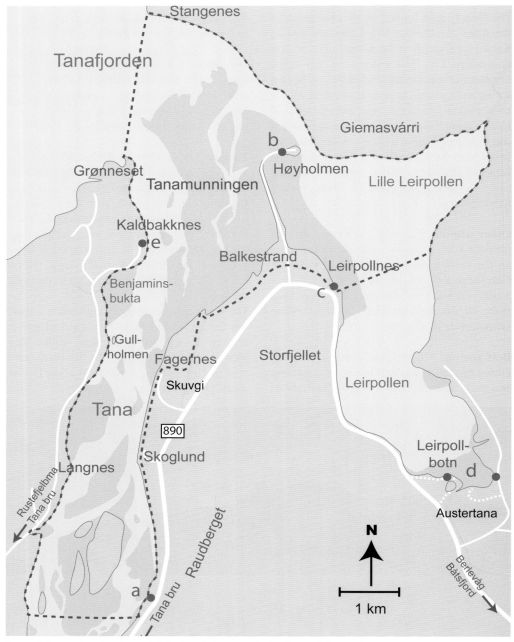

Tana delta: a. Lay-by viewpoint; **b.** Høyholmen; **c.** Leirpollnes; **d.** Leirpollbotn; **e.** Kaldbakknes. See the Varanger overview map on p. 372.

gravel and mud banks in **d.** Leirpollbotn by Austertana village.

For an overview of the Tanafjord, the river mouth and the outer parts of the Tana delta, head for **e.** Kaldbakknes headland on the west side. Follow Fv98 from Tana bridge northwards 23 km to

Rustefjelbma. If you are arriving from the west, you reach Rustefjelbma along Fv98 188 km east of Lakselv (and just over three hours' drive from Valdakmyra marsh). At the Rustefjelbma junction, turn north, signposted to Langnes, and continue just over 13 km to Kaldbakknes, where there is a rest area by the end of the road. Make stops along the way, wherever you get a view of interesting parts of the Tana river.

Distances to the bird flocks of up to a couple of km make a telescope necessary when visiting the Tana delta. On the east side, the birds are at their closest in the hours before the tide is at its highest. If you travel in a group, it can be practical if someone walks the 3 km along the beach between Høyholmen and Leirpollnes or Balkestrand and is picked up at the other end with the vehicle.

13. VARANGER NORTH

Finnmark County
GPS: 70.85194° N 29.23246° Ø (Kjølnes lighthouse)

Notable species: Tundra Bean Goose, Pintail, Scaup, Long-tailed Duck, Steller's Eider, King Eider, Rock Ptarmigan, Red-throated Diver, Black-throated Diver, White-billed Diver, Fulmar, White-tailed Eagle, Rough-legged Buzzard, Golden Eagle, Gyrfalcon, Peregrine Falcon, Dotterel, Little Stint, Temminck's Stint, Purple Sandpiper, Ruff, Red-necked Phalarope, Great Skua, Pomarine Skua, Arctic Skua, Long-tailed Skua, Glaucous Gull, Iceland Gull, Ross' Gull (rare), Arctic Tern, Common Tern, Black Guillemot, Snowy Owl (scarce), Skylark, Shore Lark, Red-throated Pipit, Dipper, Bluethroat, Arctic Redpoll, Lapland Bunting and Snow Bunting.

Description: Along the north coast of the Varanger Peninsula, facing the merciless Barents Sea, the landscape is barren, brutal and wildly beautiful. The birdlife is equally exciting, with Gyrfalcons patrolling the cliffs, White-billed Divers and King Eiders diving along the coast, and with a real possibility, although slight, of finding a Ross' Gull among the preening Kittiwakes. In Kongsfjord there are several seabird colonies on the small islands off the coast. Here, Kittiwakes and auks nest, as well as Cormorants and Shags. Arctic Terns and some Common Terns as well.

Kjølnes lighthouse near Berlevåg is the Varanger area's most strategically located seawatching site. Here, a variety of Arctic wetland birds regularly pass, such as sea ducks, divers, skuas and auks. The road out here is snow-ploughed and held open

Båtsfjordfjellet mountain plateau is an Eldorado for species such as Dotterel, Shore Lark and Rock Ptarmigan. Before the snow melts completely in late spring, most birds seek to the bare ground. An exception is grouse, which, if still in their white winter plumage, prefer to hide on the patches of snow. *Photo: Bjørn Olav Tveit.*

Gyrfalcon nests several places on the Varanger peninsula. The species mainly hunts ptarmigans inland and seabirds along the coast. The population of Gyrfalcon today is greatly reduced compared to earlier times. Human nest depredation and collision with reindeer fences are considered to be some of the biggest threats to the species here. *Photo: Tomas Aarvak.*

One of the main attractions of traveling to this part of the Varanger peninsula lies before the descent to the coast, namely up on the Kongsfjordfjellet mountain plateau. Here, the tundra landscape has interesting wetlands that are rich in birds, both in terms of the number of species and the density of individuals. The area is easily accessible and close to the more well-known birdwatching sites on the south and east side of the Varanger peninsula. The Tana delta lies as a bonus site along the route. On Kongsfjordfjellet it is freezing cold, with late snowmelt and incredibly sparse vegetation. At its most lush, you will find a few knee-high willow thickets in the lowest depressions. Here, you will find a number of breeding shorebirds, including Red-necked Phalarope, Dotterel, Purple Sandpiper, Bar-tailed Godwit, Little and Temminck's Stints

throughout the winter, which guarantees access to the seawatch in the best migration period in May, a time when the winter closed road to Hamningberg still may not have opened for the summer season. Counts from Kjølnes lighthouse in early May, include daily numbers of over 1,000 Glaucous Gulls, half a million Fulmars and several hundred thousand auks in mixed species flocks, including Brünnich's Guillemot. Similar to Slettnes and Hamningberg headlands, impressive numbers of White-billed Diver and Pomarine Skua also pass through in May. There is little vegetation near the lighthouse, but the gardens in the small town of Berlevåg and the willow bushes at Storsletta plain near the airport have great potential for passerines. At Storsletta plain there is a dense population of Red-throated Pipit. Shore Lark, Lapland Bunting and Arctic Redpoll nest here as well. Furthermore, Skylark has its northernmost European stronghold here. King Eider is regularly seen in the harbours of Berlevåg and Båtsfjord all year, but Steller's Eider has become increasingly hard to find in summer in recent years. The beaches in the area can accommodate shorebirds, particularly those in Berlevåg.

and Turnstone. A few pairs of Arctic Skua are to be found among the usually more dominant Long-tailed Skua. The occasional Pomarine Skua may also be among them – the species has bred on the Varanger peninsula at least once recently. An often reliable and accessible place for Rock Ptarmigan and Dotterel is along the side road towards Lake Oarddoskáidi on the municipal border between Berlevåg and Båtsfjord. It is also possible to find Snowy Owl on Kongsfjordfjellet mountain, but you will need a keen eye and quite a lot of luck finding it. Skånsvika inlet just past Berlevåg airport has nesting Temminck's Stint and Red-necked Phalarope, as well as regular visits by migrant shorebirds. The entire stretch from Sandfjord bay via Berlevåg harbour to Skånsvika inlet can be teeming with gulls, sometimes with five-figure numbers of both Kittiwake and Herring Gulls. Glaucous and Iceland Gulls can often be found among them, even in summer. Several records of Ross' Gull in the harbour area in Berlevåg make this probably the no. 1 site for this species in Europe.

The marshes and shallow lakes around Båtsfjord town are home to Red-necked Phalarope and other exciting wetland birds during the breeding season. In addition, the large Syltefjordstauran bird cliff is close by. This bird cliff, which is part of the Makkaurhalvøya nature reserve, is 4 km long, over 200 m high, and houses at least what to used to be one of Norway's largest colonies of Kittiwake. Razorbill, Black and Common Guillemot breed here also, and probably Puffin and perhaps still a few pairs of Brünnich's Guillemot. Outside the cliff face are several large stone pillars, which hold large numbers of nesting seabirds as well. One of these, Store Alkestauran, has housed the world's northernmost colony of Gannet since 1961. Today, the colony numbers around half a thousand pairs. Birds of prey, such as White-tailed Eagle, Peregrine Falcon and Gyrfalcon are regularly seen at the bird cliff. Snow Bunting and Rock Ptarmigan are among the few species that nest on the surrounding vegetation-poor mountain plateau. In the fjord outside the seabird cliff, there are often large numbers of ducks in summer, especially Goosanders. It is not unusual to find King Eider here.

Quite a few other rarities have been recorded along the northern part of the Varanger peninsula (west of Hamningberg, which is featured in the Varangerfjord section), including Laughing Gull, a few Ivory Gulls, Sandhill Crane (Gednje junction), Black Kite, and Norway's first records of Egyptian Vulture (Båtsfjordfjellet mountain plateau) and Yellow-browed Bunting (singing along Skånsvikelva river, west of Berlevåg).

Best season: The migration past Kjølnes lighthouse is at its most active from the end of April to the beginning of June and in September and October. The coast can be interesting here all year round. Late May to early August is best on Kongsfjordfjellet mountain plateau.

Directions: See the Varanger overview map on p. 372. From the Tana delta, continue Fv890 in the direction of Berlevåg. Both Berlevåg and Båtsfjord towns have a short-haul airport and coastal express calls. Note that there is often convoy driving over Kongsfjordfjellet and Båtsfjordfjellet mountains in bad weather in winter, sometimes well into spring.

Kongsfjord. The coast here is barren, cold and spectacular. Gyrfalcon breeds in areas like this. Islands in the fjord are important for Cormorant, Shag, and several other seabirds. *Photo: Ingar Jostein Øien.*

Oslo

Interior

Skagerrak

Western

Central

Northern

Finnmark

Svalbard

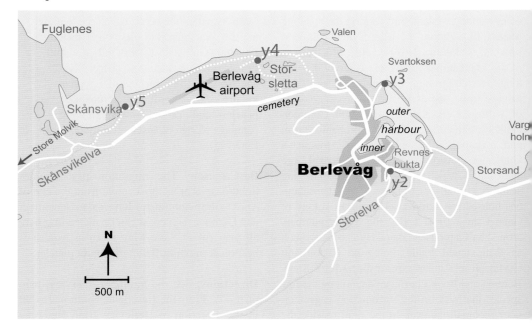

Tactics: The mountain plateaus are crossed by the Fv890 and Fv891, and a lot can be seen roadside. Stop often, scan and walk around the terrain to maximize the outcome. In spring and early summer, before all the snow has melted, bare spots concentrate the birds and may seethe with shorebirds, skuas and passerines. For the greatest chance of Dotterel and Rock Ptarmigan, you should search at night in the dim mid-night sun, when the birds are most active.

You reach Kongsfjordfjellet mountain plateau about 20 km after the exit to Høyholmen in the Tana delta, and around 60 km after you left the E6 at the bridge crossing the river at Tana bru.

Lake Stjernevannet and the upcoming lakes and waterways needs to be scanned carefully. Further down the road you reach the Gednje junction where the road splits: Fv890 continues towards Berlevåg, while Fv891 takes you to Båtsfjord and Syltefjord.

Gednje junction. This is the epicentre of birdwatching on this mountain plateau. The small ponds, marshes, and waterways here are easily scanned from the road, and most - or all - of the typical breeding species in the area may be found around this junction.

Fv890 towards Berlevåg. If you head in this direction from the Gednje junction, you pass the first 9 km through **y**. Gednjedalen valley with the Kongsfjordelva watercourse's rich wetlands. There are also several narrow gravel side roads that may be explored. Along the Fv890, Ruff often displays on the marsh 1.6 km from the Gednje junction, and Dipper breed along the river that empties Lake Buetjernet, which crosses underneath the main road 6.7 km further ahead.

x. Kongsfjord. The Fv890 takes you down towards the sea at Kongsfjord. The harbour here needs to be checked for gulls, before you drive out onto Veidnes peninsula, a few hundred metres further on. Here it is also possible to find some shorebirds in the bays on both sides of the road. Terns and Black Guillemot breed on the partially collapsed jetty in the bay by the settlement. Lake Veidnesvannet often contain some dabbling ducks and shorebirds. From Veidnes, continue Fv890 a couple of km further. Both Shag and Cormorant tend to congregate on the sea around the small, rugged islet out in the Risfjord. They can be viewed from quite a comfortable distance, if you stay inside the vehicle. The road further towards Berlevåg has a truly spectacular cliff coast where Gyrfalcon and Ring Ouzel breed. Stop and

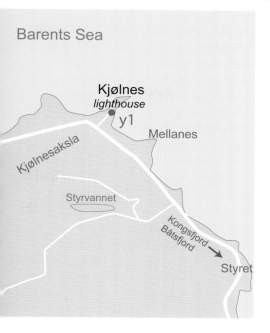

Barents Sea

Kjølnes
lighthouse
y1

Mellanes

Kjølnesaksla

Styrvannet

Kongsfjord
Båtsfjord

Styret

Berlevåg: y1. Kjølnes seawatch; **y2.** Storelva river delta; **y3.** harbour viewpoint; **y4.** Storsletta plain; **y5.** Skånsvika inlet. See the Varanger overview map, next page.

scan the fjord with the telescope at suitable places. Sandfjord bay has flocks of gulls that sometimes number in the thousands, and there are often divers and flocks of various sea ducks in the bay. The thickets at the outlet of the Sandfjordelva river are good for Arctic Redpoll. Red-throated Pipit and Lapland Bunting breed here as well.

y. Berlevåg. About 5 km before you reach the town, you see **y1.** Kjølnes lighthouse on the right. This is a top notch seawatch. However, you need to make an effort to find shelter from the wind. When you arrive in Berlevåg town, you see the **y2.** Storelva river delta on your left. This is one of Norway's most exciting gull sites. Here, Ross' Gull is almost annual while Glaucous and Iceland Gulls are regular even in midsummer. Kittiwake and Herring Gull rest in their thousands at low tide and the replacement is constant so you can study the gull gathering for hours and constantly discover new individuals. Revnesbukta inlet and **y3.** Berlevåg harbour has King Eider all year round and Steller's Eider in winter. The breakwaters can house thousands of Kittiwakes and the species breed in large numbers on buildings along the inner harbour.

y4. Storsletta plain between the settlement

in Berlevåg and the airport has fields that are good for resting geese, and nesting larks and pipis. There are several small ponds here with nesting Red-necked Phalarope and Temminck's Stint as well as willow thickets with Arctic Redpoll. Stray migrants can settle here as well. **y5.** Skånsvika inlet on the west side of Berlevåg airport has many of the same qualities as Storsletta. It is one of the more reliable sites in Varanger for Sanderling stopping by on migration.

Båtsfjordfjellet mountain plateau. Back at the Gednje junction, drive slowly along Fv891 towards Berlevåg, stopping and checking the terrain regularly. Lake Magistervannet and the bog on the north side of the road at the head of Tanadalen valley a couple of kilometres towards Båtsfjord, are particularly interesting. The Oarddoskáidi height on the south side of the road provides easy access to e.g., Rock Ptarmigan and Dotterel – a gravel road takes off from Fv891 to Båtsfjord, vis-a-vis Lake Magistervannet. You may follow this dirt road all the way to the cabin area at Lake Oarddojávri. This also provides a trailhead to Jakobsdalen valley and the central parts of the Varanger peninsula if you are up for a longer hiking trip in the wilderness. Another trailhead is from just south of Lake Stjernevannet, mentioned above.

Syltefjord. 25 km along Fv891 from Gednje junction, the road to Syltefjord takes off to the east. Follow the road to Nordfjord at the end of the road. This is the starting point for visits to the seabird cliff at Syltefjordstauran, which is preferably reached by boat.

z. Båtsfjord town. You will find interesting wetlands close to the town and a bird-rich harbour. You may rent photo hides with the possibility of King and Steller's Eider, besides Long-tailed Duck and more, in late winter and early spring.

Oslo
Interior
Skagerrak
Western
Central
Northern
Finnmark
Svalbard

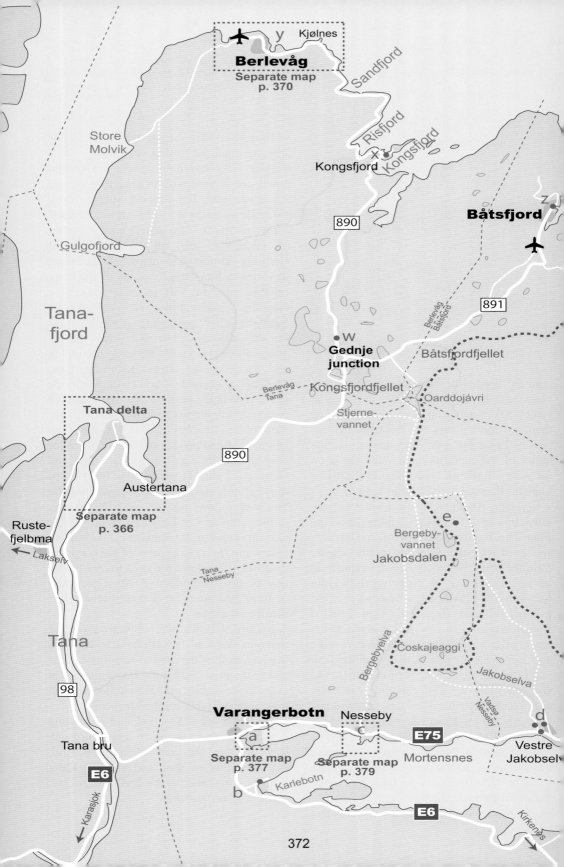

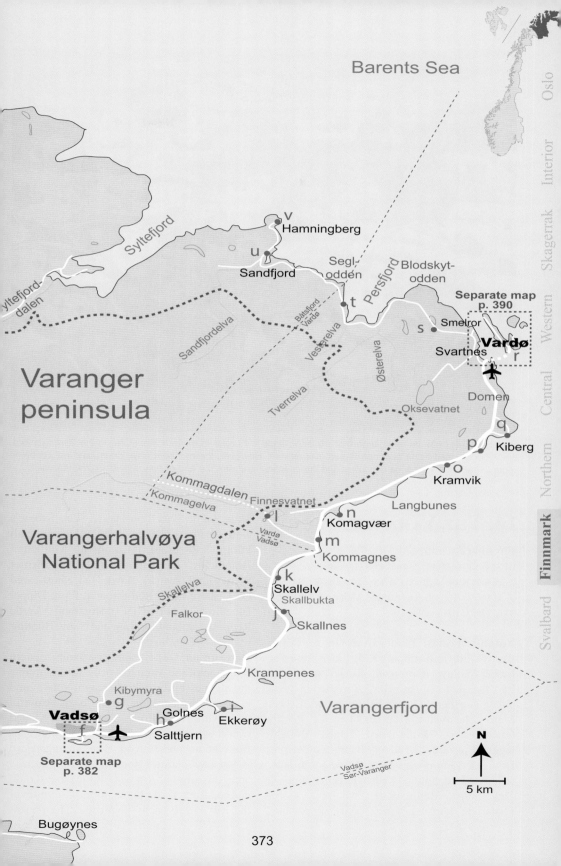

Barents Sea

Oslo

Interior

Skagerrak

Western

Syltefjord

v
Hamningberg

Sandfjord
u

Segl-
odden

Persfjord

Blodskyt-
odden

Syltefjord-
dalen

t

Smelror

s

Svartnes

Vardø

Separate map
p. 390

r

Central

Northern

Sandfjordelva

Bátsfjord
Vardø

Vesterelva

Østerelva

Domen

**Varanger
peninsula**

Tverrelva

Oksevatnet

q

p

Kiberg

o

Kramvik

Finnmark

Kommagdalen

Langbunes

Kommagelva

Finnesvatnet

l

Vardø
Vadsø

n

Komagvær

m

Kommagnes

**Varangerhalvøya
National Park**

Skallelva

k

Skallelv

Skallbukta

j

Skallnes

Svalbard

Falkor

Krampenes

Kibymyra

g

Varangerfjord

Vadsø

f

Golnes

h

i

Ekkerøy

Salttjern

Separate map
p. 382

N

Vadsø
Sør-Varanger

5 km

Bugøynes

Steller's Eider has its most important wintering area in Europe in eastern Finnmark. In the Varangerfjord, small flocks used to be common in summer as well. However, the numbers have decreased noticeably in recent years. It can now be difficult to find the species here after the end of May. *Photo: Tomas Aarvak.*

14. THE VARANGERFJORD

Finnmark County

GPS: 70.14562° N 28.86214° Ø (Nesseby church), 70.06803° N 29.74872° Ø (Vadsøya island), 70.37933° N 31.11726° Ø (Hasselneset headland, Vardø), 70.54032° N 30.62849° Ø (Hamningberg)

Notable species: Tundra Bean Goose, Taiga Bean Goose (uncommon), King Eider, Steller's Eider, Willow Ptarmigan, Rock Ptarmigan, Red-throated Diver, Black-throated Diver, Great Northern Diver, Yellow-billed Diver, Manx Shearwater (scarce), Leach's Petrel (uncommon), White-tailed Eagle, Rough-legged Buzzard, Gyrfalcon, Peregrine Falcon, Dotterel, Temminck's Stint, Little Stint, Bar-tailed Godwit, Spotted Redshank (uncommon), Jack Snipe, Red-necked Phalarope, Grey Phalarope (scarce), Great Skua, Pomarine

Skua, Arctic Skua, Long-tailed Skua, Sabine's Gull (scarce), Glaucous Gull, Iceland Gull, Puffin, Razorbill, Common Guillemot, Brünnich's Guillemot, Black Guillemot, Snowy Owl (scarce), Hawk Owl, Short-eared Owl, Shore Lark, Red-throated Pipit, Dipper, Bluethroat, Siberian Tit (uncommon), Arctic Redpoll, Lapland Bunting and Snow Bunting.

Description: The southern and eastern coastline of the Varanger peninsula is among the very best birdwatching areas in Norway and is also considered among the most beautiful and exotic in Europe. The area is visited every year by hundreds of bird enthusiasts from home and abroad. The Varangerfjord is Norway's largest fjord in terms of surface area. It is a full 55 km wide at the outer mouth, defined from Vardø to the north-eastern tip of the Russian Rybachy peninsula. The narrow inner part of the Varangerfjord turns to the southwest and extends to Varangerbotn, 118 km from the mouth of the Varangerfjord. Along the coast of the Varanger peninsula you will find several sheltered coves and hospitable shorebird sites, but the terrain becomes increasingly Arctic and inhospitable the further north you go. When the coastline at Vardø turns towards the Barents Sea, the landscape is barren and harsh, with a spectacular rugged cliff coast. The landscape on the Varanger peninsula is otherwise dominated by rounded rock formations with barren blocky land and moraine ridges, with several large marshy areas surrounded by willow marshes. This distinctive landscape is the closest you will get to Arctic tundra in Europe, which is reflected in the fact that species such as Snowy Owl, Gyrfalcon, Tundra Bean Goose, several Arctic shorebirds, Brünnich's Guillemot, Red-throated Pipit, Siberian Tit and Arctic Redpoll all breed here. In addition, the High Arctic species of Glaucous Gull, King and Steller's Eiders winter in the thousands in the Varangerfjord and contribute to giving the area its exotic feel. The occurrence of Iceland Gull is one of the strongest in Norway, with several records also of the rare subspecies *kumlieni*. Sabine's Gull and Grey Phalarope are seen almost every summer. Varanger is passed by large numbers of migrating seabirds. They can be appreciated from strategic vantage points particularly along the coast along the Barents Sea but also in the Varangerfjord. Thousands of shorebirds gather along the coast, and the area is known to harbour flocks of many tens of Red-necked Phalarope in stunning breeding plumage in early summer, not uncommonly in three-digit numbers. Their young congregate along the coast in late summer. All four skuas breed, and both Pomarine and Long-tailed Skuas can appear in significant flocks, especially in May, a time when White-billed Divers also pass by in internationally sensational numbers, with up to a quarter of a thousand individuals in a day. Great Northern Diver is also regularly encountered in the Varangerfjord, especially from May to October, so one cannot take the identification issues too lightly. Willow Ptarmigan is the common grouse along the coast. Rock Ptarmigan, as well as Dotterel and Shore Lark, can be found in the more barren areas, especially if you move a little in from the coastline. Red-throated Pipit is common in some places, while Arctic Redpoll often must be looked for among the numerically superior Common Redpoll. With a bit of luck – or by systematic searching – you can also find Siberian Tit here, preferably in sheltered, birch-covered valleys in the innermost part of the Varangerfjord and by Vadsø. Snowy Owl may be found in the central parts of the peninsula in summer, but in spring they can seek towards the snow-bare areas down by the coast. They are then seen rather frequently along the E75 between Vadsø and Vardø.

Short-eared Owl is relatively common in Varanger, and in higher-lying marsh areas and in coastal heaths in most of the country. The numbers fluctuate with the rodent populations. The species hunts in open country both night and day by flying low above the ground, much like a harrier. *Photo: Terje Kolaas.*

Sabine's Gull is an Arctic species you must look out for in the Varangerfjord. It can surprisingly easily get lost in the multitude of Kittiwakes and Arctic Terns. *Photo: Bjørn Fuldseth.*

coast. Among land-based species, it is natural to highlight the finds of Little Bustard, Steppe Eagle, Long-legged Buzzard, Roller, a couple of Red-rumped Swallows, Tawny and Olive-backed Pipits, Siberian/Amur Stonechat, Lanceolated and Paddyfield Warblers, and Yellow-breasted Bunting. Pallid Harrier has become an almost annual visitor.

Other wildlife: The area has a rich wildlife, and birdwatchers especially often encounter the many marine mammals. Several species of whales, porpoises and seals are relatively common here, including beluga and killer whales, and bearded and harp seals. The Varanger peninsula is a core area for Arctic fox, which is highly endangered in Norway. From the Varangerfjord, there are two records of polar bear (July 1953 and May 1986).

As the icing on the cake, this stretch of coast has shown enormous potential for rarities within almost all species groups. The most extreme are Europe's first records of Little Curlew, Soft-plumaged Petrel and the Pacific subspecies of Common Eider, *v-nigrum*. There are also several firsts for Norway, including Stejneger's Scoter, Greater Spotted Eagle, Semipalmated Plover, Caspian Plover, Stilt Sandpiper, Short-billed Dowitcher, Glaucous-winged Gull, Bridled Tern, White-winged Lark (three records) and Buff-bellied Pipit (*rubescens*). In addition, there are three records of Spectacled Eider. Harlequin Duck has been encountered several times, and so have Ruddy Shelduck, American Wigeon and Surf Scoter. Among the shorebirds, Pectoral Sandpiper is frequent, and there are quite a few sightings of Terek and White-rumped Sandpiper, besides the fact that that Pacific Golden Plover, Lesser Yellowlegs and Sharp-tailed, Semipalmated and Marsh Sandpipers have all been recorded at least once. The same applies to Laughing and Ring-billed Gulls. However, based on what has been implied about the occurrence of Arctic gulls above, one should perhaps expect that Ross' and Ivory Gulls were also regular visitors to the Varangerfjord, but the fact is that there are less than ten records of each along this stretch of

Best season: All year; see the When to go-section in the introduction to the Finnmark chapter (p. 336) for details.

Directions: By car, aim for Varangerbotn along the E6. Here, follow the E75 eastwards in the direction of Vadsø and Vardø. You can reach Vadsø and Vardø directly by plane via Kirkenes. The Norwegian Coastal Express calls at both Vadsø and Vardø.

Tactics: Birdwatching in this area is mainly conducted along the 160 km long road along the coast from Varangerbotn to Hamningberg. Along this road, the ornithological goldmines appear like pearls on a string. You need to allow at least 2.5 hours of efficient driving for the entire stretch one way, plus the time you need for making stops. And you should really take your time, preferably a few days or more in this area, so that you do not need to rush from place to place too quickly, ending up having spent more time inside the car than outside. A telescope is absolutely mandatory but take the time to walk along the beaches and elsewhere in the terrain. This is also a wise move whenever you take detours inland in search of mountain species. Apart from the localities that are specifically mentioned here, you can stop almost anywhere for a view over the Varangerfjord and to scan fields and pastures along the way.

Seawatching here is usually best conducted from v. Hamningberg during winds from the western and northern sectors. If the road is still winter closed, try r9. Skagodden in Vardø instead (or head for y1. Kjølnes or 11a. Slettnes). If the wind is blowing from the east, c2. Nesseby is usually the better choice, although Vadsøya, Ekkerøy and Kiberg can be good as well.

In the following, the best sites along the road from Varangerbotn to Hamningberg are described in the order in which they appear, with references to the Varanger overview map on p. 372.

a–b. Varangerbotn and Karlebotn

The innermost part of the Varangerfjord has huge, exposed mudflats at low tide, and you therefore get the best view when the tide is quite high and the shorebirds are pushed closer to land. The inner part of Varangerbotn can be viewed from a1. two hides which are reached via a boardwalk behind the outdoor exhibition of the Varanger Sami Museum. This museum is clearly signposted from the main road 100 m south of the roundabout in Varangerbotn. The access to the hides is poorly sheltered, so you

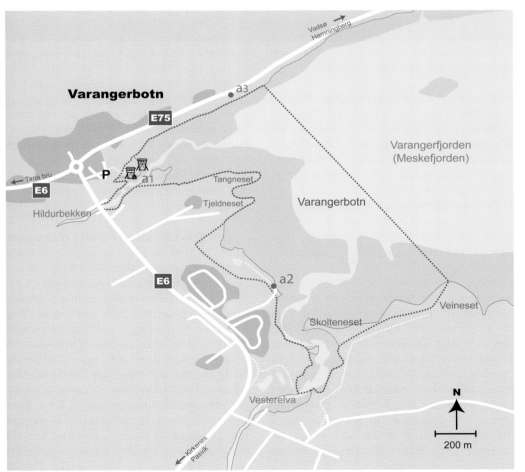

Varangerbotn: a1. Two estuary hides; **a2.** viewpoint Vesterelva river mouth; **a3.** roadside viewpoint.

Nesseby headland in the heart of the Varangerfjord is one of the Norway's best sites for both seabird migration and shorebirds. *Photo: Ingar Jostein Øien.*

must walk quietly so as not to scare away the birds that roost close to the hides. The right-hand hide offers a view of the river estuary and is a particularly good place for Temminck's Stint and other typical freshwater shorebirds. The left-hand hide is overlooking the delta which usually have more birds.

Follow the E6 for a few hundred metres in the direction of Kirkenes and take detours to the left (see map) for similarly fine tidal flats and salt meadows at the **a3.** outlet of the Vesterelva river.

Continue E6 5 km further in order to reach **b.** Karlebotn, the southern arm of the inner Varangerfjord. Karlebotn is signposted from the E6, 5 km from Varangerbotn. There are almost always shorebirds to be found here, in addition to ducks and divers on the fjord. The tidal area of Karlebotn is quite large and easiest to overlook roadside from the far end of the settlement. An alternative viewpoint is obtained by parking at the Karlebotn chapel and walking 150 m to the ridge above the shoreline, and perhaps walk even further south along the ridge from here.

Back at the roundabout in Varangerbotn, exit onto E75 towards Vadsø and Vardø and

continue 750 metres. Here is **a3.** a lay-by overlooking the outer section of the same mudflats you saw from the two hides.

Further along the E75, look for grebes, divers and fleets of sea ducks in Meskefjorden. Bluethroat, geese, White-tailed Eagle, Hawk and Short-eared Owls are often seen along the road in this area.

c. Nesseby headland and Bergeby river mouth
After 13 km on the E75, you will see the white-painted Nesseby church on a headland that juts out a few hundred metres into the fjord on the right. The promontory has nesting Shelduck, Arctic Skua, various gulls and Arctic Tern. Park at **c1.** by the church and from here, check the large tidal areas, which at low tide can be very rich in shorebirds, especially during spring migration. At high tide, the birds often sit in the rocks by the shore out on the headland or on the small Løkholmen island, 400 m from shore. Here there are also quite a few gulls, as well as occasional resting geese. Also turn the telescope south towards Skjåholmen island, even if the distance is about 2 km. This is one of the best places for geese, in May and especially August sometimes including Lesser

White-fronted Goose. Check the small pond on the headland for resting dabbling ducks and Red-necked Phalarope.

c2. At the far end of the headland, you get a good overview of the inner part of the Varangerfjord, a good place to look for migrating and resting seabirds that are funnelled into the fjord, especially in easterly winds. Pelagic species may turn up, such as Manx and Sooty Shearwaters, and Leach's Petrel. Of the latter, up to 76 in a day, which is a Norwegian record. This, of course, in addition to good numbers of more common species, such as Fulmar, auks, skuas, Kittiwake and Long-tailed Duck, most of them migrating towards the west, often at considerable heights.

Steller's Eider can often be seen at Nesseby, until recently, all year round. In recent years, they are usually only found in winter and early spring this far into the fjord. Nesseby has a long species list containing a number of rarities, including Soft-plumaged Petrel, Steppe Eagle and Buff-bellied Pipit. It is possible to find Siberian Tit in the forest around Nesseby and at feeders in the area. At the **c3.** Bergebyelva river mouth, immediately east

of Nesseby, gulls often congregate. They must be checked thoroughly for unusual species, including Sabine's Gull (up to 4 individuals have been seen at once here). There may also be shorebirds here at low tide, especially freshwater species. You have a good overview of the river basin from the lay-by along the E75 just east of the bridge. Ross' Gull, Terek and Marsh Sandpipers have been seen here – the latter two simultaneously even.

Mortensnes headland. About 5 km further east on the E75, by the *Transteinen* and the old Sami sacrifice place, you get a nice view of the fjord. There are often quite a few seagulls here. Spectacled Eider has been seen from here.

d. Vestre Jakobselv

Further 13 km east and past the border sign to Vadsø municipality, you come to Vestre Jakobselv village, where Jakobselva river flows into the Varangerfjord. The river outlet is along the main road and is a good place for resting ducks, gulls, terns and waders, especially at low tide. Turn left before the river into the dirt road signposted to *Tana–Varangerløypa* for

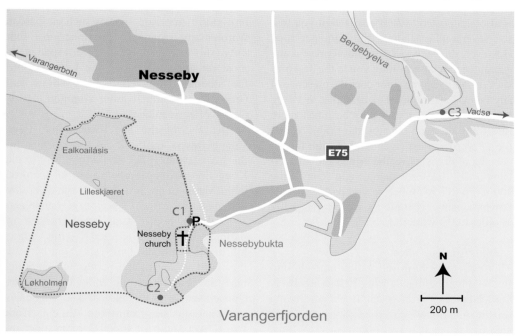

Nesseby headland: c1. Viewpoint to tidal area; **c2.** seawatch; **c3.** Bergebyelva river mouth.

Jakobsdalen valley is one of the most remote and bird-rich areas on the Varanger peninsula. Red-throated Pipit,Tundra Bean Goose, and several shorebirds are among the typical breeding species here. *Photo: Tor Olsen.*

an overview of the west side of the outlet. It is also signposted to *Bird site*, a sign you will see several places along the Varanger road. However, this signpost does not represent any guarantee that it is pointing you to the best bird sites or vantage points, nor does the presence of any fancy-looking hut. For example, in Vestre Jakobselv, at low tide, it may be much more rewarding to stand over <u>by the bridge</u> where the E75 crosses the river. And at high tide, it is better to drive across the bridge and turn left, signposted Jakobselvdalen, and continuing 400 m up this road, before you park and walk down the narrow path to <u>the eastern riverbank</u>, in order to view the shorebirds up-close.

e. Jakobsdalen valley

This extremely bird-rich valley extends into the Varanger peninsula from Vestre Jakobselv (and Nesseby). You reach it by continuing on the mentioned dirt road along the west side of <u>Jakobselva river</u> (signposted to Tana–Varangerløypa). The lower part of the valley has lush vegetation and is one of the few areas

on the Varanger peninsula where you can find more temperate species such as Garden Warbler and Dunnock. The innermost four km of this increasingly bumpy dirt road is the most reliable place along the Varangerfjord for Siberian Tit, even though Willow and Great Tits are much more common. Hawk Owl is often seen here as well.

At the end of the dirt road is the <u>Tana–Varangerløypa trailhead</u> leading further up the Jakobsdalen valley and all the way to the Kongsfjordfjellet mountain plateau, featured in the Varanger north section. Along this route you pass the large <u>Čoskajeaggi marshland</u> as well as higher-lying areas further in, particularly bird-rich between <u>Kjerringhaugen hill</u> in the south and the <u>Svanevatna lakes</u> in the north. Many of the mentioned breeding birds are found here, including Tundra Bean Goose. Access to this part of Jakobsdalen is presumably easier from the north, e.g., from Fv890 just south of Lake Stjernevannet on Kongsfjordfjellet (see the previous section, on Varanger north). You may also hike straight

Red-throated Pipit is relatively common at many places in Finnmark, especially in marshy areas along the coast. Your odds of finding it increase radically if you learn the rather discreet but characteristic call and song. *Photo: Terje Kolaas.*

Bean Goose. Look closely for other species in between, Lesser White-fronted included. You can explore this area in more detail by driving the side roads on the left side of the E75, signposted to *Golfbane*.

f. Vadsø town

When you approach Vadsø town, you see Lille Vadsøya island out in the fjord. The waters around this can often have King Eider in summer. Look for them with a telescope from a suitable lay-by 1 km east of Andersby.

Vadsø harbour. With its about 5,500 inhabitants, Vadsø is

across the Varanger peninsula, although such a trip through the vast wilderness needs to be planned and executed accordingly.

Kariel pastures. Back on the E75, the lush pastures on the left side of the road for the next 5 km, especially around the Kariel settlement, are good for resting geese, plovers, Ruff, larks and such. Harriers are sometimes seen hunting here. From early May, this is the most easily accessible site on the Varanger peninsula to see good numbers of Tundra

the largest town in the area and has most necessary facilities. Drive through the town and turn right at the roundabout signposted to *Luftskipmasta* (airship mast). This road takes you past the harbour and out onto Vadsøya island. Park in the **f1.** parking lot on the left immediately after the Vadsøya bridge. At low tide, scan the large mudflats with a telescope from here. This is the best shorebird location in Finnmark in spring, with up to a couple of thousand Dunlin on the

Vadsø town and Vadsøya island seen from the road up to Kibymyra marsh. In the background is the south-side of the Varangerfjord, west of Bugøynes. *Photo: Bjørn Olav Tveit.*

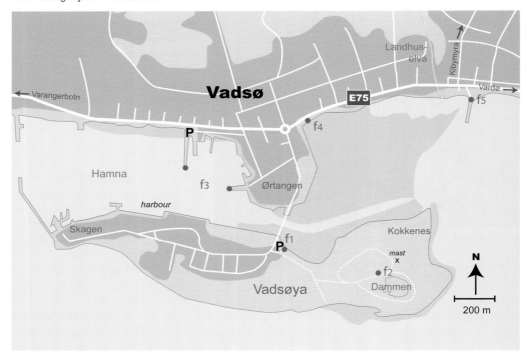

Vadsø town: f1. Viewpoint to the mudflats; **f2.** Dammen pond and grasslands on Vadsøya island; **f3.** Vadsø harbour jetties; **f4.** roadside viewpoint; **f5.** viewpoint to the Landhuselva river outlet and mudflats.

best days in May. In autumn, many hundreds can still be found here – a time of year when the Tana delta is ranked higher. The mudflats are also good for gulls. At high tide, the shorebirds usually rest on the less visible grass and pebble beaches, scattered around the entire Vadsøya island. After scanning the sound, walk through the gate and out towards the eastern tip of the island. Here stands an eye-catching mast to which the Norwegian explorer Roald Amundsen moored his airships during his voyages to the North Pole in the 1920s. On the right side of this mast is the small **f2.** Vadsøydammen pond, or just Dammen, is known for its regular occurrence and occasionally significant numbers of Red-necked Phalarope. These birds are especially here from mid-June to early August – up to 560 ind. are seen at the same time in early July. The chance of finding a Grey Phalarope among them is always present. The pond is also good for ducks and other wetland species. In this area you can also find Little and Temminck's Stints, *Tringa* waders and displaying Ruff, as well as Red-throated

Pipit and Bluethroat. During spring and autumn migration, quite a few Shore Larks can rest here. Check the beach on the seaside southeast of the pond, but be considerate of the colonies of Arctic Tern and Herring Gull at the eastern tip of the island. It can also be nice to look for seabirds from Vadsøya, although **c2.** Nesseby is preferable. It is here around Vadsø that the inner part of the Varangerfjord begins to narrow, and it is only 10 km across to Bugøynes, visible on the opposite side. Vadsø harbour has an active reception of fishing boats and is thus popular with the gulls. This is considered the best place in the Varangerfjord for Iceland Gull in winter. They may in fact be more numerous than Glaucous Gull. The fishing boats also attract a variety of sea ducks. Flocks of King and Steller's Eiders stay here from November to at least March. Towards the summer, the numbers of these species decrease significantly, but single individuals or small flocks of King Eider can usually be found even then. However, Steller's Eider has decreased in number in recent years, also in winter, and can now be difficult to find after the end of May. The birds in the harbour can be appreciated from the **f3.** harbour jetties. It is possible to rent a floating hide in Vadsø harbour, allowing for close encounters and fantastic photography of King, Steller's and Common Eiders, besides Long-tailed Duck, gulls and more.

If you continue 1 km east along the E75, you will come to the **f5.** Landhuselva river outlet, where gulls, Common Eider, Steller's Eider and shorebirds often congregate. These, and the outer mudflats and shallow seas, can be viewed from the jetty at Ytrebyen marina.

g. Kibymyra marshlands

Inland from Vadsø town are some nice wetlands. Species such as Scaup, Jack Snipe, Bar-tailed Godwit, Red-throated Pipit and Bluethroat breed here. From the E75, enter a side road on the left, signposted to *Skytebane* and *Crossbane*. Follow the road through the residential area and up into the mountain. After the asphalt paving has been replaced with gravel, take the first road to the left, then the next to the right, and finally the next to the left, and then keep straight ahead. Snowdrifts can block the roads here until well into June. Check the lakes for divers and ducks, and on the way up, keep an eye out for other of the area's specialities, such as Dotterel, Long-tailed Skua, Snowy Owl and Red-throated Pipit.

h–i. Ekkerøy area

h. Salttjern. Eastward from Vadsø, the coastline, and the E75 with it, gradually turns north. You are now driving along the outer Varangerfjord. Here are several small coves, headlands and puddles that may be worth investigating more closely for waders, gulls and more. The inlet immediately north of the Salttjern settlement, around which the road makes a small bend, is particularly interesting. Storsjødammen pond on the right side of the road, a little over a kilometre further east, is also nice. On the stretch Salttjern–Golnes there are often quite a few gulls.

Falkefjell mountain. A further couple of kilometres east, as you drive out of the Golnes settlement, a road goes to the left which takes you several kilometres inland, to areas with new opportunities for mountain species.

i. Ekkerøya peninsula. Turn right from E75 12 km east of Vadsø, signposted to Ekkerøya. The sand spit that connects the island to the mainland, and thereby robs Ekkerøya of its island status, often has significant numbers of shorebirds. Purple Sandpiper is often abundant in the rocky parts of the beach. In summer, Red-throated and Rock Pipits are often found, sometimes Shore Lark as well, especially during migration. Caspian Plover is among the rarities found here. In order to do this 3 km stretch of beach justice, you should walk along the shore, especially on the north-side of the spit and beyond. There are often divers and ducks on the sea.

Ekkerøya is home to a colony of Kittiwake, a good number of Black Guillemot and a very few pairs of Razorbill and Common Guillemot. White-tailed Eagle, Gyrfalcon and Peregrine Falcon visit the bird cliff regularly during the breeding season. Snowy Owl may

Shore Lark nests right down to the sea-level in several places along the Varangerfjord. It is elsewhere in Norway a distinctly mountain bird associated with dry tundra, often occurring side-by-side with Dotterel. *Photo: Kjetil Schjølberg.*

Ekkerøya peninsula from the beach on Yttersida, the north side of the sand spit that connects Ekkerøya with the mainland. This is a favoured site for shorebirds. *Photo: Bjørn Olav Tveit.*

occasionally drop by as well. Follow the asphalt road out to Ekkerøya harbour and turn left just before the museum and up to a car park. Here you can either walk a marked path to the top of the plateau, or you can walk the tractor path along the mountain ledge straight ahead. Both take you to the seabird cliff. This is also a good seawatching vantage point. The rocky coastline that dominates this side of Ekkerøya is good for species such as Grey Plover and Purple Sandpiper, the latter of which winters in four-digit numbers. A few pairs of Shore Lark nest on top of the plateau. Storelvneset headland. The entire coastal stretch from Ekkerøya and the 4 km northeast to Storelvneset is of interest, and may be worth walking, or at least drive down to the beach where possible. The northern part of this stretch, between Lilleelvneset and Storelvneset headlands, is particularly good for resting gulls and geese, sometimes including Lesser White-fronted Goose.

Krampenes–Skallelv

Krampenes headland, 6 km past Ekkerøya, is a good place for ducks, divers and gulls. From this point on, the landscape changes character and becomes barren and more Arctic.

Barely 2 km after the quarry at Skallneset headland, take an unmarked gravel road to the right and then down to the left towards a (private) house for a view of the two fine shorebird beaches at Nyhamn. This place represents the southernmost part of j. Skallbukta bay, which consists of a 4 km long sandy beach where small rivers and streams run out in several places, forming interesting feeding and bathing sites for waders and gulls. This area is also good for geese and other birds that are attracted to the fields along the road. At the other end of Skallbukta bay, by the Skallelva river, is Skallelv village. Stop before the bridge and check both upstream and downstream for resting ducks, gulls and waders. Then cross the bridge and immediately turn right. Look for <u>feeders</u> in the village gardens, sometimes catering to interesting species such as Arctic Redpoll and migrant passerines, besides the local House Sparrows. At the <u>river mouth</u> in Skallbukta bay, visible from the side road, ducks, gulls and shorebirds often congregate. Continue along the side road through Skallelv village and you will be soon back onto the E75 again. Follow it further towards Vardø for 400 m before turning right again and onto a dirt road that takes you across some nice fields and down to <u>Makkenes headland</u>. This is an exotic area of large sand dunes, flanked by lush pastures and shorebird-friendly beaches. Check the pastures for geese, plovers, etc., and the bays on both sides of the headland for shorebirds, ducks, White-tailed Eagle and more.

The fields on the opposite side of the E75 should also be checked. If you drive the gravel road here leading up towards <u>Holmfjellet mountain</u> and walk the last 2 km from the road barrier, you will find a relatively easily accessible tundra area with nesting Tundra Bean Goose and a nice selection of shorebirds.

Next along E75, follows a slightly less bird-rich stretch of sandy beach, perhaps apart from the kelp washed ashore by <u>Kobbeskjæret skerry</u>, which may hold Purple Sandpiper and perhaps some more. Cormorants are often seen on and around the skerry, sometimes accompanied by a few Shags and King Eiders.

l. Kommagværdalen valley

Along E75, 1 km after you have entered Vardø municipality and rounded <u>Kommagnes headland</u>, the <u>Portveien road</u> takes off to the left. This bumpy gravel road leads inland along the l. <u>Kommagdalen valley</u>, to a very interesting wetland area. At about 6 km up this road, the

Tundra Bean Goose was rather recently split from Taiga Bean Goose. Both species have almost disappeared as breeding birds in most of Norway, except in Finnmark. Researchers study the development of the populations here closely, and several of the birds are marked with collars for individual recognition. Taiga Bean Goose breeds in marshes in the forests of Pasvik and Anárjohka valleys, while Tundra Bean Goose (depicted) breeds on the Finnmarksvidda plateau and on the Varanger peninsula. Both can appear together on migration, though, and the specific identification is not always straightforward. *Photo: Tomas Aarvak.*

Oslo
Interior
Skagerrak
Western
Central
Northern
Finnmark
Svalbard

Red (Grey) Phalarope appears once in a while, normally singly, among the much more common Red-necked Phalaropes, especially after periods of strong winds from the north. *Photo: Bjørn Fuldseth.*

small Lake Soppavatnet appears on the left (south) side of the road. This lake is rich in nesting shorebirds, including Bar-tailed Godwit and Ruff, the latter often lekking roadside. Jack Snipe can sometimes be heard and seen performing its peculiar display flight over the area at night. Scan the marshland opposite the road from Lake Soppavatnet with the telescope, looking for Short-eared Owl, Hen Harrier and various wetland species by the many ponds and on the larger Lake Finnesvatnet.

From the parking area 1 km further on, a trailhead leads further into the Kommagdalen valley where there are many similarly beautiful areas. A hike from here to e.g., Svartnes near Vardø, 30 km east, is a fantastic experience which can provide immemorable encounters with species such as Dotterel, Temminck's Stint, Bar-tailed Godwit, Tundra Bean Goose and Long-tailed Skua, with a slight possibility of stumbling across a Snowy Owl.

m–n. Komagvær bay

Back onto E75, opposite the road to Kommagværdalen valley, you see **m.** Kvalnes headland with the small Lake Kvalnesvatnet. The lake often has a nice

selection of ducks and other waterbirds. The tip of the headland has a narrow pebble tongue. On the north side of the headland, Pakolabuktbekken creek meets the sea. The pebble tongue and the seashore between it and the creek outlet often have good numbers of shorebirds in summer, occasionally with some unexpected species to be found. The sea surrounding the headland often has divers, grebes and fleets of various sea ducks that require thorough scrutinizing. Black Scoter has been found here in the past.

You may scan the lake and much of the sea from the road, but to take full advantage of the potential of this site, you should park roadside by the creek and walk around Kvalnes headland on foot, bringing your telescope along with you. Check the lush willow thickets along the creek for Bluethroat, Arctic Redpoll and migrant passerines.

Along the E75 the next couple of kilomtres, you have the Komagvær bay on your right, which can hold good numbers of sea ducks, shorebirds, gulls and terns. Stop and scan the beach, sea and fields here and there. The coast outside Komagvær is one of the best places for White-billed Diver, King and Steller's Eiders.

About 600 metres after crossing the Komagelva river, a road exits to the right, leading to **n.** Komagværnes headland. This side road takes you across fine pastures which, in the right season, can be a good place to find the likes of geese, Ruff and Golden Plover. After 400 m you reach a lookout point over Komagvær bay. From here you can walk the path 1 km further across Komagværneset headland to Ytterstranda beach with a view over Trollbukta bay, a hidden gem, attractive to resting shorebirds and sea ducks.

On the opposite side of the E75, a road leads inland 2 km to Hollamyra marsh. Along this side road, you can find species such as Bar-tailed Godwit, Whimbrel, Turnstone, Shore Lark, Bluethroat and Lapland Bunting.

o–n. Langbunes–Kramvik

Leaving Komagvær on the E75 towards Vardøy, the coastline alternates between bays and barren headlands, where Long-tailed Skua may be seen and raptors may soar above the mountain ridge. About 7 km past Komagvær, the Langbunes fields, with its contrastingly inviting grazing pastures, appear on your right. Skylark, Red-throated Pipit, Lapland Bunting and variety of resting migrants may be found here.

After another 4 km, Lake Svartnesvatnet appear on your right, which must be scanned for ducks etc. from the road. Next, the Grunnesbukta fields appears, recognizable by the lush pastures on both sides of the road. Here, flocks of geese and Golden Plover often feed. The birds are easily spooked away here if you leave the vehicle. Down by the sea is a river estuary with washed-up seaweed on a pebble beach. You will have to walk along either side of the field to get access to the shorebirds often roosting here. The willow thickets along the Langbuelva river leading down to the shore are fairly reliable for Arctic Redpoll.

The next site a couple of km further on is o. Kramvik bay, a small but very interesting site. Sharp-tailed Sandpiper and White-winged Lark are among the rarities that have been found here. Start by scanning the beach from the side of the road just after the stream crossing. Then drive a little further and walk down on the seaward side of the large farm building on your right to check the two small coves, separated by a small stone breakwater. Migrant passerines often find shelter and feeding opportunities behind this farm building. You can sometimes find Shore Lark, Lapland and Snow Buntings in the meadows here.

A trip up the gravel road on the opposite side of the E75, leading into Kramvikdalen valley, with a few fields more peacefully located than those along the main road. The valley sees the possibility of Long-tailed Skua and other montane species as well.

King Eider has its most important wintering area in Europe along the coast of Finnmark, and especially in the outer part of the Varangerfjord. The species seems to prefer deeper water than Common and Steller's Eiders. While the latter is often seen far into the Varangerfjord, the King Eider is more common in the outer part of the fjord and along the coast of the Barents Sea. *Photo: Ingar Jostein Øien.*

p–q. Kiberg

A couple of km further on, you reach the settlement of Indre Kiberg. Here is an interesting beach for resting shorebirds, gulls and terns. Characteristically, one half of the beach consists of sand, the other of pebbles and washed-up seaweed. The beach is bordered by fields often holding Golden Plover, Ruff and more. Indre Kiberg is connected to neighbouring village Ytre Kiberg a little over a kilometre further north with a gravel road closed to car traffic. This road runs through fine cultural land areas that you may explore on foot.

Ytre Kiberg is otherwise reached by continuing on the E75 northwards and turning right onto the road signposted to *Bird site*. Drive all the way down to the beach, which needs to be scanned for shorebirds and gulls. The gravel road along the beach southwards is the one connecting Indre Kiberg to Ytre Kiberg. The tarmac road north along the beach leads past the

Glaucous Gull breeds in the Arctic, including on Svalbard, and winter in significant numbers on the coast of Finnmark, particularly along the Varanger peninsula. Many immature birds even spend the summer here. In the rest of Norway it is mainly a winter visitor, becoming progressively more uncommon the further south you go. *Photo: Kjetil Schjølberg.*

Kibergselva river estuary, which often has a nice gathering of gulls. There is a fancy-looking hide here but by entering it, you risk spooking all the gulls. It might be better to scan the estuary from inside the vehicle or through the telescope at a safer distance.

Continue into **q**. Kiberg harbour. It is often packed with gulls, and Brünnich's Guillemot is sometimes seen here. The coast between Kramvik and Kiberg is home to thousands of Common Eider in summer. It is also here that you find the largest flocks of King Eider at that time of year, but the distance is usually great, so a telescope is absolutely necessary. Ytre Kiberg is one of the very best sites for exciting gulls in Varanger. You can usually find Glaucous and a few Iceland Gulls here all year round, in March with up to three-digit numbers of the former. The first Glaucous-winged Gull in Norway was found here. At the end of the harbour, after the last house on the right, a dirt road goes uphill and out onto Kibergneset headland. This is a good vantage point for seabird watching. Also check the lakes on the headland for Red-necked Phalarope, Scaup, Long-tailed Duck and more. The gravel road leads out onto the E75.

Domen mountain pass

Domen (The Dome) is the high-lying section that the E75 takes you across on the way from Kiberg to Svartnes/Vardø. It provides easy access to habitat favoured by the likes of Dotterel, Rock Ptarmigan, Shore Lark and Snow Bunting. Snowy Owl is seen here from time to time. In order to explore Domen, either park near the Red Cross cabin, visible on the left side of the road, and do a bit of walking in the terrain from here, or you can drive a bit further and turn right along the gravel road leading up to the antenna installations near the summit. Domen has also proven to be a good vantage point for watching raptors migrating along the Varanger coast. Gulls and ducks often roost in the lakes here.

Rock Ptarmigan is quite common in Finnmark but still one of those bird species that can sometimes be challenging to find. In southern Norway, it lives in the high mountains while in Finnmark, where the terrain is right, you may find them down to sea level. In spring, listen for the male's peculiar call. *Photo: Bjørn Fuldseth.*

r. Svartnes

Svartnes marks the crossroads where the E75 continues out to Vardø and the road out to Hamningberg turns left. Here you will also find Vardø Airport Svartnes, which has a runway surrounded by short-cut grass, which plovers, larks, pipits and a number of other birds may find alluring. A vagrant Caspian Plover has been seen here. Svartnesbukta bay is an industrial deep harbour, surrounded by gravel fields and stone jetties. The **r1.** Storelva river outlet in the middle of the bay is a popular bathing and resting place for gulls. You may find three-digit numbers of Glaucous Gull here in March. Shorebirds and dabbling ducks also tend to gather here. You get a good overview from the roadside. Avoid stepping out of the vehicle close to the outlet, you will only scare the gulls away. There is an active **r2.** fishing reception in the harbour, often attracting large gatherings of gulls. Along the sandy beaches north to Skyttargamnes headland, there are also often several gulls and geese resting. You get a nice view of the beach from the **r3.** northern jetty. From the tip of either of the two jetties, you can see across Busseundet sound between Svartnes and Vardøya island, where there

are often good numbers of sea ducks, divers, auks, and more. Inland along the beach and sand dunes are plains with short straw and heather vegetation which have breeding Little Stint, Shore Lark and a good population of Red-throated Pipit. This is an exciting place to look for resting passerines during migration. So are the willow thickets here. The small **r4.** Svartnesskogen forest, a spruce plantation just southwest of the exit to Hamningberg is one of the very few places out here where forest-dwelling species can find something close to natural shelter. Take a detour through the small settlement of **r5.** Smelror and follow the road down to the sea. From here you have a view over to Tjuvholmen island where there are almost without exception King Eider to be found among the Common Eiders in summer. The road to Lake Oksevatnet, which exits by the northern jetty, provides easy access to tundra habitat with breeding Dotterel, Long-tailed Skua, Lapland Bunting, and more.

r. Vardø island

Vardø town marks the end of E75 and is reached via an underwater tunnel. The town is situated on two interconnected islands 1.5 km from the mainland. In terms of population, Vardø is about half the size of Vadsø. The facilities are therefore somewhat more frugal. The road network is surprisingly complex for a town of this size, so keep a keen eye on the map when navigating here. Vardø island is situated at the mouth of the Varangerfjord and has a lot to offer birdwatchers.

Vardø harbour has an active fish reception which attracts gulls and sea divers. This is one of Norway's best sites for Glaucous Gull. Here you can find pure flocks containing several hundred individuals in winter, and up to 2,000 have been counted here. You will always find the species here, even in midsummer. The harbour also has the

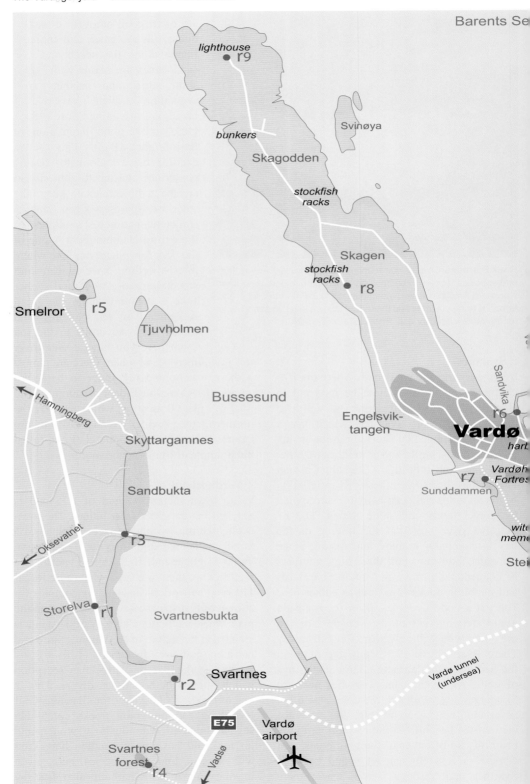

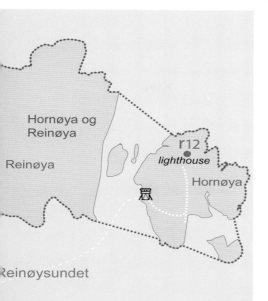

Hornøya og Reinøya

Reinøya

r12

lighthouse

Hornøya

Reinøysundet

selneset

r11

Vardøya

r10

Østervågen

Gullringnes
lighthouse

N

200 m

opportunity for King Eider all year round, and they can appear in their thousands during winter. Steller's Eider is often encountered in winter as well, with a maximum count of 900 in the harbour. Look for Brünnich's Guillemot and even more exotic species as well. Harlequin Duck has been recorded in the harbour, as well as a flock of three Spectacled Eiders. In recent years, the number of birds in the harbour has decreased somewhat, probably because of reduced fishing activity. Large numbers of Kittiwake breed on the harbour buildings and there is a mixed colony of Arctic and Common Terns on the roof of Vardø Hotel. The gardens of Vardø are for the most part meagre but should still be checked for migrant passerines. The small **r6. Sandvika inlet** by the western breakwater may hold a few shorebirds, including Red-necked Phalaropes in late summer.

r7. Sunddammen lagoon by Vardøhus fortress is a nice little shorebird site, often holding a surprisingly wide selection of shorebird species in spring and summer. Short-billed Dowitcher and Semipalmated Plover are just two of many examples of vagrants that have turned up here. It is also a favoured roosting site for gulls, the smaller species in particular. Because of the tall grass lining the lagoon and thereby blocking the view, you need to walk carefully down to the shore and scan the lagoon from there. Be sure to have scrutinized the more shy gulls from a distance first. Species such as Red-throated Pipit, Bluethroat, Sedge Warbler and Arctic Redpoll can usually be found in the surrounding area, and you may find unexpected passerines during migration. The site is most easily accessed from the western end of Birger Dahls gate street.

Steilneset headland south of Sunddammen has grazing pastures and horse stables with a potential for pipits, larks and buntings. A

Svartnes and Vardø island: r1. Storelva river outlet; **r2.** fishing reception; **r3.** northern jetty viewpoint; **r4.** Svartnes forest; **r5.** Smelror viewpoint; **r6.** Sandvika inlet in Vardø harbour; **r7.** viewpoint Sunddammen lagoon; **r8.** Skagen, stockfish racks and grassland; **r9.** seawatch Skagodden; **r10.** seawatch Gullringnes; **r11.** viewpoint Hasselneset; **r12.** Hornøya lighthouse.

Oslo

Interior

Skagerrak

Western

Central

Northern

Finnmark

Svalbard

Hornøya island provides close encounters with large numbers of nesting seabirds, a truly breathtaking experience. Vardø island can be seen in the background. *Photo: Bjørn Olav Tveit.*

hut provides shelter with a view to the sea, which here often has good good numbers of Common and King Eiders. The scrapyard to the east is situated on top of a small, now filled wetland – Skagnesdammen lagoon – which should be restored.

r8. Skagodden headland is rather dry and covered in heather, grass, and low willow thickets. The headland is reached by continuing the road north past the fortress. The area is good for plovers, pipits, larks, buntings, Twite, and other species that thrive in open terrain. Norway's second White-winged Lark was encountered here, just 15 km from the first record at Kramvik. Look for migrant passerines on and around the old stockfish racks out here, which are the closest the birds come to forest-like habitat. The shoreline can hold shorebirds and gulls, and there are often many sea ducks out in Bussesundet sound between Vardøya island and the mainland, often including King Eider.

r9. Skagodden lighthouse at the end of the Skagodden road, provides a good viewpoint for seabird migration, at least if the wind is not too strong, because finding shelter is a challenge here. This is an alternative to Hamningberg, not least before the road there opens in spring (normally around May 15).

r10. Gullringnes is also a place with seabird potential, located with a view to the south. To get here from the tunnel, bear right twice and continue out to the small lighthouse. An advantage here is that you have a view from the car, practical when it rains.

r11. Hasselneset headland. Vardø is the most reliable and accessible place in Norway to find Brünnich's Guillemot. Just direct your telescope towards the large numbers of auks that roost and pass out in the strait between Vardø and the breeding colonies on Hornøya island, about a kilometre further out. Razorbill and Common Guillemot are the dominant species, but you should be able to pick out a few Brünnich's by looking carefully through flocks, prioritizing the closest birds. This is also a very educational exercise that will come in handy when you

Brünnich's Guillemot formerly nested sparsely in bird cliffs south to Runde island. It is now probably gone elsewhere in Norway, apart from the northernmost bird cliffs along the mainland, besides Svalbard. Vardø is the most reliable and easiest place in mainland Norway to see this species. *Photo: Terje Kolaas.*

want to pick out a Brünnich's Guillemot among passing auks at your local patch back home. There are often flocks of King Eider here, even in summer. From the tunnel, turn right and then immediately left, and continue to Hasselneset at the end of the road, where a shelter is provided.

r12. Hornøya island. About a kilometre off Vardø are two islands, Hornøya and Reinøya, that both house large colonies of seabirds. Most of the species there can be found scanning through a telescope from Vardø but it is highly recommended to take the trip out to the easily accessible Hornøya island during the breeding season to see the hordes of seabirds up close. The island can be reached with a 15-minute express boat trip from Vardø harbour. Contact Vardø Port for information about boat trips. In the period May 1–August 15 there is a traffic ban on the islands, with the exception of the disembarkation area at Hornøya, around the lighthouse there, as well as along the almost 1 km long path between the two places. At the seabird cliffs on the west side of Hornøya, large numbers of Kittiwake and Puffin nest, as well as several Razorbills, Common Guillemots and a few Brünnich's Guillemots. The latter tends to stay in the upper part of the cliff face. The neighbouring and less accessible Reinøya

island, holds one of Europe's largest colonies of Herring Gull. There are several Great Black-backed Gulls nesting here as well, and a few Great Skuas among them. In addition, many Common Eiders nest on the two islands, as well as Cormorant, Shag, Black Guillemot and probably Leach's Petrel. Single pairs of Fulmar have tried to breed here in recent years. Incredibly, both Black-browed Albatross and Tufted Puffin has been recorded here in summer. Hornøya is Norway's easternmost point and is strategically located for resting land birds during migration. Records of species such as Turkestan Short-toed Lark and Paddyfield Warbler show that the potential for vagrant passerines is very much present out here.

s–v. Hamningberg road
The almost 40 km long stretch of road along the coast of the Barents Sea from Svartnes to Hamningberg is narrow and winding. The road is closed and not ploughed in winter. It is usually opened around May 15, depending on the snow conditions. Do not attempt to drive out before it opens even if the road is passable; overzealous seabird watchers have become stuck in Hamningberg after being surprised by snowfall in May. For seabird watching when the road is still closed, try **r9.** Skagodden near Vardø.

After the village of Smelror, the road takes you across a marsh area covered by the Barvikmyran and Blodskuttodden nature reserve, with nesting ducks, divers and shorebirds including Little Stint, along the many lakes. In addition, a variety of tundra species are to be found in this area. A **s.** wind shelter with a view of some of the most bird-rich marsh lakes is available on the left side of the road, 2.5 km after the exit to Smelror. These may also be scanned from the parking area by the road. There is a traffic ban in part of the nature reserve, in a marked area here, around Grøhøgdmyra marsh, between May 15 and July 31.

Oslo

Interior

Skagerrak

Western

Central

Northern

Finnmark

Svalbard

The Hamningberg road takes you past Grøhøgdmyra marsh, the most bird-rich part of Barvikmyran and Blodskytodden nature reserve. Domen mountain can be seen in the background, on the left. *Photo: Bjørn Olav Tveit.*

The road further on leads you through a wild and spectacular landscape of distinctive rock formations and craggy cliffs that plunge down towards the inhospitable Barents Sea. Species such as Gannet and King Eider are often found along this coastline, especially in the large **t.** Persfjorden bay. In summer, there are also large flocks of moulting Goosander, Red-breasted Merganser and various eiders and other sea ducks here, sometimes Little Auk as well. Scan the sea from several roadside stops. In Persfjorden, there may be shorebirds and dabbling ducks to be found at Østerelva and Vesterelva river outlets. A number of passerines can also be found in this area. Be sure to stop by at the Vesterelva garden on the south side of the road from the Vesterelva river estuary, which is the second of the two river outlets. As always, respect private property. There is willow scrub to be found underneath the cliffs further along the road, with a potential for resting migrants here as well. Black Stork, Harlequin Duck, Stejneger's Scoter and Nightingale are among the rarities found at Persfjorden bay.

u. Sandfjorden bay. Five to six km before you reach Hamningberg, you pass Sandfjorden, a wide bay with a magnificent sandy beach, although, luckily, usually without a single bather.

Instead, you often see flocks of resting Eiders, Goosanders, Red-breasted Mergansers and gulls on the beach, often including some gulls in the white-winged category. White-billed Diver often dive outside the surf. There may be dabbling ducks to be found in the estuary as well, and, to a lesser extent, shorebirds. Dense willow thickets line the riverbanks, more lush than elsewhere along this stretch of coast. Thus, Sandfjorden is an interesting site for passerines. Bluethroat is common, and this is one of Varanger's best places to look for Arctic Redpoll. A startlingly barren and rocky valley road stretches further inland and forms an Eldorado for Rock Ptarmigan. A dead-end side road leads into the valley.

v. Hamningberg is a small, windswept ghost village that was wrested from its permanent residents many years ago. Nowadays, most of the buildings have been restored and are used as summer holiday homes. For birdwatchers, Hamningberg is primarily known for both its exceptionally good migration of Arctic seabirds and for a number records of rare birds. The latter includes finds of Stone-curlew, Little Curlew, Stilt Sandpiper, Bridled Tern and White-winged Lark. Look for resting migrants in the grasslands around the settlement and along the beach, and west past the cemetery to

Arctic Redpoll is often encountered in Sandfjorden. *Photo: Bjørn Fuldseth.*

the seaweed accumulations down in <u>Skjåvika bay</u>, where there may be some gulls and shorebirds as well. In Hamningberg, you won't find lush gardens, as one might have hoped,

but a few of the <u>gardens</u> are provided with gnarled bushes that can provide some shelter for passerines. You must return to Sandfjorden to find scrub vegetation of any importance. For observing seabird migration, an <u>observation hut</u> is available at the headland beyond the harbour. As a seabird migration location, Hamningberg has many of the same qualities as Slettnes headland and Kjølberg lighthouse (which see). Slettnes tends to have an even higher number of passing Pomarine Skuas, while Hamningberg generally has higher counts of White-billed Diver, Arctic Tern and partly Long-tailed Skua. Of the day-counts at Hamningberg, 100,000 Fulmars, 279 White-billed Divers, 874 Pomarine Skuas and a startling 6,840 Long-tailed Skuas can be mentioned. Hamningberg is a fairly reliable location for King Eider and Glaucous Gull in midsummer, and it is usually easy to find Red-throated Pipit here during the breeding

Sandfjorden bay just east of Hamningberg has significant amounts of willow scrub, in contrast to the more vegetation-free Hamningberg a little further west. Arctic Redpoll and Bluethroat breed. *Photo: Bjørn Olav Tveit.*

Hamningberg has not much to offer migrant passerines seeking shelter, save for a few rather meagre gardens. This, however, serves only to make such birds easier to locate. *Photo: Bjørn Olav Tveit.*

season, e.g., around the cemetery. There are few practical facilities in Hamningberg but you will find a public restroom at the road junction by the old school and a public fresh water tap 100 m further on. There is a cafe here as well, although open only on some days during mid-summer.

15. KIRKENES–
VARANGERBOTN ROAD

Finnmark County

GPS: 69.70213° N 29.38824° Ø (Neiden chapel), 69.97169° N 29.63152° Ø (Bugøynes)

Notable species: Ducks, divers, shorebirds and gulls; Tundra Bean Goose, Taiga Bean Goose, White-tailed Eagle, Hen Harrier, Golden Eagle, Rough-legged Buzzard, Gyrfalcon, Knot, Broad-billed Sandpiper, Short-eared Owl, Lesser Spotted Woodpecker, Yellow Wagtail, Bluethroat, Sedge Warbler, Arctic Warbler, Lapland Bunting and Little Bunting (scarce).

Description: Many birdwatchers visiting Finnmark fly in to Kirkenes, hire a car and drive to Varangerbotn to concentrate on the Varanger peninsula. Along the way, you pass several sites well worth stopping by.

This area is known to be perhaps the most reliable breeding site in Norway for Arctic Warbler, conveniently located along the E6 by the Neiden village. The occurrence of Arctic Warbler is variable, with up to 12–15 singing individuals along Munkelva river and in the lush birch forest around Neiden chapel. Little Bunting may also be found breeding in these areas, although it is much scarcer.

The outlets of the two rivers Neidenelva and Munkelva form nutrient-rich deltas. At low tide, large mudflats are exposed, worth checking for gulls and shorebirds. In May, four-digit numbers of Knot may stage here. At Mikkelsnes by the Neidenelva river mouth you will find salt meadows which are good for resting geese, often with Bean Geese of both species and with a certain possibility of Lesser White-fronted Goose. During the migration periods, especially in the spring, the fjord can hold several hundred Black-throated Divers and many Red-throated Divers as well. And it is not uncommon to find White-billed and Great Northern Divers here. Considerable numbers of ducks also take advantage of the favourable conditions here, and there are often hundreds of King Eiders in the Munkefjord in winter. Neiden has a winter record of a vagrant Naumann's Thrush.

Best season: Arctic Warbler and Little Bunting arrive around mid-June and are often in active song until early July, before leaving the breeding grounds in late summer. Resting ducks and divers can be seen all year round, shorebirds mainly from late spring to early autumn.

Directions: The sites covered here are all situated along the E6 between Kirkenes and Varangerbotn. At Munkefjord and Neiden, see the map below.

Tactics: Keep your telescope ready and available in the car – you will need it regularly. Head for the E6 in the direction signposted to Tana bru. Be on the look-out for Hawk Owls on roadside poles and treetops along the way, and for raptors soaring overhead.

Munkefjord. The E6 runs along the eastern bank of the Munkefjord in its entire length, starting about 10 km west of Kirkenes Airport Høybuktmoen. The fjord usually holds good numbers of divers and sea ducks and can be explored with a telescope from the road, although there are few safe places to stop along the E6. The best (but not the first) place is at **a.** Sandneset, 17 km west of the airport, recognized by a sand quarry on the left. Here,

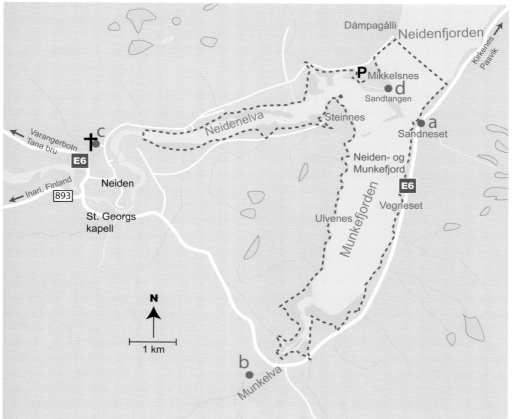

Neiden: a. Sandneset viewpoint; **b.** Munkelva river and **c.** Neiden chapel are both fairly reliable sites for Arctic Warbler in season; **d.** Mikkelsnes headland.

Arctic Warbler is a regular Norwegian breeding bird only in Sør-Varanger municipality. It arrives at the breeding grounds very late in spring, usually not before mid-June. Dense, thin-stemmed birch forest in damp areas is the favorite habitat (see photo on the facing page). The most reliable sites for this species are in Pasvik valley as well as at Neiden and other fjord bottoms east to Grense Jakobselv at the border with Russia. The species may show up in some years in similar habitat further west and south in northern Norway. At migration sites elsewhere in the country, Arctic Warbler is a major rarity. *Photo: Terje Kolaas.*

you get a view of the outer part of the Munkefjord and the inner part of the Neidenfjord. Færdesmyra marsh is located along the E6 a couple of km west of Neiden, seen on the left when you arrive from Kirkenes. Stop at the signposted rest area on the right side of the road and scan the vast marsh with the telescope. Here you can find resting geese, divers and other wetland birds, as well as hunting Hen Harrier and Short-eared Owl. Broad-billed Sandpiper and several shorebird and passerine species nest out in the marsh, but you will probably need to walk out onto the marsh in order to find them. Elk is commonly seen on the marsh, occasionally brown bear as well.

Bugøyfjord reveals itself on the right side of the E6 a further 16 km west. Here at Gjerdebukta, the head of the fjord, is a nice tidal area where shorebirds often gather. Further out in the fjord there are usually sea ducks and divers to be seen. You get the best view of the tidal area by exiting onto the slip road signposted to the memorial site of the Sami artist John Savio. Yellow-breasted Bunting and Steppe Eagle are among the vagrants encountered in this area.

Continue along the E6, and stop here and there to scan the hills on the left for Golden Eagle and other raptors.

Bugøynes is reached via a 40 km long detour eastwards from the E6, clearly signposted 15 km further past Bugøyfjord (72 km west of Kirkenes Airport and 40 km from Varangerbotn). This promontory is located along the southern coast of the Varangerfjord and houses the fjord's second largest Kittiwake colony. The Fuglefjellet seabird cliff is located in Ranvika at the mouth of the Bugøyfjord and can be reached by a 90-minute walk on a clearly signposted and well-marked path south from Bugøynes village. Bugøynes itself

there is an information sign concerning the Mikkelsnes nature reserve, which is visible across the fjord. Scan the fjord and sand banks for a variety of waterbirds.

Further along the E6, at the head of the Munkefjord, there are exposed mudflats at low tide, but again it is difficult to find safe stopping places with a view. From here, you can hike upstream along **b.** Munkelva river. Drive a few km further and turn right, signposted to Neiden chapel and Mikkelsnes. After 1.6 km, turn left to **c.** Neiden chapel. Look and listen for Arctic Warbler in the birch forest around the chapel and in similar habitat nearby. Note that the chapel is in active use as a church. Also beware that there may be many birdwatchers visiting here every day in the high season, so be especially considerate not to stress the birds with song playback.

Continue the gravel road past the chapel to reach **d.** Mikkelsnes nature reserve at the Neidenelva river mouth. A stroll onto the headland might prove rewarding. If you continue the road 400 m past the parking area, has great potential for migrating seabirds in summer and autumn, and resting shorebirds and passerines during migration, but it is not as well visited and well explored as the sites along the north coast of the Varangerfjord. Along the road here, Gyrfalcon is often seen. In Norway, the critically endangered flower boreal Jacobs-ladder can only be found in and around Bugøynes cemetery.

Nyelv is reached along E6 20 km past the exit to Bugøynes. The small settlement is located at the mouth of the Nyelva river, which often has a small gathering of gulls that may be worth looking through. The outlet is visible from the E6. Nyelv also marks the beginning of the Veidnesfjord, which is a small branch of the Varangerfjord, often having quite a few divers and ducks.

Sabahuset viewpoint. Further west along E6, 8 km from Nyelv, you get a view of a shallow part of the inner Varangerfjord with the possibility of shorebirds, gulls and White-tailed Eagles. The area is visible from the E6, but it is a more traffic-safe option to scan the area from Sabahuset, a restored old house, clearly signposted from the E6. Park in the staging area and walk the few metres to Sabahuset, from where you get an overview of a nice part of southern Varangerfjord. In the surrounding bushes you can find vegetation-demanding passerines such as iron sparrows, species that are more difficult to find out on the Varanger Peninsula.

The next site coming up along the E6, is Karlebotn – here featured as a part of Varangerbotn in the Varangerfjord section, which see.

16. PASVIK VALLEY

Finnmark County

GPS: 69.44909° N 29.92531° Ø (Nittisekshøgda hill), 69.1239° N 29.1684° Ø (Blankvassåsen hill)

Notable species: Whooper Swan, Taiga Bean Goose, Shoveler, Smew, Red-throated Diver, Black-throated Diver, White-tailed Eagle, Hen Harrier, Goshawk, Golden Eagle, Osprey, Willow Ptarmigan, Capercaillie, Black Grouse, Hazel Grouse, Crane, Lapwing, Broad-billed Sandpiper, Temminck's Stint,

Typical Arctic Warbler habitat consisting of dense and thin-stemmed birch forest, usually in moist terrain with meadow-like undergrowth. Here from Lake Svanevatn in the Pasvik valley. *Photo: Terje Kolaas.*

Great Grey Owl is an uncommon breeding bird in the Pasvik valley. Ural Owl is more irregular here. Look for them at Skrøytnes and wherever you are able to scan large marshes, meadows and other clearings in the forest. They may hunt from an exposed perch or fly around searching for rodents in the manner of a Short-eared Owl or harrier. *Photo: Kjetil Schjølberg.*

Green Sandpiper (scarce), Spotted Redshank, Bar-tailed Godwit, Jack Snipe, Little Gull, Great Grey Owl, Ural Owl (scarce), Hawk Owl, Short-eared Owl, Tengmalm's Owl, Wryneck, Black Woodpecker, Great Spotted Woodpecker, Lesser Spotted Woodpecker, Three-toed Woodpecker, Yellow Wagtail, Waxwing, Dipper, Red-flanked Bluetail (rare), Redstart, Whinchat, Mistle Thrush, Arctic Warbler, Siberian Tit, Great Grey Shrike, Siberian Jay, Pine Grosbeak, Parrot Crossbill, Two-barred Crossbill (scarce), Arctic Redpoll, Rustic Bunting (scarce) and Little Bunting (variable).

Description: Pasvik is a forested valley with large marshy areas and rolling hills that gently lead the water down to the Pasvik river in the valley floor. The valley runs south from Kirkenes town and is sandwiched between Finland in the west and Russia in the east. This is among the most exotic and exciting locations in Norway and one of the few places where you can with a high degree of probability find Smew, Great Gray Owl, Arctic Warbler and Little Bunting during the breeding season. In addition, Pasvik is home to several other uncommon bird species, as listed above. Several of the birds here are species that have their distribution eastwards in the conifer forest, and which barely touch this part of northern Norway, such as Hazel Grouse and Crane as well as more sporadically occurring species such as Green Sandpiper, Black Woodpecker, Mistle Thrush, Rustic Bunting and - even less often – Red-flanked Bluetail. From 1999, Little Gull has established itself with dozens along the Pasvik river, and this waterway thus forms the core area for the species in Norway. In Pasvik valley, Three-toed is the most common woodpecker, and Hawk and Short-eared are the most common owls. The owl occurrences vary in line with the rodent populations, and the Great Gray Owl can apparently be absent in some years, while in good rodent years the Ural Owl may also be found in the valley.

In the summer months, the waterways, and the large marshy areas in the Pasvik valley are important resting and nesting places for several shorebirds and other wetland species. The shorebirds often display most intensely in the bright summer nights, contributing to producing an exotic, Arctic soundscape. The Pasvik river has its source in Lake Enare in Finland. Parts of its course marks the border between Norway and Russia. The most bird-rich stretches are the shallow areas with lush riverbanks. However, seven power stations regulate the river and limit the natural spring flood, which negatively affects the edge vegetation. The least affected areas, such as Lake Fjærvatn, are still of great importance to birds. Large numbers of Whooper Swans, geese and ducks use these areas as long as there are open, not frozen, sections to be found. There have on several occasions

flocks counting around 150 Smew in Lake Fjærvatn in September.

In 1987, 20–25 territories with Little Bunting were recorded on the 13 km long stretch from Noatun in Lake Fjærvatn south to Kneppåsen hill, but this was probably an exceptionally good year. Usually, Little Bunting is scarcer, but in season, you will usually be able to find it in the areas around Svanvik village, the eastern parts of Nilamyra marsh, south of Lake Vaggetem or at Gjeddebekken creek. In recent years, a small population of Rustic Bunting seems to have been established in the valley. Red-flanked Bluetail was confirmed breeding at Kjerringneset headland in 2011, for the first time in Norway. Breeding has been confirmed or suspected at this site in a few subsequent years as well and is well worth keeping on the radar. The species is still regarded a rarity but is easily overlooked due to its secretive lifestyle and with a song that may pass as just an odd Redstart. Arctic Warbler is often found in some traditional places, such as at Bjørnevatn, Ryeng, Svanvik

Pasvik valley, here from Blankvassåsen hill far up the valley, a place providing a bit of a view in otherwise fairly flat terrain. At viewpoints like this, scan the surroundings for large owls, raptors and bears. This habitat is typical for Siberian Tit. *Photo: Terje Kolaas.*

and Nyrud–Hestefossen, but it too can appear in different places in the valley from one year to the next. Arctic Warbler prefers dense, usually thin-stemmed birch forest close to water. Little Bunting prefers moist areas with dense undergrowth of willows or wild rosemary and more scattered birch trees than Arctic Warbler.

In upper Pasvik valley you will find large areas of untouched, old-growth pine forest. Here, the primeval forest provides nesting sites for birds of prey, woodpeckers and cavity-nesting ducks and owls. The forest may not seem so rich in birds at first glance but with a little patience, you can find many of the most sought-after forest species. The forest areas in Pasvik are connected to large wilderness areas on both the Russian and Finnish sides of the border.

Large parts of the inner valley are protected in the form of *Øvre Pasvik National Park*, *Øvre Pasvik landscape protection area* and *Pasvik nature reserve*. These protected areas join the *Vätsäri wilderness area* on the Finnish side and *Pasvik Zapovednik* on the Russian side of the border.

Pasvik's nature and location give a justified hope of finding rare species from the east. Several eastern celebrities has already been recorded here, such as Ruddy Shelduck, Pacific Golden Plover, Calandra Lark, Red-rumped Swallow *japonica* (Europe's first), Siberian Stonechat (mixed pair breeding with Whinchat), Lanceolated Warbler, several Nuthatch *asiatica*, Black-headed Bunting and Yellow-breasted Bunting. Long-billed Dowitcher, Ivory Gull and much more have also been found here.

If you intend to drive all the way to the head of the Pasvik valley, you will be thankful for driving a car with high ground clearance and four-wheel drive. Be sure to fill up the tank before leaving Hesseng. Even though you can buy fuel at the grocery store in Svanvik, the prices here are rather steep. You might want

Waxwing is to many people in Norway a familiar winter visitor, arriving in large flocks in irruption years, eating berries in gardens and city parks. The breeding areas, however, are in the taiga, the coniferous forest belt that stretches from Norway eastwards through Russia. Even though the species also breeds scattered in higher-altitude conifer forest further south in Norway, it is here in Pasvik valley that it has the strongest population. *Photo: Terje Kolaas.*

to buy groceries in Hesseng as well, where there is a larger selection of goods.

Other wildlife: Pasvik valley has a brown bear population counting a few dozen animals that roam freely across the country borders. The best way to find them is to scan large marsh and logging areas. The bears in Pasvik are relatively fearless of humans. Be careful if you are lucky enough to come across one, especially if it has cubs or is about to eat a carcass. At the exit to Ryeng (see below) there is an old bear den that you can safely crawl into. Among other mammals you can come across in Pasvik are domestic reindeer, elk, lynx, wolverine, marten, and mink. Muskrats are also common along the Pasvik river, as the only watercourse in Norway where you can expect to find them. You may notice the small huts they make along the riverbanks. Wolves roam the valley, and rarely stray raccoon dogs.

Best season: You will find the greatest species diversity in Pasvik at the end of June and the first half of July, but several conifer woodland species stay here all year round. For more details, see the When to go-section in the introduction to the Finnmark chapter (p. 336) for details.

Directions: From the E6 at the petrol stations at Hesseng, 3 km south of Kirkenes town and 1 km east of Kirkenes Airport Høybuktmoen, turn south on Fv885 signposted Pasvikdalen. It is almost 100 km from here all the way to Øvre Pasvik National Park with plenty of sites along the way.

Tactics: If you are short on time and only want a taste of Pasvik, for instance because you are in a hurry to get to Varanger, concentrate on the Svanvik–Skrøytnes area, with stops along the way and detours on some of the forest roads in this area. You may be lucky and find the most important species during a day trip

and you don't necessarily have to drive the entire long and bumpy road to the innermost part of the valley. Still, the area is large and as in most forest areas, birds can be few and far between. You really need to slow down and move calmly and deliberately through the sections you want to prioritize in order to maximize your Pasvik experience. Preferably, you should spend at least a couple of days in the valley. Or nights, rather – in late spring and summer, the birds are at their most active from around midnight until early morning. When driving, make frequent stops at places that look exciting, especially at larger forest clearings and where the road passes open waterbodies. Do not hesitate to take walks into the terrain. This is often the best way to find species such as woodpeckers and Siberian Jay. But be careful in this impregnable terrain.

The Pasvik watercourse is a dominant feature of the Pasvik valley, here from a section at Tangenfossloken near Gjeddebekken creek, along the road to Grensefoss. Birdwatching in Pasvik in late spring and summer is at its best at night, like here, when the terrain is bathed in the midnight sun. *Photo: Bjørn Olav Tveit.*

Oslo

Interior

Skagerrak

Western

Central

Northern

Finnmark

Svalbard

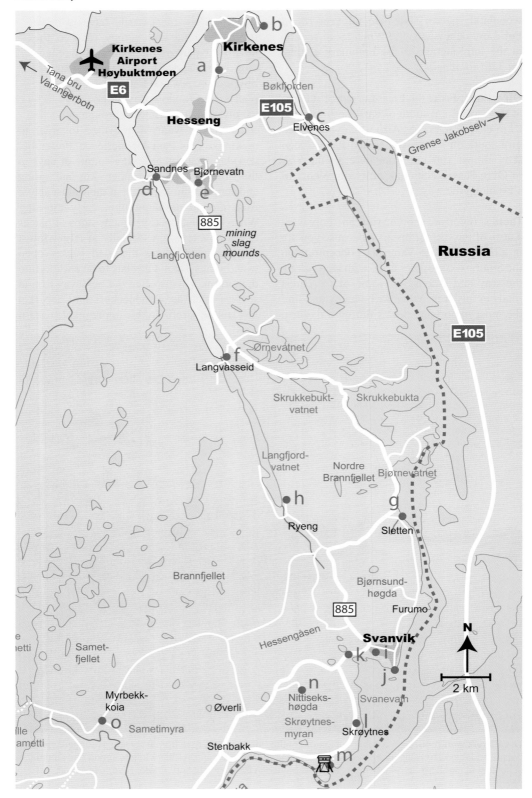

Pasvik valley

Pasvik valley

Kirkenes

Neiden

See map p. 404

Svanvik

Finland

Skogfoss

Russia

See map p. 408

Vaggetem

Nyrud

See map p. 410

Pasvik valley, north: a. Kirkenes lakes; **b.** Prestøya headland, viewpoint Bøkefjord north; **c.** Elvenes, viewpoint Bøkefjord south; **d.** Sandnes, viewpoint Langfjord; **e.** Sandnesmyra marsh, Arctic Warbler habitat; **f.** Langvasseid; **g.** Mortensenbekken creek outlet and Sletta farmlands; **h.** Ryeng, Arctic Warbler habitat and viewpoint Lake Langfjordvatnet south; **i.** Svanhovd, fields and feeders, information centre; **j.** Utnes, viewpoint Lake Svanevatn; **k.** Lake Loken, inner and outer; **l.** Skrøytnes, lakeside fields and woodland; **m.** Skrøytnes tower hide; **n.** Nittisekshøgda hill, viewpoint, Siberian Tit habitat; **o.** Sametimyra marsh, forest species.

You should bring a map and compass but be aware that there is a lot of local magnetism in the terrain some places, so you cannot blindly trust the compass.

Look for active feeders near houses and cabins, they may be visited by Siberian Jay, Pine Grosbeak or Siberian Tit – but please respect private property. Scan lakes, marshes, and the horizon regularly, preferably with your telescope close at hand.

The best sites are revealed from north to south in the following order (see the maps on this and the following pages):

Kirkenes town: Before you enter the valley,

you can do some shopping and scan **a.** the three lakes appearing on your right on the approach. Little Gull breed on the marsh between the two lakes closest to Kirkenes. Drive through the town and onto **b.** Prestøya headland. Park in the bend by the address Bøkfjordveien 39, and walk the path down to the fjord. This site may hold a few migrants and you can scan the Bøkefjord from here. The fjord often holds a selection of ducks, divers and Little Gulls. A different section of the Bøkefjord can be accessed at **c.** Elvenes, if you head back towards Hesseng and exit onto E105 towards Murmansk (Russia). This is also the road to Grense Jakobselv, which see. Head back to Hesseng and exit onto Fv885, signposted to Pasvikdalen.

d. Langfjorden. After just over 2 km on the Fv885 from Hesseng, Langfjordveien road turns to the right, leading down to the Langfjord. Scan the fjord from the bridge. In early spring, continue across the bridge and keep right towards a farm. Check the area around the horse stables for early migrant passerines etc.

e. Bjørnevatn. Back on the Fv885, continue 1 km and turn left onto Grubeveien road, signposted to Bjørnevatn. From this junction and a few hundred metres on, you pass through typical Arctic Warbler birch-forest habitat. Look and listen for the species in similar habitat further into Pasvik valley (and elsewhere) as well.

f. Langvasseid. After another 8 km on the Fv885, turn right, signposted to Langvasseid. Just as you enter the settlement situated on the narrow isthmus between the Langfjord and Lake Langfjorden, a track leads to the right. The track pass by a nice marshy pond, often attracting shorebirds and ducks. The fields and woodlands around this settlement typically holds species like Woodpigeon, Woodcock, Dunnock and Chaffinch, species not so common further up the valley. Tengmalm's Owl is sometimes heard calling in spring and Dipper may be found along the stream. In winter and early spring, feeders here often attract Siberian Jay, Siberian Tit and Pine Grosbeak.

Tengmalm's Owl may be found in the Pasvik valley. However, they usually only call in late winter and early spring. If you are here in summer, you might find it by scratching carefully on the tree trunk underneath Black Woodpecker holes or large nest boxes, and seeing what peeks out to investigate the intruder. *Photo: Terje Kolaas.*

<u>Sletten</u>. After another 13 km on the Fv885, turn left, signposted to Furumo. The first kilometre of this side road leads through the inviting <u>Sletta farmlands</u> and past the **g.** <u>Mortensenbekken creek outlet</u>, worth checking for shorebirds and ducks. If you continue further along this road, you will eventually end up in Svanvik village, passing some interesting, flooded forest habitat and a few riverside fields along the way. You might want to save this stretch for the return trip.

h. <u>Ryeng</u>. Back onto Fv885, continue 3 km and turn right, signposted to Strand. This road leads to Ryeng village and Lake Langfjord–vatn, a classic area for Arctic Warbler and with the possibility of Little Bunting. From outside *Sør-Varanger Museum*, you get a view to the inner section of <u>Lake Langfjordvatn</u>, usually holding a collection of ducks.

Back up by the highway, there is a sign pointing you to *Bjørnehi*, an old bear den. You reach the den after a short walk on a well-marked path. If the barrier on this forest road is open, the road can be driven through nice marshland and forest. By keeping to the left in junctions, this road leads back onto Fv885 further south, past the <u>Grasmyra marsh</u> and surrounding fields 1.5 km north of Svanvik village. If the barrier is closed, you may still head for the Fv885 and drive almost 4 km to enter this forest road from the opposite direction.

i. <u>Svanvik village</u>. This is the main settlement in the Pasvik valley, reached a few km further south along the Fv885. Here, a research station and Øvre Pasvik National Park visitor centre is co-located at the <u>Svanhovd</u> conference centre. Here, they have on display an exhibition about the fauna in the national park, as well as a cafe and shop. There is a botanical garden and an active feeder outside, however, normally without celebre jays, grosbeaks nor tits. The fields around the buildings may have several Golden Plovers, thrushes and more, especially when the fields are bare and moist from rain or meltwater. From Svanhovd you can walk down to the bird-rich Lake Svanevatn and get a view of its north-western section. The swampy birch-forest around Lake Svanevatn is worth exploring with sought-after woodpeckers, warblers and buntings in mind.

If you drive a little east past Svanhovd and turn right towards **j.** Utnes, you will come to a stone jetty at the end of the road that gives a view of the central part of Lake Svanevatn. Back at Fv885, continue south, but note: The grocery store you leave behind at the junction here is the only one in the valley.

k. Lake Loken. The first kilometre or so along Fv885 after leaving Svanvik is worth checking carefully for Arctic Warbler and Little Bunting. And the whole area around Svanvik and the upcoming Skrøytnes may hold the odd pair of Rustic Bunting, although these birds are usually extremely skulky. Any finds are usually kept secret. 1 km from the Svanvik junction, the outer Lake Loken appear on your left, followed by the larger inner Lake Loken on the right. The outer lake, which should be scanned from the car in order not to spook the birds, often holds Smew and a variety of other ducks. You may park by the inner Lake Loken and walk along a wooden boardwalk semi circling the lake. Listen for warblers, buntings and look for ducks on the lake.

l. Skrøytnes is clearly signposted to from Fv885 just south of Loken. This side road loops around the Skrøytnes marsh, joining Fv885 again further down the road. The first kilometres on the Skrøytnes road goes across cultivated land with small meadows and fields surrounded by birch and pine forests. This area is perfect for Taiga Bean Goose, Crane, Golden Plovers, a variety of passerines, Hen Harrier, the odd Marsh or Pallid Harrier besides Hawk and Short-eared Owls. There is also a real potential of finding Great Grey Owl here. Elk is a rather

common sight, and you may stumble across a brown bear. You can see some of the fields from the road, while others you can reach by trying some of the side roads. After the fields and small settlement of Skrøytnes, the road leads onto the large Skrøytnes marsh. Stop at suitable places and scan from the side of the road.

A few km after the road heads out over the marsh, a marked path, in part paved with wooden boards, leads down to a **m.** tower hide. This is located on the southern bank of Lake Svanevatnet with a view across to the small Lille Skogøy island. Here you often find good numbers of Whooper Swans and dabbling ducks, often including Shoveler. Especially at low water levels, when large mud banks are exposed, there is a good chance of mixed flocks of shorebirds with Broad-billed Sandpiper included. Smew and Little Gull are often seen far out on the lake.

Continue along the Skrøytnes road, scrutinizing the upcoming fields and the Pasvik river. Be sure to step out of the car here and there to listen for Little or even Rustic Buntings.

When the road meets Fv885, you can then continue south, or drive north for 4.5 km to get to Nittisekshøgda hill and Sametimyra marsh.

n. Nittisekshøgda hill. This was originally a military guard post which now has a cafe in the summer. The side road leading up this hill is signposted *96-høyden* (96 height) from Fv885, 5 km south of Svanvik village. From behind the trees, you can catch a glimpse of Skrøytnesmyra marsh. You may also see the Russian, heavily polluted industrial town of Nikel in the distance. Pine-clad hills such as Nittisekshøgda are good places for Siberian Tit. Look for the species around nest boxes or natural cavities here.

o. Sametimyra marsh. The forest road that runs west from Fv885, 2.2 km south of Nittisekshøgda hill, takes you into an exciting, wooded landscape around Sametimyra marsh with opportunity of encountering many of the valley's specialities.

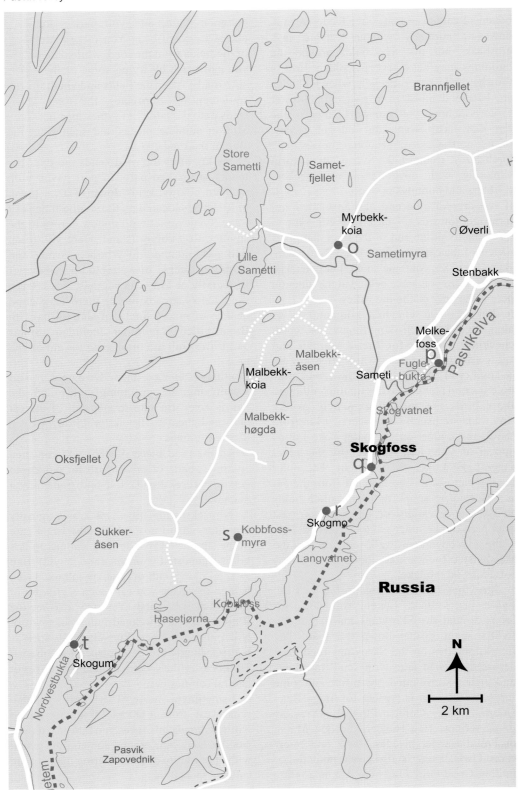

Brannfjellet

Store
Sametti

Samet-
fjellet

Myrbekk-
koia

o

Sametimyra

Øverli

Stenbakk

Lille
Sametti

Melke-
foss

p

Pasvikelva

Malbekk-
åsen

Malbekk-
koia

Sameti

Fugle-
bukta

Skogvatnet

Malbekk-
høgda

Skogfoss

q

Oksfjellet

r

Skogmo

Sukker-
åsen

s

Kobbfoss-
myra

Langvatnet

Russia

N

Kobbfoss

Hasetjørna

t

Nordvestbukta

Skogum

2 km

Pasvik
Zapovednik

Pasvik valley, middle section: o. Sametimyra marsh, forest species; **p.** Melkefoss dam, migrants; **q.** Skogfoss cove, photo opportunities from the car; **r.** Skogmo, viewpoint Lake Langvatnet, lush farmland; **s.** Kobbfossmyra marsh, sheltered and moist fields; **t.** Elgbekken creek, small wetland.

p. Melkefoss dam. Melkefoss is signposted from the Fv885 almost 3 km south of the Skrøytnes road junction. It provides a view to a slow-flowing stretch of the Pasvik river. Swallows and Sand Martins often gather here, as does a variety of other passerines along the riverbank. A road runs parallel to the river north from Melkefoss dam, past Birk Husky with often active feeders and further on past several riverside fields, often good for Black Grouse and Taiga Bean Goose. This road leads back to Skrøytnes marsh.

From Melkefoss, continue further south on the Fv885. You will from time to time see the Pasvik river with associated backwaters and lakes on your left. Capercaillie and Black Grouse can often be seen eating roadside gravel.

Make sure to make a stop just after Skogfoss power plant where the road crosses the **q.** Skogfoss cove and you see water on both sides of the road. Smew is often seen here. Continue south for 3 km and you get a view of the larger Lake Langvatnet on your left, which should be scanned for ducks and divers. Here is also a horticulture outlet, signposted *Garneri*, and the upcoming few hundred metres have the lush **r.** Skogmo fields on your right, sometimes holding resting migrant passerines and shorebirds. Next up, 5 km further ahead, is **s.** Kobbfossmyra marsh. The fields behind the farm buildings often hold flocks of Taiga Bean Goose, Crane and displaying Black Grouse. Yellowhammer and Whinchat breed in the area. Elk and the occational brown bear is sometimes seen here.

A nice little site 9.5 km further on is the little cove in the innermost part of Lake Nordvestbukta, **t.** Elgbekken creek, worth checking for wetland species.

Pine Grosbeak is a sparse breeder in the Pasvik valley, in coniferous forest with a rich juniper undergrowth. In winter and spring, it can often be found at feeders in the valley. *Photo: Kjartan Trana.*

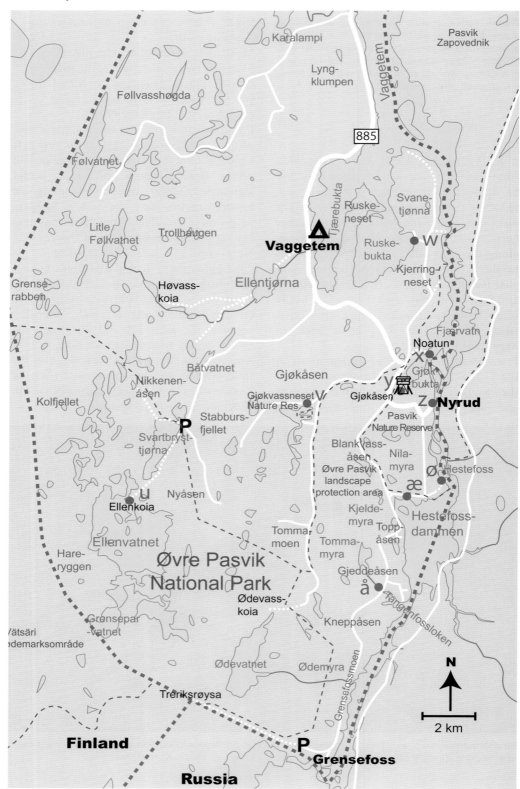

Pasvik valley, south: u. Ellenkoia hut, open to accommodation, forest species; **v.** Gjøkvassneset NR and Gjøkåsen hill, unspoiled forest; **w.** Kjerringneset headland, viewpoint to Ruskebukta cove; **x.** border post viewpoint to Lake Fjærvatn and Gjøkbukta cove, north; **y.** Gjøkbukta hide, view to the cove from the south; **z.** Nyrud, viewpoint to the Pasvik river; **æ.** Hestefoss dam, viewpoint to lake and marshland; **ø.** Hestefoss, viewpoint to the Pasvik river; **å.** Gjøkbekken creek, Little Bunting habitat.

Hawk Owl is often seen near Vaggetem village a little further south, and the <u>Vaggetem campsite</u> has usually an active and rewarding feeder.

<u>Upper Pasvik National Park</u>. From Fv885, just under 2 km south of Vaggetem village, a forest road enters on the right. It is closed to public traffic from October to May. This forest road leads almost 9 km to the car park by Lake Svartbrysttjønna. From here, a 4 km long marked path runs along the east bank of the lake up to **u.** <u>Ellenkoia hut</u> within the national park. The hut is simple and unattended but open and can be freely used for accommodation. You can enter the national park from the east side as well, see directions for Nilamyra marsh below.

v. <u>Gjøkvassneset peninsula and Gjøkåsen hill</u>. Back on Fv885, continue south for a further 5 km, and turn onto the forest road on the right. This road is closed from October to May as well. After 4 km on this forest road, keep to the left and park at Gjøkvasskoia hut and explore the site on foot. Gjøkvassneset is a peninsula that protrudes 600 metres into the small Lake Gjøkvatnet. The birds here are basically the same as in the surrounding area but here you get to see them in primeval pine and birch forest, protected as a nature reserve. The lake can be viewed from the forest road continuing along the northern bank. The <u>Gjøkåsen hill</u> on the opposite side of the forest road from Lake Gjøkvatnet, is also well worth exploring on foot in search for the forest species.

<u>Kjerringneset headland</u>. Continue just 200 m on the Fv885 and turn left onto a gravel road. This road leads 9 km, or nearly all the way, out on Kjerringneset headland. The headland represents some of the best forest areas and is surrounded by a part of the Pasvik watercourse that constitutes one of the most important and bird-rich wetlands in Pasvik. It is also known for being one of the most reliable places to se brown bear, although you will still need a lot of luck. The stretch of forest road from Lake Koietjernet (3.5 km from the main road) and further out on Kjerringneset is regarded as fairly reliable for Capercaillie, Pine Grosbeak, Parrot Crossbill and Siberian Jay, and Red-flanked Bluetail has been found nesting here a few times in recent years.

Parts of **w.** <u>Ruskebukta cove</u> can be viewed by following the forest roads out onto Kjerringneset, taking detours to the left. At Jordanfossen on the opposite side of the headland and at Lake Svanetjønna, you may find Little Gull.

For a view of the very bird-rich <u>Lake Fjærvatn and Gjøkbukta cove</u>, park along the Kjerringneset road by the barrier blocking the side road exiting to the right just 200 m from the Fv885. Walk this road and later the path past the residential buildings at Noatun farm and further out to **x.** <u>the border post</u> at the far end of the headland, about 1 km from the barrier. Please respect private property along the way, and be particularly considerate at night, and do not enter the farmyard without a permission obtained in advance from the landowner.

An alternative view of <u>Gjøkbukta cove</u> can be obtained from the **y.** <u>Gjøkbukta hide</u> on the south side. You reach this site by continuing 1.2 km on Fv885. Here, on the left side of the road, signposted *Fuglebu* (bird hide), is a designated parking area and a path leading the 200 m to a shelter. Trees block much of the view from the hut. You can solve this by manoeuvring a bit along the ridge or continuing down to the water's edge.

z. <u>Nyrud</u>. The Fv885 ends just 3 km further on, by Nyrud police station. Park roadside outside the fencings and walk into the yard in order to obtain a view of a nice stretch of the Pasvik river. Bird ringing has been conducted here in early autumn the last several years, encountering quite a few Rustic Buntings and

Long-tailed Duck and other sea ducks often gather on the sea in good numbers at Grense Jakobselv. Many may also pass by. *Photo: Kjetil Schjølberg.*

Bunting even in years when it is absent from other parts of the valley. Upper Pasvik valley. You can find several more of the Pasvik specialties along the road to Grensefoss, from where a marked path leads the 4.5 km hike through old-growth forest to Treriksrøysa. The road is quite rough some places and even with four-wheel drive and good ground clearance, you must expect a good hour's steady drive from the Hestefoss dam to Grensefoss. Treriksrøysa – or the Three-Country Cairn – marks the point at which the borders of Russia, Finland and Norway meet. Note that it is strictly forbidden to circle around the cairn, and thereby entering Russia without permission, a provision that is actively enforced and results in substantial fines. The forest road from Rv885 to Grensefoss is closed in winter but if you happen to be here in spring before it opens, you can drive 100 m in this road and park by the garage and have a good chance of finding Siberian Tit on the hill behind it.

Red-flanked Bluetails, possibly as a result of breeding nearby.
Nilamyra, Hestefoss dam and Tommamyra. Together, this inner part of Pasvik constitutes a vast and very rich marshland. You can find a number of wetland species here, in addition to Little Bunting, large owls and brown bear.
From Nyrud, drive back along the Fv885. After 2.5 km, the road into this area is on your left, discreetly signposted to *Treriksrøysa*. The road leads 15 km to Grensefoss. Along the way you pass through pine and birch forest alternating with marshlands. Drive slowly and stop regularly to walk a bit or to scan the surroundings wherever it looks interesting. Blankvassåsen, Toppåsen and other pine-clad hilltops are the most reliable places to find Siberian Tit in late spring and summer. Look out for Pine Grosbeak and Siberian Jay along the road as well. At æ. Hestefoss dam, you can scan the lake and surrounding marshlands from on top of the dam and from the roads. From the east end of the Hestefoss dam, you can follow the track to ø. Hestefoss, where you will find good habitat for Arctic Warbler and Little Bunting. Here is also an alternative road back to Fv885 close to Nyrud, running parallel to the Pasvik river, although this road is even more bumpy than the main forest road.
å. Gjeddebekken creek tends to hold Little

18. GRENSE JAKOBSELV

Finnmark County
GPS: 69.79118° N 30.79445° Ø (sea viewpoint)

Notable species: Resting and migrating ducks, Cormorant, Shag, skuas, gulls and auks. Resting shorebirds and passerines on migration. Willow Ptarmigan, White-tailed Eagle, Rough-legged Buzzard, Golden Eagle, Bluethroat and Arctic Warbler.

Description: Grense Jakobselv is a small settlement and a military base at the border with Russia. This is the most easily accessible place near Kirkenes and Pasvik valley with a clear view to the open sea and thus with a potential for resting and migrating seabirds. Grense Jakobselv is located along the outer Varangerfjord, between the Varanger Peninsula and the Russian Rybachy peninsula. The place has not been well explored as a migration site, and it is therefore unclear to what extent it is affected by the massive migration of Arctic seabirds that passes along the outer coast of Finnmark. It should be well worth trying out though, especially in winds from the northern sector.

At the outlet of the border river there is a lagoon usually holding several gulls, shorebirds and large flocks of Goosander and Red-breasted Merganser. Also check the edge vegetation and the surrounding beaches, dunes and willow scrub for resting migrants. Up along the river valley you will find both White-tailed and Golden Eagles. The small village has no facilities for visitors apart from a public restroom and rain shelters at the rest area near the end of the road. The drive from Hesseng to Grense Jakobselv takes you through interesting habitats, including montane areas with associated species, and dense birch forest with a proven potential for Arctic Warbler.

Best season: Migration seasons. Arctic Warbler from mid-June to mid-July.

Other wildlife: Beluga is sometimes seen in summer.

Directions: From the E6 at Hesseng, take the exit towards Grense Jakobselv and keep to the left just before the border crossing to Russia. The trip takes about one hour. Look and listen for Arctic Warblers in suitable habitat, particularly along the fjords and lakes you pass by. In Grense Jakobselv, continue 1 km past Oscar II's chapel and park at the turning point. You get the best view of the sea by walking up onto the small knoll towards the sea. It can be a challenge to find sufficient shelter here in strong winds. The last 30 km of the road to Grense Jakobselv is closed in winter, between October and May – the exact date varies, depending on the snow conditions.

Grense Jakobselv has exciting habitats for migratory birds, both on sea and land. On the fjord beyond, considerable numbers of seabirds can roost and migrate. The headland just across the bay belongs to Russia. *Photo: Bjørn Olav Tveit.*

SVALBARD

S valbard is an exotic part of the Kingdom of Norway. A trip up here is something every birdwatcher should have on their bucket list. The archipelago is located in the Arctic, just 1,100 km from the North Pole. The landscape and fauna are strongly characterized by the northern location, and plant life is dominated by moss and ankle-high plants. Nevertheless, Svalbard is not quite as inhospitable as the latitude would suggest, thanks to the Gulf Stream which carries warmer water from the south. In a few short, hectic summer months, there is a bustling birdlife here. The breeding fauna is made up of only around 40 species. However, among these you will find celebrities such as King Eider, Red (Grey) Phalarope, Sabine's Gull,

Ivory Gull is one of Svalbard's most desirable species. You might find it in Longyearbyen but you have an even better chance if you take a trip further afield. *Photo: Eirik Grønningsæter / WildNature.no*

Ivory Gull, Brünnich's Guillemot and Little Auk, in addition to the fact that some of the more common species appear as subspecies and varieties that are different from what you may be used to further south: Many male Common Eiders here have mustard yellow colour on their bills and can raise small, white scapular sails on their backs, clear signs that they originate from the Arctic subspecies *borealis*. The Rock Ptarmigans are of the subspecies *hyperborea*, colloquially called Svalbard Ptarmigan. They

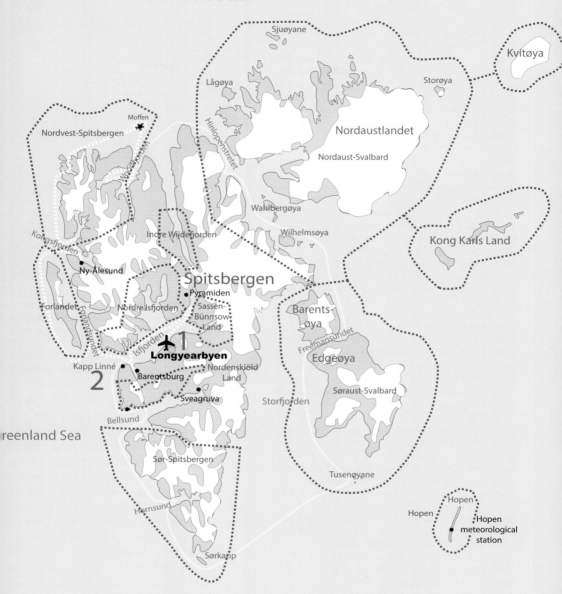

Arctic Ocean

Kvitøya

Sjuøyane

Lågøya

Storøya

Nordaustlandet

Nordaust-Svalbard

Moffen

Nordvest-Spitsbergen

Wahlbergøya

Wilhelmsøya

Kong Karls Land

Indre Wijdefjorden

Ny-Ålesund

Spitsbergen

Pyramiden

Barents-øya

Forlandet

Nordre Isfjorden

Sassen-Bünnsow-Land

Edgeøya

Longyearbyen

Nordenskiöld Land

Søraust-Svalbard

Kapp Linné

2

Barentsburg

Sveagruva

Storfjorden

Greenland Sea

Bellsund

Sør-Spitsbergen

Hornsund

Tusenøyane

Hopen

Hopen

Hopen meteorological station

Sørkapp

Barents Sea

N

50 km

Norwegian Sea

Map base © Cappelen Damm

Some practicalities

Longyearbyen is usually shrouded in snow from October to May, and even in midsummer you can be surprised by snow that can lie several inches thick for days. As a rule, the temperature is around 5 degrees in July, and there is often high humidity and fog. Prepare a trip to Svalbard in summer as you would prepare a trip to mainland Norway in January – full winter gear, in other words. Robust waterproof footwear is important, as many places are wet and muddy. Mosquitoes and midges occur in Svalbard, especially in the Isfjord area, but usually not in troublesome quantities. At Longyearbyen, there is a midnight sun with associated benefits and challenges between April 20 and August 22. It is mostly too dark for effective birdwatching from early November to mid-February.

A number of special traffic rules apply in Svalbard, and you would do well to familiarize yourself with them before you set off. The regulations are enforced by the Governor of Svalbard, who supervises from boats and helicopters. You can travel freely in what is called Zone 10, an area that includes Longyearbyen, Adventdalen valley and large parts of Nordenskiold Land. Traffic outside Zone 10 must be reported to the Governor, who may also require special devices, such as emergency beacons and insurance in the event of rescue operations. You can safely drink freshwater in most places on Svalbard, with the exception of water that flows directly from glaciers and water between Longyearbyen and Barentsburg. There are sibling voles here, which have brought with them a tapeworm whose final host is the Arctic fox, but which also poses a risk to humans. Therefore, do not drink water from streams in the areas where sibling voles are widespread. There is mobile phone coverage in Longyearbyen, most of Isfjorden and in Barentsburg, but you cannot find coverage outside the built-up areas.

Longyearbyen has most facilities, a rich nightlife and plenty of shopping opportunities. There is no VAT on goods sold on Svalbard, so the prices of most things are significantly lower than in mainland Norway. However, fresh foods are more expensive due to high shipping costs. It is common practice in Svalbard to take off your shoes in hotels, museums and in some shops, so thick socks or slippers will come in handy here. You will find on offer a wide range of organized trips with dog sleds, snowmobiles, horses and on foot, as well as short and long boat trips.

Ross's Gull is the crown jewel among the birds frequenting Svalbard. However, you may need to travel to the north and east side of the archipelago in late summer. Even there, the species is uncommon. *Photo: Steve Baines.*

are significantly larger and, in summer plumage, warmer brown than those on the mainland. The Black Guillemots are of the subspecies *mandtii*, nesting higher up on the mountain sides and further inland than mainland birds. On closer inspection, they have larger white wing patches and a narrower bill than its relatives in the south. The Puffins are large and grey-cheeked, and most Fulmars are of the dark form. In addition, there is a greater chance at Svalbard than anywhere else in Europe to come across a Ross's Gull, one of the world's most enigmatic and beautiful birds. Mammals are a prominent element in the fauna, led by Arctic species such as Svalbard reindeer, Arctic fox, polar bear, walrus, beluga whale and a number of others. Even the soundscape up here seems foreign and exotic to first-time visitors: Red-throated Diver, Arctic Tern, Kittiwake, Glaucous Gull, Little Auk, Black Guillemot and Snow Bunting fill the cool air with peculiar calls that help make your maiden trip to Svalbard unforgettable.

Longyearbyen is the administrative centre of the archipelago. It is a distinctive small town with just over 2,600 inhabitants of many nationalities. The society was built around mining and still bears the mark of this, although research and tourism are increasingly important industries. The total land area on Svalbard is 62,229 km². More than half of the land areas are covered by glaciers, several of which extend out over the sea and provide an additional area of firm ground all year on parts of the surrounding sea. As much as 65% of the land area on Svalbard is protected, most of it in the form of national parks.

Around the solid ice cap that surrounds the North Pole, there is a belt of drift ice. This is a very important factor for the fauna in the Arctic, and it is here that you find the richest diversity of both birds and marine mammals. The further north you go, the denser the drift ice packs together. The belt of such pack ice varies seasonally and from one year to the next. In winter, the pack ice belt creeps south and encloses most of Svalbard. In summer it retreats north again. In some years it retreats no further up than that it blocks boats intending to sail around the archipelago. In other summers, the pack ice and drift ice retreat so far north that you cannot reach the desirable pack ice belt except on the most far-reaching boat cruises. The mountain formations on Svalbard are spectacular and alien compared to those you find in mainland Norway. At Longyearbyen,

they are characterized by dark, sedimentary rocks that lie horizontally in the terrain and form large flat mountain plateaus. From these plateaus, the mountains plunge straight down, exposing layers of rocks of varying hardness. These have weathered irregularly to form jagged rock ledges with huge piles of gravel below. Meltwater has carved deep furrows in these piles. The snow at the bottom of the furrows often does not melt during the summer, so they appear as narrow white streams down the lead-black mountain sides. Elsewhere on Svalbard, especially in the west, the mountains have the character of pointed Alpine peaks. From this, the largest island, Spitsbergen, got its name – a name some use as a collective term for the entire archipelago. The two highest mountain peaks, Newtontoppen and Perriertoppen, are located on Spitsbergen and both rise to 1,717 metres above sea level. Bjørnøya island, which lies by itself approximately midway between Spitsbergen and Finnmark, is considered a part of Svalbard. Jan Mayen is another isolated island, positioned closer to Greenland, about halfway between Spitsbergen and Iceland. It is part of Norway and faunistically has a lot in common with Bjørnøya,

Fuglefjella by Isfjorden and the entrance to Bjørndalen, just outside Longyearbyen. *Photo: Bjørn Olav Tveit.*

417

Polar bears are rather uncommon around Longyearbyen. Still, if you go hiking on your own, special precautions must be followed to ensure that both you and the bear escape any encounter unscathed. *Photo: Eirik Grønningsæter / WildNature.no*

but Jan Mayen is not considered part of the Svalbardarchipelago. Although some cruise ships stop by, Bjørnøya and Jan Mayen are in practice inaccessible to most birdwatchers. Longyearbyen, on the other hand, is easily accessible by regular flights from mainland Norway, and almost any species you can hope to find in the European part of the Arctic, you may find here.

THE BEST SITES

Fortunately, the best birdwatching sites on Svalbard are the most easily accessible, namely the areas around Longyearbyen. All the nesting Svalbard specialties can be seen here, and the area is one of the most important roosting places in Svalbard for geese and shorebirds. You may even find stray passerines from the mainland. You should try to arrange for a boat trip into the Isfjord, to Barentsburg, Ny-Ålesund or preferably even further. However, a long trip is a time-consuming and very expensive pleasure, and the return is meagre in terms of encountered species, compared to a stay of a similar duration in Longyearbyen. On the other hand, a boat trip up to the drift ice does give an increased chance

of finding Ross's Gull, besides Ivory Gull and certain other species, as well as polar bears, walruses and more. Your encounters with the huge seabird cliffs and the impressive nature on such a trip will certainly give value for money.

1. LONGYEARBYEN AREA
Spitsbergen, Svalbard
GPS: 78.25173° N 15.41624° Ø (Vestpynten headland)

Notable species: Staging and passing migrants; Pink-footed Goose, Barnacle Goose, Brent Goose, Common Eider (*borealis* and *mollissima*, or a mix), King Eider, Long-tailed Duck, Rock Ptarmigan *hyperborea* (Svalbard Ptarmigan), Red-throated Diver, Fulmar (dark phase), Greater Ringed Plover, Pectoral Sandpiper (uncommon), Sanderling, Purple Sandpiper, Dunlin *arctica*, Grey Phalarope, Pomarine Skua, Arctic Skua, Long-tailed Skua, Great Skua, Kittiwake, Sabine's Gull, Glaucous Gull, Iceland Gull, Ivory Gull (uncommon), Arctic Tern, Little Auk, Puffin (*"naumanni"*), Black Guillemot *mandtii*, Brünnich's Guillemot and Snow Bunting.

Polar bear precautions

Experiencing polar bears up close is one of the most beautiful and exciting nature experiences Svalbard has to offer. You have the best chance of finding it by taking part in one of the boat trips, but you can also find polar bears at Longyearbyen, especially in early spring. The polar bears depend on sea ice to catch seals, which are the most important component of their diet. They are therefore mainly linked to the drift ice belt around the North Pole, and Svalbard is thus part of the species' normal range. There is no clearly defined, indigenous population on Svalbard, as the animals migrate over large areas. In summer, when the drift ice moves north, most of the polar bears follow. Some individuals may still remain on the islands during the summer.

Polar bears generally avoid people, but they are curious by nature and may walk towards you to investigate you more closely. It is especially young bears that have recently been rejected by their mother who can then pose a threat to humans. In order to avoid provoking confrontations, it is forbidden on Svalbard to actively seek out polar bears or to lure them to you.

Anyone who moves outside the settlements must have suitable means of scaring and chasing away polar bears. The governor also recommends carrying a firearm. A signal gun is considered a more effective means of scaring away polar bears than, for example, to fire rifle shots into the air. Weapons for polar bear protection can be rented from authorized weapons dealers in Svalbard. There can be up to a month processing time for an application to hire a weapon. Private individuals over the age of 18 who have a valid firearms license can hire a rifle without applying to the Governor.

Never try to run from a polar bear. Then the hunting instinct can be awakened so that it takes up the hunt. Instead, calmly walk away facing it. If it continues to come at you, fire shots into the air or the ground with the flare gun. Shoot repeatedly if the bear does not immediately turn sideways and run away from you. Only in extreme cases will it be necessary to shoot the bear with a rifle. If you spend the night in a tent, you should set up trip flares around the tent or let the tour participants sit on polar bear watch in turn. Provisions and toilets should be located at least one hundred metres from the tent, visible from the tent opening.

Description: Longyearbyen town is located in the Adventfjord, a small side fjord to the large Isfjord that cuts deep into the western side of Spitsbergen. Not only is Longyearbyen the most easily accessible place on Svalbard, it is also one of the most ornithologically interesting. With hardly any special preparations, and without renouncing comfort, a visit to Longyearbyen can give you almost all the specialties in the European part of the Arctic.

The terrain around the city is spectacular and the bird life exotic, with Snow Buntings in the streets, Glaucous Gull as the dominant large gull in the harbour and with Little Auks breeding in the mountain slopes above the town. Birdwatching in the Longyearbyen area is centred on the Adventfjord and the Advent delta, as well as the tundra and wetlands in the Adventdalen valley.

The dominating species in and around the town, are Pink-footed and Barnacle Geese, sometimes by the thousands, Common Eider, Long-tailed Duck, Svalbard Ptarmigan, Red-throated Diver, Fulmar, Purple Sandpiper, Dunlin, Kittiwake, Glaucous Gull, Arctic Tern, Brünnich's and Black Guillemots, Little Auk, Puffin and Snow Bunting. More sparse, but still common, is King Eider, Grey Phalarope and Arctic Skua, which all breed. Great Skua is sometimes seen here as well. Ivory Gull is regular in April and May, although they may visit this area throughout summer from the breeding grounds in other parts of the island. During migration, the most common visitors are Brent Goose, Pintail, Teal, Tufted Duck, Greater Ringed and Golden Plovers, Turnstone, Sanderling, Red-necked Phalarope and a few other shorebirds, Pomarine and Long-tailed Skuas, Sabine's, Herring, Iceland, Lesser and Greater Black-backed gulls, besides Wheatear. Pectoral Sandpiper is almost annual and several observations of displaying birds indicate that it may breed nearby sometimes. Ross's Gull and Snowy Owl are highly unusual visitors to this part of Svalbard. Gyrfalcon is rarely seen but when it first appears, it is usually in a beautiful, pale morph. Snow and Canada Geese that are sometimes seen here are regarded as wild birds from North America, and should be

Little Auk has long been one of the most numerous bird species in the Arctic, but populations have declined dramatically in recent years, like so many other seabirds. The species nests in colonies in a number of places on Svalbard, including near Longyearbyen. *Photo: Eirik Grønningsæter / WildNature.no*

well documented. Especially in the autumn, migrants from the mainland can stray up here, with Fieldfare and Redwing being the most regular. Other passerines that have ended up here include River, Marsh, Yellow-browed and Wood Warblers and Rustic Bunting. A number of other vagrants have been found on Svalbard, including Norway's first Solitary and Spotted Sandpipers, as well as Blue-winged Teal, Bittern, Baird's Sandpiper, Ring-billed and Bonaparte's Gulls. Many of these records are from Longyearbyen.

Other wildlife: Svalbard has a rich wildlife, and many of the species can be found around Longyearbyen. Most visible in the terrain is the Svalbard reindeer, which graze individually or in small groups on the tundra. Arctic fox is a common sight, often close to the settlements. In the mountain slopes below the seabird cliffs between Longyearbyen and Barentsburg you find sibling vole, an introduced species from Russia. Polar bears are seen annually near Longyearbyen. Out on the fjord, especially in spring before the ice disappears completely, there can be many other marine mammals to see. Bearded seal is the most common of the seals, but ringed and harp seals and walrus also occur. Among the whale species, beluga is the most common, sometimes appearing in large flocks. Whales in general, including blue and fin whales, can be seen from Vestpynten headland, usually far out in the fjord. It is best to scan with a telescope on calm days in the evening when you have some backlighting, and the blow is easily visible. Even better is taking part in boat trips out of the Adventfjord.

Best season: The seabirds normally arrive around the last week of March. From being completely empty, within a few days there can be hundreds of thousands of birds in the fjords and in the colonies. The first Snow Buntings arrive in mid-April. From May onwards, the King Eiders are also in place, but after the first week of June they spread out into the nesting areas on the tundra and can become more difficult to find. For shorebirds and geese, it usually peaks between the end of May and mid-June. Grey Phalarope arrive at the turn of May–June. In the last week of

May, the snow is still melting, and the birds are then concentrated around bare spots along the road into Adventdalen valley. For ducks and geese this is the best time, while for shorebirds it is probably the first two weeks of June that have the greatest species diversity. Otherwise, the birds are here throughout June and July, but the seabirds leave the colonies at the turn of July–August. Only the occasional Grey Phalarope is seen after the first week of August. In late autumn and winter, there are very few bird species to be found, but the Ptarmigans hold their ground, and you can find seabirds in ice-free parts of the Isfjorden when the light conditions permit. Many of the mammals can be seen here then as well. Resident birdwatchers look for birds in the dark from the floodlights on the harbour and shine flashlights under the quays, which is often the last place the ice cover forms.

Directions: By plane to Longyearbyen, 3 hrs direct flight from Oslo and 1.5 hrs from Tromsø.

Tactics: You can hire a car or bicycle to get around, but you will also be able to manage well on foot, possibly with the help of a taxi if you want to move further out of the town. In summer, it can be nice to look for birds and also photograph in the light of the midnight sun, when few people are out and Arctic foxes and geese can wander about in the middle of the town. Within Longyearbyen town you will find Little Auk and Svalbard Ptarmigan up in the mountain sides. It is particularly nice to look for the ptarmigan by walking into Longyeardalen from where the asphalt road ends in Nybyen, but they can be difficult to find as they are well camouflaged and confiding, and thus difficult to flush. Often, they are detected by the call.

a. The Sjøskrenten ridge overlooking the Adventfjord and the Advent delta is equipped

Adventdalen valley seen towards Fivelflya in August, with Glaucous Gulls, the most common large gull in the area. *Photo: Bjørn Olav Tveit.*

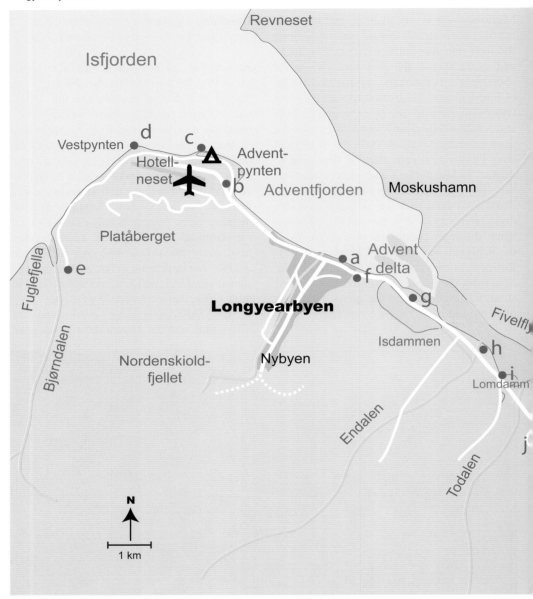

Longyearbyen area: a. Sjøskrenten ridge, viewpoint with warming shelter; **b.** horse stable; **c.** Laguna lagoon; **d.** Vestpynten seawatch; **e.** Bjørndalen valley; **f.** dog yard by the Advent delta; **g.** Lake Isdammen; **h.** The old northern lights observatory; **i.** Lake Lomdammen; **j.** scrap yard; **k.** Dog yard, viewoint to Bolterdalen valley; **l.** Breinosa mountain.

with a warming shelter (*LoFF-huset*), open to all in the summer season. The mudflats here are very good for shorebirds and gulls. Out on the fjord there are usually eiders, auks and divers, and there may be movements of gulls, terns, skuas and other seabirds. In the bay 300 m from the shelter, there is a discharge attracting birds, not least when the fjord is otherwise iced over.

b. The horse stable by the airport (on the right, 1 km before the terminal building) keep animals which in turn attract insects. The dung heaps here represent a welcome oasis for warblers and other stray insectivores.

c. Laguna is an artificial lagoon with a few islands at the Hotellneset headland. Barnacle Goose, Arctic Tern, Arctic Skua and Purple

Operafjellet

Adventdalen

k

Gruve 7

EISCAT

Kjell Henriksen
Observatory

Breinosa

d. <u>Vestpynten headland</u> with the characteristic lighthouse is located at the mouth of the Adventfjord. This is the best place in the area for migrating seabirds. Here you have an unobstructed view of <u>the Isfjord</u>. Many of the birds pass at close range. Here, there is an opportunity for Sabine's Gull, Pomarine Skua and more. Besides Common Eiders, there are often King Eiders on the sea, especially in spring.

e. <u>Bjørndalen valley</u> at the end of the road have nice areas for ptarmigans as well as some Little Auks at close range, in addition to foxes and reindeer. Check the outlet of <u>Bjørndalselva river</u> for resting ducks, gulls and shorebirds. Brent Geese often roost here at the beginning of June and in the transition from August to September.

<u>Adventdalen valley</u> stretches from Longyear-byen to the south and consists of tundra with a mosaic of rivers, streams, and ponds. Follow the road south from the Sjøskrenten ridge and scan the terrain regularly at suitable places. Feel free to take detours on foot into the damp, muddy terrain or up into the side valleys, but please pay attention to the nesting birds. A scant kilometre outside the town are two large enclosures with sled dogs, often referred to as the **f.** <u>dog yards</u>. Dog yards like this can be found elsewhere in the Adventdalen as well, and they all tend to attract birds. The dog yards have traditionally been the most reliable places for Ivory Gulls, which in summer often appear here at night. The dog yards must also be checked for unusual passerines that may seek shelter here and benefit from the insects that abound in the enclosures. Up to 200 pairs of Common Eider nest near the enclosures at the dog yards closest to the town, as the dogs provide some protection against Arctic foxes. Grey Phalarope nests in the area as well.

g. <u>Lake Isdammen</u> is the large body of water on the south side of the road just beyond the first dog yards. Here you will often find flocks of Barnacle and Pink-footed Geese, as well as ducks and Red-throated Divers. In most years, a few pairs of King Eider make nesting attempts here, especially in the parts farthest from roads and buildings. On the opposite side of the road from Lake Isdammen is Lake Tuedammen with great habitat for shorebirds, geese and dabbling ducks. At the **h.** <u>old northern lights observatory</u>, you have

Sandpiper breed. This is one of the best places for resting gulls as well, with the possibility of Sabine's Gull. King Eider and other ducks often stop by. Like Sjøskrenten at the Advent delta, Laguna should be checked often, since the turnover of birds can sometimes be high. Arctic foxes are regularly seen here. From Longyearbyen you reach Laguna by taking the road north towards the airport and turning right on the road to Bjørndalen, 1.2 km before the terminal building.

a view with a telescope to the particularly bird-rich <u>Fivelflya plateau</u> on the opposite side of the Adventelva river. Try to find something here that can give you some height above the flat landscape. Further up in the valley, on the left side of the road, is **i.** <u>Lake Lomdammen</u>, where there are often ducks and Red-throated Diver to be found. After the ice has gone in June, there are often small flocks of King Eider here. A little further into Adventdalen, at the end of a side road up the mountainside on the right, is a fenced **j.** <u>scrap yard</u>. which used to be a good place for gulls but now

Fulmar populations have a higher proportion of dark individuals, like this one, the further north you go. *Photo: Eirik Grønningsæter / WildNature.no*

containing scraps that provide shelter to stray passerines. The marshy area between the waste facility and the main road below is one of the best places for shorebirds in Adventdalen valley, with nesting Dunlin and Purple Sandpiper.

On the hill just behind the waste facility and along the road down in the **k.** <u>Bolterdalen valley</u> are a couple of dog yards that also need to be checked. The wetlands here, where Adventdalen and Bolterdalen meet, lie undisturbed by people and become ice-free early in the spring. The area is therefore particularly important for resting geese, as well as for ducks and shorebirds. Scan the area from the main road. Follow the road all the way up the mountain to **l.** <u>Breinosa mountain</u> and Gruve 7, where there are often Ptarmigan and where you have a fantastic view of Adventdalen.

2. BEYOND THE ADVENTFJORD

Svalbard
GPS: 78.92788° N 11.93668° Ø (Ny-Ålesund harbour)

Notable species: All species mentioned for Longyearbyen above. Greater chance of finding, e.g., Brent Goose, Sanderling, Pomarine Skua, Long-tailed Skua, Ross's Gull (scarce), Sabine's Gull, Ivory Gull, Common Guillemot and Razorbill.

Description: There is no road connection between the towns on Svalbard and no regular public transport either. You can explore the island on your own and at your own risk on foot, by boat, or snowmobile before the snow melts. However, most people choose to take part in an organized ship cruise along the west side of Spitsbergen (marked with a thick, white dashed line on the overview map, p. 415) or further around the entire archipelago (thin dashed line). These tours not only give you the opportunity to see the above-mentioned species, but you get up close and personal with the large seabird cliffs and many marine mammals as well. In addition, they take you through a spectacular scenery, with nature experiences that far exceed those you can expect to get at Longyearbyen. The shorter cruises usually last five days. If you are going around the entire archipelago, it will take about ten days.

In addition, there is a varied offer of day trips from Longyearbyen. The trips typically go to the Russian settlement of Barentsburg, Esmarkbreen or Pyramiden. Barentsburg has sites for stray passerines and resting gulls and ducks. An interesting offer is a day trip with a fast-moving RIB via some of the bird cliffs to the walrus colony at Poolepynten in Forlandsundet strait. In spring and early summer, there may still be ice in the fjords, which provides a much better birdwatching outcome than after the ice has melted. Check with the tour operators about the ice conditions and their specific tour options. The adventurous can rent a kayak or inflatable boat with an outboard motor in Longyearbyen and explore the Isfjorden or beyond on their own. An early summer trip along the west side of Spitsbergen, before the drift ice recedes too far to the north, can certainly offer many exciting experiences. Such an expedition is of course associated with considerable risk and requires thorough preparation. Note that all nature reserves on Svalbard have a traffic ban on land and at sea within 300 m of land or reefs at the lowest water level from May 15 to August 15. In some places, disembarkation is prohibited all year, such as on King Karl's Land. See details and updated information about the conservation regulations in the areas you plan to visit on the Governor's website. Here you will also find a lot of information for practical planning.

Other wildlife: In addition to the species mentioned under Longyearbyen, boat trips around the archipelago give an increased chance of finding more rare or locally occurring species, such as hooded seal, white-beaked dolphin, bowhead whale and narwhal. The walrus population has increased greatly in recent years, and the species has become more common in Isfjorden. The day trips by boat here offer a good chance to see it, and day trips to walrus colonies (Poolepynten) are now arranged as well. Blue and fin whales have also become increasingly common, and are now regularly seen on fjord cruises in outer parts of Isfjorden, e.g. the trips headed for Barentsburg. June–July is the best time for these species here.

Best season: The organized boat trips typically run in June, July and August. The ice situation set the conditions for accessibility and vary greatly from on year to the next. The chance of the ships coming all the way around the archipelago is greatest late in the season. Late summer is also the best time for Ross's Gull.

Directions: By air to Longyearbyen. Further transport mainly takes place by boat. On the day trips you can often choose between fast-moving open boats and slower-moving vessels where you spend the whole day at sea.

Tactics: The guides on board the organized boat cruises are often more concerned with finding polar bears and other marine mammals than birds. You must therefore be prepared to spot

From Longyearbyen you can take boat trips to several destinations on the archipelago, either on a long trip by ship or on shorter excursions by fast, rigid inflatable boats (RIBs). *Photo: Bjørn Olav Tveit.*

the birds yourself. Stand far forward on the boat, as high as possible and out of the wind, and use the binoculars diligently. Ross's Gull, Pomarine Skua and many other sought-after species do not normally follow ships, so keep a close eye out for birds passing on the horizon. To maximize the chances of finding Ross's Gull at Spitsbergen, you should go north past 80° N, while on the eastern side of Svalbard you can encounter it in the drift ice south to 78° N. Sabine's Gull nests in several places around Svalbard, and you don't necessarily have to travel far from Longyearbyen to find it. Look especially for them in large tern colonies as well as in aggregations of foraging Arctic Terns, Kittiwakes and Fulmars that often occur at glacier fronts and river deltas. Bring a telescope on the boat trip for observations when you are ashore.

These are some of the most interesting sites along the west and north coast of Spitsbergen: Isfjorden's many seabird cliffs, such as the Tempelet in the innermost part of the fjord, where Fulmar, Kittiwake, Brünnich's and Black Guillemot, Little Auk and Puffin dominate. On the trips to Barentsburg, the boat passes colonies of Little Auk in Fuglefjella and Grumant. Look for polar bears along the seashore. Barentsburg is one of the most exciting places in Svalbard for searching for stray migrants, as well as gulls and ducks. Concentrate on the sewer outlet, the stable and the garbage dump to the south of the settlement. The Pyramid has a colony of Kittiwake and a spectacular glacier. On the boat trip there, they often stop at a colony of Brünnich's Guillemot in the seabird cliff on Diabasodden. Kapp Linné by the mouth of Isfjorden is well situated for seabird migration. There is also a nice lagoon here with the possibility of resting shorebirds and more, as well as smaller colonies of Black Guillemot and Little Auk. The former manned radio station (Isfjord radio), which is now automated, is located here. On the opposite side of the mouth of Isfjorden rises the impressive rock formation Alkhornet. Large numbers of Brünnich's Guillemot, Kittiwake and Fulmar nest here.

Outside of Isfjorden, turning north along Spitsbergen's west coast, there is a small colony of walruses at Poolepynten in Forlandsundet strait. Fuglehuken is located at the northern tip of Prins Karl's Foreland and is a seabird cliff with considerable dimensions, although the population trend has

Walruses are among the many marine mammals you meet if you embark on a boat trip around the archipelago, such as here at Hornsund in the southwest of Spitsbergen. *Photo: Eirik Grønningsæter / WildNature.no*

been negative in recent years. Fuglehuken has Svalbard's only known colony of Common Guillemot, apart from Bjørnøya island. On the west side of the mountain, Brent Geese sometimes roost, and Norway's first record of the subspecies *nigricans* (Black Brant) was made here, in addition to the country's only summer record of Buff-bellied Pipit.

Ny-Ålesund is a small, Norwegian settlement in Kongsfjorden on the west side of Spitsbergen. It is originally an old mining community which has been the starting point for a number of adventurers on their way to the North Pole. It is now an international research centre. The place has an airport, and all the longer boat cruises stop by. However, the town is closed to tourists who are not participants in organized tours. Lake Solvannet above the quay and the lagoon 1 km northwest of the town are favourable locations for shorebirds, gulls, geese and ducks. Ivory Gull is more reliable in Ny-Ålesund than in Longyearbyen, and a number of vagrants have been found here.

Blomstrand (also called *London*) in Kongsfjorden opposite Ny-Ålesund, is a displaced settlement based on the extraction of marble. A few pairs of Long-tailed Skua nest here, as one of the very few places on Spitsbergen. A number of islets and reefs inside the Kongsfjord ensure a bustling birdlife, with several smaller seabird colonies. The seabird cliff by the 14th July Glacier in the side fjord Krossfjorden has Puffins. There are also hanging gardens here, with many unusual plants. Fuglesongen is a low-lying and easily accessible colony of Little Auk on the northwest corner of Spitsbergen. Sit quietly among the boulders and the auks will soon fly in and settle around you.

Moffen is a 2x3 km, flat gravel island at 80° N on the north side of Spitsbergen. It is considered to be among the most reliable places to see Sabine's Gull, Grey Phalarope and walrus. There are also a few dozen pairs of Brent Goose nesting here. The trip up here can be hindered by ice, especially early in the season. The island is protected as a nature reserve with a traffic ban, so observations must be made from the ship at some distance. Sjuøyane islands in the northernmost part of Svalbard, with Phippsøya island as the northernmost accessible of these, is a favoured stop by birdwatchers. On the north-east side of Lågøya island, which lies a little further to the south-west, there is a small puddle which is a favourite site for Grey Phalarope, but here there is now a disembarkation ban in the period May 15 to August 15. Hamilton Bay at the far end of the Raudfjord has a large seabird cliff that cruise ships often visit. There are also exciting islets with terns, Barnacle Geese and eiders as well as 1–2 pairs of Great Skua. The trollish Alkefjellet mountain in the north of the Hinlopen strait, is home to a large seabird colony.

Near the southwestern tip of Spitsbergen you will find the Hornsund fjord complex. Several seabird cliffs are located here, including Ariekammen, which is probably Svalbard's largest Brünnish Guillemot colony, counting several hundred thousand pairs. Dunøyane and Isøyane islands are important nesting sites for Common Eider, Pink-footed and Barnacle Geese, and Great Skua, and is an important staging ground for ducks and geese. Hornsund is known for having a high density of polar bears, partly because this is an important casting area for ringed seals. The occurrence of both ringed seals and polar bears depends on the ice conditions. This is also a good spot for Ivory Gull, whose presence is often a sign that there are polar bears nearby.

The large Bellsund fjord is Isfjorden's neighbour to the south. There are several seabird cliffs here, with Kittiwake, Brünnich's Guillemot and Little Auk as well as a few pairs of Razorbill. Midterhuken and especially Ingeborgfjellet are among the largest seabird cliffs in Svalbard. The latter has a six-figure number of nesting pairs. The grasslands below the bird cliffs are important grazing areas for geese and Svalbard reindeer. In Vårsolbukta on the north side of the fjord, Norwegian researchers study staging geese in spring. On Eholmen at the mouth of the Van Keulenfjord, approx. 2,500 pairs of Eider breed in good years. Bellsund has a good population of Arctic fox, and a number of marine mammals exploit the fjord. This also provides good conditions for polar bears. Although most polar bears migrate eastwards and away from the area in late spring. A dozen individuals may linger here throughout the summer.

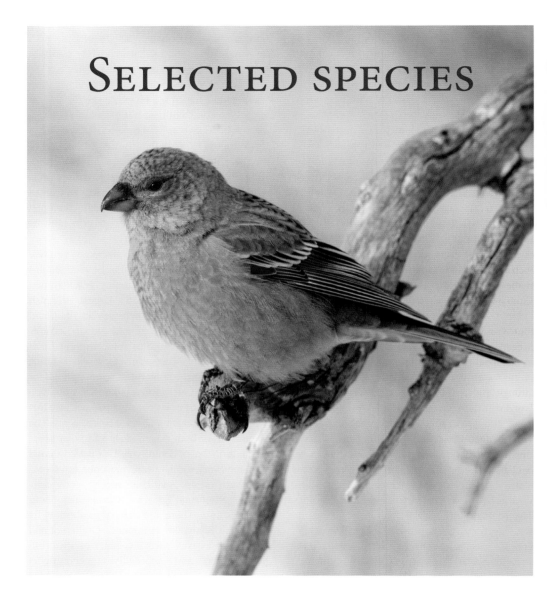

SELECTED SPECIES

The following is a selection of species sought after by many birdwatchers visiting Norway from abroad, with suggestions on where, when and how they may be found. In addition, see the list of common bird species on p. 12 and the *Notable species* section in each site account.

Brent Goose *Branta bernicla*
Pale-bellied subspecies *hrota* is a common migrant along the coast and is best seen from headlands from Andøya to the outer Oslofjord. Occurrence is limited to a restricted time slot: in spring during

Pine Grosbeak in its golden female-type plumage. *Photo: Kjartan Trana.*

the last few days of May, and in autumn around the third week of September. Dark-bellied *bernicla* is an uncommon but regular visitor along the coasts of central and southern Norway, mostly late September–November.

Barnacle Goose *Branta leucopsis*
Breeds on Svalbard. Common during migration (May and September) along the coast of mainland Norway south to Sogn og Fjordane. Large resting

areas in Lofoten–Vesterålen, along the coast of Helgeland and around the Trondheimsfjord. A feral population has become numerous in south-eastern Norway in recent decades.

Taiga Bean Goose *Anser fabalis*
Uncommon but regular breeder in Finnmark. In the breeding season, the species is associated with wetlands in the forested taiga zone such as Pasvik valley. Common in the past, but now a rare breeder further south, e.g., in the Børgefjell mountain area. Regular on staging grounds a few places, such as in the Vorma–Glomma confluence in the Oslo area in March and early September. The two bean geese often occur together during migration and in winter, such as at Jæren in southern Norway. Identification challenges demands good views and preferably photo documentation.

Pink-footed Goose *Anser brachyrhynchus*
Breeds on Svalbard. Common during migration in mainland Norway, passing over the Oslo area in flocks of many hundreds from mid-March through April. Large resting areas around the Trondheimsfjord and in Lofoten–Vesterålen.

Tundra Bean Goose *Anser serrirostris*
Uncommon but regular breeder in Finnmark in tundra habitat, such as on the Finnmarksvidda plateau and the central Varanger peninsula. The Varanger peninsula and Valdakmyra are reliable sites in late spring and late summer. This species also breeds with a few pairs on the Hardangervidda plateau in the interior south of Norway. Tundra and Taiga Bean Geese were until rather recently regarded as subspecies of the same species, Bean Goose. They occur together during migration and in winter, e.g., at Jæren in southern Norway.

Lesser White-fronted Goose *Anser erythropus*
Formerly common in the Scandinavian mountains but now nearly extinct. Only to be expected at the regular resting site of Valdakmyra marsh in the Porsangerfjord (and to a much lesser extent around the Varangerfjord), from May to early June and from mid-August to mid-September. These birds breed on secretly kept locations in the nearby mountains. Very rare elsewhere in Norway. Records in southern Norway are suspected to be of birds from the Swedish reintroduction scheme.

Bewick's Swan *Cygnus columbianus*
Uncommon but singles regularly found among wintering Whooper Swans, particularly at Lista and Jæren (November–early April). Has declined in recent years, and is now more likely to turn up during the migration periods, alone or together with Whooper Swans.

Whooper Swan *Cygnus cygnus*
Uncommon but regular breeder in small lakes in woodland and montane areas of south-eastern and northern Norway. Numbers have increased in recent years. Winters in sometimes large flocks on freshwater lakes and rivers as long as the ice conditions allow. Migrates to Denmark or Great Britain during cold spells.

Pintail *Anas acuta*
Regular but uncommon breeder in nutrient-rich lakes and marshes at higher elevations in southern and central Norway. In Finnmark, however, it is among the commoner dabbling ducks. During migration throughout Norway, most are seen in gatherings with other ducks in April–June and August–October.

Scaup *Aythya marila*
A sparse breeder. In southern Norway mostly in mountain lakes while in the north, it breeds regularly down to sea-level as well, such as in Lofoten. Winters uncommonly in coastal fjords and lakes, with Hafrsfjord near Stavanger as the single most important wintering site.

Steller's Eider *Polysticta stelleri*
Has its most important wintering area in Europe in eastern Finnmark. In the Varangerfjord it may be found all year, with up to several thousand in late winter. It used to be common during here in summer as well, and breeding has been suspected. Increasingly rare further south along the coast. Numbers have decreased in recent years, and it is now difficult to find in Finnmark in summer, after the end of May.

King Eider *Somateria spectabilis*
King Eider breeds in small numbers on Svalbard, for instance in the Advent delta by Longyearbyen. The species' most important European wintering area is along the coast of northern Norway south to

Vega, particularly in Finnmark and the outer section of the Varangerfjord. The species seems to prefer deeper water than Common and Steller's Eiders. King Eider is worth looking for in gatherings of Common Eider further south in Norway as well.

Common Eider *Somateria mollissima*
Common along the coast, often in large flocks. In decline everywhere, except for in the Oslofjord, where numbers are increasing. Subspecies *borealis* breeds in Svalbard and can be found among birds belonging to the nominate form along the coast of northern Norway, mainly in winter.

Velvet Scoter *Melanitta fusca*
Sparsely distributed breeder of inland freshwater lakes, mainly on higher ground and in the mountains. Common along the coast outside breeding season, mainly along coast of central and southern Norway. Immature and moulting birds can be found along the coast in summer, for instance at Mølen along costal Skagerrak.

Common Scoter *Melanitta nigra*
Sparsely distributed breeder of inland freshwater lakes, mainly on higher ground and in the mountains. Common along the coast outside breeding season. Migrates in large and dense flocks. The male's soft whistle (similar to Bullfinch) can sometimes be heard at night from flocks migrating overland.

Long-tailed Duck *Clangula hyemalis*
Common breeder in freshwater lakes with small islands in Finnmark and other places in northern Norway, inland and near the coast. In Svalbard, breeds along the seashore as well. Uncommon breeder in similar habitat in the mountains of southern Norway. Common along the coast outside breeding season, particularly in northern Norway. The species' presence is sometimes first noted by its peculiar and melodic call.

Goldeneye *Bucephala clangula*
Very common in most parts of the country, mainly in nutrient-poor inland lakes and tarns. Winters along the coast in sheltered fjords with deep and clear water.

Smew *Mergellus albellus*
Uncommon breeder, mainly in Finnmark. It breeds in large nest boxes and natural cavities in mature coniferous woodland, mostly in Pasvik but also in Anárjohka valleys and probably underneath boulders in the birch zone of south-western Finnmarksvidda plateau. In recent years, it has also been found breeding in Vikna archipelago in central Norway. In southern Norway, Smew is uncommon in winter, although regular at Lista and Jæren.

Goosander *Mergus merganser*
Common breeder in deep, clear inland lakes and slow-flowing sections of rivers, preferably with access to natural cavities in trees, such as old Black Woodpecker holes. It may also use nest boxes. Tens of thousands gather in sheltered fjords such as in the Tana delta in Finnmark in late summer.

Red-breasted Merganser *Mergus serrator*
Common breeder both inland and along the coast over much of the country. Common along the coast in winter, in some places numerous.

Hazel Grouse *Tetrastes bonasia*
Quite common in boreal forest in many parts of Norway, including in the woodland surrounding Oslo, but the species leads a secretive life and can be very hard to find. Search for it in shaded valleys dominated by mature spruce mixed with deciduous trees like alder and birch. Imitating its call with a recording or a specially designed flute may increase your chances of finding it, although it is not always effective.

Willow Ptarmigan *Lagopus lagopus*
Common in birch woods and the willow zone at high elevations in southern Norway and down by the coast further north. Numbers vary from one year to another. In Tromsø, the species is common in recreational areas in and around the city.

'Smøla Ptarmigan' *Lagopus l. variegata*
The only endemic bird taxon in Norway, found on islands off central Norway (mainly Smøla, Hitra and Frøya), characterized by not turning completely white in winter.

Rock Ptarmigan *Lagopus muta*
Found in mountains, high above the tree line in areas with scree, boulders and exposed rock. Some places, particularly in northern Norway,

suitable habitat like this is found along the coast as well. In areas where the two species of ptarmigan occur together, Willow Ptarmigan usually stick to the willow region while Rock Ptarmigan is found in dried-out riverbeds and other areas of grey, exposed rock and gravel. Willow Ptarmigan is usually easy to find, whereas finding Rock Ptarmigan often takes a lot of walking in difficult terrain. Identification must always be based on call or visual appearance – the habitat only gives you an indication of the species.

'Svalbard Ptarmigan' *Lagopus m. hyperborea*
Svalbard Ptarmigan is the subspecies *hyperborea* of Rock Ptarmigan, endemic to Svalbard and Franz Josef Land in Russia. It is noticeably larger in size and warmer brown than Rock Ptarmigans on the mainland and is common many places in Svalbard, including near Longyearbyen. It prefers dry, south-facing slopes where the snow melts early in summer. Its presence is often revealed by the male's rasping call. The calling bird can be difficult to locate visually, however, because it is well-camouflaged, more confident and thus harder to flush than other grouses. It is largely white in winter. The hen moults into her rusty brown summer plumage in April–May while the cock retains the white winter plumage until July, moulting into full summer plumage by mid-August. In September, both sexes moult back to winter plumage.

Capercaillie *Tetrao urogallus*
Has its Norwegian stronghold in the conifer forests of southern and central Norway, mainly in old-growth woods dominated by pine. It can be found locally in other areas across the country as well, including in Pasvik valley. Performs its lekking display in late winter and spring in open forest, usually just a few males together. Vocalizations can be heard only a few hundred metres. Capercaillies, as well as other species of grouse, can sometimes be seen roadside filling their crop with gravel in the morning.

Black Grouse *Lyrurus tetrix*
Common in woodland up to the willow zone in most parts of the country and also on heathland along the coast. Males gather on marshes and clearings in the forest throughout spring (sometimes in other seasons as well), performing a lekking display accompanied by vocalizations audible several kilometres.

Quail *Coturnix coturnix*
Uncommon in cultivated fields in southern Norway (mid-May–August). Performs far-carrying calls at night.

Nightjar *Caprimulgus europaeus*
Locally common in open pine forest with little undergrowth in south-eastern Norway, mainly in southern Østfold County, such as on the Hvaler archipelago. In the Oslo area, it can be found in the hills around the Øyeren delta and near Lake Fiskumvannet. Arrives in late May and calls actively in the middle of the night until early July. Leaves in August.

Water Rail *Rallus aquaticus*
Leads a secretive life in reedbeds and is usually only heard calling. Can occasionally be seen along the edges of the reedbed in early morning. Lake Østensjøvannet in Oslo and Presterødkilen inlet in Vestfold County is among the best sites to get a glimpse of this species. Several Water Rails winter in coastal marshes of southern Norway.

Corncrake *Crex crex*
After many years of near extinction because of industrialized farming, this species i slowly returning to farmlands and lush meadows in the lowlands of southern and central Norway. Like other rails, the Corncrake is rarely seen. Its loud and repetitive call is performed in early summer nights from May to July.

Crane *Grus grus*
Breeds on marshes in taiga forest and mountains mainly in southeast and central Norway. The population has increased in recent decades, and it is now found several places in western and northern Norway. Migrant, arriving in April and leaving in September. Gather in groups or even large flocks during migration, often resting on remote cultivated fields. Some traditional staging areas include Løten east of Hamar, Meldal west of Lake Orkelsjøen and in Namdalen valley in central Norway.

Red-necked Grebe *Podiceps grisegena*
Winters in small numbers along the Norwegian coast, most commonly along the coast of Møre, such as near Giske and Vigra islands. Breeding has occurred a few times.

Slavonian Grebe *Podiceps auritus*
The core breeding areas of Norway are in Troms and the northern part of Nordland and the Trondheimsfjord area of central Norway. Has spread to Finnmark and Southeastern Norway in recent years. The species prefers densely vegetated and nutrient-rich lakes in the lowlands. Stays often well hidden but the breeding pair often meet openly out on the lake in the evenings. Regular along the coast in winter, such as at Lista, Jæren and Møre.

European Golden Plover *Pluvialis apricaria*
Common and conspicuous breeder on mountain plateaus and tundra in summer. Its vocalizations are a dominating ambient element of the soundscape in the mountains of southern Norway and in similar habitat all across northern Norway. Nesting birds can be seen performing distraction displays, hoping to draw your (and other suspected predators) attention away from its eggs or young by pretending to be injured. Common at migration sites, often resting in cultivated fields and skerries alongside equally common Grey Plover along the coast.

Little Ringed Plover *Charadrius dubius*
Prefers dry, open areas close to fresh water, such as gravel islet in rivers and the shorelines of lakes with artificially lowered water level. Industrial areas are also among its favourite habitats. Arrives in late March, leaving by September.

Dotterel *Eudromias morinellus*
A sparse breeder on dry moraine hills above the tree line. The species is often very approachable, and as such, can be difficult to spot. In May, when the birds arrive and wait for the snow and ice to melt in the breeding areas, they can gather in fields in the valleys leading up towards the mountains. Dotterels are sometimes found on recently ploughed and fallow fields in the lowlands as well, such as near Kurefjorden bay, on Lista and Jæren, usually in May.

Whimbrel *Numenius phaeopus*
Breeds on marshes in open terrain mainly in central and northern Norway, with a few pairs also in western Norway. Rather common on migration. Most are seen May–September.

Bar-tailed Godwit *Limosa lapponica*
Breeds on tundra marshes mainly in Finnmark, arriving in May, for instance on the Varanger peninsula. A few pairs also breed east of Femundsjøen lake in Hedmark. Common migrant along coastal Norway. A few winter.

Black-tailed Godwit *Limosa limosa*
A very local breeder (April–September). A small population of the nominate form breeds in moist grassy fields on Jæren, and a similarly small population of the northern race *islandica* breeds in northern Norway along shallow lakes lined with grassy marshes, mainly in Lofoten, Vesterålen and on Vannøya island near Tromsø, but look out for it in similar habitat other places in the north.

Turnstone *Arenaria interpres*
Breeds along the outer coast in areas with little vegetation, most commonly from central Norway and northwards. Common along the coast during migration, often together with Purple Sandpipers. A few winter.

Red Knot *Calidris canutus*
Common migrant at typical shorebird sites along the coast, often in rocky habitats together with Purple Sandpiper and Turnstone. The Balsfjord and the Porsangerfjord are reckoned to be the most important resting areas for the Greenland and North American populations. In May, Knots in brick-red summer plumage may gather in the thousands at these sites.

Ruff *Calidris pugnax*
Formerly a rather common breeder on marshes in much of the country, included in salt marshes along the coast in southern Norway. In recent years, populations have decreased significantly, particularly in southern Norway, where it nowadays is easiest to find at Nekmyrene marshes. The population in Finnmark is much healthier, and lekking males are regularly seen in marshes and pastures along the roads in

Varanger in late May and June. The species can be common in freshwater wetlands and along the coast during migration.

Broad-billed Sandpiper *Calidris falcinellus*
Prefers to nest in inaccessible areas of wet and dangerous black bogs where it can be found by scanning the marshes with a telescope. However, it is more often discovered by its characteristic calls during display flights over the breeding grounds. The stronghold in Norway for this species is Finnmark, particularly on the Finnmarksvidda plateau. A few pairs still remain in a few places in southern Norway, particularly at the Nekmyrene and Langsua marshes in the interior south. Uncommon at migration sites in May–June and in August. Kurefjorden bay is one of the most reliable sites during migration in southern Norway, while Valdakmyra marsh is quite reliable in the north.

Temminck's Stint *Calidris temminckii*
Breeds along the shores of mountain lakes in southern Norway, and also at river mouths with dry gravel banks along the Trondheimsfjord and further north. During migration, tends to prefer resting on mud in freshwater or brackish environments rather than along the seashore.

Sanderling *Calidris alba*
Common migrant along the coast, mainly on sandy beaches. Uncommon breeding bird on Svalbard.

Purple Sandpiper *Calidris maritima*
Regular but rather uncommon breeder in the mountains of southern Norway, such as Hardangervidda and Valdresflya plateaus, and along the coast in northern Norway. Winters along the coast north to Finnmark, preferring rocky shores and skerries. A few tend to winter on the skerries off Bygdøy peninsula in Oslo.

Little Stint *Calidris minuta*
Breeds on the tundra at several places in Finnmark, such as on the Varanger peninsula, Slettnes headland and the Lille Porsangen wetlands. Common migrant along the Norwegian coast in flocks of other shorebirds.

Pectoral Sandpiper *Calidris melanotos*
Breeds in North America and in Siberia but its occurrence at Norwegian migration sites (mainly May–September) has increased in recent years to such a degree that the species is suspected to breed in or near the country. Displaying birds have been encountered several times in marshlands on the Varanger peninsula and on Svalbard.

Jack Snipe *Lymnocryptes minimus*
Uncommon breeder in vast and very wet marshes, mainly in Finnmark. It usually reveals its presence by performing display flights while calling. Uncommon in wet marshes and ditches along the coast in autumn, often not flushing until you almost tread upon it. A few winter in similar habitat.

Great Snipe *Gallinago media*
A declining species in Norway due to the intensification of agriculture. However, in the montane areas of southern and central Norway, it is still rather common and with stable populations in marshy areas with willow thickets, for instance at Dovrefjell and the mountains surrounding the Trondheimsfjord. Like Capercaillie, Black Grouse and Ruff, Great Snipe gather in spectacular lekking displays in spring, for instance in Folldal and Rindal valleys.

Red-necked Phalarope *Phalaropus lobatus*
Breeds in shallow ponds in montane marshlands of southern Norway, such has on the Hardangervidda plateau, Dovrefjell and at nekmyrene marshes, and along the coast as well in northern Norway, such as Seinesodden peninsula in the Bodø area. Most common, however, in Finnmark, for instance along the northern coast of the Varangerfjord where dozens famously may gather in the Vadsøydammen pond. Most are seen from May–August and are sometimes seen on pelagic trips until September, rarely later.

Red (Grey) Phalarope *Phalaropus fulicarius*
Breeds in shallow marshes and pools in Svalbard. Is usually easy to find in Adventdalen near Longyearbyen, May–August. Rarely seen in mainland Norway, except almost annually along the Varanger coast. A few are encountered along the coast in late autumn after severe storms and on pelagic trips in October–November.

Green Sandpiper *Tringa ochropus*
Fairly common by small lakes in conifer forests over much of the country, mainly April–August. Common at freshwater sites during migration, usually appearing as singles or a few together.

Wood Sandpiper *Tringa glareola*
Breeds on marshes mixed with woodland mainly in the eastern half of southern Norway but also several places in the north. Common during migration (May–August), mainly in freshwater environments, often in flocks.

Spotted Redshank *Tringa erythropus*
Breeds almost exclusively in woodland marshes in Finnmark, such as in Pasvik valley. It is fairly common on migration across the country, with a few in May and most in July–August.

Kittiwake *Rissa tridactyla*
Breeds in colonies on steep cliffs along the outer coast, a few on and around Utsira, but most from Runde and northwards. Some also breed on walls of seaside buildings in harbours, such as in Ålesund and Vardø. In some places, such as on Røst and in Berlevåg, "Kittiwake hotels" are built to offer the birds an alternative to breeding on the regular buildings.

Ivory Gull *Pagophila eburnea*
Breeds uncommonly in cliffs protruding from glaciers in Svalbard. Although populations have decreased in recent years, it is still reliable near Longyearbyen, and even more so in Ny-Ålesund, particularly in spring. It is often attracted to polar dog cages and polar bears, munching food scraps and faeces. Rare in mainland Norway, only occasionally appearing in fishing harbours or other places where gulls gather, most often in the north.

Sabine's Gull *Xema sabini*
Breeds locally in Svalbard, often in colonies of Arctic Tern. You may find it by carefully scanning gatherings of Kittiwakes and terns near Longyearbyen. Your chances increase if joining boat trips along the coasts of Svalbard. In mainland Norway, it is uncommonly encountered during seawatching in autumn. Almost annually seen in spring in Kittiwake gatherings along the Varangerfjord, most often around the Bergebyelva river mouth at Nesseby.

Little Gull *Hydrocoloeus minutus*
Has colonised eastern Finnmark in recent years and is now common along the Pasvikelva watercourse in summer (mid-May–July). Also increasingly regular in lakes in the eastern parts of the country further south, such as at Lake Vesle Sølensjøen near Nekmyrene marshes. Is regularly encountered during migration at Jæren and wetland sites such as Øyeren delta in the Oslo area in May and August.

Ross's Gull *Rhodostethia rosea*
Rare and unreliable in the waters around Svalbard, mainly in late summer. You may encounter it during seawatching at Longyearbyen, but the further north and east you can come, the greater are the odds of finding it. Extremely rare along the coast of mainland Norway. However, it is almost annually encountered in Kittiwake gatherings along the Varanger peninsula, particularly in Berlevåg.

Glaucous Gull *Larus hyperboreus*
Breeds commonly in Svalbard. Winters regularly along coast of mainland Norway, mainly in the north. Sometimes encountered in the hundreds in Varanger, where a few regularly spend the summer as well.

Iceland Gull *Larus glaucoides*
Winters regularly in Norway, mainly in the north, such as in Lofoten, Vesterålen, Andøya and along the Varangerfjord. Usually greatly outnumbered by Glaucous Gulls.

Lesser Black-backed Gull *Larus fuscus*
Race *intermedius* rather common in summer (late March–September) along the coasts of southern Norway. Race *fuscus* is a local breeder from central Norway and northwards (May–September), reliably found for instance from the ferry in the Froan archipelago. The situation from Lofoten to Troms is complex, because paler-backed individuals of unknown heritage are mixed into the colonies. Western race *graellsii* is regular but uncommon in Norway, and may be involved in the northern colonies.

Great Skua *Stercorarius skua*
Uncommon breeder along the outer coast, usually near seabird cliffs. Breeds regularly for instance at Runde, Røst and in western Finnmark. About 350 pairs breeds on Bjørnøya (Svalbard).

Pomarine Skua *Stercorarius pomarinus*
Migrates past the coast of northern Norway in vast numbers, mainly in May. Slettnes headland is the prime location, but any headland along the outer coast will do. Less common in the south. Scarce breeder on the Varanger peninsula in recent years.

Arctic Skua *Stercorarius parasiticus*
Common breeder, usually in small numbers along the outer coast, more common towards north. Mainly seen May–September. The largest colony is at Slettnes headland in Finnmark.

Long-tailed Skua *Stercorarius longicaudus*
Breeds on tundra and mountain plateaus, fluctuating with the rodent populations. Rather common in Finnmark, for instance on the Varanger peninsula. Mainly seen May–July. Sometimes encountered in large numbers at migration sites in Finnmark. In southern Norway it is easiest to find at Valdresflya.

Little Auk *Alle alle*
Numerous in Svalbard, where it also breeds on the cliffs above Longyearbyen. Seen regularly on migration in late autumn in winter along the coast of mainland Norway in variable numbers.

Brünnich's Guillemot *Uria lomvia*
Numerous in Svalbard. A few pairs bred formerly in seabird colonies south to Runde. Now only found in the northernmost colonies. Hornøya island off Vardø is the most reliable and accessible site along the coast of mainland Norway, and the species may even be seen from Vardø town.

Guillemot *Uria aalge*
Formerly numerous in seabird cliffs along the outer coast. The populations have decreased significantly in recent years, however. It is still easy to find, though, at Runde, Røst and several other places. It can be found anywhere along the coast outside the breeding season. In some years, influxes occur in the Oslofjord in late autumn and it can then be seen along the docks of Oslo. After severe storms, a few are drifted to inland lakes, where they may linger for a while.

Razorbill *Alca torda*
Breeds in large colonies along the outer coast, such as at Runde and Røst. Can be seen at many sites along the coast outside the breeding season, even in the Oslofjord during influxes in late autumn and winter.

Black Guillemot *Cepphus grylle*
Common along the outer coast of mainland Norway, often breeding underneath seaside boulders. Becoming commoner further north. In Svalbard, race *mandtii* breeds higher up on the cliffs, sometimes a bit inland.

Puffin *Fratercula arctica*
About 35 colonies are distributed along the Norwegian coast, from Utsira to Vardø. The colony at Røst is the largest, with about 300,000 pairs, even after the dramatic decrease in numbers during the last decades, in part caused by industrial fishing. Rather uncommonly encountered along the coast outside the breeding season. They breed in large colonies among boulders or in burrows dug out in grassy slopes.

Red-throated Diver *Gavia stellata*
Uncommon breeder in southern Norway, more common in the north and on Svalbard. Breeds on small tarns and marsh pools. Rather common along the coast in winter. Migrates in the hundreds past headlands such as Kråkenes and Slettnes, mainly in May and September–October.

Black-throated Diver *Gavia arctica*
Breeds on inland lakes in most parts of the country, commoner in the east. Migrates inland and is the commonest diver in the Oslo area, mainly in April and May. Very few pass coastal headlands compared to Red-throated. Gather in the hundreds in sheltered fjords of Finnmark in spring. A few winter along the Skagerrak coast, Lista and southern Jæren.

Great Northern Diver *Gavia immer*
Winters sparsely along the Norwegian coast, except off Lista and Jæren in the southwest, where it is quite common. Regular in the Varanger area. Breeding has been suspected several times at Bjørnøya (Svalbard), which is perhaps the easternmost frontier of its breeding range.

White-billed Diver *Gavia adamsii*
Winters uncommonly along the Norwegian coast, commoner in the north. The Trondheimsfjord,

Balsfjord and Varangerfjord are considered to be among the most reliable sites in winter. White-billed Diver migrates along the coast in spring and autumn, particularly in mid-May. The prime spot is Slettnes headland in Finnmark, but headlands in the south, such as Skogsøy near Bergen and Kråkenes, can be productive as well.

Storm Petrel *Hydrobates pelagicus*
A highly pelagic species only coming ashore to breed in late summer, and then only at night. It is quite often encountered on pelagic trips in late summer and is caught in nets for ringing purposes along the coast, attracted by recordings of its calls. It can often be seen on boat trips among the islands of Røst in late summer and from the ferry between Røst and Bodø.

Leach's Petrel *Hydrobates leucorhous*
Highly pelagic, uncommonly seen along the coast in stormy weathers. Breeds in small numbers on remote islands along the coast. Is sometimes caught in nets together with Storm Petrels at night along the coast in late summer but rarely more than a few individuals (while Storm can be caught in the dozens). Is occasionally seen on boat trips off Røst in late summer. Rather regular on migration at Nesseby in the Varangerfjord in early summer.

Gannet *Morus bassanus*
Commonly seen along the coast. Breeds in a few colonies in northern Norway, such as at Runde and off Lofoten, Vesterålen and Finnmark. Some of the colonies tend to move around from one year to the next.

Cormorant *Phalacrocorax carbo*
Race *carbo* common in northern Norway, migrating in large flocks overland exploiting thermal winds alongside raptors and Cranes. Commonly seen migrating over the Oslo area in late March and April. Migrates along west coast as well. Continental race *sinensis* has colonized Øra delta and other places along Skagerrak coast during the last decades and is seen increasingly in other parts of southern Norway as well, often in lakes inland. Mixed breeding with *carbo* noted in outer Oslofjord.

Shag *Gulosus aristotelis*
A common resident along most of the Norwegian coast, except in the Oslofjord. It has recently colonized the Skagerrak coast and is in recent years reaching the Oslofjord as well. It seems more tied to the outer coast compared to Cormorant.

Osprey *Pandion haliaetus*
Osprey is a regular breeder in areas of mature forests mixed with lakes, mainly in south-eastern Norway, a few also in central Norway and in Finnmark. The southern part of Østfold County and Setesdalen in Agder are regarded as being of international importance to the species. Ospreys are commonly seen at migration sites around the Oslofjord in spring, and the species is reliable at the Øyeren delta, Lake Borrevannet and several other lake-side and coastal sites in this part of the country from April to September.

Honey Buzzard *Pernis apivorus*
Breeds mostly in deciduous forests mixed with open fields and lakes in south-eastern Norway south to Agder County. It is very specialized in its feeding habits, arriving late in spring (early May to mid-June) and leaving early in autumn (August–September). It is more retiring than Common Buzzard, which often sits openly exposed on fence poles or soar above the breeding grounds, making Honey Buzzard harder to find. Lake Gjennestadvannet is one of the traditional sites for this species. Easily confused with Common Buzzard.

Golden Eagle *Aquila chrysaetos*
Sparsely distributed across Norway year-round, mainly in montane areas but also in forests. It is particularly common on the larger islands along the coast of Troms and western Finnmark. In southern Norway it is fairly reliable in southern Østfold County in winter and at southern Jæren all year. Attracted to carcasses.

Goshawk *Accipiter gentilis*
Breeds in woodland all over Norway, including the outskirts of Oslo. Significantly less common than Sparrowhawk. Although shy, it often enters suburban and even urban areas to hunt in winter.

Hen Harrier *Circus cyaneus*
An uncommon breeding bird of vast marshlands at higher elevations in the woodlands and mountains in the interior of southern and central Norway, at Finnmarksvidda and Pasvik. Regular but uncommon at migration sites such as Borrevannet in spring, mainly in April and early May, and at Lista and Jæren in September–October. A few winter regularly at Lista and Jæren.

Pallid Harrier *Circus macrourus*
Formerly regarded a vagrant but has increased in recent decades. Now an uncommon migrant from mid-April through May and in September, primarily at Lista, Jæren and in southeast Norway, but regular in Finnmark as well. Anticipated to establishing itself as a breeding bird in the future in somewhat similar habitat as Hen Harrier. Montague's Harrier is still regarded as quite rare.

White-tailed Eagle *Haliaeetus albicilla*
Common, even numerous in places along the coast from Møre to Finnmark. It is quite common south to Lista as well, particularly in winter. In recent years it has re-established itself as a breeding bird in the outer Oslofjord, with one pair in the Drøbak sound in the Oslo area as well. It is regularly encountered along watercourses inland.

Rough-legged Buzzard *Buteo lagopus*
Common in the mountains of southern Norway and in most parts of northern Norway. Numbers vary with rodent populations and can be quite numerous in Varanger some years. Summer migrant, arriving in April–May, most leaving in September–October. At migration sites in the Oslo area, this species arrives 2–3 weeks later than Common Buzzard and is usually in the minority.

Tengmalm's Owl *Aegolius funereus*
Rather common in variable numbers, mainly in conifer forest with access to abandoned Black Woodpecker nest cavities. The core breeding range in Norway is in the southeast but it can be found in conifer forests in all parts of the country, although uncommonly in western Norway. Its calls can be heard in the darkest hours of the night, mainly in March and April. The species is difficult to see but during daytime, you can scratch the tree trunk underneath Black Woodpecker holes or nest boxes and hope a Tengmalm's Owl will peak out to see who is knocking. In some years, many are caught at migration sites during nighttime ringing in October.

Hawk Owl *Surnia ulula*
Breeds in open conifer and birch woodland in many parts of the country, most commonly in the north. Is often active during daylight and may sit exposed in treetops. Is in some irruption years encountered at migration sites in autumn.

Pygmy Owl *Glaucidium passerinum*
Rather common in woodland areas in the eastern parts of south Norway and northwards to Saltfjellet mountain. In particular it fancies patches of mixed forest in areas with cultivated meadows and fields. Territorial calls in spring can easily be imitated by whistling, often resulting in a prompt response.

Snowy Owl *Bubo scandiacus*
Uncommon and very variable in mountains and on tundra, dependent on good rodent populations. May be encountered in the mountains of southern and central Norway, such as at Børgefjell mountain. More regular in Finnmark, where Magerøya island and the Nordkinn and the Varanger peninsulas are among the most reliable places to search for it. Are sometimes found along the coasts outside the breeding season, most often along the coast of the Varangerfjord.

Eagle Owl *Bubo bubo*
Uncommon and local in mature conifer forests with cliffs and on heathland along the outer coast. Most easily found near Halden, in the hills south of Jæren, at Hitra and at the Solvær archipelago in central Norway. The owls can be hard to spot but can be heard calling at dusk and dawn mainly in late winter and early spring.

Ural Owl *Strix uralensis*
A regular but uncommon breeder in the easternmost taiga forests of southern Norway, particularly in the Finnskogene forests. May also be encountered in Lierne and Pasvik valleys, particularly in years with plenty of rodents. Breeds in semi-cavities, e.g., broken treetops and in large nest boxes. Look and listen for it by driving around at dusk in areas of mature conifer forest mixed with fields, clearings and marshes. It is heard calling in the darkest hours, mainly in early spring.

Great Grey Owl *Strix nebulosa*
A regular but uncommon breeder in Pasvik and Anárjohka valleys, easiest to find in years with good rodent populations. In such years, may also breed in other areas of mature and open taiga forest, such as in Troms and Lierne. In recent years, the population has increased in southern Norway, along the border with Sweden. Its calls do not carry very far and is often heard at dusk and dawn. Look and listen for it by driving around at night in areas of fields, clearings and marshes. the species hunts by flying low over the ground like a Short-eared Owl or by sitting exposed on the edges of marshes and cultivated fields, looking for rodents and shrews. Can wander quite a bit outside the breeding season and is sometimes encountered at unlikely sites. Reveals its presence during winter by leaving a characteristic pattern ("angels") in the snow, from plunging down when hunting rodents.

Wryneck *Jynx torquilla*
Breeds in sunny and dry woodlands with meadows and clearings with good supplies of ants over much of the country, mainly south of Saltfjellet, most commonly along the Skagerrak coast and in southern Rogaland County. Declining and often hard to find. Arrives in late April and leaves by September. Often discovered by the call, which, however, is easy to confuse with the calls of Lesser Spotted Woodpecker.

Three-toed Woodpecker *Picoides tridactylus*
Prefers mature coniferous or mixed forest with a high percentage of dead or dying trees, and can be found in such habitat in many parts of the country. It can be difficult to find and is usually outnumbered by Great Spotted Woodpeckers. Three-toed may be seen in the woodlands surrounding Oslo, although it is more reliable in old-growth forests such as Trillemarka. Its presence is often revealed by the characteristic lines of small holes it pecks in the trunks of conifer trees, in order to obtain sap.

Lesser Spotted Woodpecker *Dryobates minor*
Breeds in deciduous woodland in most parts of the country, although not very commonly and in declining numbers. Still reliable e.g., at Lake Dælivannet in the Oslo area.

White-backed Woodpecker *Dendrocopos leucotos*
Rather uncommon along the coast from Telemark to the southern part of Trøndelag counties. It prefers sunny slopes with deciduous woods, particularly aspen, with a high percentage of dead and dying trees. Reliable for instance at Lista, near Ålesund and in Sengsdalen valley in central Norway. Formerly fairly common in the eastern parts of southern Norway as well, but it is now almost extinct here because of habitat loss.

Black Woodpecker *Dryocopus martius*
A large, loud and conspicuous inhabitant of mature conifer and mixed woodlands of central Norway and in the eastern and southern parts of southern Norway. It makes spacious cavities in tree trunks and often only uses them for one season. Afterwards, these cavities are vital to other species of birds and mammals, including Goldeneye, Goosander, Stock Dove, Kestrel, Tengmalm's Owl and Redstart, in some parts of the country Smew and Hawk Owl as well. Pine marten and red squirrel may use them for shelter in winter. It is always exciting to scratch carefully on the trunk underneath a Black Woodpecker hole to see what peaks out to investigate the intruder.

Grey-headed Woodpecker *Picus canus*
Quite common on sunny slopes along the fjords of the coast of western Norway. Rather uncommon in deciduous or mixed woods in central and southern Norway. Vocal at all times of year. Some of the calls are easily confused with the calls of Green and Black Woodpeckers. Grey-headed Woodpecker can be encountered along the outer coast in autumn, often far from trees.

Merlin *Falco columbarius*
Common in montane areas covered with willow and birch. Also, in marshy areas in conifer forests. Regular at migration sites. Mainly seen April–October.

Hobby *Falco subbuteo*
Uncommon breeder in the lowlands of southeast Norway, mainly in areas of mixed woods, wetlands and cultivated fields. Reliable sites include Kynndalen in Finnskogene forest, Lake Gjølsjøen and Lake Gjennestadvannet. Often seen hunting small birds and insects over lakes

and wetlands. Uncommon at migration sites mainly along the Oslofjord, Skagerrak coast and Rogaland County. Migrant arriving late April and leaving by early October.

Gyrfalcon *Falco rusticolus*
Sparsely distributed across Norway, primarily in montane areas, preying on ptarmigans inland and seabirds along the coast. Core areas include Dovre and Finnmark. The species is threatened by international collectors of eggs and young. Thus, the exact breeding locations are kept secret and known breeding sites kept under surveillance. Collision with man-made obstructions such as reindeer fences is also a major threat. The population of Gyrfalcon in Finnmark today is less than 20 % of what it used to be 150 years ago. In winter it is regularly seen at Lista, Jæren, Ørlandet and other places where large numbers of gulls and sea ducks gather.

Peregrine Falcon *Falco peregrinus*
Was on the brink of extinction in Norway up until the 1970s because of human prosecution and environmental pollutants. In recent years, the species has recovered remarkably, and it can now be encountered along the entire coast and also several places inland. It is often attracted to gatherings of shorebirds, gulls and other larger prey, even Feral Pigeons in downtown Oslo. Most birds migrate south, returning in April, but a few winter along the southern coast. Fairly reliable at Lista and Jæren in autumn.

Red-backed Shrike *Lanius collurio*
A regular but uncommon breeder in south-eastern Norway, present mainly in May–September. It is usually found in open, dry and sunny areas with patches of dense scrub, such as in forest clearings and beneath electric wires through woodlands, for instance in Maridalen and Sørkedalen valleys near Oslo. It habitually perches exposed on top of a bush or a fence post. However, it can be rather skulky at times, particularly during breeding. It is uncommon at migration sites along the coast of southern Norway.

Great Grey Shrike *Lanius excubitor*
Uncommon breeder in montane birch woodland and on marshes and clearings in forested areas.

Winters in open country, often cultivated fields, and along the coast. Is fairly reliable e.g., in the farmlands around Ås south of Oslo in winter.

Siberian Jay *Perisoreus infaustus*
Rather common in montane conifer forests across the country, apart from western Norway. However, the species may be rather difficult to find when you need them. Even when visiting core areas, like Valdres, Lesja or Pasvik valleys, you will most likely need to walk quite a bit through appropriate habitat in order to find them. They often respond to playback of their calls. In winter, the species is attracted to feeders. Siberian Jays are often remarkably approachable.

Nutcracker *Nucifraga caryocatactes*
Nests in woodlands of south-eastern Norway, in sheltered fjords along the west coast, and in the woods surrounding the Trondheimsfjord. It is a common visitor to the city parks of Oslo in late summer and early autumn, including Nordre and Vestre cemeteries.

Waxwing *Bombycilla garrulus*
Is most often seen in large flocks during irruptions, feeding on berries in the lowlands across the country in late autumn and winter. Uncommon breeder in mature taiga forest, mainly in northern Norway, for instance in Pasvik. Often very approachable, particularly outside the breeding season.

Crested Tit *Lophophanes cristatus*
Common resident in conifer forest in southern and central Norway, mainly in areas with pine. Uncommon in northern Norway.

Siberian Tit *Poecile cinctus*
Very local. Rather common resident in Pasvik valley and several other places in Finnmark, usually in dry, open pine forest. It may be found in similar habitat in Troms and Lierne, and in the Folldal valley and east of Lake Femunden in the southern interior. The species' chocolate-brown cap is its safest distinction from the always much more numerous Willow Tit. Its vocalizations, once learned, are often a good first clue of the species presence.

Marsh Tit *Poecile palustris*
Common resident in several areas of central, western and south-eastern Norway, some places along the coast of southern Norway. Seems to be declining, particularly in the south-east. Prefers moist deciduous woods, such as in river deltas and sheltered valleys.

Willow Tit *Poecile montana*
Common resident in conifer forests in most parts of the country. Also, in birch woodland at higher elevations.

Bearded Tit *Panurus biarmicus*
Uncommon and very local breeder, mainly at the Øra delta along coastal Skagerrak, in Lake Slevdalsvannet on Lista and Lake Søylandsvannet at Jæren. More unreliable in other areas of vast reedbeds, such as at Fornebu near Oslo and at Presterødkilen inlet in Tønsberg, where the birds often arrive in late autumn and stay through the winter. Even large flocks may stay well hidden in the reeds, but their presence is often revealed by their characteristic calls.

Woodlark *Lullula arborea*
Breeds in open and dry pine woodland with little undergrowth in the southern part of Østfold County, or in similar terrain a few other places in southeast Norway.

Shore Lark *Eremophila alpestris*
Breeds mainly on dry tundra habitat, often in the same areas as Dotterel. In southern Norway such habitat is found in the mountains, e.g., on Hardangervidda, Valdresflya and Dovre. In northern Norway it can be found at the coast as well, such as along the Varangerfjord.

Long-tailed Tit *Aegithalos caudatus*
In Norway, the white-headed nominate form occurs. A resident species mainly found in deciduous woodland, for instance at Lake Dælivannet near Oslo.

Wood Warbler *Phylloscopus sibilatrix*
A sparse breeder in deciduous forest and mixed woodland in the lowlands of south-eastern and central Norway. It is a summer migrant, arriving in late April in the Oslo area. Lake Dælivann is one of several reliable sites in the Oslo area.

Yellow-browed Warbler *Phylloscopus inornatus*
This Asian species is annually recorded on migration, mainly along the coast from mid-September to mid-October. From being a genuine vagrant a hundred years ago, umbers have steadily increased since, and the species have now reached day-counts of up to 160 individuals at Værøy island in the Lofoten archipelago, the highest in Europe.

Arctic Warbler *Phylloscopus borealis*
Breeds regularly in eastern Finnmark, and more sporadically in other locations in Finnmark and other parts of northern Norway. It is a summer migrant arriving very late in spring, usually around or after mid-June. It prefers dense and thin-stemmed birch forest in damp areas. The Pasvik valley and Neiden are traditionally the most reliable sites to find it. At migration sites along the coast, Arctic Warbler is regarded a genuine rarity, mainly recorded in September–October.

Blyth's Reed Warbler *Acrocephalus dumetorum*
Rare in Norway, but singing, often long-staying males annually encountered in late spring and early summer in south-eastern Norway. The species prefers open country with scrub and dense undergrowth, often in drier terrain than Marsh Warbler. Has bred a few times. Sometimes difficult to identify from Marsh and Reed Warblers. Records should be documented and submitted to the regional rarities committee (LRSK).

Marsh Warbler *Acrocephalus palustris*
Common in agricultural areas of south-eastern Norway, along the Skagerrak coast and on Lista and Jæren from late May to early September. Most are found singing on quiet early summer nights from wet ditches along the fields. Sings at daytime as well but is then more easily confused with Icterine Warbler.

Icterine Warbler *Hippolais icterina*
Rather common in mature deciduous forest in southern and central Norway, mainly mid-May to August. They can be hard to see high up in the canopies but are fortunately easy to hear when singing. Song is performed at daytime and may be confused with that of Marsh Warbler but note differences in habitat.

Barred Warbler *Curruca nisoria*
Used to be an uncommon and local breeder along the Skagerrak coast, mainly found late May to September. However, the species seem to have vanished completely as a regular breeding bird. Still, be on the lookout for it, particularly at the last regular breeding sites of Stråholmen and Jomfruland island and at Mølen headland in spring. The species prefer open, dry terrain with patches of dense scrub and scattered deciduous trees, a habitat often shared with Red-backed Shrike and Common Rosefinch. The song is similar to that of Garden Warbler, only fuller and deeper. Garden Warbler tends to prefer areas of taller deciduous trees. In late August through September, the species is regular but uncommon at coastal migration sites.

Mistle Thrush *Turdus viscivorus*
Rather uncommon summer migrant. Mainly found in open pine forest in the southeast but also further north, such as in Lierne and Pasvik valleys.

Redwing *Turdus iliacus*
Numerous and conspicuous in most kinds of woodland across the country, mainly April–November. Redwings have a multitude of dialects, clearly audible in the first half of the song. Dialects may vary greatly from one hill to the next. Redwings gather in large flocks in autumn before migrating across to Britain and Continental Europe. The characteristic call is commonly heard at night during migration.

Fieldfare *Turdus pilaris*
Numerous and conspicuous in most kinds of woodland across the country, mainly March–November. Gathers in large flocks in autumn before migrating south across the North Sea.

Ring Ouzel *Turdus torquatus*
Breeds in areas of cliffs and boulders. Most migrate and are regularly seen at migration sites from April to mid-May, more uncommonly in September–October.

Bluethroat *Luscinia svecica*
Summer migrant visiting Norway mainly in May–September. It is typically found in willow scrub, in southern Norway in the mountains, in northern Norway near the coast as well. During migration, it is often found in reedbeds in the lowlands.

Thrush Nightingale *Luscinia luscinia*
Summer migrant, arriving in early May. A secretive species but its loud and characteristic song is performed actively both day and night. Most individuals stop singing after mid-June. However, the species can still be found on the breeding areas until the end of August, particularly if you have learned its characteristic contact calls. The species prefers dense and moist deciduous woods, mainly around the outer Oslofjord. Particularly common at Lake Borrevannet and the Tjøme archipelago in Vestfold County and also around Kurefjorden bay in Østfold County.

Red-breasted Flycatcher *Ficedula parva*
Rare in Norway but singles are almost annually found singing in deciduous and mixed woodland around the Oslofjord in May and June. It is an uncommon migrant at coastal sites, such as Utsira island, in late September and early October.

Stonechat *Saxicola rubicola*
Uncommon breeding bird, mainly on heathland with clusters of taller bushes or trees along the west coast, such as at Kråkenes and Stad headlands.

Dipper *Cinclus cinclus*
Norway's national bird. It is found along fast-running streams and whitewater in most parts of the country, in winter as well.

Yellow Wagtail *Motacilla flava*
Breeds in open, moist terrain in most of the country, including in the mountains. The grey-headed race *thunbergi* is the commoner subspecies. Nominate *flava* and *flavissima* are uncommon breeders mainly at Lista and Jæren.

Red-throated Pipit *Anthus cervinus*
Common many places in Finnmark, less common in marshy tundra terrain other places in northern Norway, for instance around Tromsø. The densest population is thought to be in the Čoskajeaggi marshes on the Varanger peninsula. Usually easy to find around Vardø. Uncommon at migration sites elsewhere, mainly in late May and September. Your odds of encountering it increases significantly if you are aware of its discrete but characteristic calls.

Brambling *Fringilla montifringilla*
A common breeder in high altitude woodlands across the country. It is a short-distance migrant, gathering with Chaffinches in the lowlands before heading south in flocks numbering tens or even hundreds of thousands of individuals. Some spend the winter here as well.

Hawfinch *Coccothraustes coccothraustes*
Is usually found in mature deciduous forests. It has become increasingly common in south-eastern Norway in recent years, particularly in the Oslo area and along coastal Skagerrak. The species may have a flashy visual appearance but may still be hard to find due to its usual shyness and preferences to spend most of its time high up in the canopies of large, broad-leaved trees. It often reveals its presence by uttering its subtle calls, which are fairly distinctive once learned.

Pine Grosbeak *Pinicola enucleator*
A scarce breeder in high altitude coniferous forest with a rich juniper undergrowth, mainly in the eastern parts of Norway, and in Lierne, Anárjohka and Pasvik valleys. You may well discover them by their loud whistled calls and song while you walk or drive slowly with your windows open through these vast forests. Pine Grosbeaks are attracted to feeders within their range. In some irruption years, the species turn up in forests and even parks in other parts of the country in late autumn and winter.

Bullfinch *Pyrrhula pyrrhula*
Common resident in woodland in much of the country. An irruption of birds probably originating from the east occurred in autumn 2004, bringing birds with a call sounding like a small toy trumpet, somewhat similar to Two-barred Crossbill and distinctly different from the soft plaintive whistled call of the native Scandinavian birds (the latter's song, however, may include nasal sounds similar to the calls of the supposed eastern Bullfinches). Some of these have lingered ever since, although less commonly as time goes by.

Common Rosefinch *Carpodacus erythrinus*
Has been an uncommon breeding bird in Norway since the first confirmed breeding in 1970. It is usually found in the lowlands in the south-eastern parts of the country, preferring open, sunny terrain with scattered bushes and trees, often close to water. It is a migrant arriving in mid-May and are often noticed by the male's pleasant whistle: "I'm pleased to meet you!".

Twite *Linaria flavirostris*
Traditionally, Twite has been a regular breeding birds in montane areas of southern Norway and numerous along the west and north coasts. However, in recent years it has suffered a noticeable decline throughout its range. The global distribution of the nominate subspecies is mainly confined to Norway, with a small and declining population in Great Britain as well.

Arctic Redpoll *Acanthis hornemanni*
Breeds in northern Norway only but can be found in winter in the south. Becomes increasingly regular the further north in Norway you get. It is quite common in Finnmark. Even here, however, it is outnumbered by Common Redpoll, and the two species can be very difficult to separate in the field. They often occur in mixed flocks. The willow scrub in Sandfjorden bay just east of Hamningberg in Varanger, is considered to be one of the most reliable sites for Arctic Redpoll.

Parrot Crossbill *Loxia pytyopsittacus*
Common in pine dominated forest across Norway. It can be hard to identify with confidence though, because it cannot be reliably identified on habitat alone. The size of the bill and the calls are the only safe means of field identification, and both are subtle and variable. Usually, flocks contain only one species but sometimes all three crossbills occur together, at least at migration sites.

Common Crossbill *Loxia curvirostra*
Common in conifer forest across the country. It is the dominant species of the two common crossbills in most parts of the country, in both spruce and mixed spruce/pine forests. Never trust habitat as an identification feature.

Two-barred Crossbill *Loxia leucoptera*
A scarce breeding bird and uncommon and irruptive migrant, sometimes encountered in flocks of other crossbills. It prefers larch trees not native to Norway and should be looked for in larch plantations in arboretums and parks.

Lapland Bunting *Calcarius lapponicus*
Rather common, particularly in the willow region just above the tree line in many parts of Norway. It has been declining in some areas in recent years, such as on the Hardangervidda plateau. Uncommon at migration sites, favouring open country. A very few winter, mainly on Lista and Jæren.

Snow Bunting *Plectrophenax nivalis*
Rather common breeding bird in arid, montane areas all across Norway, including Svalbard. Are found at coastal sites in northern Norway and at Svalbard. It is one of the earliest migrants in spring, often appearing in large flocks at migration sites in March and early April. In northern Norway flocks in the tens of thousands are regularly seen. A few winter along the coast north to Finnmark.

Ortolan Bunting *Emberiza hortulana*
Critically endangered in Norway and may disappear completely in the next few years. The species is found mainly on peat bogs, clearings and old forest-fire areas in woodland areas in south-eastern Norway, such as at Meløyfloen and Starmoen and in suitable habitat further south along the Glomma watercourse. Controlled forest fires have been started in several places in order to provide Ortolan Buntings with suitable habitat.

The species is an uncommon visitor at migration sites in southern Norway, primarily at Utsira and similar islands in May and September.

Little Bunting *Emberiza pusilla*
A Siberian breeder just barely including Norway in its range. It is an uncommon but probably annual breeding bird in eastern Finnmark, primarily in damp birch forest. The Norwegian population is estimated to no more than 50 pairs in good years. The best places to look for it are in Pasvik valley, but it may be found in suitable habitats at the Finnmarksvidda plateau or in the Porsangerfjord or other places in the northern parts of Norway. During autumn migration it can be encountered in other parts of Norway, most regularly at Utsira in September–October.

Rustic Bunting *Emberiza rustica*
Scarce breeding bird, associated with flooded conifer forests along streams lined with deciduous trees, mainly in the southern interior of Norway. Arrives mid-May and leaves in August but can be extremely hard to find after the male stops singing in mid-June. Migrates to the east and is only rarely encountered on migration sites in Norway. After a serious decline for many years, the species now seem to slowly recover.

REFERENCES

Anker-Nilssen, T. 2006. The avifaunal value of the Lofoten Islands in a World Heritage perspective. *NINA Report 201.*

Anker-Nilssen, T. and Aarvak, T. 2006. Tidsseriestudier av sjøfugler i Røst kommune, Nordland. Resultater med fokus på 2004 og 2005. *NINA Rapport 133.*

Bangjord, G. 2009. Fuglelivet i Longyearbyen og nærområdene. Longyearbyen feltbiologiske forening.

Bangjord, G., Hübner, C. and Soot, K. M. 2007. Faunaregistreringer ved Sørkapp september 2007. *Arbeidsrapport 3-2007.* Longyearbyen feltbiologiske forening.

Bangjord, G., Frantzen, B. O., Hammer, S. and Hagen, O. 2006. Registreringer av fugl sør på Sørkappland august–september 2006. *Arbeidsrapport 1-2006.* Longyearbyen feltbiologiske forening.

Bekken, J. 2001. Fugler og pattedyr i 18 våtmarksreservater i Hedmark. Fylkesmannen i Hedmark, Miljøvernavdelingen. *Rapport nr. 8/2001.*

Bell, J. Natural Born Birder. http://www.naturalbornbirder.com/

Borch, H. 2006. Nytt Rusasetvatn. Plan for restaurering av vatnet – Ørland kommune. *Bioforsk Rapport vol. 1 nr. 78, 2006.*

Bratlid, H. 2000. Biologisk mangfold i Inderøy kommune. Norsk institutt for jord- og skogkartlegging.

Byrkjeland, S., Chapman, E., Falkenberg, F., Hansen, T., Heggland, T. H., Heggøy, O., Helland, O. B., Kjærandsen, J., Lislevand, T., Mjøs, A. T. and Måge, I. Fuglar i Hordaland. http://fuglar.no/lokaliteter/

Bøhler, T. 2010. Fuglelivet i Asker og Bærum 2010.

Dale, S. 2001. Hortulanen i Oslo og Akershus – fra karakterfugl til sjeldenhet. *Toppdykker'n* 24: 152–161.

Dale, S. and Manceau, N. 2003. Habitat selection of two locally sympatric species of *Emberiza* buntings (*E. citrinella* and *E. hortulana*). *Journal für Ornithologie* 144 (1): 58–68.

Dale, S., Andersen, G. S., Eie, K., Bergan, M. and Stensland, P. 2001. *Guide til fuglelivet i Oslo og Akershus.* NOF, avd. Oslo og Akershus.

Direktoratet for naturforvaltning. Elvedeltadatabasen. www.elvedelta.no

Dyresen, A. 2008. Nattsangere i Østfold 1991–2008. Syngende/spillende individer. Internett.

Fylkesmannen i Nordland 2008. Forvaltningsplan for Børgefjell/Byrkije nasjonalpark.

Fylkesmannen i Oppland 2009. Verneforslag og konsekvensutredning, høring. Utvidelse av Ormtjernkampen nasjonalpark med tilgrensende landskapsvernområder og naturreservater. Del A og C – Innledning, bakgrunn og forslag til verneplan. *Rapport nr. 7/2009.*

Gjershaug, J. O., Thingstad, P. G., Eldøy, S. and Byrkjeland, S. (red.) 1994. *Norsk fugleatlas.* NOF, Klæbu.

Grønningsæter, E. 2007. *Hvor finner man fugler og dyr i Møre og Romsdal.* Eget forlag.

Günther, M. 2003. Dvergmåke 2003. Resultater fra årets registreringer i Sør-Varanger og Nordvest-Russland. Svanhovd Miljøsenter.

Günther, M. 2004. *Field Guide to Protected Areas in the Barents Region.* Svanhovd Environmental Centre.

Günther, M. (red.) 2004. Protected Areas in the Barents Region. Svanhovd Environmental Centre.

Günther, M. Birding between borders. Internett.

Günther, M. and Thingstad, P.G. 2001. Vannfuglregistreringer i Pasvik naturreservat og omkringliggende våtmarksområder. Resultater fra 2000 og 2001. Oppsummering av prosjektarbeidet i perioden 1996–2001 samt statusoversikt for vannfuglfaunaen i Pasvik. Svanhovd Miljøsenter.

Haftorn, S. 1971. *Norges fugler.* Universitetsforlaget, Oslo.

Harrop, H. 2007. Birding Varanger in winter. *Birding World* 20: 517–525.

Haugerud, R. E and Gabler, H. M. 1995. Reisadalen. *Info-hefte.* Fylkesmannen i Troms. Bioforsk Svanhovds nettsted www.pasvik.no

Haugskott, T. 1997. Indre deler av Trondheimsfjorden – fugleområde av internasjonal betydning. *Vår Fuglefauna* 20: 8–13, med oppdatering på NOF avd. Nord-Trøndelags nettsider.

Heggland, A., Blindheim, T. and Olsen, K. M. 2006. Naturverdier i Sørkedalen. Siste Sjanse-notat 2006-2.

Heggland, T. H. Sævarhagsvikjo. Internett.

Hellan, M. E. 2004. Fugletaksering i Øvre Forra Naturreservat 2003. Bacheloroppgave. Høgskolen i Nord-Trøndelag.

Hintikka, J. 2006. Birding in Longyearbyen and Spitsbergen 12th–27th June 2006. Egen utgivelse. Reiserapport på http://www.tarsiger.com

444

Hofton, T. H. 2003. Trillemarka–Rollagsfjell: en sammenstilling av registreringer med hovedvekt på biologiske verdier. Siste Sjanse rapport 2003-5.

Holmström, N. Seawatching at Slettnes. Internett.

Husby, M. 2000. Fuglene i Levanger. HiNT Utredning nr. 14. Steinkjer.

Isaksen, K. 2008. Kartlegging av nattravn i Oslo og Akershus i 2007. Rapport, NOF, avd. Oslo og Akershus.

Jacobsen, K.-O. and Bjerke, J.W. 2006. Reguleringsplan for havne- og næringsformål på Tønsnes, Tromsø. Konsekvensutredning, deltema zoologi og vegetasjon. NINA Rapport 182.

Jacobsen, K.-O., Øien, I. J., Steen, O. F., Oddane, B. and Røv, N. 2008. Hubroens bestandsstatus i Norge. *Vår Fuglefauna* 31: 150–158.

Jakobsen, C. A. 2004. Akerøya Ornitologiske Stasjon – for nye og gamle skådere. *Natur i Østfold* 23 (1–2): 34–42.

Kolaas, T. Fugler i Træna. Træna kommunes Internett-sider.

Kovacs, K. M and Lydersen, C. (red.) 2007. Svalbards fugler og pattedyr. *Polarhåndbok nr. 13.*

Larsen, B. H. 2003. Vårrastende sangsvaner i Indre Østfold – viktige lokaliteter, regulerende faktorer og habitatbruk. *Natur i Østfold* 22 (1–2): 47–58.

Larsen, T. 2005a. Sjøfuglteljingar i Sogn og Fjordane i 2003 og 2004. Hekkefuglteljingar i sjøfuglreservata. Fylkesmannen i Sogn og Fjordane. Rapport nr. 6-2005.

Larsen, T. 2005b. Sjøfuglteljingar i Sogn og Fjordane i 2005. Hekkefuglteljingar i sjøfuglreservata. Fylkesmannen i Sogn og Fjordane. Rapport nr. 11-2005.

Lislevand, T. 2000. Viktige fugleområder i Europa er kartlagt. *Vår Fuglefauna* 23: 101–105.

LRSK/Østfold. Fugleområder i Østfold. Internett.

Norderhaug, M. 1989. *Svalbards fugler.* Dreyer.

NOF, avd. Oppland. 1998. *Fugler i Oppland.*

Nygård, T., Einvik, K. and Røv, N. 2006. Sklinna – Fugleøya lengst ut i havet. Fylkesmannen i Nord-Trøndelag, Miljøvernavdelingen. Rapport 6-2006. 44 s.

Overrein, Ø., Henriksen, J. and Johansen, B. F. 2007. Cruisehåndbok for Svalbard (nettutgave). Norsk Polarinstitutt.

Pedersen, T. 2000. Birding the Varanger Peninsula. Internett.

Sandvik, J. 1998. Konsekvenser av veibygging og hogst i Seterseterdalen i Hemne kommune, Sør-Trøndelag. NOFs rapportserie. Rapport nr. 3-1998.

Shimmings, P. 2004. Vindmøllepark i Solværøyan/Sleneset, Lurøy kommune: Konsekvensutredning av tema Fugle- og dyreliv. Rapport til Nord-Norsk Vindkraft as. Planteforsk Tjøtta fagsenter.

Solbakken, K. Aa., Rudolfsen, G. and Myklebust, M. 2002. Kartlegging av hvitryggspett i Trøndelag 1999. NOFs Rapportserie nr. 3-2001.

Strann, K.-B. and Nilsen, S. 1996. Verneverdige myrer og våtmarker i Finnmark. Fylkesmannen i Finnmark, miljøvernavdelingen. Rapport nr. 3 – 1996.

Strann, K.-B., Rae, R., Francis, I., Nilsen, S. Ø. and Johnsen, T. 2005. Hekkende vadefugl i Goatteluobbal med særlig vekt på fjellmyrløper. En pilotstudie i 2004. NINA Rapport 70.

Størkersen, Ø. R. 1993. Guide til fuglelokaliteter ved Trondheim og andre nærliggende lokaliteter. *Vår Fuglefauna* 16: 34–40, med oppdatering på NOF avd. Sør-Trøndelags Internett-sider.

Suul, J. (red.) 2007. Vegafuglene. Oversikt over fuglelivet i Vega. Vegaøyans Venner and NOF.

Svorkmo-Lundberg, T., Bakken, V., Helberg, M., Mork, K., Røer, J.E. and Sæbø, S. 2006. *Norsk vinterfuglatlas.* NOF, Klæbu.

Systad, G. H., Øien, I. J. and Nilsen, S. Ø. 2003. Zoologisk kartlegging innenfor utvalgte områder på Varangerhalvøya, Finnmark. Rapport til Fylkesmannen i Finnmark.

Sørmo, G. Naturvernområder i Verdal. Verdal kommune. Internett.

Thingstad, P. G. and Frengen, O. 2005. Restaureringsprosjekt Tautra. Del 1: Status vannfugler. NTNU Vitenskapsmuseet Zoologisk Notat 2: 1–32.

Tveit, B. O. 1991. Gode fuglelokaliteter i Oslo og Akershus. *Vår Fuglefauna* 14: 91–99.

Tveit, B. O. 2005. Fugletrekk på tvers – noen betraktninger om landfuglenes trekkretninger og nedslagsområder på Utsira. *Utsira Fuglestasjons Årbok* 2000/2001: 93–110.

Tveit, B. O., Mobakken, G. and Bryne, O. 2004. *Fugler og fuglafolk på Utsira.* Utsira fuglestasjon.

Værnesbranden, P. I. 2004. Fuglelivet i Stjørdal kommune – status pr. 01.09.2004. NOF avd. Stjørdal.

Våge, H. and Baines, S. 1993. *Fugler midt i Lofoten.* NOF Vestvågøy lokallag.

Wilson, J., Dick, W. J. A., Frivoll, V., Harrison, M., Soot, K. M., Stanyard, D., Strann, K-B., Strugnell, R., Swinfen, B., Swinfen, R. and Wilson, R. 2008. The migration of Red Knots through Porsangerfjord in spring 2008: a progress report on the Norwegian Knot Project. *Wader Study Group Bull.* 115(3).

Øien, I. J. and Aarvak, T. 2007. Overvåking av dverggås og sædgås i Norge i 2007. NOF rapport 6-2007.

Aarvak, T. and Øien, I. J. 2006. Kartlegging av hekkende havsvaler og stormsvaler på Hernyken, Røst, oktober 2006. NOF rapport 7-2005.

Aarvak, T. and Øien, I. J. 2009. Monitoring of Bean Goose in Finnmark County, Norway – results from 2008. NOF rapportserie report No 2-2009.

INDEX
OF SITES AND PLACES

K

Notes

Please, report your sightings in Norway onto Artsobservasjoner.no (English) or eBird.org.

460